D1121721

Macromolecules · 1

Structure and Properties

Macromolecules

VOLUME 1 • Structure and Properties

VOLUME 2 • Synthesis and Materials

Macromolecules · 1

Structure and Properties

Hans-Georg Elias
Midland Macromolecular Institute
Midland, Michigan

Translated from German by ***John W. Stafford***

PLENUM PRESS · NEW YORK AND LONDON

Library of Congress Cataloging in Publication Data

Elias, Hans-Georg, 1928-
Macromolecules.

Translation of Makromoleküle.
Includes bibliographical references and indexes.
CONTENTS: v. 1. Structure and properties. – v. 2. Synthesis and materials.
1. Macromolecules. I. Title. [DNLM: 1. Macromolecular systems. QD381 E42m]
QD381.E4413 574.1'924 76-46499
ISBN 0-306-35111-0 (v. 1)

A Division of Plenum Publishing Corporation
227 West 17th Street, New York, N.Y. 10011

Printed in the United States of America

Acknowledgments

The permission of the following publishers to reproduce tables and figures is gratefully acknowledged:

Part I

Academic Press, London: D. Lang, H. Bujard, B. Wolff, and D. Russell, *J. Mol. Biol.* **23,** 163 (1967) (Fig. 4-10).

Akademie Verlag, Berlin: H. Dautzenberg, *Faserforsch. Textiltech.* **21,** 117 (1970) (Fig. 4-18).

American Chemical Society, Washington, D.C.: P. Doty and J. T. Yang, *J. Am. Chem. Soc.* **78,** 498 (1956) (Fig. 4-24); M. Goodman and E. E. Schmitt, *J. Am. Chem. Soc.* **81,** 5507 (1959) (Fig. 4-20); S. I. Mizushima and T. Shimanouchi, *J. Am. Chem. Soc.* **86,** 3521 (1964) (Table 4-4).

American Institute of Physics, New York: W. D. Niegisch and P. R. Swan, *J. Appl. Phys.* **31,** 1906 (1960) (Fig. 5-15).

Butterworths, London: A. Nakajima and F. Hameda, *Macromolecular Chemistry 8 and 9, Proceedings,* IUPAC (1972), p. 1 (Fig. 5-21).

W. H. Freeman and Company, Publishers, San Francisco: M. F. Perutz, *Sci. Am.* **1964** (Nov.), 71 (Fig. 4-19).

Gazzetta Chimica Italiana, Rome: G. Natta, P. Corradini, and I. W. Bassi, *Gazz. Chim. Ital.* **89,** 784 (1959) (Fig. 4-5).

Interscience Publishers, New York: T. M. Birshtein and O. B. Ptitsyn, *Conformation of Macromolecules,* p. 34 (Fig. 4-2); P. H. Lindenmeyer, V. F. Holland, and F. R. Anderson, *J. Polym. Sci. C* **1,** 5 (1963) (Figs. 5-16, 5-17, and 5-19); P. J. Flory, *Statistical Mechanics of Chain Molecules* (1969), Chapter V, Fig. 9 (Fig. 4-14); J. Berry and E. F. Casassa, *J. Polym. Sci. D* **4,** 33 (1972) (Fig. 4-15); P. Pino, F. Ciardelli, G Montagnoli, and O. Pieroni, *Polym. Lett.* **5,** 307 (1967) (Fig. 4-21); A. Jeziorny and S. Kepka, *J. Polym. Sci. B* **10,** 257 (1972) (Fig. 5-4); H. D. Keith, F. J. Padden and R. G. Vadimsky, *J. Polym. Sci. A* **2** (4), 267 (1966) (Fig. 5-22).

Kogyo Chosakai Pub. Co., Tokyo, Japan: M. Matsuo, *Japan Plastics,* **1968** (July), 6 (Fig. 5-31).

Pergamon Press, New York: J. T. Yang, *Tetrahedron* **13,** 143 (1961) (Fig. 4-23).

Societa Italiana di Fisica, Bologna: G. Natta and P. Corradini, *Nuovo Cimento Suppl.* **15,** 111 (1960) (Fig 5-9).

D. Steinkopff Verlag, Darmstadt: A. J. Pennings, J. M. M. A. van der Mark, and A. M. Keil, *Kolloid-Z.,* **237,** 336 (1970) (Fig. 5-28).

Textile Research Institute, Princeton, N. J.: H. M. Morgan, *Textile Res. J.,* **32,** 866 (1962) (Fig. 5-33).

Verlag Chemie, Weinheim/Bergstrasse: L. Pauling, *Die Natur der chemischen Bindung,* p. 80 (Table 2-2); H. Staudinger and E. Husemann, *Liebigs Ann. Chem.* **527** (1937) 195 (Table 1-3).

Part II

Akademie-Verlag, Berlin: K. Edelmann, *Faserforsch. Textiltech.* **3,** 344 (1952) (Fig. 7-6).

American Chemical Society, Washington, D.C.: K. G. Siow, G. Delmas, and D. Patterson, *Macromolecules* **5,** 29 (1972) (Fig. 6-10).

The Biochemical Journal, London: P. Andrews, *Biochem. J.* **91,** 222 (1964) (Fig. 9-18).

Butterworths, London: H. P. Schreiber, E. B. Bagley, and D. C. West, *Polymer* **4,** 355 (1963) (Fig. 7-7).

The Faraday Society, London: R. M. Barrer, *Trans. Faraday Soc.* **35,** 628 (1939) (Table 7-2); R. B. Richards, *Trans. Faraday Soc.* **42,** 10 (1946) (Fig. 6-19).

General Electric Co., Schenectady: A. R. Schultz, *General Electric Report 67-C-072* (Fig. 6-15).

Carl Hanser Zeitschriften Verlag, München: G. Rehage, *Kunststoffe* **53,** 605 (1963) (Fig. 6-11).

Institution of the Rubber Industry, London: G. Gee, *Trans. Inst. Rubber Ind.* **18,** 266 (1943) (Fig. 6-1).

Interscience Publishers, New York: T. G. Fox, *J. Polym. Sci. C* **9,** 35 (1965) (Fig. 7-8); G. Rehage and D. Möller, *J. Polym. Sci. C* **16,** 1787 (1967) (Fig. 6-14); Z. Grubisic, P. Rempp, and H. Benoit, *J. Polym. Sci. B* **5,** 753 (1967) (Fig. 9-19).

Journal of the Royal Netherlands Chemical Society, 's-Gravenhage: D. T. F. Paals and J. J. Hermans, *Rec. Trav.* **71,** 433 (1952) (Fig. 9-25).

Pergamon Press, New York: H. Hadjichristidis, M. Devaleriola, and V. Desreux *Eur. Polym. J.* **8,** 1193 (1972) (Fig. 9-27).

Springer Verlag, New York: H.-G. Elias, R. Bareiss, and J. G. Watterson, *Adv. Polym. Sci.* **11,** 111 (1973) (Fig. 8-6).

Verlag Chemie, Weinheim/Bergstrasse: H. Benoit, *Ber. Bunsenges* **70,** 286 (1966) (Fig. 9-5); G. V. Schulz, *Ber. Dtsch. Chem. Ges.* **80,** 232 (1947) (Fig. 9-1).

Part III

American Institute of Physics, New York: H. D. Keith and F. J. Padden, Jr., *J. Appl. Phys.* **30,** 1479 (1959) (Fig. 11-15); R. S. Spencer and R. F. Boyer, *J. Appl. Phys.* **16,** 594 (1945) (Fig. 11-17).

Badische Anilin- & Soda-Fabrik AG, Ludwigshafen: *Kunststoff-Physik im Gespräch,* pp. 103, 107 (Figs. 11-1 and 11-2).

Butterworths, London: A. Sharples, *Polymer* **3,** 250 (1962) (Fig. 10-7); A. Gandica and J. H. Magill, *Polymer* **13,** 595 (1972) (Fig. 10-9).

Engineering, Chemical & Marine Press, Ltd., London: R. A. Hudson, *Brit. Plast.* **26,** 6 (1953) (Fig. 11-12).

The Faraday Society, London: L. R. G Treloar, *Trans. Faraday Soc.* **40,** 59 (1944) (Fig. 11-5).

General Electric Co., Schenectady: F. E Karasz, H. E. Bair, and J. M. O'Reilly, *General Electric Report 68-C-001* (Fig. 10-4).

Interscience Publishers, New York: N. Berendjick, *in: Newer Methods of Polymer Characterization* (B. Ke, ed.) (1964) (Fig. 13-1); J. P. Berry, *J. Polym. Sci.* **50,** 313 (1961) (Fig. 11-14); O. B. Edgar and R. Hill, *J. Polym. Sci.* **8,** 1 (1952) (Fig. 10-17); K. V. Fulcher, D. S. Brown, and R. E. Wetton, *J. Polym. Sci. C* **38,** 315 (1972) (Fig. 10-10); H. W. McCormick, F. M. Brower, and L. Kin, *J. Polym. Sci.* **39,** 87 (1959) (Fig. 11-16); N. Overbergh, H. Bergmans, and G. Smets, *J. Polym. Sci. C* **38,** 237 (1972) (Fig. 10-12); G. Rehage and W. Borchard, *in: The Physics of the Glassy State* (R. N. Haward, ed.), p. 54 (Fig. 10-2); P. I. Vincent, *Encyclopedia of Polymer Science Technology,* Vol. VII, p. 292 (Fig. 11-10); A. Ziabicki, *in: Man-Made Fibers* (H. Mark, S. M. Atlas, and E. Cernia, eds.), Vol. I (1967), pp. 17, 21 (Figs. 12-4 and 12-5).

Japan Synthetic Rubber Co., Tokyo: anon., *Japan Synthetic Rubber News* **10** (1) (1972) (Fig. 10-18).

Verlag B. M. Leitner, Wien: F. Patat, *Allg. Prakt. Chem.* **18,** 96 (1967) (Fig. 13-5).

McGraw-Hill Book Co., New York: A. X. Schmidt and C. A. Marlies, *Principles of High Polymer Theory and Practice* (1948), p. 66 (Fig. 10-6).

Research Group of Polymer Physics in Japan, Tokyo: H. Tadokoro, Y. Chatani, M. Kobayashi, T. Yoshihara, S. Murahashi, and K. Imada, *Rep. Prog. Polym. Phys. Jpn* **6,** 305 (1963) (Fig. 10-15).

Society of Plastics Engineers, Greenwich, Conn.: J. D. Hoffman, *SPE Trans.* **4,** 315 (1964) (Fig. 10-11).

Springer-Verlag, Berlin: H. Mark, *in: Die Physik der Hochpolymeren* (H. A. Stuart, ed.), Vol. IV (1956), p. 630 (Table 12-5).

Dr. Dietrich Steinkopff Verlag, Darmstadt: G. Kanig, *Kolloid-Z.* **190,** 1 (1963) (Fig. 10-22).

Van Nostrand Reinhold Company, New York: R. C. Bowers and W. A. Zisman, *in: Engineering Design for Plastics* (E. Baer, ed.), p. 696 (Fig. 13-4).

Verlag Chemie, Weinheim/Bergstrasse: K.-H. Illers, *Ber. Bunsenges.* **70,** 353 (1966) (Fig. 10-20); G. Rehage, *Ber. Bunsenges.* **74,** 796 (1970) (Fig. 10-3).

Part IV

Akademische Verlagsgesellschaft, Leipzig: G. V. Schulz, A. Dinglinger, and E. Husemann, *Z. Phys. Chem.* **B43,** 385 (1939) (Fig. 20-7).

American Chemical Society, Washington: P. J. Flory, *J. Am. Chem. Soc.* **63,** 3083 (1941) (Fig. 17-5); H. P. Gregor, L. B. Luttinger, and E. M. Loebl, *J. Phys. Chem.* **59,** 34 (1955) (Fig. 23-4); G. V. Schulz, *Chemtech.,* **1973** (April), 224 (Fig. 18-3), 221 (Fig. 20-3).

Butterworths, London: C. E. H. Bawn and M. B. Huglin, *Polymer* **3,** 257 (1962) (Fig. 17-6); D. R. Burfield and P. J. T. Tait, *Polymer* **13,** 307 (1972) (Figs. 19-1 and 19-2); I. D. McKenzie, P. J. Tait, and D. R. Burfield, *Polymer* **13,** 307 (1972) (Fig. 19-3).

The Chemical Society, London: W. C. Higginson and N. S. Wooding, *J. Chem. Soc.* **1952,** 774 (Table 18-1).

Chemie-Verlag Vogt-Schild AG, Solothurn: G. Henrici-Olivé and S. Olivé, *Kunstst.-Plast.* **5,** 315 (1958) (Figs. 20-4 and 20-5).

The Faraday Society, London: F. S. Dainton and K. J. Ivin, *Trans. Faraday Soc.* **46,** 331 (1950) (Table 16-9).

Interscience Publishers, New York: E. J. Lawton, W. T. Grubb, and J. S. Balwit, *J. Polym. Sci.* **19,** 455 (1956) (Fig. 21-1).

The Royal Society, London: N. Grassie and H. W. Melville, *Proc. R. Soc. (London) A* **199,** 14 (1949) (Fig. 23-6).

Verlag Chemie, Weinheim/Bergstrasse: F. Patat and Hj. Sinn, *Angew. Chem.* **70,** 496 (1958) (Eq. 19-11); G. V. Schulz, *Ber. Dtsch. Chem. Ges.* **80,** 232 (1947) (Fig. 20-6); J. Smid, *Angew. Chem.* **84,** 127 (1972) (Fig. 18-1); K. J. Ivin, *Angew. Chem.* **85,** 533 (1973) (Fig. 16-3).

Part V

Academic Press, New York: R. S. Baer, *Adv. Protein Chem.* **7,** 69 (1952) (Fig. 30-3).

W. H. Freeman and Company Publishers, San Francisco: Hans Neurath, "Protein-digesting enzymes," *Sci. Am.* **1964** (December), 69 (Fig. 30-1).

Interscience Publishers, New York: J. F. Brown, Jr., *J. Polym. Sci. C* **1,** 83 (1963) (Fig. 33-3).

Verlag Chemie, Weinheim/Bergstrasse: E. Thilo, *Angew. Chem.* **77,** 1057 (1965) (Fig. 33-4).

Didici in mathematicis ingenio, in natura experimentis, in legibus divinis humanisque auctoritate, in historia testimoniis nitendum esse.

G.W. Leibniz

(I learned that in mathematics one depends on inspiration, in science on experimental evidence, in the study of divine and human law on authority, and in historical research on authentic sources.)

Preface

Like so many of its kind, this textbook originated from the requirements of teaching. While lecturing on macromolecular science as a required subject for chemists and materials scientists on the undergraduate, graduate, and postgraduate levels at Swiss Federal Institute of Technology at Zurich (1960–1971), I needed a one-volume textbook which treated the whole field of macromolecular science, from its chemistry and physics to its applications, in a not too elementary manner. This textbook thus intends to bridge the gap between the often oversimplified introductory books and the highly specialized texts and monographs that cover only parts of macromolecular science. This first English edition is based on the third German edition (1975), which is about 40% different from the first German edition (1971), a result of rapid progress in macromolecular science and the less rapid education of the writer.

This text intends to survey the whole field of macromolecular science. Its organization results from the following considerations.

The chemical structure of macromolecular compounds should be independent of the method of synthesis, at least in the ideal case. Part I is thus concerned with the chemical and physical structure of macromolecules.

Properties depend on structure. Solution properties are thus discussed in Part II, solid state properties in Part III. There are other reasons for discussing properties before syntheses: For example, it is difficult to understand equilibrium polymerization without knowledge of solution thermodynamics, the gel effect without knowledge of the glass temperature, etc.

Part IV treats the principles of macromolecular syntheses and reactions. The emphasis is on general considerations, not on special mechanisms, which are treated in Part V. Part V is a surveylike description of important polymers, especially the industrially important ones. It also contains information about industrial monomer syntheses and selected biopolymers.

Undergraduate-level knowledge of inorganic, organic, and physical chemistry is assumed for the study of certain chapters. Whenever possible, all treatments and derivations were developed step-by-step from basic phenomena and concepts. In certain cases, I found it necessary to replace rigorous and mathematically complex derivations by simpler ones. I very much hope that this makes the book suitable for self-study.

Physical quantities are expressed in SI units. In many cases, the gram and not the kilogram was chosen as the more convenient unit and commas to group numbers have been retained, in addition to decimal points. Most symbols for physical quantities correspond to those recommended by the International Union of Pure and Applied Chemistry, although sometimes others had to be chosen for the sake of clarity.

A textbook must, of necessity, rely heavily on secondary literature available as review articles and monographs. Although I have consulted more than 5000 original papers before, during, and after the compilation of the individual chapters, I have (with the exception of one area) not cited the original literature. The exception is in the historical development of the subject, and this exception has been made because I was unable to find an accessible, balanced account treating macromolecular science in terms of the development of its ideas and concepts. I believe also that reading these old original works rewards the student with an insight on how a better understanding of the observed phenomena developed from the difficulties, prejudices, and ill-defined concepts of the times. However, because of the width and diversity of the field, a fully comprehensive and historically sound treatment of the development of its ideas and discoveries is beyond the scope of this work. Thus, since I have not been able to give due recognition to the work of individual chemists and physicists, I have only used names in the text when they have become *termini technici* in relation to methodology, phenomena, and reactions (for example: Ziegler catalysis, Flory–Huggins constant, Smith–Harkins theory, etc.). The occasional use of trade names cannot be taken to mean that these are free for general use.

In writing this book, I have tried to follow the practice of Dr. Andreas Libavius,* who had

> principally taken, from the most far-flung sources, individual data from the best authors, old and new, and also from some general texts, and these were then, according to theoretical considerations and the widest possible experience, carefully interpreted and painstakingly molded into a homogeneous treatise.

The reader may judge how much of this is true of this book.

* *Alchemia*, chemistry textbook from the year 1597; new edition in German, Gmelin Institute, 1964.

It is a pleasure for me to thank all of my colleagues, especially Drs. R. Bareiss, Mainz, P. Plesch, Keele, as well as Profs. G. Rehage, Clausthal, H. Ringsdorf, Mainz, and G. Zuber, Zurich, for the stimulating discussions we had prior to the publication of the first German edition. My special thanks go to the translator, Dr. John W. Stafford, Basel, and to my colleagues at Midland Macromolecular Institute, who kindly agreed to check all chapters: Dr. Mary Exner (first draft), Drs. Robert J. Kostelnik (Chapters 1–3, 29–33), Dale J. Meier (Chapters 4, 6–9, 11–15), Robert L. Miller, Sr. (Chapters 5, 10), John Semen (Chapters 16–28), and Karel Solc (Chapters 4, 6–9). I take, of course, the blame for factual and conceptual errors, and for the transcription of the Irish "Queen's English" translation of a Swiss–German text into German–American and any resulting mixup.

Midland *Hans-Georg Elias*

Contents

Vol. 1. Structure and Properties

Chapter 5. Supermolecular Structures 155

Part II. Solution Properties

Chapter 6. Solution Thermodynamics 203

Part III. Solid-State Properties

Contents

Vol. 2. Synthesis and Materials

Chapter 17. Polycondensations 595

Chapter 18. Ionic Polymerization 623

Chapter 23. Reactions of Macromolecules 799

Part V. Polymers

Chapter 31. Polysaccharides ... 1067

Notation

As far as possible, the abbreviations have been taken from the "Manual of Symbols and Terminology for Physicochemical Quantities and Units," *Pure and Applied Chemistry* **21**(1) (1970). However, for clarity, some of the symbols used there had to be replaced by others.

The ISO (International Standardization Organization) has suggested that all extensive quantities should be described by capital letters and all intensive quantities by lower-case letters. IUPAC does not follow this recommendation, however, but uses lower-case letters for specific quantities.

The following symbols are used above or after a letter:

Symbols Above Letters

— signifies an average, e.g., $\overline{M}$ is the average molecular weight; more complicated averages are often indicated by $\langle\ \rangle$, e.g., $\langle R_G^2 \rangle_z$ is another way of writing $(\overline{R_G^2})_z$

~ stands for a partial quantity, e.g., $\tilde{v}_A$ is the partial specific volume of the compound A; V_A is the volume of A, whereas $\tilde{V}_A^m$ is the partial molar volume of A

Superscripts

$^\circ$ pure substance or standard state

∞ infinite dilution or infinitely high molecular weight

m molar quantity (in cases where subscript letters are impractical)

- (q) the q order of a moment (always in parentheses)
- ‡ activated complex

Subscripts

- 0 initial state
- 1 solvent
- 2 solute
- 3 additional components (e.g., precipitant, salt, etc.)
- am amorphous
- B brittleness
- bd bond
- cr crystalline
- crit critical
- cryst crystallization
- e equilibrium
- E end group
- G glassy state
- i run number
- i initiation
- i isotactic diads
- ii isotactic triads
- is heterotactic triads
- j run number
- k run number
- m molar
- M melting process
- mon monomer
- n number average
- p polymerization, especially propagation
- pol polymer
- r general for average
- s syndiotactic diads
- ss syndiotactic triads
- st start reaction
- t termination
- tr transfer
- u monomeric unit
- w weight average
- z z average

Prefixes

at atactic
ct *cis*-tactic
eit erythrodiisotactic
it isotactic
st syndiotactic
tit threodiisotactic
tt *trans*-tactic

Square brackets around a letter signify molar concentrations. (IUPAC prescribes the symbol c for molar concentrations, but to date this has consistently been used for the mass/volume unit.)

Angles are always given by °.

Apart from some exceptions, the meter is not used as a unit of length; the units cm and mm derived from it are used. Use of the meter in macromolecular science leads to very impractical units.

Symbols

A absorption (formerly extinction) ($= \log \tau_i^{-1}$)
A surface
A Helmholtz energy ($A = U - TS$)
A^m molar Helmholtz energy
A preexponential constant [in $k = A \exp(-E^{\ddagger}/RT)$]
A_2 second virial coefficient
a exponent in the property/molecular weight relationship ($E^{\ddagger} = KM^a$); always with an index, e.g., a_η, a_s, etc.
a linear absorption coefficient, $a = l^{-1} \log(I_0/I)$
a_0 constant in the Moffit–Yang equation
b_0 constant in the Moffit–Yang equation
C heat capacity
C^m molar heat capacity
C_{tr} transfer constant ($C_{tr} = k_{tr}/k_p$)
c specific heat capacity (formerly: specific heat); c_p = specific isobaric heat capacity, c_v = specific isochore heat capacity
c "weight" concentration (= weight of solute divided by volume of solvent); IUPAC suggests the symbol ρ for this quantity, which

could lead to confusion with the same IUPAC symbol for density

$\hat{c}$ speed of light in a vacuum
$\hat{c}$ speed of sound
D diffusion coefficient
D_{rot} rotational diffusion coefficient
E energy (E_k = kinetic energy, E_p = potential energy, $E^{\ddagger}$ = energy of activation)
E electronegativity
E modulus of elasticity, Young's modulus ($E = \sigma_{ii}/\varepsilon_{ii}$)
E general property
$\mathbf{E}$ electrical field strength
e elementary charge
e parameter in the Q–e copolymerization theory
e cohesive energy density (always with an index)
F force
f fraction (excluding molar fraction, mass fraction, volume fraction)
f molecular coefficient of friction (e.g., f_s, f_D, f_{rot})
f functionality
G Gibbs energy (formerly free energy or free enthalpy) ($G = H - TS$)
G^m molar Gibbs energy
G shear modulus ($G = \sigma_{ij}$/angle of shear)
G statistical weight fraction ($G_i = g_i/\sum_i g_i$)
g gravitational acceleration
g statistical weight
g *gauche* conformation
g parameter for the dimensions of branched macromolecules
H enthalpy
H^m molar enthalpy
h height
h Planck constant
I electrical current strength
I radiation intensity of a system
i radiation intensity of a molecule
J flow (of mass, volume, energy, etc.), always with a corresponding index
K general constant
K equilibrium constant
K compression modulus ($p = -K\,\Delta V/V_0$)
k Boltzmann constant

k rate constant for chemical reactions (always with an index)
L chain end-to-end distance
L phenomenological coefficient
l length
M "molecular weight" (IUPAC molar mass)
m mass
N number of elementary particles (e.g., molecules, groups, atoms, electrons)
N_L Avogadro number (Loschmidt's number)
n amount of a substance (mole)
n refractive index
P permeability of membranes
p probability
p dipole moment
$\mathbf{p}_i$ induced dipolar moment
p pressure
p extent of reaction
Q quantity of electricity, charge
Q heat
Q partition function (system)
Q parameter in the Q–e copolymerization equation
Q polymolecularity index ($Q = \overline{M}_w/\overline{M}_n$)
q partition function (particles)
R molar gas constant
R electrical resistance
R_G radius of gyration
R_n run number
R_ϑ Rayleigh ratio
r radius
r_0 initial molar ratio of reactive groups in polycondensations
S entropy
S^m molar entropy
S solubility coefficient
s sedimentation coefficient
s selectivity coefficient (in osmotic measurements)
T temperature
t time
t *trans* conformation
U voltage

U internal energy
U^m molar internal energy
u excluded volume
V volume
V electrical potential
v rate, rate of reaction
v specific volume (always with an index)
W weight
W work
w mass fraction
X degree of polymerization
X electrical resistance
x mole fraction
y yield
Z collision number
Z z fraction
z ionic charge
z coordination number
z dissymmetry (light scattering)
z parameter in excluded volume theory

α angle, especially angle of rotation in optical activity
α cubic expansion coefficient $[\alpha = V^{-1}(\partial V/\partial T)_p]$
α expansion coefficient (as reduced length, e.g., α_L in the chain end-to-end distance or α_R for the radius of gyration)
α degree of crystallinity (always with an index)
α electric polarizability of a molecule
$[\alpha]$ "specific" optical rotation
β angle
β coefficient of pressure
β excluded volume cluster integral
Γ preferential solvation
γ angle
γ surface tension
γ linear expansion coefficient
δ loss angle
δ solubility parameter
δ chemical shift
ε linear expansion ($\varepsilon = \Delta l/l_0$)
ε expectation

ε_r relative permittivity (dielectric number)
η dynamic viscosity
$[\eta]$ Staudinger index (called J_0 in DIN 1342)
Θ characteristic temperature, especially theta temperature
θ angle, especially angle of rotation
ϑ angle, especially valence angle
κ isothermal compressibility $[\kappa = V^{-1}(\partial V/\partial p)_T]$
κ electrical conductivity (formerly specific conductivity)
κ enthalpic interaction parameter in solution theory
λ wavelength
λ heat conductivity
λ degree of coupling
μ chemical potential
μ moment
$\boldsymbol{\mu}$ permanent dipole moment
ν moment, with respect to a reference value
ν frequency
ν kinetic chain length
ξ shielding ratio in the theory of random coils
Ξ partition function
Π osmotic pressure
ρ density
σ mechanical stress (σ_{ii} = normal stress, σ_{ij} = shear stress)
σ standard deviation
σ hindrance parameter
τ relaxation time
τ_i internal transmittance (transmission factor) (represents the ratio of transmitted to absorbed light)
ϕ volume fraction
$\varphi(r)$ potential between two segments separated by a distance r
Φ constant in the viscosity–molecular-weight relationship
$[\Phi]$ "molar" optical rotation
χ interaction parameter in solution theory
ψ entropic interaction parameter in solution theory
ω angular frequency, angular velocity
Ω angle
Ω probability
Ω skewness of a distribution

Part I
Structure

Chapter 1
Introduction

1.1. Micro- and Macromolecular Chemistry

Macromolecules are chemical compounds built from a large number of atoms. They are therefore of high molecular weight. Macromolecules can be of natural origin (e.g., cellulose, proteins, natural rubber) or they may be synthetically produced [e.g., poly(ethylene), nylon, silicones]. As in the case of low-molecular-weight compounds, macromolecular substances can be classed as "organic" and "inorganic" macromolecules. All macromolecules, both organic and inorganic, consist of at least one chain of atoms bonded together and running through the whole molecule. This chain forms, so to speak, the backbone of the macromolecule. It can contain carbon–carbon bonds, e.g.,

$$\mathrm{R{-}CH_2{-}CH_2{-}(CH_2)_m{-}CH_2{-}CH_2{-}CH_2{-}R'}$$

poly(methylene)

or carbon–oxygen bonds, e.g.,

$$\mathrm{R{-}CH_2{-}CH_2{-}O{-}(CH_2{-}CH_2{-}O)_m{-}CH_2{-}CH_2{-}OR'}$$

poly(ethylene oxide) or poly(oxyethylene)

or carbon–nitrogen bonds, e.g.,

$$\mathrm{R{-}NH{-}CHR'{-}CO{-}NH{-}CHR''{-}CO \cdots NHCHR^{(x)}{-}CO{-}R^{(y)}}$$

polypeptides

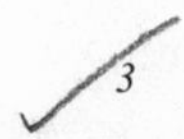

or contain no carbon atoms, e.g.,

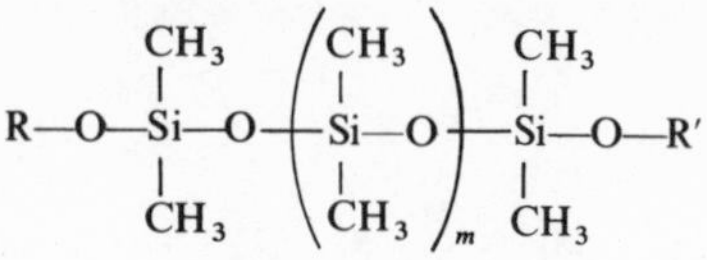

$$R—O—\overset{CH_3}{\underset{CH_3}{Si}}—O—\left(\overset{CH_3}{\underset{CH_3}{Si}}—O\right)_m—\overset{CH_3}{\underset{CH_3}{Si}}—O—R'$$

poly(dimethylsiloxane); silicon

Instead of purely covalent compounds, chelates are possible, as for poly[nickel-bis(8-hydroxyquinoline)],

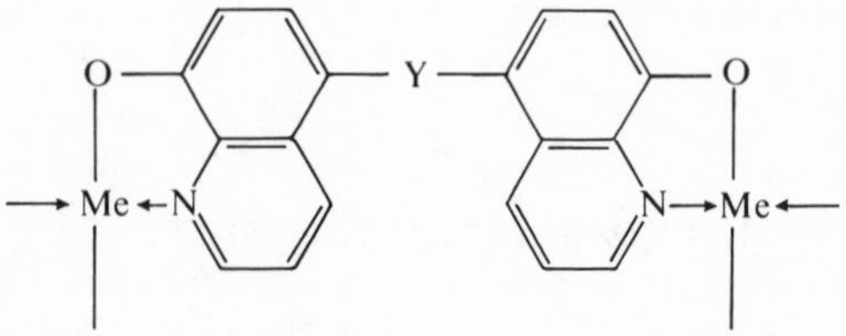

O Y O

→ Me ← N N → Me ←

or else exclusively electron-deficient bonds can occur, as for beryllium hydride,

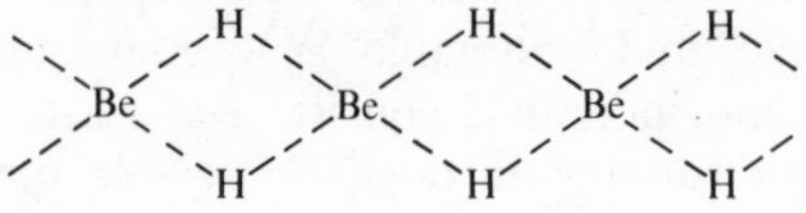

H H H

Be Be Be

H H H

(see Section 2.2.2). Chains composed entirely of like atoms are called isochains and those of unlike atoms, heterochains.

The tendency to form iso- or heterochains varies for different elements (see Section 2.2). Carbon is particularly prone to self-bonding or to bonding with the so-called heteroatoms of organic chemistry. The predominance of such "organic" chains with covalent bonds led in the past to the conception that macromolecular science is a small part of organic chemistry.

The opinion, forcefully defended by H. Staudinger in particular, that only chains with covalent bonds represent genuine macromolecules is historically reasonable and was necessary as a concept in the development of the idea of macromolecular character (see Section 1.3). The division between the true macromolecule and a coordination lattice can be quite fluid for inorganic compounds, so that the macromolecular character of such compounds cannot always be easily recognized or defined. Ion lattices, such as those present in common salt crystals,

$$\begin{array}{cccccc} Na^+ & Cl^- & Na^+ & Cl^- & Na^+ & Cl^- \\ Cl^- & Na^+ & Cl^- & Na^+ & Cl^- & Na^+ \\ Na^+ & Cl^- & Na^+ & Cl^- & Na^+ & Cl^- \\ Cl^- & Na^+ & Cl^- & Na^+ & Cl^- & Na^+ \end{array}$$

are not, however, considered to be macromolecular structures (see Section 1.2).

An organic macromolecule with covalently bonded atoms represents an ideal case of a macromolecular structure for which the concept of a macromolecule can be defined in a particularly unequivocal way and proven by means of unambiguous experiments. The existence of inorganic macromolecules was hesitantly acknowledged since many findings seemed to speak against it. The success of the formulation of *reactions* via ions led to the erroneous conclusion that the corresponding chemical *bonds* are entirely polar. From this it has been further concluded that there can be no inorganic macromolecules in the solid state. In fact, the purely ionic structure is an ideal case which is seldom realized. The bonds between two ions may contain some covalent character in addition to the electrostatic interactions. The formulation of ionic reactions does not exclude the existence of genuine inorganic macromolecules.

The impossibility of transferring these solid-state structures into a fluid state was also an apparent argument against inorganic macromolecules. The existence of inorganic macromolecules in the crystalline state is also difficult to prove decisively, since a transformation from a molecular to an ionic lattice can occur with temperature. Germanium telluride, for example, is present in an arsenic-type structure below 670 K, whereas above 670 K, it exists as a rock-salt structure. The former is a molecule lattice, the latter an ion lattice. The same substance could, therefore, be defined as macromolecular or nonmacromolecular according to its state.

Macromolecules are sometimes also called polymers. A polymer, in a strict sense, is a substance containing many identical monomeric units. Such a monomeric unit is the ethylene oxide $\text{—}(CH_2CH_2O)\text{—}$ unit in poly(ethylene oxide). In the strictest sense, a polymer is built solely of the monomeric units and possesses no end groups. In poly(ethylene oxide) formulas such end groups can be, for example, R = OH and R′ = H, so that the molecule then possesses two hydroxyl end groups.

Instead of monomeric units, one frequently talks of base units or mers. These terms always relate to the origin of the building block in the chain (process-based nomenclature). The smallest structure-based unit is called,

Table 1-1. Monomeric Units and Constitutional Repeating Units of Various Macromolecules

Initial Monomers	Monomeric units	Repeating units
$CH_2{=}CH_2$	$—CH_2—CH_2—$	$—CH_2—$
CH_2N_2	$—CH_2—$	$—CH_2—$
$Cl(CH_2)_2Cl$ + $Na(CH_2)_2Na$	$—CH_2—CH_2—$	$—CH_2—$
$Cl(CH_2)_2Cl$ + $Na(CH_2)_3Na$	$—CH_2—CH_2—$ $—CH_2—CH_2—CH_2—$	$—CH_2—$
$NH_2(CH_2)_6NH_2$ + $HOOC(CH_2)_4COOH$	$—NH(CH_2)_6NH—$ $—CO(CH_2)_4CO—$	$—NH(CH_2)_6NH—CO(CH_2)_4CO—$

by contrast, a repeating unit. A repeating unit can be larger than, smaller than, or equal to the monomeric unit (Table 1-1). The repeating unit was formerly called a structural element. Repeating units can be classified as constitutional, configurational, or conformational repeating units.

The term "poly(ethylene oxide)" and similar terms such as "poly(ethylene)," "poly(styrene)," etc., leave the nature of the end groups open. This designation is justified, since the mass contribution of the end groups to the mass of the whole macromolecule is slight, especially at high molecular weights. Conversely, the chemical structure of the end groups is difficult to identify at high molecular weights. However, in spite of this small contribution, they are still quite frequently able to influence the properties of macromolecules, especially electrical properties and resistance to degradation.

Thus, strictly speaking, a polymer can only be a macrocyclic structure. For example, the cyclic oligomers of ε-aminocaprolactam,

$$\begin{array}{c} \ulcorner\!\!-\!\![NH\!\!-\!\!(CH_2)_5\!\!-\!\!CO]_{n-1}\!\!-\!\!\urcorner \\ \llcorner\!\!-\!\!CO\!\!-\!\!(CH_2)_5\!\!-\!\!NH\!\!-\!\!\lrcorner \end{array}$$

can be included among rings of this type, and yet these molecules are of relatively low molecular size. Polymers with a small number of monomeric units (mostly 2 to ~20) with unknown or unspecified end groups are called *oligomers*. On the other hand, *telomers* are oligomers formed by transfer reactions (see Section 20.3.3) where the end groups consist of fragments of the chain transfer agent. *Telechelic polymers* are low-molecular-weight compounds with known functional end groups.

Macrocyclic compounds in the proper sense, on the other hand, are present in a few naturally occurring polymers, as in the deoxyribonucleic acids (DNA) of the bacteriophage ϕX 174, or of a polyoma virus. These macromolecules are true macrocyclic compounds; they are, however, not polymer rings. That is to say, a true polymer contains only one type of monomeric unit. Macromolecules of this kind are also called *homopolymers* (previously, unipolymers).

Macromolecules with several types of monomeric units are called *copolymers*. Copolymers can be classified according to the number of different types of monomeric units in each macromolecule, e.g., bipolymers, terpolymers, quaterpolymers, etc. Periodic bipolymers with the sequence ABABAB are also called alternating copolymers. Conversely, aperiodic copolymers show no order in their sequences of monomeric units.

The degree of polymerization X of a compound is defined as the number of monomeric units of molecular weight M_u joined together in a macromolecule of molecular weight M. For high molecular weights (the molecular weight of the end groups M_E making a negligible contribution), X is obtained by dividing the macromolecular weight M by the monomeric-unit molecular weight M_u, i.e.,

$$X = (M - M_E)/M_u \approx M/M_u \tag{1-1}$$

(The correct term now for "molecular weight" is "molar mass." The old name "molecular weight" has, however, been retained because of its familiarity and in order to avoid misunderstanding.)

Most macromolecular substances consist of a mixture of macromolecules of differing degrees of polymerization: They are *polymolecular*. Molecular weights or degrees of polymerization of such macromolecular substances are thus always averages. According to which statistical weight forms the basis of the averaging (see Chapter 8), one can distinguish between, for example, the number average

$$\langle X_n \rangle = \frac{\sum_i n_i X_i}{\sum_i n_i}, \qquad \langle M_n \rangle = \frac{\sum_i n_i M_i}{\sum_i n_i} \tag{1-2}$$

and the weight average of the molecular weight or degree of polymerization

$$\langle X_w \rangle = \frac{\sum_i W_i X_i}{\sum_i W_i}, \qquad \langle M_w \rangle = \frac{\sum_i W_i M_i}{\sum_i W_i} \tag{1-3}$$

where n_i and W_i are the moles and the weights of the species i of molecular weight M_i.

Most authors use the word "polydisperse" instead of "polymolecular." Polydispersity refers, however, to the dispersion of properties of any particles, whereas polymolecularity refers to the distribution of molecular weights only. A distinction is necessary because a given polymolecularity may lead to different polydispersities in aggregating systems.

Macromolecules are sometimes divided into natural and synthetic macromolecules. Synthetic macromolecules are normally composed of one to three different types of monomeric units, whereas natural macromolecules or biopolymers may contain many different types. Four main classes of biopolymers exist: polyprenes, polysaccharides, nucleic acids, and proteins. Polyprenes (see Section 25.3.2) contain only one type of monomeric unit: They are homopolymers and correspond in constitution and configuration to the synthetic polymers. Polysaccharides (see Chapter 31) are either homopolymers or copolymers with up to five different types of monomeric units. Proteins (see Chapter 30) may consist of up to 20 different monomeric units per macromolecular chain. Nucleic acids are composed of relatively few types of monomeric units (see Chapter 29). The combination of units of these four types of biopolymers may lead to other biopolymer classes, e.g., nucleoproteins, glycoproteins, etc.

There is, however, one significant difference between synthetic and biological polymers. Biological copolymers very often possess an aperiodic arrangement of the monomeric units in a given chain similar to many synthetic polymers. In contrast, the arrangement of the monomeric units in biopolymers is the same for all chains of the biopolymer, whereas synthetic copolymers normally exhibit a distribution from chain to chain. The biopolymers are thus molecularly homogeneous, i.e., their chains possess the same composition and sequence. Such molecularly homogeneous but aperiodic chains can be synthesized *in vitro* only with enormous difficulty, if at all. The aperiodic sequence of the enzymes is the key to their importance as very specific and selective catalysts. In nucleic acids, the aperiodicity serves as a carrier of genetic information.

The significance of synthetic polymers, on the other hand, lies more in their mechanical, electrical, or optical properties. Their use as raw materials for plastics, elastomers, or synthetic fibers is especially important. The chemical structure of these substances plays a subordinate role for their application, and it is desirable that such substances are as chemically inert as possible: otherwise the useful properties could change unfavorably with time. Since the properties of plastics depend on physical quantities, the chemistry of synthetic macromolecules is bound inseparably to the physics and physical chemistry of such substances. Thus, a clear-cut division into pure preparative chemistry and pure physics of macromolecules is inadvisable. Macromolecular science represents a true "interdisciplinary science."

A comparison of the industrial preparation and use of high- and low-molecular-weight substances illustrates this point. In the chemistry of small molecules the technology of reactions and products can be clearly distinguished. If the reaction conditions (pressure, temperature, catalyst concentration, reactant concentration, etc.) in low-molecular-weight chemistry are altered, then the yields of the primary and secondary products change, although as a rule the same products are obtained. In macromolecular chemistry, on the other hand, small changes in reaction conditions lead not only to variations in yield, but generally also to a different product composition since by-products in this case can occur as an inseparable part of the macromolecular chain (i.e., extra branching, different degrees of stereoregularity, changes in the molecular weight, changes in the molecular-weight distribution, etc.). Compositional alterations of this kind lead to different properties and, hence, to other areas of application. Very often these compositional alterations can only be detected with difficulty by quantitative chemical and physical methods. In some cases, chemical and physical techniques do not even give qualitative evidence, and the compositional change can only be detected by the variation of properties in actual use of the polymer. Knowledge of the synthetic process and its mechanism frequently provides an insight into these compositional changes.

The great influence exerted by the structure of macromolecular compounds on properties, in contrast to low-molecular-weight structures, becomes obvious in a comparison of the two 1,2-dimethylethylenes with the 1,4-poly(butadienes). The melting point T_M of the *trans*-butene-2 is about 34°C higher than that of the *cis*-butene-2, while the boiling points T_{bp} only differ by about 3°C. While butene-2 shows two isomers, poly(butadiene) can occur in five isomeric forms.

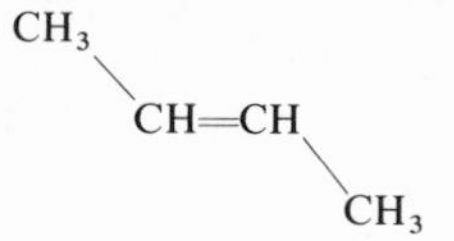

CH_3—CH═CH—CH_3 (trans)

trans-butene-2
(*trans*-1,2-dimethyl ethylene)
$T_M = -105.8°C$
$T_{bp} = +1°C$

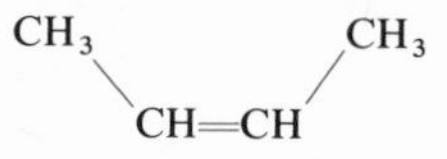

CH_3—CH═CH—CH_3 (cis)

cis-butene-2
(*cis*-1,2-dimethyl ethylene)
$T_M = -139.3°C$
$T_{bp} = +3.7°C$

The 1,4-poly(butadienes) corresponding to the butene-2 compounds, however, differ in their melting points by almost 140°C (Table 1-2). 1,4-*cis*-Poly(butadiene) is an elastomer; 1,4-*trans*-poly(butadiene) is thermoplastic. The other isomeric poly(butadienes) likewise show a marked difference in their properties. Aromaticized 1,2-poly(butadiene), although not an isomer of poly(butadiene), shows semiconductor properties because of its conjugated double bonds.

In practice, *cis*- and *trans*-poly(butadiene) can be subjected to the same reactions, although, of course, the reaction rates are different. In a similar way, high-molecular-weight poly(ethylene) differs from low-molecular-weight paraffins only in its mechanical properties and not in its chemical behavior.

The essential difference between low-molecular-weight and macromolecular chemistry lies not in the type of bond or in chemical structure and reactions, but only in the size of the molecule and its related properties. In one sense, low-molecular-weight substances can be considered as point molecules and are a limiting case for macromolecular compounds of very low molecular weight. Because of the great chain length of high polymers,

Table 1-2. Isomeric Poly(butadienes)

Designation	Structural formula	Melting point T_M, °C	Properties
1,4-*cis*	$—CH_2$ $CH_2—$ CH=CH	2	Elastomer
1,4-*trans*	$—CH_2$ CH=CH $CH_2—$	140	Thermoplastic
1,2-isotactic	$/CH_2\backslash CH/CH_2\backslash CH/$ CH (under each CH) ‖ CH_2 CH_2	126	Thermoplastic
1,2-syndiotactic	CH_2 ‖ CH (above second CH) $/CH_2\backslash CH/CH_2\backslash CH/$ CH (under first CH) ‖ CH_2	156	Thermoplastic
1,2-cyclic	$/CH_2\backslash CH/CH_2\backslash CH/$ | | $\backslash CH_2/CH\backslash CH_2/CH\backslash$	—	Thermoset[a] (insulator)
1,2-aromatic[b]	/CH=C/CH=C/ | | \CH=C\CH=C\	—	Thermoset[a] (electrical semiconductor)

[a] Insoluble, since the cyclization reaction also leads to intermolecular bonds (cross-linking).
[b] No isomer.

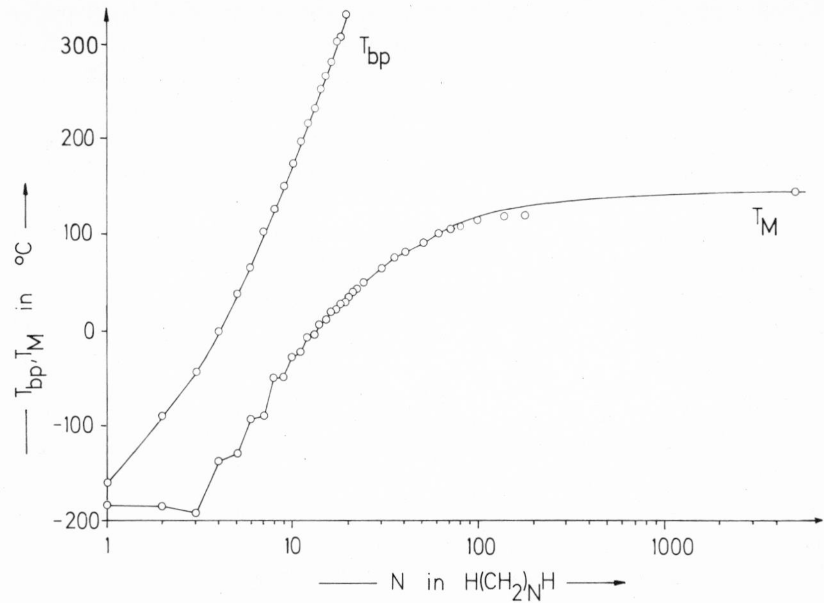

Figure 1-1. Variation of boiling temperature T_{bp} and melting temperature T_M with molecular weight for a homologous series of alkanes.

many different arrangements of chain segments are possible in space, leading to a significant role of entropic effects and thus of statistics.

The dominant role of statistics is shown not only in the number of possible spatial arrangements of the chain atoms, but also in the reaction mechanisms (reactions leading to macromolecules are called polyreactions). Only under stringent conditions during synthesis can one obtain molecularly homogeneous compounds, i.e., compounds in which each molecule shows the same overall composition, the same chain sequence, and the same molecular weight. If the macromolecules are of identical chemical structure, but differ in molecular weight (ignoring end groups), then one speaks of a *polymer homologous* series. A mixture of a few members of the same, or different, molecularly homogeneous polymer series is termed *paucimolecular*, while one containing a large number of such polymer homolog is *polymolecular*. Thus polymolecular materials possess a molecular weight distribution.

The homologous series concept thus differs for low-molecular-weight chemistry and for polymers. In the former, for example, one refers to aliphatic, unbranched alcohols as homologous:

$$CH_3OH, \quad CH_3{-}CH_2OH, \quad CH_3{-}CH_2{-}CH_2OH, \quad \text{etc.}$$

The homology here is the hydroxyl group, one of the two end groups in the sense of macromolecular chemistry. In the homologous polymer series concept it is the central methylene groups which are decisive. They define the series, and the chemical nature of the end groups is of secondary interest.

Chemically, no sharp distinction exists between the small molecules (oligomers) and large molecules (polymers) representing a homologous polymer series. A compound can be considered macromolecular if the end groups, or the substitution of *one* representative group by another, no longer have any significant influence on the properties of this compound. This definition is not the same as the one often used, which states that properties should not change when the number of monomeric units is increased by one. In fact, there are many properties in a homologous polymer series which alter continually with the degree of polymerization, while others become virtually constant above a certain degree of polymerization. For example, with alkanes, the boiling point increases with molecular weight until at high molecular weights, they decompose before volatilizing, but the melting point is practically independent of molecular weight at very high molecular weights (Figure 1-1). However, with alkanes, the melt viscosity increases exponentially with the molecular weight.

1.2. The Molecular Concept

In low-molecular-weight chemistry, a molecule is seen as a physical unit with a fixed spatial order. A unit of this kind can only exist if the bonding energy is greater than the energy of thermal movement kT. Since the Boltzmann constant has a value of $k = 1.3807 \times 10^{-23}$ J/K, the bonding energy at room temperature (293 K) per mole must be greater than 2430 J/mol bond. Thus a structure considered as a molecule at a low temperature can lose its identity at high temperatures and change to a new structure which can again be called a molecule. An example of this is the dissociation of N_2O_4 to 2 NO_2.

The concepts of a "physical unit" and "bonding energy" are closely linked in low-molecular-weight compounds. It is appropriate here to differentiate between chemical and physical bonds. Chemical bonds can be considered those bonds with a bonding energy greater than about 4.2×10^5 J/mol, while physical bonds possess a lower bonding energy. The greater energy content of a chemical bond means a shorter bonding distance. Thus in low-molecular-weight chemistry, a chemical molecule can also be defined as a group of bonded atoms, which can be separated in principle from any other such unit. If such a molecule is in a fluid phase, i.e., gas or liquid, then true chemical molecules are always present, at least at infinite dilution. In the gaseous phase, this molecule remains a stable unit since its bonding energy is by definition greater than the energy of thermal motion.

In solution, the relationships are more complicated. In short, for low-molecular-weight substances the bonding energies are greater than the energies of the physical bonds between the solute and solvent molecules.

In the crystalline state the situation is unequivocal as long as the compounds finally crystallize into molecular lattices. This crystallographic definition means no more than that the atomic distances within the so-defined molecule are smaller than those outside. Expressed another way: Chemical bonds in molecular lattices should not overlap the boundaries of the unit cell (for definition of unit cell, see Section 5.3.2).

At finite concentrations of low-molecular-weight molecules in fluid phases, intermolecular physical bonds can effectively bring molecules together as large aggregates, i.e., "physical molecules." Associations of this type are still called "polymers" in many areas of chemistry. This designation is correct in the historical sense, as well as the strictly verbal sense, since it deals with an association of atoms that develops from a simple plurality of one unit, a mer. However, since true chemical molecules and not physical aggregates are considered polymers, these types of "polymers" will be referred to in this book as *associates.* In principle, associates can be distinguished from macromolecules by studying the effect of different solvents on aggregate size as the inter- and intramolecular energy interactions for the free macromolecule are then varied. Since these interaction energies are mostly temperature-dependent, a conclusion can be reached by determining molecular weight at different temperatures. The decision as to whether the colloidal properties of many substances arise from an association of small particles or from true macromolecular structures marks the beginnings of macromolecular chemistry (see Section 1.3).

Thus, from the definition of a molecule as a physical unit with fixed bond angles and bond lengths, it follows that it must be possible to isolate a single species. In low-molecular-weight compounds this can be achieved at least in theory by working in the vapor or solution state. Macromolecules, in contrast, cannot be vaporized, since the energy of vaporization is given by the total of all the cohesion energies for each polymer molecule. Thus the macromolecule will be thermally degraded even below its boiling point.

Macromolecules that are separated from one another by a large distance can, therefore, only be found in solution. In a solid, the molecules are closely packed, so that the differences in distances between atoms belonging to a specific molecule and those from different molecules are small. The question of whether certain atoms belong to a specific molecule thus becomes more difficult. It has led certain authors to consider that three-dimensional, insoluble macromolecules (e.g., phenolic resin, vulcanized rubber, and also diamond and quartz) are no longer macromolecules. There were essentially two reasons given for this suggestion, beginning with the insolubility of the compound. According to one group of authors, these structures were too large to be macromolecular, while the other group considered irregular three-dimensional compounds (phenolic resins, etc.) as macromolecules, but not those with regular structures (diamond, etc.).

Both viewpoints are untenable, since they arise from an experimental aspect, solubility. Experimental convenience and the skill of an investigator can, however, never provide the basis for a scientific definition. In addition, both opinions confuse cause and effect. Insolubility is a consequence and not a cause of high molecular weight. In principle it is impossible to give either an upper limit for molecular weight or an upper degree of order above which structures can no longer be considered macromolecules. Solubility decreases with increasing size of the highly ordered, annular, aromatic hydrocarbons, i.e., benzene, naphthalene, anthracene, through perylene, coronene, to graphite. However, this does not enable one to draw a sharp line between the "molecules" benzene, naphthalene, anthracene, perylene, etc., and the "nonmolecular" graphite. The two-dimensional graphite is thus just as much a macromolecule as one-dimensional poly(ethylene) or the three-dimensional diamond.

Macromolecules are referred to as one, two, or three dimensional when there are two, three, or four chain bonds per chain atom. This form of reference, based on the constitution, gives no indication of the arrangement of the chains in space, which is determined by the conformation (see also Section 2.4).

Macromolecular character can easily be recognized in one-dimensional molecules with atoms bonded covalently in the main chain. Such molecules occur, for example, in the chain molecules of poly(ethylene), nylon, poly(dimethylsiloxane), etc. (see Section 1.1). In the chains of isotactic poly(styrene), the length of the carbon–carbon bond is 0.154 nm, while the distance from one carbon atom to another in a different chain is about 0.35 nm, and solvent molecules are about 0.4–0.5 nm apart. Therefore, the bond distances between adjacent chain atoms are always substantially shorter than the distances between the chain atoms and those of the surroundings.

It is not as easy to draw a sharp dividing line between associations of atoms thought of as molecules and their surroundings when the intra- and intermolecular distances are comparable. The covalent bond is most pronounced for carbon. The contribution of covalency to the bond character decreases for other elements, whereas the bond length increases. Inorganic macromolecules are therefore often difficult to recognize because they cannot always be transferred unchanged into the solution state due to strong interactions with the solvent.

The definition of a molecule as a physical unit with a stable arrangement in space is very broad and includes, theoretically, both a common salt crystal and a piece of iron. From the chemist's point of view it is appropriate to limit this definition a little more by taking into account the nature of the bonds. In this book a compound will be considered a macromolecule when its main-chain atoms are bonded by directed valences with the bond electrons shared by all chain atoms involved in the bonding.

A definition of this kind confines bond types to covalent bonds, coordinate bonds, and electron-deficient bonds. Metallic bonds and ionic bonds are excluded. Associations of atoms built from metallic bonds are not considered to be macromolecules since the electrons could be shared by all atoms, but the bonds are nondirectional. Likewise, ionic crystals such as common salt are not considered macromolecules because, for the ideal ionic bond, the electrons are not common to both atoms and the bond is likewise nondirectional. On the other hand, compounds held together by hydrogen bonds, such as the double helix of DNA, are considered macromolecules and not associates since the electrons are shared by the hydrogen bonds, and the hydrogen bond is directed.

1.3. Historical Development

There is no remembrance of former things; neither shall there be any remembrance of things that are to come with those that shall come after.

Ecclesiastes 1.11

Since ancient times, naturally occurring polymers have been used by mankind for various purposes. Meat (protein, see Chapter 31) and corn (polysaccharide, see Chapter 32) are important sources of food. Wool and silk, both proteins, serve as clothing. Wood, the main component of which is cellulose, a polysaccharide, is used for building and fire-making. Amber, a high-molecular-weight resin, was worn by the Greeks as a jewel. The use of asphalt as an adhesive is mentioned in the Bible.

In 1839, Simon[(1)] observed that styrene ($CH_2{=}CHC_6H_5$) changed on heating from a clear liquid to a solid, transparent mass—to a poly(styrene) in the modern sense. Since the overall composition of carbon and hydrogen remained constant, Berthelot[(2)] called this event a polymerization. So the word polymerization originally meant simply that several molecules formed one larger association without alteration of the overall composition. This left open to question how one was to imagine such a combination (polymerization in the modern sense) and the structure of the products obtained. Berthelot also noticed that depolymerization of the solid mass back to styrene occurred at even higher temperatures. This simple conversion of styrene $\rightleftharpoons$ poly(styrene) $\rightleftharpoons$ styrene, obtainable only through temperature change, later formed an apparently reliable basis for the micelle theory of such substances.

Before Berthelot, Wurtz[3] had already converted ethylene oxide into low-molecular-weight poly(ethylene oxides) (using the modern terminology):

$$n\,\underset{\diagdown\ \mathrm{O}\ \diagup}{CH_2{-}CH_2} \longrightarrow -\!\!\left(CH_2{-}CH_2{-}O\right)\!\!-_n \tag{1-4}$$

Lourenço[4] carried out this transformation in the presence of ethylene halides and isolated substances up to $n = 6$ from the reaction medium. He established that the overall structure for compounds of this kind approached that of pure ethylene oxide, although the properties of these substances were different from those of ethylene oxide. Lourenço also noticed that this compound, which was a liquid at room temperature, showed increased viscosity with an increasing degree of polymerization n and he proposed a chain formula for the products.

Shortly after this study, Graham[5] discovered that certain substances, such as lime, diffuse much more slowly than, say, common salt, and do not readily permeate a membrane. Since this behavior was characteristic of limelike, noncrystalline substances, whereas the crystalline substances known at that time all diffused and permeated rapidly, Graham differentiated between crystalloids and colloids (from the Greek $\kappa o\lambda\lambda\alpha$ = lime). He thus attributed the colloidal behavior to the structure of the colloids and not to their state.

Further subdivision of the colloids concerned many researchers. For example, studies of coagulation processes lead Müller[6] to connect suspensions with physical disintegration processes and large molecules with chemical precipitation methods. He designated as "high molecular" such substances as albumin and colloidal silica. A later classification by Staudinger[7] into colloidal dispersions, micellar colloids (association colloids), and colloidal molecules (macromolecules), proved to be very suitable and forms the foundation of modern textbooks on colloid science.[8] It was later found that inorganic substances can also behave as colloids, e.g., the basic oxides of iron and aluminum. In their capacity for reaction these colloidal substances showed no great difference from the chemistry of similar, but crystalloid, substances. The correct conclusion was obviously that all substances can be transferred into the colloidal state under suitable conditions and could also revert to the noncolloidal state (e.g., styrene). Thus the colloidal state is, in general, a possible *state* for materials (Ostwald,[9] von Weimarn[10]). Von Weimarn also disputed Graham's thesis on crystalloid and colloidal *materials*, since he was able to change crystalline substances into the colloidal state. The correct conclusion, that all low-molecular-weight substances can be transferred into the colloidal state, led to a false antithesis to this principle, namely, that all colloidal particles or aggregates are therefore composed of smaller molecules, i.e., physical polymers.

In the years between Graham's discovery and Ostwald's postulation, the modern idea of true macromolecules was very much alive. For example, in 1871 Hlasiwetz and Habermann[11] assumed that proteins and polysaccharides were macromolecules, but with the methods available at that time they could not prove their thesis. In particular, the methods failed to prove the high molecular weight which they suggested.

The possibility of such proof became available from laws governing the relation between vapor pressure and mole fraction, or between osmotic pressure, concentration, temperature, and molecular weight, which were discovered by Raoult[12] (1882–1885) and van't Hoff[13] (1887–1888). With these methods, very high molecular weights (between 10,000 and 40,000) were subsequently obtained for rubber, starch, and cellulose nitrate. Other authors found similarly high values for the same materials; e.g., Gladstone and Hibbert[14] found 6000–12,000 for rubber, and Brown and Morris[15] obtained cryoscopically about 30,000 for a product of starch obtained by degradation hydrolysis.

To most research workers at that time, however, these high molecular weights appeared untrustworthy. That is to say, the same methods used on covalently structured crystalloids gave molecular weights which agreed satisfactorily with the chemical formula weights. But since formula weights for colloids could not be found unequivocally, the molecular weights obtained by physical methods also appeared suspect.

In addition, the proportionality between vapor pressure and concentration (Raoult's law) and that between osmotic pressure and concentration (van't Hoff's law) had to be satisfied. Both requirements were adequately fulfilled within the limits of experimental error by the covalent crystalloids then studied, but not by the colloids. This "error" concerning the two laws made the high molecular weights of the colloids also seem suspect. However, we know today that both laws are only limiting laws for infinite dilution. A molecular weight apparently dependent on concentration, i.e., calculated from the limiting laws, is also the rule rather than the exception for low-molecular-weight substances. This effect, dependent on the interaction between the molecules in the solution, was known as early as 1900 from ebullioscopic measurements by Nastukoff,[16] who also proposed an extrapolation to zero solute concentration. Caspari[17] obtained a molecular weight of 100,000 for rubber using osmotic measurements by a similar extrapolation procedure.

At that time, however, the formulation of the laws of Raoult and van't Hoff as limiting laws seemed unacceptable. The apparent unlimited validity of the laws for covalent crystalloids and the retention of identity of these low-molecular-weight compounds in the colloid contradicted this theorem. In addition, colloids were not the only class of compound which showed a marked deviation from Raoult's law. Similar discrepancies were found in electrolytes. Since the electrolytes known at that time were solely inorganic compounds and could theoretically be formed into colloids, this was coming close to the notion that some peculiar forces were involved.

The results of the first great advances in full-fledged modern organic chemistry also spoke against the assumption that macromolecules are held together by covalent bonds. The great success of classical organic chemistry was based mainly on three principles: the formulation of a reaction in terms of the smallest constitutional change occurring; the use of elemental analysis as a basic test of a proposed constitutional formula; and, in particular, the possibility of crystallizing pure substances. At that time, however, colloids could not be crystallized. Indeed, even in low-molecular-weight organic chemistry there were substances that were difficult to crystallize, such as alcohol or sugar, but these were considered primarily as inexplicable exceptions. In addition, colloids lacked a broad purity criterion as used in organic chemistry. A substance was considered "pure" when it showed a definite structural formula with a definite molecular weight. For one class of colloids known at that time, however, there were obviously similar structural formulas with varying molecular weights.

The nature of the peculiar binding forces exhibited by such colloids was therefore important. The existence of intermolecular forces was known form the study of gases.(18) Similar forces could exist in solution. For organic molecules the law of partial valence seemed appropriate and could be effective, according to Thiele,(19) in substances with conjugated double bonds. This thesis seemed established since molecular compounds such as chinhydrones existed.(20)

The hypothesis of partial valence gave a convenient explanation for the behavior of natural rubber. The overall formula C_5H_8 already proposed by Faraday in 1826(21) pointed to one double bond per unit. Harries(22) confirmed this conclusion by ozonolysis of natural rubber and subsequent hydrolysis of the ozonides. Since he also found C_5H_8 to be the overall formula, he did not think that he would have to consider any "end groups." From the observed low molecular weights, he concluded initially that there were rings of two isoprene units:

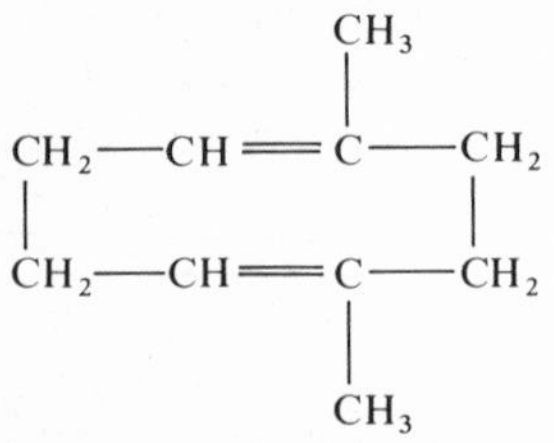

```
                    CH3
                    |
CH2——CH══C——CH2
|                   |
CH2——CH══C——CH2
                    |
                    CH3
```

Later he concluded that there were five to seven isoprene units per cyclic molecule.

The fact that rubber cannot be distilled also seems to suggest low-molecular-weight, cyclic compounds held together by partial valence. It was known that associated substances have a much higher boiling point than nonassociated ones. Pickles,[23] on the other hand, suggested a chain structure for rubber. As proof of constitution he carried out the first relevant polymer modification, namely the addition of bromine across the double bonds of rubber. Since the bromine addition did not alter the molecular size, Pickles considered natural rubber to be a true molecule. However, his conclusions were not generally accepted.

Rings similar to those of rubber were then proposed for numerous organic colloids. For example, the structural formula for cellulose (see Section 31.5) was written as

$$\left(\begin{array}{l} \text{OH} \\ | \\ \text{CH—CH—CH—OH} \\ | \qquad \diagdown \text{O} \;\; \diagdown \text{O} \\ | \qquad \diagup \qquad \diagup \\ \text{CH—CH—CH}_2 \\ | \\ \text{OH} \end{array} \right)_n$$

With its three hydroxyl groups and the hemiacetal group, the formula satisfactorily reproduced the chemical nature of cellulose. The colloidal character was explained to a large extent by an association of many cyclic compounds of this kind. The fact that no end groups were found was also in agreement with the assumption of cyclic compounds. We know today that the contribution made by end groups to the high molecular weight was much too small to be detected by the methods then in use.

Association was affirmed by a further observation: The rotation of optically active diamylitaconate was, in fact, more or less equal for monomer and "polymer."[24] Otherwise, differences were found for substances which varied constitutionally.

Around 1910–1920, all facts seemed to support the theory that organic colloids were physical groups of particles and not true covalent macromolecules. Organic colloids had the same overall composition and reactivity as their known noncolloidal basic components. Furthermore, they could often be converted back to the noncolloidal material (as in the styrene study by Berthelot) and were not crystallizable. Anomalies appeared during molecular-weight determinations (with regard to a concentration dependence of the apparent molecular weight). Yet all these phenomena were already known in inorganic colloids. These, too, could only be crystallized, if at all, with the loss of colloidal character. In addition, since no end groups were found, all observations seemed to confirm the assumption of low-molecular-weight rings. Theoretically, the colloidal character was easily explained: It was held together by van der Waals forces, e.g., by Thiele's partial valence.

Staudinger, however, disputed the assumption of molecular complexes in organic colloids. In his studies on ketenes,[25] he obtained "polymeric" products which he considered to be cyclobutane derivatives. Since another author[26] considered these dimers to be molecular complexes, Staudinger compiled all the arguments for covalent bonds in a work which has since become classic.[27] End groups, necessary from valence considerations but not found experimentally, did not seem to contradict this idea, since it was generally accepted at that time that the reactivity of a group decreased with increasing molecular weight.

In his later work, Staudinger tried to establish experimentally his concept of organic colloids as true macromolecules. For this, it was first necessary to refute the idea of the so-called first micellar theory, which states that in organic colloids small rings were held together by partial valence. In 1922, Staudinger and Fritschi[28] hydrogenated rubber. Since the hydrogenated rubber no longer contained any double bonds, it should according to the micellar theory, no longer show any colloidal properties. Yet in actual fact the colloidal properties were retained, as Pickles had already found in his bromination study. The hydrogenation of poly(styrene) to poly(vinyl cyclohexane) also excluded colloids based on Thiele's partial valence. Staudinger therefore concluded that these organic colloids consisted of many atoms joined together by covalent bonds, i.e., true "macromolecules."[29]

Since bonds consisting of van der Waals forces are much weaker than covalent bonds, covalently bonded molecular colloids, unlike associated colloids, should retain their colloidal character in all solvents.[30] However, this discovery was not accepted as proof by most scientists. Cryoscopic molecular-weight studies on natural rubber in camphor, for example, gave molecular weights of 1400–2000,[31] whereas Staudinger has found values of 3000–5000 for hydrogenated rubber. The X-ray studies carried out on such compounds seemed to contradict Staudinger's idea of molecular colloids by showing that a large section of the organic colloids gave X-ray diagrams more like those of liquids than those of low-molecular-weight crystalloids. Further, the more crystal-like X-ray pictures for organic colloids revealed a relatively small unit cell. Yet it was known from measurements on homologous series of low-molecular-weight substances that the size of the unit cell is directly proportional to the molecular weight. Since it was impossible to imagine that such a proportionality should not be true for all molecular-weight ranges, it was concluded that colloids were composed of low-molecular-weight compounds.

The X-ray measurements also contradicted the existence of small rings and supported the idea of chain structures,[32] since rings did not correspond with the elementary cell structure that had been found. An analysis of the X-ray diffraction pattern for rubber revealed crystal lengths of about 30–60 nm. Assuming that the crystal length is identical to the molecular length, K. H. Meyer and H. Mark thus found molecular weights of about 5000–10,000 for natural rubber. The much greater molecular weights of about 150,000–380,000 found for rubber in solution was interpreted as the weight of the solvated chain and, later, also by the supposition that the micelles found by X-ray crystallography were present in solution as associated structures. The second micellar theory, contrary to the first, assumed chains instead of rings, as well as higher molecular weights. To explain the very high molecular weights obtained experimentally, it was assumed that these chains associated to form colloidal aggregates.

In contrast, H. Staudinger and R. Signer* stressed that the crystal length need not have anything to do with the molecular length. Since the crystal structure was essentially dependent on the constitution of the compound, Straudinger tried to prove his theory by polymer-analogous reactions. In reactions of this type, the side groups of a compound are replaced without attacking the bonds of the main chain. Unbranched poly(vinyl acetate) could be converted by saponification to poly(vinyl alcohol), and then by esterification back to poly(vinyl acetate) again:

$$\underset{\substack{|\\ O\\ |\\ CO\\ |\\ CH_3}}{\left(CH_2{-}CH\right)_n} \xrightarrow[-CH_3COOH]{+H_2O} \underset{\substack{|\\ OH}}{\left(CH_2{-}CH\right)_n} \xrightarrow[H_2O]{+CH_3COOH} \underset{\substack{|\\ O\\ |\\ CO\\ |\\ CH_3}}{\left(CH_2{-}CH\right)_n} \qquad (1\text{-}5)$$

The fact that no degradation appeared in polymer-analogous reactions of this kind could be proved by molecular-weight determinations on the initial and conversion products. If the same degree of polymerization is obtained for a single polymer analog in different solvents, then, because of differing polymer–solvent interactions, it is quite improbable that association colloids are present. For example, in a study of this kind made on amylopectin, a polysaccharide with three hydroxyl groups per monomeric unit, the same degree of polymerization was always found, thus excluding the possibility of association colloids[34] (Table 1-3).

Staudinger formulated his ideas from studies of natural products (amylose, cellulose) and compounds obtained by addition polymerization. According to these ideas, addition polymerization is an event which causes

* See, e.g., the description of the historical development given by Mark.[33]

Table 1-3. Degree of Polymerization $\overline{X}_n$ from Osmotic Measurements During Polymer-Analogous Reaction of Amylopectin Fractions[a]

Fraction number	$\overline{X}_n$: Amylopectin (initial product) in formamide	Amylopectin triacetate: in acetone	Amylopectin triacetate: in $CHCl_3$	Amylopectin (regenerated) in formamide
1	185	190	190	185
2	380	390	390	—
3	560	540	540	570
4	940	960	960	870

[a] From Staudinger and Husemann.[34]

many monomer units to come together in the course of forming covalent bonds, thus giving a macromolecule which has the same overall elemental composition as that of the initial monomer (Carothers[35]), e.g., in styrene:

$$n\mathrm{CH_2{=}CH(C_6H_5)} \longrightarrow \mathrm{\left(CH_2{-}CH(C_6H_5)\right)_n} \tag{1-6}$$

During this process the monomer molecules are attached one after the other, i.e., sequentially, onto the active growing end of the polymer chain[45]:

$$\mathrm{R\left(CH_2{-}CH(C_6H_5)\right)_n CH_2CH^*(C_6H_5) + CH_2{=}CH(C_6H_5) \longrightarrow R\left(CH_2{-}CH(C_6H_5)\right)_{n+1} CH_2CH^*(C_6H_5)} \quad \text{etc.}$$

The chemical nature of this active growing end was not known. A radical chain mechanism was proposed,[46,47] but the initiation or starting mechanism was not clarified. A starting mechanism involving radical transfer from free radicals to the monomer[48] or a process involving an active complex between styrene and, for example, dibenzoyl peroxide[49] was discussed. The problem was eventually solved by appropriate labeling of the initiator.[50-52] It could then be shown that labeled initiator fragments were incorporated into the polymer as chain ends. At that time the addition polymerization mechanism leading to compounds used by Staudinger as model substances was anything but clear. For the second group of compounds that he used—natural macromolecular products—even less was known about the mechanism of formation. Carothers[36] therefore decided to build macromolecular compounds, stepwise, using condensation reactions familiar in low-molecular-weight organic chemistry, e.g., reacting diols with dicarboxylic acids:

$$\text{HO—R—OH} + \text{HOOC—R}'\text{—COOH} \xrightarrow{-\text{H}_2\text{O}} \text{HO—R—OCO—R}'\text{—COOH}$$

$$\text{HO—R—OCO—R}'\text{—COOH} + \text{HO—R—OH} \xrightarrow{-\text{H}_2\text{O}} \qquad (1\text{-}8)$$

$$\text{HO—R—OCO—R}'\text{—COO—R—OH} \quad \text{etc.}$$

In contrast to addition polymerization, every step must be separately initiated in *polycondensation* (condensation polymerization).

Carothers was able to show that many macromolecules could be built, not only through some mysterious process, but also with the known methods of organic chemistry. His work yielded further proof for the formation of organic molecular colloids by covalent bonds, and led to the first synthetic fiber produced industrially on a large scale, nylon 6,6 [poly(hexamethylene-adipamide)]. This polymer is obtained from hexamethylenediamine and adipic acid:

$$n\text{H}_2\text{N—(CH}_2)_6\text{—NH}_2 + n\text{HOOC}\!\left(\text{CH}_2\right)_4\!\text{COOH} \longrightarrow$$

$$\text{H}\!\left[\text{NH—(CH}_2)_6\text{—NHCO}\!\left(\text{CH}_2\right)_4\!\text{CO}\right]_n\!\text{OH} + (2n-1)\text{H}_2\text{O} \qquad (1\text{-}9)$$

Further questioning of the micellar theory came when Sumner[37] succeeded in crystallizing the enzyme urease in 1926, and when Northrop[38] crystallized the enzyme pepsin in 1930. Thus, the theory that colloids could only be crystallized by losing their colloidal properties was refuted. Then, over the period 1927–1940, Svedberg,[39] using the ultracentrifuge that he had invented (see Section 8.7), showed that a colloidal solution containing proteins, when ultracentrifuged at various temperatures and in different salt solutions, proved to be uniform with regard to molecular weight. This discovery also contradicted the idea of colloidal association. Using the electrophoresis which he developed,[40] Tiselius showed conclusively that a specific protein always had the same charge per mass, and this, too, was contradictory to inorganic colloidal association. Thus, at the beginning of the 1930s, it was realized that true macromolecules were being dealt with in organic colloids.

The inability of many synthetic organic macromolecules to crystallize was traced back in the 1940s to the irregularity in configuration of connecting monomeric units. In 1948, Schildknecht and co-workers[41] found that vinyl ether gave polymers with different physical properties according to the catalyst used. Poly(vinylmethylethers) proposed via free radical polymerizations were amorphous, whereas cationic polymerizations at low temperatures yielded crystalline material. This discovery was correctly interpreted as being dependent on differences in the steric makeup of the polymer. However, these results were not much heeded because no other stereoregular polymers can be generated in this way.

The key to the desired synthesis of stereoregular polymers was first provided by the discovery of Ziegler catalysts. Ziegler found that catalyst systems of aluminum alkyls and titanium tetrachloride can polymerize ethylene to poly(ethylene) even at room temperature and normal pressure.[(42)] Up to this time, poly(ethylene) had been made exclusively by radical polymerization of ethylene at high pressure. Natta and his co-workers[(43)] observed that these catalysts enabled α-olefins to be converted into stereoregular polymers that could often be crystallized. Later alterations to the Ziegler catalysts also made the polymerization of other monomers possible, so that today a large number of stereoregular polymers is known.

Chain-forming macromolecules contain many bonds in the main chain. The individual chain atoms can therefore assume many different arrangements in space relative to each other. Kuhn[(44)] already knew in the 1930s that the problem of the spatial arrangement of chain macromolecules can be solved particularly well by statistical analysis and statistical calculation methods. Generally speaking, statistics plays a large role in macromolecular chemistry, as was shown especially by P. J. Flory in a masterly manner.[(53)]

1.4. Nomenclature

> "When *I* use a word," Humpty Dumpty said in a rather scornful tone "it means just what I choose it to mean—neither more nor less."
>
> "The question is," said Alice, "whether you can make words mean so many different things."
>
> "The question is," said Humpty Dumpty, "which is to be master—that is all."
>
> Lewis Carroll, *Through the Looking-Glass*

In chemistry, materials and reactions are usually classified according to three principles: *Phenomenological* definitions are constructed from external features (overall composition of a compound, elimination of groups during reactions, etc.). *Molecular* definitions are based on chemical structure or the reaction mechanism, i.e., bonds and changes occurring in these bonds. *Operative* definitions consider everything with regard to the suitability, e.g., the magnitude of the desired molecular weight, the speed of the reaction, the mechanical properties required, etc. All three types of definition are used to describe macromolecular materials.

In the early days of macromolecular chemistry, synthetic polymers were simply labeled according to the initial monomer. Thus, ethylene polymers became poly(ethylenes), styrene polymers became poly(styrenes), those from lactams became poly(lactams), etc. In other cases the choice of name was provided by a characteristic group in the finished polymer. Thus polymers from diamines and dicarboxylic acids were given the name of polyamides.

These phenomenological labels must fail whenever more than one kind of monomeric unit can arise from the monomer. An example of this is the polymerization of butadiene (see Table 1-2).

Naturally occurring macromolecules generally have trivial names. These trivial names are often derived from their function (e.g., catalase) or their chemical character (e.g., nucleic acids), or describe their origin (e.g., cellulose).

On the other hand, modern nomenclature is based on the chemical structure of the macromolecules. The name of a polymer of unspecified degree of polymerization consists of the prefix "poly" and the name of the smallest repeating unit. With unbranched polymers, the smallest repeating unit is a diradical. The name of this diradical is the same as that met with in the nomenclature for low-molecular-weight organic diradicals. Thus, the group —CH_2— is called "methylene" and the corresponding polymer is called poly(methylene) (example 1 in Table 1-4). The diradicals —CH_2—CH_2— and —CH_2—$CH(CH_3)$— do not have simple definitive names; the trivial names "ethylene" and "propylene" are retained. Examples of names for other diradicals are as follows:

—CH_2—	—CO—	—CH=CH—
methylene	carbonyl	vinylene
—O—	—S—	—NH—
oxy	thio	imino
1,4-phenylene	1,4-cyclohexylene	4,6-quinolinidyl

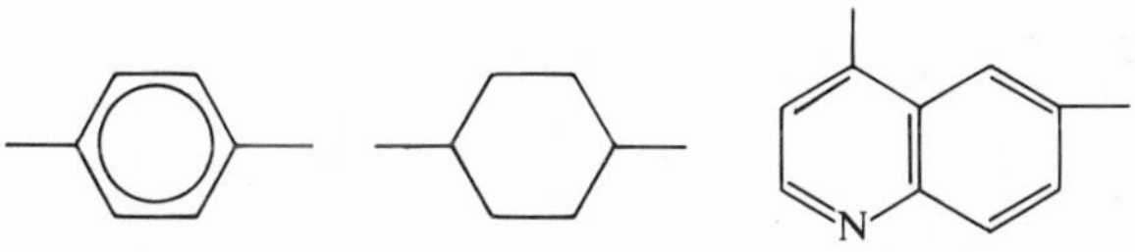

Substituents are generally given before the name of the diradical. Poly(vinyl alcohol) (example 4, Table 1-4) is thus correctly called "poly-(hydroxyethylene)." Another example is "poly(1-oxotrimethylene)" (example 11, Table 1-4). Note that the name of the diradical is always given in parentheses or other brackets. All examples given in Table 1-4 have structural names which imply a direction in the polymer chain. This direction is not always the same as the direction of chain propagation in polyreactions (see examples 5, 6, and 9 in Table 1-4).

For complicated structures, the structural name generally consists of the names of simple diradicals given in sequence. Examples of combina-

Table 1-4. Trivial and Structural Names of Polymers

Structure number and formula	Trivial name (based on the structure of the monomers)	Structural name
1 $\text{+CH}_2\text{+}_n$	Poly(methylene)	Poly(methylene)
2 $\text{+CH}_2\text{CH}_2\text{+}_n$	Poly(ethylene)	Poly(ethylene)
3 $\text{+CH(CH}_3\text{)—CH}_2\text{+}_n$	Poly(propylene)	Poly(propylene)
4 $\text{+CH(OH)—CH}_2\text{+}_n$	Poly(vinyl alcohol)	Poly(hydroxyethylene)
5 $\text{+CH(C}_6\text{H}_5\text{)—CH}_2\text{+}_n$	Poly(styrene)	Poly(phenylethylene)
6 $\text{+C(CH}_3\text{)(COOCH}_3\text{)—CH}_2\text{+}_n$	Poly(methyl methacrylate)	Poly[1-(methoxy-carbonyl)-1-methyl ethylene]
7 $\text{+CH=CHCH}_2\text{CH}_2\text{+}_n$	Poly(butadiene)	Poly(1-butenylene)
8 $\text{+OCH}_2\text{+}_n$	Poly(formaldehyde)	Poly(oxymethylene)
9 $\text{+OCH}_2\text{CH}_2\text{+}_n$	Poly(ethylene oxide), poly(ethylene glycol)	Poly(oxyethylene)
10 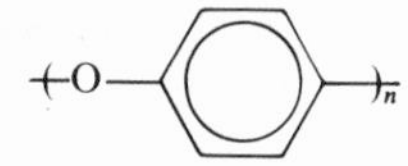 $\text{+O—C}_6\text{H}_4\text{+}_n$	Poly(phenylene oxide)	Poly(oxy-1,4-phenylene)
11 $\text{+CO—CH}_2\text{CH}_2\text{+}_n$	Poly(ethylene-co-carbon monoxide)	Poly(1-oxotrimethylene)
12 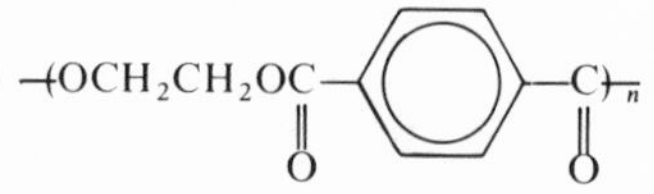 $\text{+OCH}_2\text{CH}_2\text{OC(=O)—C}_6\text{H}_4\text{—C(=O)+}_n$	Poly(ethylene terephthalate)	Poly(oxyethylene-oxyterephthaloyl)
13 $\text{+NHC(=O)(CH}_2\text{)}_4\text{C(=O)NH(CH}_2\text{)}_6\text{+}_n$	Poly(hexamethylene diamine-co-adipic acid); nylon 6,6	Poly(iminoadipoyl-iminohexamethylene)
14 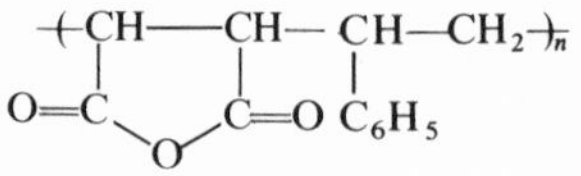 +CH—CH—CH(C$_6$H$_5$)—CH$_2$+$_n$ (first two CH joined via O=C–O–C=O ring)	Poly(maleic anhydride-co-styrene)	Poly[(tetrahydro-2,5-dioxo-3,4-furandiyl)(1-phenylethylene)]

tions of two simple diradicals are poly(oxymethylene) and poly(oxy-1,4-phenylene) (examples 8 and 10, respectively, Table 1-4).

In many cases, trivial names are retained for certain combinations of simple diradicals. Thus, the unit $\text{—CO—C}_6\text{H}_4\text{—CO—}$ is called "terephthaloyl" and not "carbonyl-1,4-phenylene-carbonyl" (example 12 in Table 1-4). The unit $\text{—CO—CH}_2\text{CH}_2\text{—}$ is called "1-oxotrimethylene" and not "carbonylethylene." However, carbonyl groups bound directly to heterogroups are called "acyl groups." The compound $\text{+NH—CO—CH}_2\text{CH}_2\text{+}_n$

is thus called "poly(iminocarbonylethylene)" and not "poly[imino(1-oxo-trimethylene)]."

Sequence rules have to be observed in naming polymers with two or more diradical components in the structural element. The component with highest priority is given first. Components that follow are then given in order of decreasing priority. Heterocyclic rings have highest priority; linear groups containing heteroatoms follow; next come carbocyclic rings; then, finally, there are the linear groups containing carbon atoms only. Substitution does not alter this order of priority. In each component group, the following rules are observed in the nomenclature:

1. In each ring system, bonding to the main chain is designated by a numbering system utilizing the shortest distance between the atoms involved in this bonding.

2. When two rings of the same kind are present, the following rules are used: The ring with most substituents has highest priority. When the number of substituents per ring is the same, the ring with the lowest position numbers for bonding to the main chain has highest priority. If the rings have the same number of substituents and the same position numbers, the highest priority is given to the ring with substituents the first letter of whose name comes earliest in the alphabet.

3. With different rings as components, the highest priority is given to (a) the component with the largest number of rings, (b) the largest individual ring, and (c) the least hydrogenated ring.

4. Heteroatoms are given priority in the order: O, S, Se, Te, N, P, As, Sb, Bi, Si, Ge, Sn, Pb, B, and Hg.

5. If, for example, aliphatic groups occur between heteroatoms or rings, the structural elements are given such that the shortest distance occurs between components with the highest priorities. Example: The structure $\text{+O—CH}_2\text{—NH—CHCl—CH}_2\text{—SO}_2\text{—(CH}_2)_6\text{+}$ is written in the order given, since this represents the shortest distance between O (highest priority) and S (second highest priority). When written as $\text{+O—(CH}_2)_6\text{SO}_2\text{—CH}_2\text{—CHCl—NH—CH}_2\text{+}$, the correct order of priority between O, S, and N is retained, but the distance between O and S (six atoms) is greater than in the case where it is written correctly with four atoms, i.e., $\text{—CH}_2\text{—NH—CHCl—CH}_2\text{—}$ between O and S.

6. Heterocyclic rings are given the following priority ratings: (a) The largest nitrogen-containing ring system (independent of the number of nitrogens in the ring) has highest priority; (b) with rings of the same size, the ring system with the largest number of nitrogen atoms has highest priority; and (c) the ring system with the largest number of other heteroatoms has highest priority. Thus, with rings N has the highest priority, followed by O, S, Se, Te, P, As, etc. (see also point 4).

Double-strand or ladder polymers have four bonding positions. The positional relationship between the bonds is designated by a colon between the two sets of main chain bonds. Examples of this system are:

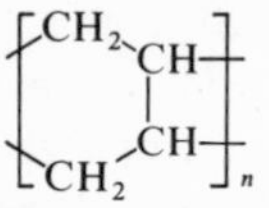

poly(1,4:2,3-butanetetrayl) 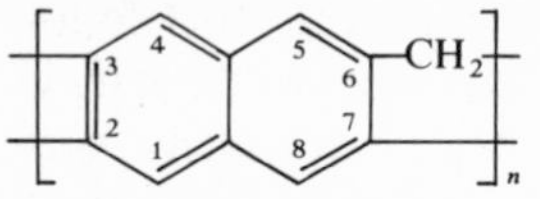 poly(2,3:6,7-naphthalenetetrayl-6-methylene) 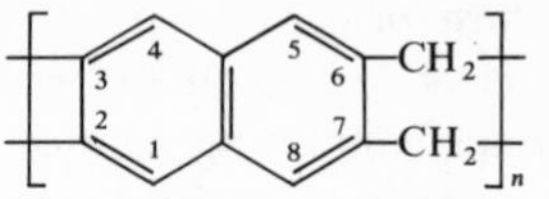 poly(2,3:6,7-naphthalenetetrayl-6,7-dimethylene)

When there are tetravalent radicals as well as diradicals in the structural element (e.g., in spiro polymers), the tetravalent radicals take a higher priority than the diradical components:

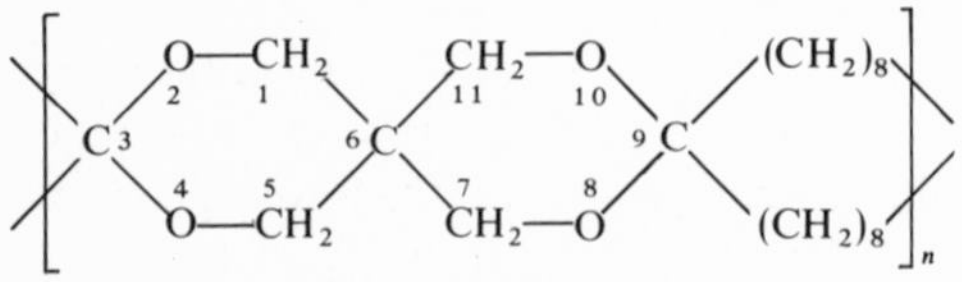

poly[2,4,8,10-tetraoxaspiro(5,5)-undecane-3,9-diylidene-9,9-bis(octamethylene)]

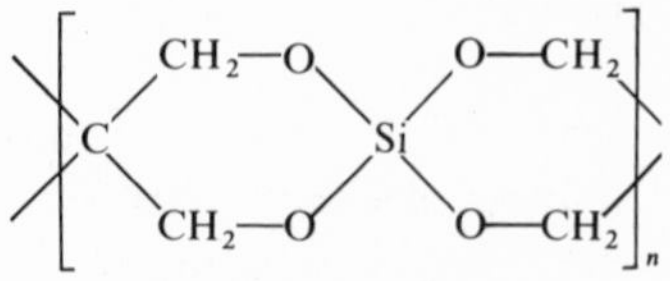

poly[1,3-dioxa-2-silacyclohexane-5,2-diylidene-2,2-bis(oxymethylene)]

End groups are not normally specified in polymers. However, when they are known, they should be prefixed by the Greek letters α and ω and written before the name of the polymer. An example is

$$Cl\text{-}(CH_2)_n\text{-}CCl_3$$

α-chloro-ω(trichloromethy')poly(methylene)

The nomenclature rules say that the trivial names of common polymers need not necessarily be replaced by structural names. Therefore, both trivial and structural names of polymers will be used in this book. Standard abbreviations of trivial names (see Table A1 in the Appendix to this book) will generally only be used in diagrammatic illustrations. For further information, a special table (Table A2 in the Appendix) gives trade names of some important plastics, elastomers, and fibers.

1.5. Commercial Classification and Significance

Commercially available polymers are generally classified according to their field of application or the process used for their production.

Classification according to production processes is based, on the one hand, on the consideration of the chemical equations describing the overall reaction, and, on the other hand, on the practical operational process used to make the polymer. This classification procedure (although it does give useful technological information on the nature of end groups, unreacted monomer content, etc.) has been superseded by one based on mechanisms. The mechanistic classification describes classes such as addition polymerizates, polycondensates, and adduct polymerizates according to whether the polymer is produced by addition polymerization, polycondensation, or adduct polymerization.

In this scheme a reaction in which monomers react with the elimination of low-molecular-weight compounds is called a *polycondensation.* An example of this is the formation of polyamides from diamines and dicarboxylic acids [see equation (1-9)].

In contrast to polycondensation, no low-molecular-weight components are eliminated during *adduct polymerization*, e.g., by the reaction of diisocyanates with diols to form polyurethanes:

$$n\text{HO—R—OH} + n\text{OCN—R'—NCO} \longrightarrow$$
$$\text{H}\!\left(\text{O—R—O—CO—NH—R'—NH—CO}\right)_{n-1}\text{O—R—O—CO—NH—R'—NCO} \qquad (1\text{-}10)$$

Many authors consider the transfer of H atoms from one monomeric unit to another as a further characteristic of adduct polymerization. Many also include the formation of poly(ethylene oxide) from ethylene oxide among adduct polymerizations, although there is no H transfer:

$$n\underset{\text{O}}{\text{CH}_2\text{—CH}_2} \longrightarrow \left(\text{CH}_2\text{CH}_2\text{O}\right)_n \qquad (1\text{-}11)$$

As in adduct polymerization, there is no elimination during *addition polymerization* [see e.g., equation (1-4)], but in addition polymerization the overall composition of the monomeric units is not altered.

Synthetic polymers are further classified as plastics (thermoplastics, plastomers), thermosets, elastomers (rubbers), and fibers. Materials that can be deformed reversibly at constant temperature and under normal conditions are called elastomers. The best-known example is vulcanized (cross-linked) natural rubber. Materials that are deformed under stress, but do not revert when the stress ceases, are known as plastics. Masticating rubber and poly(ethylene) belong to this group. The materials can be brittle, i.e., they do not deform when a short, sharp stress is applied but break into smaller pieces, or tough, i.e., only their shape is altered under stress. Thermosets are substances that during or after their manufacture, change irreversibly from the plastic state into a strong, three-dimensional, cross-linked form and can no longer be changed by temperature or stress into any other form. They are, therefore, only slightly plastic. Fibers are distinguished by a preferential orientation in the arrangement of molecules in the orientation in which they have high tensile strength and for the most part low extensibility.

Plastics, elastomers, and fibers thus differ essentially in their modulus of elasticity and by their extensibility (Table 1-5). Tensile strength and elongation at break can be used for further distinction.

The properties desirable for a certain application can also be considerably influenced by temperature. The mobility of chain segments is low at low temperatures. At higher temperatures the segments can rotate more easily. Therefore, increasing temperature means a corresponding increase in elasticity and plasticity at the expense of the brittleness. Thus a certain non-cross-linked macromolecule can be brittle, viscoelastic (rubbery), or plastic, depending upon the temperature:

$$\underset{\text{(plastic)}}{\text{Brittle state}} \xrightarrow[\text{temperature}]{\text{glass transition}} \underset{\text{(elastomer)}}{\text{Viscoelastic state}} \xrightarrow{\text{flow temperature}} \underset{\text{(viscous liquid)}}{\text{Viscofluid state}}$$

The transitions between the three states occur at well-defined temperatures. Plastics, elastomers, and viscous liquids can be changed into thermosets. They can be changed back to their original state if the cross-linking results from ionic bonds. This is also possible if reversible covalent cross-linking takes place.

The first artificially produced and industrially applied macromolecules were modified natural products. The vulcanization of natural rubber was

Table 1-5. Classification of Macromolecules According to Application Properties

Type	Elastic modulus, N/cm^2	Elongation, %	Substances	Tensile strength, 10^4 kg/m^2	Elongation at break, %
Elastomers	$10–10^2$	1000	Natural rubber	230–320	470–600
			GRS	280	580
Plastics	$10^3–10^4$	100–200	Low-pressure poly(ethylene)	200–400	15–100
			Nylon 6,6	500–800	90
			Poly(vinyl chloride)	350–630	2–40
Fibers	$10^5–10^6$	10–30	Wool	1550	10–50
			Silk	3550	25
			Cotton	2100–8400	5–8
			Low-pressure poly(ethylene)	3400–6100[a]	10–20[a]
			Nylon 6,6	4600–8700[a]	19–32[a]
			Poly(vinyl chloride)	3400–3800[a]	14–20[a]

[a] Depends on draw ratio.

introduced in 1839, and hard rubber was produced in 1851. In 1859 it was noticed that cellulose treated with $ZnCl_2$ gave a material, vulcan fiber, that was mechanically very resiliant. Cellulose nitrate plasticized with camphor was used to produce celluloid in 1865. The protein casein cross-linked with formaldehyde led to the introduction of galalith in 1897.

In contrast, the first completely synthetic macromolecules had a much longer interval between discovery and technical application. Indeed, poly(styrene) was discovered as early as 1839, but only manufactured around 1930. The first polyamides were observed in 1907, but the pioneer work of W. H. Carothers (1929–1935) was needed to make possible the production of nylon yarn in 1937. This time lag was a result of the big difference between laboratory-scale synthesis and the plant production of a plastic for industrial application. Tests must be carried out on properties such as tensile strength, etc., and on the modification of these properties by producing polymers that differ in regard to molecular weight, molecular-weight distribution, stereoregularity, etc. Industry must consider the marketability of the polymer, technical production problems, and, finally, the most important "physical quantity," the price. Thus, for every 6000 or so plastics developed in the laboratory, only one goes into production. (In pharmaceuticals the proportion is only 3000:1.) Development costs are correspondingly high. At the E. I. du Pont de Nemours factory the development of nylon 6,6 cost $27 million, that of poly(acrylonitrile) fiber $60 million and that of poly(formaldehyde) $52 million. Another company spent $30 million on the technical development of poly(propylene), even though the patent and some of the know-how were purchased.

Spending sums of this kind, however, is justified since the development is frequently for a much-needed, mass-production article, or for a special product that sells at high prices. The rapid increase in production can be seen in Table 1-6. At an average price of 25¢/lb, the 1962 production shows a value of 6×10^9 dollars. The production of 10 million tons of plastics per year corresponds to 1/35 by weight, 1/5 by volume, and 1/6 by value of the yearly production of steel (350 million long tons). In contrast to a slowly increasing steel production, the polymer industry shows the fantastically high growth rate of 16%. Should this increase continue—as all the signs indicate—then in 1980 the production of polymers will equal that of steel in volume and value.

The strong increase in the production of synthetic fibers, particularly polyester fibers, is especially striking. Other important man-made fibers are nylon and the acrylates [poly(acrylonitrile)]. Among actual plastics the main bulk is likewise principally made up of a few types. In 1970, poly(ethylene) made up 29%, poly(vinyl chloride) 22%, and poly(styrene) 15% of total plastic production.

Polymers required for the electrical industry as a whole play a sub-

Table 1-6. Annual World Production of Macromolecules (in millions of tons)[a]

Type	1950	1960	1970
Thermoplasts and thermosets	2	6.7	31.0
Synthetic elastomers	1	2.6	4.7
Natural rubber	1.8	2.2	2.7
Synthetic fibers	0.07	0.7	4.8
Natural fibers and fibers from naturally occurring raw materials			
Cotton	7.0	10.4	11.6
Other cellulosic fibers	1.6	2.6	3.6
Wool	1.8	1.5	1.6
Naturally occurring macromolecules			
Cellulose	100,000	100,000	100,000
Chitin	1,000	1,000	1,000

[a] Data for the naturally occurring materials cellulose and chitin refer to their production in nature.

ordinate role. Most of the macromolecules used in the electrical industry are employed as insulators, and polymer-based semiconductors are only in the experimental stage.

The small portion of polymers used for their chemical reactivity are limited by the need for special chemical properties. Important in this context are ion-exchange resins and solid rocket fuels.

Among biopolymers, properties are distributed a little differently. Macromolecules of natural origin serve as catalysts (enzymes), as genetic messengers (nucleic acids), as transport agents (certain proteins), as construction materials (cellulose, collagen), and as foodstuffs and food reserves (glycogen, starch), or else they are quite simply metabolic substances generated in nature, e.g., natural rubber. The production is also high among industrially useful biopolymers. In the United States in 1959, for example, the production of collagen products (leather, gelatine, animal glue) had a value of 4×10^9 dollars, that of keratin products (hair, nails, feathers, wool) 1.5×10^9 dollars, and raw silk 500,000 dollars.

Literature

Periodicals

Journal of Polymer Science, Vols. 1–62 (1946–1962); since 1963 divided into three parts: *A* (General Papers), *B* (Letters), and *C* (Symposia); since 1966 divided into *A-1* (Polymer Chemistry), *A-2* (Polymer Physics), *B* (Polymer Letters), *C* (Polymer Symposia); since

1970, *D* is Vol. 4 of the earlier Macromolecular Reviews series; from October 1972 changed to *Polymer Chemistry Edition, Polymer Physics Edition, Polymer Letters Edition, Polymer Symposia*, and *Macromolecular Reviews*

Journal of Applied Polymer Science, Vol. 1ff. (1959ff.)

Journal of Applied Polymer Science, Applied Polymer Symposia, Vol. 1ff. (1965ff.)

Journal für Makromolekulare Chemie (Vol. 1 = 1944, after Vol. 2 discontinued)

Die Makromolekulare Chemie, Vol. 1ff. (1947ff.)

Die Angewandte Makromolekulare Chemie, Vol. 1ff. (1967ff.)

Journal of Macromolecular Science (from 1967), *A* (Chemistry), *B* (Physics), *C* (Reviews in Macromolecular Chemistry), and *D* (Polymer Processing and Technology). Introduced as: *Journal of Macromolecular Chemistry* (1 Vol. 1966) and *Reviews in Macromolecular Chemistry* (1st vol. 1966, issued later as a journal in the *C* Series). Since 1974, *D* series as *Polymer-Plastics Technology and Engineering*.

British Polymer Journal, Vol. 1ff. (1969ff.)

European Polymer Journal, Vol. 1ff. (1965ff.)

International Journal of Polymeric Materials, Vol. 1ff (1971ff.)

Polymer, Vol. 1ff. (1960ff.)

Polymer Journal (Vol. 1 from 1970)

Kolloid-Z. und Z. für Polymer [under this title since 1962, previously Kolloid-Z. (1906)]; since 1974 as *Colloid and Polymer Science*

Journal of Materials Science, Vol. 1ff. (1966ff.)

The Chemistry of High Polymers (Japan), Vol. 1ff. (1944ff.)

Vysokomol. Soyed. (High-molecular-weight compounds) (USSR); in English translation as *Polymer Science USSR* (1960ff.)

Macromolecules, Vol. 1ff. (1968ff.)

Polymer Preprints, Vol. 1ff. (1960ff.)

Biopolymers, Vol. 1ff. (1963ff.)

Biochemistry, Vol. 1ff. (1964ff.)

Biophysics, Vol. 1ff. (1957ff.); English translation of *Biofizika*

Biochimica et biophysica acta, Vol. 1ff. (1947ff.)

Archives of Biochemistry and Biophysics, Vol. 1ff. (1952ff.)

Materials Science and Engineering, (1966ff.)

Soviet Plastics, Vol. 1ff. (1961ff.); English translation of *Plasticheskie Massy*

Plastics and Polymers (Journal of the Plastics Institute), Vol. 1ff. (1933ff.)

Reports on Progress in Polymer Physics in Japan, Vol. 1ff. (1958ff.)

Polymer Engineering and Science (from 1965, previously *Society of Plastics Engineers Transactions)*

Reviews

Advances in Polymer Sciences—Fortschritte der Hochpolymerenforschg., Vol. 1ff. (1958ff.); Vols. 1–3 as *Fortschr. Hochpolymeren-Forschg.—Adv. Polymer Science*

Macromolecular Reviews, Vol. 1ff. (1967ff.); from Vol. 4 appears as *Journal of Polymer Science D*

Reviews in Macromolecular Chemistry, Vol. 1ff. (1967ff.)

Progress in Polymer Science, Japan, Vol. 1ff. (1971ff.)

Progress in Polymer Science, Vol. 1ff. (1967ff.)

Advances in Macromolecular Chemistry, Vol. 1ff. (ca. 1968ff.)

Critical Reviews in Macromolecular Science, Vol. 1ff. (1972ff.)

Rubber Chemistry and Technology, Vol. 1ff. (1928ff.)

Advances in Materials Research, Vol. 1ff. (1967ff.)

Macromolecular Science (Vol. 8 of the *Physical Chemistry Series* of the *MTP International Review of Science)*, Series 1ff. (1972ff.)

Information Retrieval Sources

Chemical Abstracts, Macromolecular Sections
Polymer News (from 1970)
Polymers (Quarterly Reports), 1969ff.
Literaturschnelldienst (Rapid literature survey service) *Kunststoffe und Kautschuk,* Vol. 1ff. (1955ff.)

Bibliography

E. R. Yescombe, *Sources of Information on the Rubber, Plastics and Allied Industries,* Pergamon, Oxford, 1968

Handbooks and Series

Houben-Weyl, *Methoden der organischen Chemie* (E. Müller, ed.), G. Thieme Verlag, Stuttgart, Vol. XIV, *Makromolekulare Stoffe,* Part 1 "Polymerisate," 1961, Part 2 "Polykondensate, Reaktionen an Polymeren," 1963; Vol. XV *Makromolekulare Naturstoffe* (planned)

R. Vieweg, ed., *Kunststoff-Handbuch,* Carl Hanser Verlag München, 1963ff. (12 vols. planned)

A. D. Jenkins, ed., *Polymer Science—A Materials Science Handbook,* North-Holland Publ., Amsterdam, 1971

H. Mark, N. G. Gaylord, and N. M. Bikales, eds., *Encyclopedia of Polymer Science and Technology,* J. Wiley, New York, 1966ff. (approx. 10 vols.)

W. J. Roff and J. R. Scott, *Handbook of Common Polymers,* Butterworths, London, 1971, and The Chemical Rubber Co., Cleveland, 1971

J. Brandrup and E. H. Immergut, *Polymer Handbook,* Wiley–Interscience, New York, 1965; second ed., 1974

B. Carroll, *Methods in Macromolecular Analysis,* M. Dekker, New York, 1969ff.

Practical Books and Books on Preparative Macromolecular Chemistry

S. H. Pinner, *A Practical Course in Polymer Chemistry,* Pergamon Press, New York, 1961

G. F. D'Alelio, *Kunststoff-Praktikum,* Carl Hanser Verlag, München, 1952

E. M. McCaffery, *Laboratory Preparation for Macromolecular Chemistry,* McGraw-Hill, New York, 1970

D. Braun, H. Cherdron, and W. Kern, *Praktikum der makromolekularen Chemie,* Verlag Hüthig, Heidelberg, 1966; second ed., 1971; English translation: *Techniques of Polymer Synthesis and Characterization,* Wiley, New York, 1972

I. P. Lossew and O. Ja. Fedotowa, *Praktikum der Chemie hochmolekularer Verbindungen,* Akademische Verlagsgesellschaft, Geest & Portig K.-G., Leipzig, 1962

W. R. Sorensen and T. W. Campbell, *Preparative Methods of Polymer Chemistry,* Interscience Publishers, New York, 1961

C. G. Overberger, ed., *Macromolecular Syntheses,* Wiley, New York, Vol. 1, 1963 (continuing series); from Vol. 2, under various editors

E. A. Collins, J. Bares, and F. W. Billmeyer, Jr., *Experiments in Polymer Science,* Wiley-Interscience, New York, 1973

A. M. Toroptseva, K. V. Belogorodskaya, and V. M. Bondarenko, *Laboratory Practical Course in the Chemistry and Technology of Macromolecular Compounds,* Khimiya, Leningrad, 1972 (in Russian)

S. R. Sandler and W. Karo, *Polymer Syntheses,* Academic Press, New York, Vol. 1, 1974

Section 1.3. Historical Development

1. E. Simon, *Ann. Chim. Phys.* **31**, 265 (1839).
2. M. Berthelot, *Bull. Soc. Chim. France* **6** (2), 294 (1866).

3. A. Wurtz, *Compt. Rend.* **49**, 813 (1859); **50**, 1195 (1860).
4. A.-V. Lourenco, *Compt. Rend.* **49**, 619 (1859); **51**, 365 (1860); *Ann. Chim. Phys.* **67** (3), 273 (1863).
5. T. Graham, *Philos. Trans. R. Soc. London* **151**, 183 (1861); *J. Chem. Soc. (London)* **1864**, 318.
6. A. Müller, *Z. Anorg. Chem.* **36**, 340 (1903).
7. H. Staudinger, *Organische Kolloidchemie,* Vieweg, Braunschweig, Germany, first ed. 1940, third ed. 1950.
8. J. Stauff, *Kolloidchemie,* Springer, Berlin, Germany, 1960.
9. W. Ostwald, *Kolloid-Z.* **1**, 291, 331 (1907).
10. P. P. v. Weimarn, *Kolloid-Z.* **2**, 76 (1907/1908).
11. H. Hlasiwetz and J. Habermann, *Ann. Chem. Pharmacol.* **159**, 304 (1871).
12. F. M. Raoult, *Compt. Rend.* **95**, 1030 (1882); *Ann. Chim. Phys.* **2** (6), 66 (1884); *Compt. Rend.* **101**, 1056 (1885).
13. J. H. van't Hoff, *Z. Phys. Chem.* **1**, 481 (1887); *Philos. Mag.* **26** (5), 81 (1888).
14. J. H. Gladstone and W. Hibbert, *J. Chem. Soc. (London)* **53**, 679 (1888); *Philos. Mag.* **28** (5), 38 (1889).
15. H. T. Brown and G. H. Morris, *J. Chem. Soc. (London)* **55,** 462 (1889).
16. A. Nastukoff, *Ber. Dtsch. Chem. Ges.* **33**, 2237 (1900).
17. W. A. Caspari, *J. Chem. Soc. (London)* **105**, 2139 (1914).
18. J. D. van der Waals, *Die Kontinuität des gasförmigen und flüssigen Zustands,* Thesis, Leiden, 1873; second ed., J. A. Barter, Leipzig, 1895 and 1900; *Die Zustandsgleichung,* Nobelpreisrede, Akad. Verlagsgesellschaft, Leipzig, 1911.
19. J. Thiele, *Liebigs Ann. Chem.* **306**, 87 (1899).
20. P. Pfeiffer, *Liebigs Ann. Chem.* **404**, 1 (1914); **412**, 253 (1917).
21. M. Faraday, *Q. J. Sci.* **21**, 19 (1826).
22. C. Harries, *Ber. Dtsch. Chem. Ges.* **37**, 2708 (1904); **38**, 1195, 3985 (1909).
23. S. S. Pickles, *J. Chem. Soc. (London)* **97**, 1085 (1910).
24. P. Walden, *Z. Phys. Chem.* **20**, 383 (1896).
25. H. Staudinger, *Die Ketene,* F. Enke, Stuttgart, Germany, 1912, p. 46.
26. G. Schroeter, *Ber. Dtsch. Chem. Ges.* **49**, 2697 (1916).
27. H. Staudinger, *Ber. Dtsch. Chem. Ges.* **53**, 1073 (1920).
28. H. Staudinger and J. Fritschi, *Helv. Chim. Acta* **5,** 785 (1922).
29. H. Staudinger, *Ber. Dtsch. Chem. Ges.* **57**, 1203 (1924).
30. H. Staudinger, *Ber. Dtsch. Chem. Ges.* **59**, 3019 (1926); H. Staudinger, K. Frey, and W. Starck, *Ber. Dtsch. Chem. Ges.* **60**, 1782 (1927).
31. R. Pummerer, H. Nielsen, and W. Gündel, *Ber. Dtsch. Chem. Ges.* **60**, 2167 (1927).
32. K. H. Meyer and H. Mark, *Ber. Dtsch. Chem. Ges.* **61**, 593, 1939 (1928).
33. H. Mark, *Physical Chemistry of High Polymeric Systems,* Interscience, New York, 1940.
34. H. Staudinger and E. Husemann, *Liebigs Ann. Chem.* **527**, 195 (1937).
35. W. H. Carothers, *Chem. Rev.* **8**, 353 (1931).
36. H. Mark and G. S. Whitby, eds., *Collected Papers of W. H. Carothers,* Interscience, New York, 1940.
37. J. B. Sumner, *J. Biol. Chem.* **69**, 435 (1926).
38. J. H. Northrop, *J. Gen. Physiol.* **13**, 739 (1930).
39. T. Svedberg and K. O. Pedersen, *The Ultracentrifuge,* Oxford University Press, London and New York, 1940.
40. A. Tiselius, *Kolloid-Z.* **85**, 129 (1938).
41. C. E. Schildknecht, S. T. Gross, H. R. Davidson, J. M. Lambert, and A. O. Zoss, *Ind. Eng.* **40**, 2104 (1948).
42. K. Ziegler, Folgen und Werdegang einer Erfindung, *Angew. Chem.* **76**, 545 (1964).
43. G. Natta, P. Pino, P. Corradini, F. Danusso, E. Mantica, G. Mazzanti, and G. Moraglio, *J. Am. Chem. Soc.* **77**, 1708 (1955); see also G. Natta, Von der stereospezifischen Poly-

merisation zur asymmetrischen autokatalytischen Synthese von Makromolekülen, *Angew. Chem.* **76**, 553 (1964) (historical review).
44. W. Kuhn, *Ber. Dtsch. Chem. Ges.* **65**, 1503 (1930).
45. H. Staudinger and E. Urech, *Helv. Chim. Acta* **12**, 1107 (1929).
46. W. Chalmers, *J. Am. Chem. Soc.* **56**, 912 (1934).
47. H. Staudinger and W. Frost, *Ber. Dtsch. Chem. Ges.* **68**, 2351 (1935).
48. H. W. Melville, *Proc. R. Soc. London* **A163**, 511 (1937).
49. G. V. Schulz and E. Husemann, *Z. Phys. Chem.* **B39**, 246 (1938).
50. C. C. Price, R. W. Kell, and E. Kred, *J. Am. Chem. Soc.* **63**, 2708 (1941); **64**, 1103 (1942).
51. W. Kern and H. Kämmerer, *J. Prakt. Chem.* **161**, 81, 289 (1942).
52. P. D. Bartlett and S. C. Cohen, *J. Am. Chem. Soc.* **65**, 543 (1943).
53. P. J. Flory, *Principles of Polymer Chemistry*, Cornell University Press, Ithaca, New York, 1953; P. J. Flory, *Statistical Mechanics of Chain Molecules*, Wiley-Interscience, New York, 1969.

Section 1.4. Nomenclature

IUPAC Macromolecular Nomenclature Commission, Nomenclature of regular single strand organic polymers, *Macromolecules* **6**, 149 (1973).

IUPAC, Macromolecular Division, Tentative nomenclature of regular single strand organic polymers, *J. Polym. Sci., Polym. Lett. Ed.* **11**, 389 (1973).

IUPAC, Abbreviated nomenclature of synthetic polypeptides (polymerized amino acids): revised recommendations (1971), *Biopolymers* **11**, 321–327 (1972).

K. L. Loenig, W. Metanomski, and W. H. Powell, Indexing of polymers in chemical abstracts, *J. Chem. Doc.* **9**, 248 (1969).

M. L. Huggins, G. Natta, V. Desreux, and H. Mark, Report on nomenclature dealing with steric regularity in high polymers, *Pure Appl. Chem.* **12**, 645 (1966).

J. W. Breitenbach *et al.*, Richtlinien für die Nomenklatur auf dem Gebiet der makromolekularen Stoffe, *Makromol. Chem.* **38**, 1 (1960).

Chapter 2
Constitution

2.1. The Concept of Structure

The concept of "structure" has different meanings in chemistry and in physics. Constitution and configuration are considered to be part of the chemical structure, while orientation and crystallinity come under physical structure. Conformation can be associated with both the physical and chemical structure.

Under *constitution* are included the type and arrangement of the chain atoms, the kinds of substituents and end groups, the sequence of monomeric units, type and length of branching, the molecular weight, and the breadth of the molecular-weight distribution. Since molecular weights are almost exclusively evaluated from solution properties, these will be discussed in Part II.

The *configuration* (Chapter 3) refers to the arrangement in space of the substituents around a particular atom. It is only possible to change one configuration into another by destroying and reforming chemical bonds. Macromolecular compounds show special configurations because of the many monomeric units combined in one chain. The configuration of the substituents themselves is identical to that of low-molecular-weight compounds and need not be discussed here.

The *conformation* (constellation or configuration to the physicist) is determined by the rotation of atomic groups about single bonds to preferred positions (Chapter 4). Conformations can interchange without the destruction of chemical bonds. In macromolecular chemistry the overall conformation—the shape—of the whole macromolecule is important, as well as the conformation about single bonds. Similarly, because of the very large number of possible conformations, preferred conformations are determined by statistical laws. The overall conformations in the solid state and in solution can differ considerably.

Orientation (Chapter 5) refers to the preferred positioning of parts or groups of molecules without the establishment of large-scale order.

Crystallinity (Chapter 5) presupposes not only a three-dimensional preferential arrangement of the chains, but definite interrelationships between the lattice points. In macromolecular science, the chain atoms can be considered as lattice points. In polymers, however, in contrast to low-molecular-weight substances, it is not only the mutual arrangement of the lattice points of various molecules which must be considered, but also the arrangement of the lattice points of an individual macromolecule relative to the other lattice points of the same molecule.

Thus, in essence, the concept of chemical structure embraces the construction of the individual molecule, and the concept of physical structure refers to the arrangement of groups or molecules relative to each other.

2.2. *Atomic Structure and Polymer Chain Bonds*

Isochains consist of identical main-chain atoms, while heterochains consist of different main-chain atoms. The capacity of an atom to form iso- and heterochains is strongly dependent on its position in the periodic table. In general, only a small number of elements can form macromole-

Table 2-1. Periodic Position of the Elements and Their Ability to Form Isochains[a]

IIIB $2s, 1p$	IVB $2s, 2p$	VB $2s, 3p$	VIB $2s, 4p$	VIIB $2s, 5p$
5 B, ~*5*	6 C, ∞	7 N, ∞?	8 O, ∞?	9 F, *2*
13 Al, *1*	14 Si, *45*	15 P, >*4*	16 S, *30,000*	17 Cl, *2*
31 Ga, *1*	32 Ge, *6*	33 As, *5*	34 Se, ?	35 Br, *2*
49 In, *1*	50 Sn, *5*	51 Sb, *3*	52 Te, ?	53 I, *2*
81 Tl, *1*	82 Pb, *2*	83 Bi, ?	84 Po, ?	85 At, *2*

[a] The figures at the right are the highest chain-link numbers so far observed for isochains after isolation.

cules, mainly the elements of groups IVB, VB, and VIB of the periodic system (see Table 2-1). For inorganic macromolecules it is often said that isochains result from catenation, whereas heterochains come from alternation. However, since the main-chain atoms of a heterochain need not necessarily be arranged alternately, the concept of alternation is too limited to describe heterochains. The isochain corresponds to the homopolymer of organic macromolecular chemistry, and the heterochain to the heteropolymer.

The expression "homopolymer" was formerly used for chains from a single type of base unit [e.g., polyethylene or poly(ethylene oxide)], whereas the expression "heteropolymer" was used for what is now known as a copolymer.

Iso- and heterochains can be substituted or unsubstituted. Unsubstituted isochains are, in the strictest sense and in inorganic nomenclature, macromolecules such as polymeric sulfur, while silanes, $H(SiH_2)_nH$, belong to the substituted chains. In organic chemistry, hydrogen-substituted main chains are, however, considered to be unsubstituted, since it is not diamond or graphite which is considered to be the parent structure, but the aliphatic hydrocarbon. According to this definition, polymethylene, $H(CH_2)_nH$, is an unsubstituted isochain.

2.2.1. *Isochains*

Isochains are formed from elements of groups IVB, VB, and VIB of the periodic table, as well as from boron (Table 2-1). All elements for which isolatable compounds containing at least three successive identical atoms in a chain exist are considered to be capable of forming such isochains. The number of atoms joined together in the chain is called the chain-link number. The highest chain-link numbers are shown by the elements of the first row. Within each group, the numbers decrease with increasing atomic number. Thus, in group IVB, carbon (first row) practically forms isolatable chains of infinite length (alkanes). Silanes (second row) can only be obtained with up to 45 chain links, germanes (third row) with only up to 6, and stannanes (fourth row) with no more than 5. All the numbers quoted relate to compounds that can be isolated, i.e., compounds that can be obtained as single molecules in a fluid form, e.g., as a gas, in solution, or even as a melt.

Third Group. Elemental boron is present in polymeric form in the solid state. With the boranes (borohydrides), main-chain bonds are only partly B—B; they are also partly B—H—B in character.

Fourth Group. In the last analysis, the whole of organic chemistry depends on the capacity of carbon to form isochains. Diamond and graphite are polymeric forms of carbon.

Silicon also occurs as a polymer in the solid state. Its ability to form isolatable isochains, however, is considerably lower than that of carbon. Nevertheless, silanes, $H(SiH_2)_nH$, can be isolated up to chain-link numbers of $n = 45$. Silanes of the $(SiH)_n$ type are also known, and probably exist in the form of a graphite-type lattice with hydrogen substitution:

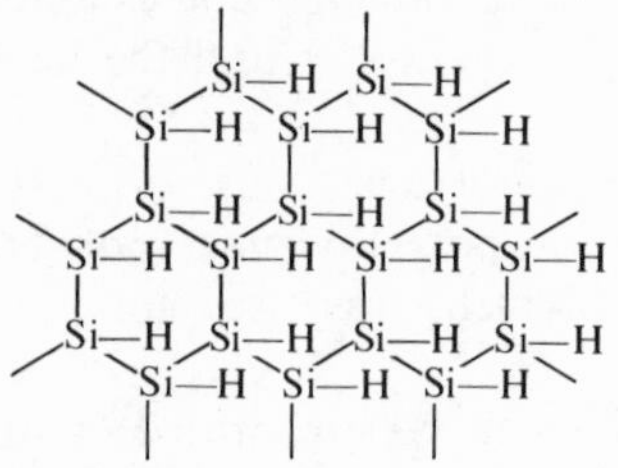

```
             |       |
            Si—H   Si—H
        \  /    \ /    \ /
        Si—H   Si—H   Si—H
         |       |       |
        Si—H   Si—H   Si—H
    \  /    \ /    \ /    \ /
    Si—H   Si—H   Si—H   Si—H
     |       |       |       |
    Si—H   Si—H   Si—H   Si—H
   /    \ /    \ /    \ /    \
        Si—H   Si—H   Si—H
         |       |       |
```

The tendency to form isochains decreases still further with the germanes, $H(GeH_2)_nH$, and stannanes, $H(SnH_2)_nH$. Elemental germanium occurs in polymeric form, while tin exists as a polymer in one modification, whereas in another it exhibits metallic bonding.

Fifth Group. Nitrogen isochains probably exist in a product, $(NH)_n$, which is obtained as a blue substance by decomposing hydrogen azide at 1000°C and chilling the reaction product with liquid nitrogen. The product converts to ammonium azide, NH_4N_3, at −125°C. Nitrogen isochains substituted with organic end groups can be obtained with up to six nitrogen atoms joined together.

Phosphanes, arsanes, and stilbanes, i.e., hydrogen compounds of phosphorus, arsenic, and antimony, are known in the form of short isochains or rings. The elements phosphorus, arsenic, and antimony exist polymerically with different chain atom arrangements in what are known as their allotropic modifications. The best known example is black phosphorus.

Sixth Group. An ozone modification occurring at low temperatures probably consists of oxygen isochains. Selenium and tellurium form chain polymers in the solid state, and sulfur does so in the melt within a certain temperature range. In contrast to the elements of group V, the tendency to form sulfanes, etc., is distinctly decreased.

Isochains from elements of other groups are unknown. The experimental observation that only a quite distinct and limited number of elements close together in the periodic table can form such isochains can be explained as follows: Elements of the first row do not possess available d orbitals. They can therefore form no more than four σ bonds per atom, which corresponds to sp^3 hybridization. Only carbon and the elements to the right of it have enough electrons to contribute at least one electron to each com-

Table 2-2. Bond Energies (10^{-5} *J/mol bond) for Bonds between Like Elements*[a]

Bond	Energy	Bond	Energy	Bond	Energy	Bond	Energy
C—C	3.5	N—N	1.6	O—O	1.6	H—H	4.3
Si—Si	1.8	P—P	2.1	S—S	2.1	F—F	1.6
Ge—Ge	1.6	As—As	1.3	Se—Se	1.8	Cl—Cl	2.4
Sn—Sn	1.4	Sb—Sb	1.2	Te—Te	1.4	Br—Br	1.5
						I—I	1.5

[a] After L. Pauling.

plete bond (σ, π_x, π_y). The elements to the right of carbon thus show a lower bonding energy than carbon (Table 2-2), and acting as electron donors with the corresponding electron acceptors, they form heterochains particularly readily. Nitrogen, oxygen, and fluorine, when compared with their successors in the same group, have too low a bonding energy, and this, according to K. S. Pitzer, results from the strong mutual repulsion of the free electron pairs.

The elements to the left of carbon, on the other hand, have less available electrons than unoccupied orbitals. Since the atoms tend to fill their energetically more accessible outer orbitals, multicenter bridge bonds, involving hydrogen, methyl groups, etc., will occur with boron and beryllium (cf. Figure 2-1).

Elements of the second row possess *d* orbitals of sufficiently low energy to participate in bonding. In general, however, *d* orbitals are not used for σ bonding, but, via hybridization, for π bonding. This hybridization increases the stability of the molecule. Hybridization is most marked in silicon. phosphorus, and sulfur. These elements are therefore polymeric in the solid state, and even less condensed states show to some extent a high chain-link number.

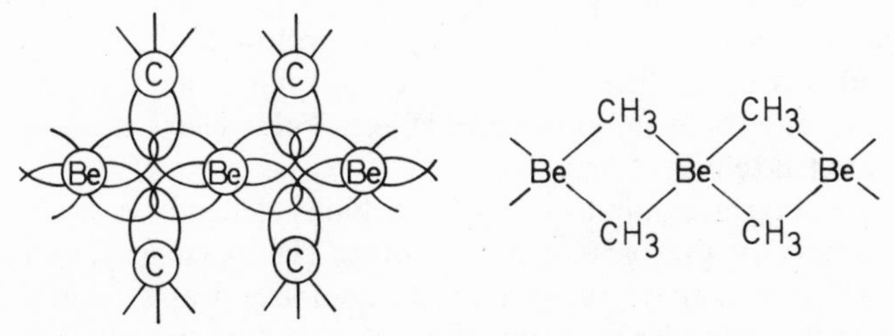

Figure 2-1. Dimethyl beryllium. *Left*: Schematic representation of the overlapping orbitals, *Right*: Normal valence bond method of writing the structure.

In the fourth row the *d* orbitals are used more for σ bonding than for π bonding. As would be expected, the bonding capacity of the third-row elements lies between that of the second- and fourth-row elements.

It is therefore to be expected that the bond energy within each period falls with increasing atomic number, and likewise within each row (cf. Table 2-2). The higher the atomic number of the element, the less marked the macromolecular character of the inorganic polymer will be.

The bond energies reproduced in Table 2-2 are largely what one would expect. Admittedly they are not entirely unambiguous, as, on the whole, they are not values for chain structures, but were measured on low-molecular-weight compounds containing only one isobond. The bond strengths of such bonds, however, are influenced by the nature of the substituents. The bond energies of Table 2-2 are thus "average" bond energies from a numerous set of compounds, and are not genuine bond dissociation energies. Complications can be expected even in macromolecules themselves, since their one-, two-, or three-dimensional structures are strongly affected to a varying extent by polarization effects, and therefore exhibit different bond stabilities.

Thus, only a few elements form *isolatable* isochains. However, for all elements for which crystallographic data are available, only 11 do not have at least one macromolecular form in the crystalline state. In a study involving about 1200 crystallographically investigated compounds of two elements, only 5% were nonmacromolecular, and of the others, 1.5% were linear, 7.5% were parquet, and 86% were layer polymers.

2.2.2. *Heterochains*

All elements that form isochains can also form heterochains with other elements. In addition, simple heterochains can sometimes be formed by the next higher row of the periods III, IVB, and VB, namely aluminum, iron, and bismuth. Compounds with multicenter bonds are also formed by beryllium and a few elements from higher rows, e.g., niobium and vanadium. In these compounds, as well as the OH and H groups, the elements of period VIIB can also occur as the central atoms in multicenter bridge bonding, that is, in borohydrides, niobium iodide, etc. These elements, however, do not represent true chain atoms.

The bonding and stability of heterochains depends basically on the electronegativity of the participating atoms. The electronegativity E is a measure of the capacity of the element to compete at any instant with the other elements for the larger proportion of the electronic charge. Electronegativity cannot be measured directly, but is estimated from ionization potentials, atomic radii, bond force constants, or bond energies. The

Table 2-3. Heterochain-Forming Elements and Their Relative Electronegativities

IIIB	IVB	VB	VIB	VIIB	0
				1 H, *2.1*	2 He
5 B, *2.0*	6 C, *2.5*	7 N, *3.0*	8 O, *3.5*	9 F, *4.0*	10 Ne
13 Al, *1.5*	14 Si, *1.8*	15 P, *2.1*	16 S, *2.5*	17 Cl, *3.0*	18 A
31 Ga, *1.6*	32 Ge, *1.8*	33 As, *2.0*	34 Se, *2.4*	35 Br, *2.8*	36 Kr
49 In. *1.7*	50 Sn, *1.8*	51 Sb, *1.9*	52 Te, *2.1*	53 I, *2.5*	54 Xe
81 Tl, *1.8*	82 Pb, *1.8*	83 Bi, *1.9*	84 Po, *2.0*	85 At, *2.2*	86 Rn

Pauling electronegativity series based on bond energies is the best known.

In Pauling's electronegativity series, the most electronegative element, fluorine, acts as a standard or reference element. In this scheme, carbon has a value of $E = 2.5$. Every combination of elements with electronegativity higher than 2.5 with those of electronegativity below 2.5 leads, within the observance of certain selection rules, to heterochains. Oxygen (3.5), nitrogen (3.0), and sulfur (2.5) thus form heterochains with boron (2.0), aluminum (1.8), silicon (1.8), germanium (1.8), lead (1.8), titanium (1.5), zirconium (1.4), phosphorus (2.1), arsenic (2.0), antimony (1.9), bismuth (1.9), and vanadium (1.6).

On the basis of the electronegativity value, it is to be expected that sulfur and selenium are very similar to carbon in their tendency to form chains (cf. Table 2-3). Equally, $(C—S)_x$ chains are also relatively stable.

The tendency to form multiple bonds competes with the tendency to form heterochains. In multiple bonds σ bonds are also present with π bonds. A π bond possesses a bond order of two. Bond orders can be calculated from force constants, which, in turn, can be obtained from vibrational spectra (infrared and Raman spectroscopy).

From data of this kind, the following conditions were drawn up empirically for π-bond occurrence:

1. Both atoms between which the bond is formed must be electron deficient.
2. The sum of the Pauling electronegativities for both bond partners must be at least five.
3. The difference in the Pauling electronegativities for both bond partners must be as low as possible, i.e., less than 1.5.

The significance of these rules for polymer chemistry can be demonstrated by the following example.

The nitrogen atom has an electronegativity of 3.0 (Table 2-3). Thus for the nitrogen–nitrogen bond, the sum of the electronegativities $\sum E$ is 6.0, whereas the difference ΔE is zero. Two nitrogen atoms therefore form a very stable multiple (triple) bond and polymeric nitrogen is not stable under normal conditions.

In hydrogen cyanide, H—C≡N, the values for the carbon–nitrogen bond are $\sum E = 5.5$ and $\Delta E = 0.5$. The triple bond is thus stable, but weaker than the nitrogen–nitrogen triple bond. Correspondingly, hydrogen cyanide is also known in polymeric form.

In the boron–nitrogen compounds, $\sum E$ falls to 5.0, while ΔE rises to 1.0. HBNH no longer exists as a monomer but is exclusively polymeric.

Too great a difference in electronegativity leads to ionic bonding, and therefore no macromolecular character. Correspondingly, it is found that bond energies between elements of the first and second row, which have fully occupied orbitals, increase to a first approximation (Table 2-4) with the electronegativity difference between the bonded atoms. In contrast, the bond energies between boron and carbon (440 kJ/mol bond) and boron and nitrogen (830 kJ/mol bond) can be traced back to the electronic structure of boron. In comparisons of this kind, therefore, allowance must always be made for the position of the element in the periodic table, i.e., for the "ionic bond contribution."

Bond energies (actually bond dissociation energies) primarily indicate the susceptibility of the bond to thermal scission, and therefore give evidence of the thermal stability of the macromolecule. The vulnerability of a bond to other reagents, on the other hand, depends mainly on the ionic character of the bond and whether there remain unoccupied orbitals or electron pairs in the molecule, as these will lower the activation energy of reaction with the reagent. The resistance to reduction, oxidation, hydrolysis, etc., decreases with increasing atomic number in every period. Thus hydrocarbons, C_nH_{2n+2} are not hydrolyzed, but silanes, Si_nH_{2n+2}, certainly are.

Table 2-4. Bond Energies and Differences in Electronegativity

Bond	Bond energy, 10^{-5} J/mol bond	Difference in electronegativity
C—S	2.6	0
C—N	2.9	0.5
C—Si	2.9	0.7
C—O	3.5	1.0
Si—O	3.7	1.7

In the latter case there are only four completely filled orbitals of a possible maximum coordination number of six.

In bonds between carbon and nitrogen, phosphorus, oxygen, sulfur, selenium, or the halogens, the carbon is made more positive ($C^{\delta+}$—$E^{\delta-}$) according to the position of the element in the periodic table, and is thus more easily attacked by nucleophilic reagents. If, on the other hand, the bond partner of carbon is a metal atom, then the now negative carbon ($C^{\delta-}$—$Me^{\delta+}$) can only be attacked by electrophilic reagents. Thus all macromolecules with heteroatoms in the main chain react more readily than pure carbon chains. Under most conditions, they undergo exchange equilibration, and are readily attacked chemically. The vulnerability of carbon is further dependent on its substituents. These act as electron donors (e.g., methyl groups) or as electron acceptors (e.g., halogens) and can either stabilize or activate the main-chain bonds once they have been formed.

Similar considerations apply to carbon bonded to noncarbon atoms in the main chain. The —Si—CH_3 is not so polar as —SiH_3, and so the dimethyl derivative is sufficiently stable to enable poly(dimethylsiloxanes), $-[Si(CH_3)_2-O]_n-$, to be produced commercially. By contrast, Ti—C and Al—C bonds are sensitive to oxygen and hydrogen (electronegativity of metals!). Thus organic groups can only be bonded with these metal atoms via ether or carboxyl groups.

In boron and beryllium there are fewer electrons than available orbitals. The outer orbitals can be filled, for example, in dimethyl beryllium, by a mechanism whereby one carbon atom orbital overlaps with an orbital from each of two neighboring beryllium atoms (Figure 2-1). The resulting high-molecular-weight chain of dimethyl beryllium thus has three center bonds, which leads to an absurd structural formula when valences are represented in the normal way. Boron hydrides and some aluminum compounds have similar structures.

Main and subgroup elements of the earlier rows often exhibit a dynamic equilibrium between different coordination numbers. Elements such as fluorine, chlorine, oxygen, etc., can donate one or two free electron pairs to vacant lower energy orbitals of these metal atoms. Fluorine can therefore act as a bifunctional bridging atom, and oxygen can even be mono- to tetrafunctional, according to its bond partner. In all these cases the coordination number is increased above normal. In the case of fluorine, fluorine bridge bonds exist, for example, in the anion of the complex between thallium fluoride and aluminum fluoride:

$$\sim\left(-F_2AlF_3-\right)^{-2}\left(-F_2AlF_3-\right)^{-2}\sim$$

where every chain unit carries a double-negative charge. Chlorine bridging occurs, for example, in palladium (II) chloride, and iodine bridging atoms in niobium (II) iodide:

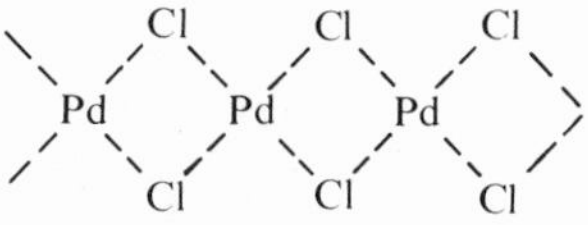

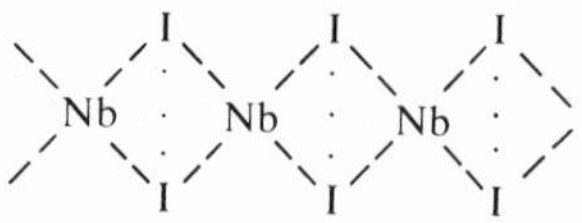

Cl Cl Cl
Pd Pd Pd
Cl Cl Cl

I I I
Nb Nb Nb
I I I

All of these compounds have many vacant orbitals at their disposal. They are therefore readily attacked, and in all of the common solvents they decompose into smaller units. For this reason they were previously not even considered to be macromolecules in the solid state.

An example of an atom-bridged compound which is soluble without loss of macromolecular structure is the basic aluminum soap:

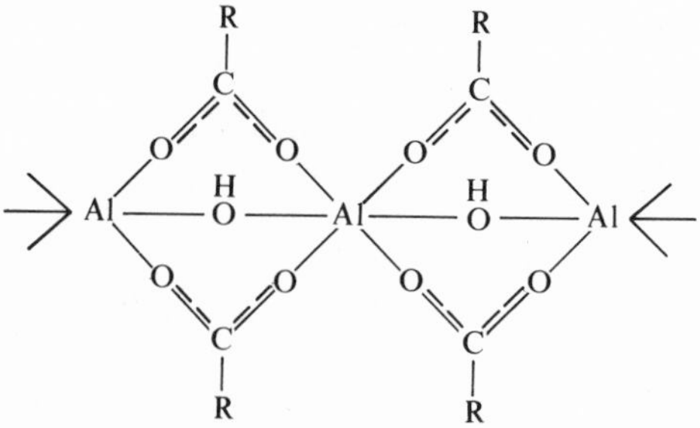

R R
C C
O O O O
H H
Al O Al O Al
O O O O
C C
R R

The compound can have a molecular weight running into the millions. It is soluble in hydrocarbons and is used as a thickener for gasoline.

Other high-molecular-weight structures can, by means of π complexing, be formed by inorganic compounds. Examples of these are the ferrocene sandwich complexes and copper acetylide. The acetylene π orbitals of the copper acetylide, $Cu_2C_2 \cdot 2H_2O$, donate electrons to unoccupied copper orbitals:

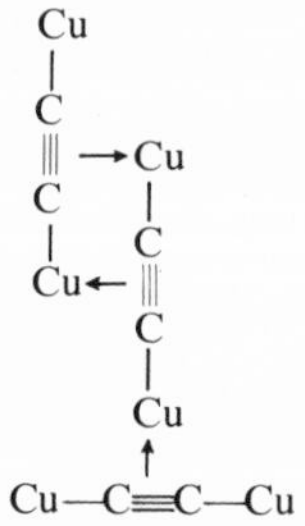

Cu
C
C → Cu
C
Cu ← C
C
Cu
↑
Cu—C≡C—Cu

2.3. Monomeric Unit Bonding

2.3.1. Unipolymers

In polycondensation and ring-opening polymerization, the monomers are joined together in an unequivocal manner. In the polymerization of lactones,

$$\left.\begin{array}{l} (CH_2)_x\begin{array}{c} CO \\ | \\ O \end{array} \\[2em] HO{-}(CH_2)_x{-}COOH \xrightarrow{-H_2O} \end{array}\right\} \longrightarrow -\!\!\left[O{-}(CH_2)_x{-}CO\right]\!\!- \qquad (2\text{-}1)$$

as in the polymerization of α,ω-hydroxycarboxylic acids, polyesters are obtained. The monomeric unit structure does not alter on being converted into polymer, and they are always joined in the "head-to-tail" position—that is, no peroxide or diketo ("head-to-head" or "tail-to-tail") structures are formed.

The addition polymerization of monomers is often less straightforward, since on the one hand the structure of the monomeric unit can alter under certain circumstances, and on the other hand, head-to-head or tail-to-tail structures are possible. For example, under the action of strong bases, acrylamide does not yield poly(acrylamide), but isomerizes to give poly-(β-alanine). Styrene-p-sulfonamide behaves similarly:

$$CH_2{=}CH(C_6H_4SO_2NH_2) \begin{cases} \xrightarrow{\text{radical}} & -\!(CH_2{-}CH(C_6H_4SO_2NH_2))_x\!- \\ \xrightarrow{\text{anionic}} & -\!(CH_2{-}CH_2{-}C_6H_4{-}SO_2{-}NH)_x\!- \end{cases} \qquad (2\text{-}2)$$

Isomerization can also occur in cationic polymerizations, for example, in the polymerization of 4,4-dimethyl-1-pentene:

$$CH_2{=}CH{-}CH_2{-}C(CH_3)_3 \xrightarrow{+AlCl_3} \begin{cases} \xrightarrow{-130^\circ C} & \underset{(I)}{-\!(CH_2{-}CH(CH_2{-}C(CH_3)_3))\!-} \\ \xrightarrow{0^\circ C} & (I) + -\!(CH_2{-}CH_2{-}CH(C(CH_3)_3))\!- \end{cases} \qquad (2\text{-}3)$$

or even in the polymerization of propylene to "poly(ethylene)," with certain catalysts. The altered structure of the monomeric unit can be determined in these cases by spectroscopic means (infrared, nuclear magnetic resonance), or even chemically, as in the case of acrylamide or styrene-*p*-sulfonic acid, by hydrolysis of the polymer. Such polymers with no monomer similar to the monomeric unit are therefore also called "phantom" or "exotic" polymers.

A polymeric product of unknown structure, in some cases with monomeric impurities, etc., is called a polymerizate, whereas a polymer is a macromolecular compound of known structure.

In the cases described, the polymerization can be controlled (by choice of catalyst) in such a way that one or another structure is the almost exclusive result. Under normal polymerization conditions, however, allowance must be made for the few monomer units that react anomalously. In the free-radical polymerization of methacrylonitrile, for example, the majority of monomer molecules polymerize via the carbon double bond, whereas a limited amount of polymerization occurs via the nitrile group, as can be shown spectroscopically:

$$CH_2{=}C(CH_3)(C{\equiv}N) \longrightarrow {+}CH_2{-}C(CH_3){=}C{=}N{+}_x \tag{2-4}$$

In the polymerization of styrene, degradation studies on the polymer product suggest that reaction takes place via the phenyl nucleus to a very small extent:

$$CH_2{=}CH(C_6H_5) \longrightarrow {+}CH_2{-}CH_2{-}C_6H_4{+}_x \tag{2-5}$$

The chemical groups that result from side reactions of this kind are generally less stable than those that normally occur. Therefore, they are "active sites" or "weak links" in the polymer chain, which can frequently be recognized in degradation studies. Because they are normally present in low concentrations only, their constitution is often impossible to elucidate. Thus, as a rule, their existence can only be deduced from degradation studies, which certainly are not very conclusive.

Weak links can be formed through a variation in the linkages between monomeric units as well as through monomer isomerization. For example, in the polymerization of vinyl compounds, $CH_2{=}CHR$, a certain proportion of head-to-head and tail-to-tail linkages will occur along with the normal head-to-tail bonding:

$$CH_2{=}CH(R) \begin{cases} \xrightarrow{1,2\ \text{addition}} -CH_2-CH(R)-CH_2-CH(R)-CH_2-CH(R)- & \text{head-to-tail} \\ \xrightarrow{1,1\ \text{addition}} -CH_2-CH(R)-CH(R)-CH_2-CH_2-CH(R) & \text{head-to-head or tail-to-tail} \end{cases} \quad (2\text{-}6)$$

The first studies of this problem were carried out on poly(styrene). For example, in the pyrolysis of radically polymerized poly(styrene), 1,3-diphenylpropane and 1,3,5-triphenylpentane were recovered, but no diphenylethane. Small proportions of 1,1 linkages could not be found since the pyrolysis products could not be recovered quantitatively, and the method of determination was not very sensitive.

Small quantities of 1,1 linkages in poly(vinyl alcohol) were readily recognized by selective oxidation of the diols. Head-to-tail structures can be oxidized to acetic acid by CrO_3, while the diols resulting from 1,1 linkages are not attacked:

$$-CH_2-CH(OH)-CH_2-CH(OH)- \longrightarrow -CH_2-C(=O)-CH_2-C(=O)- \longrightarrow CH_3COOH \quad (2\text{-}7)$$

The diols from 1,1 linkages are, however, selectively oxidized to succinic and oxalic acids by H_5IO_6:

$$-CH_2-CH(HO)-CH(OH)-CH_2-CH_2-CH(HO)-CH(OH)-CH_2- \longrightarrow \begin{matrix} HOOC-COOH \\ + \\ HOOC-CH_2-CH_2-COOH \end{matrix} \quad (2\text{-}8)$$

Normally, poly(vinyl alcohol) contains about 1–2% head-to-head linkages. In these studies, it was found that the proportion of head-to-head links formed increased with polymerization temperature, as would be expected on steric and energetic grounds.

Chemical methods of investigation of this problem are only unequivocal when the attacking reagent is a selective reagent for one of the structures involved (i.e., H_5IO_6 for diols). If the reagent also reacts with the other structures, the random nature of the reaction has to be taken into account in the evaluation of the results (cf. also Section 23.4.4.1).

The proportion of head-to-head structures can be considerable in some cases. For example, in free-radical-polymerized poly(vinylidene fluoride), $\text{+}CH_2-CF_2\text{+}_n$, there are about 10–12% of such links according to nuclear magnetic resonance measurements (^1H and ^{19}F). The proportion is as high as 6–10% in poly(vinyl fluoride). Admittedly, these physical methods are not sensitive enough for the determination of small amounts of "wrong" structures.

Larger proportions of head-to-head or tail-to-tail structures must always be expected when steric effects are small and the resonance stabilization of the growing end (free radicals, cations, anions) is low. The atomic radius of the fluorine atom is less than that of the hydrogen atom, which helps to explain the high proportion of head-to-head structures in poly(vinyl fluoride). Similar effects lead to the 40% head-to-head structures that occur in the polymerization of propylene oxide with diethyl zinc and water as initiator.

The "weak links" so far discussed have been, so to speak, inherent in the monomer and/or the method of polymerization. In addition, weak links can also result from the incorporation of small amounts of impurities into the chain. Small amounts of oxygen can be unintentionally built into chains as comonomer in free-radical polymerizations. The peroxy groups can subsequently decompose with degradative oxidation of the polymer and/or the formation of keto groups in the polymer. The keto groups can render the polymer more susceptible to photolysis, etc. (cf. Section 24.3.1).

2.3.2. *Copolymers*

Copolymers can vary in their average composition and mean monomeric unit sequence length, as well as in the distributions in composition and sequences. It is useful to consider sequences of polymers with a small number of different monomeric units (e.g., proteins). Since the determination of sequence is very specific of multipolymers with many different monomeric units, this will be discussed under the relevant polymer.

In bipolymers (copolymers from two different monomeric units A and B), there is the further classification into alternating, random, graft, and block copolymers:

```
—A—B—A—B—A—B—A—B—A—B—A—B—     Alternating copolymer

—A—B—A—A—B—A—A—A—B—B—A—B—     Random copolymer

—A—A—···A—A—B—B—···B—B—       Block copolymer

—A—A—A—A—A—A—A—A—A—A—A—       Graft copolymer
     |                 |
     B                 B
     |                 |
     B                 B
     |                 |
     B
     |
     B
     |
```

Block and graft copolymers are obtained from two consecutive and separate polyreactions, and so are called multistep polymers. These concepts can, in principle, also be extended to multipolymers (terpolymers and quaterpolymers, etc.).

2.3.2.1. Composition

The average composition of copolymers can be determined most readily for the case where the monomeric units can be isolated and identified by suitable scission or degradation reactions. This is the usual method for the elucidation of protein structure. The proteins are hydrolyzed by acids and/or bases in an automated apparatus. The resulting amino acids are chromatographically separated and assayed quantitatively via the color reaction with ninhydrine in so-called amino acid analyzers (Section 30.2.1).

Such mild degradative reactions are usually not successful with synthetic polymers and are completely inapplicable for carbon chains. When synthetic polymers are pyrolyzed under controlled conditions, a gas-chromatographic analysis of the degradation products provides a kind of fingerprint for the composition and sequence of the polymer in question. Since the method is rapid but not absolute, it is preferentially used for industrial quality control.

The composition of carbon-chain polymers with monomeric units having widely differing analytical composition, characteristic elements or groups, or radioactive labels can be readily determined. Chemical (microanalysis, functional group determination, etc.) and spectroscopic methods (infrared, ultraviolet, nuclear magnetic resonance, etc.), as well as the determination of radioactivity, yield the average composition of the polymer. The mean composition can also be determined from the refractive indices of solid samples. The composition can be calculated from the principle that the copolymer is considered to be a solution of one unipolymer (from one of the monomeric units) in the other. The composition can also be found by means of the refractive index increment dn/dc in solution, which gives the variation in refractive index with concentration. The mass fraction w_A of the monomeric unit A can be calculated from

$$\left(\frac{dn}{dc}\right)_{\text{copolymer}} = \left(\frac{dn}{dc}\right)_A w_A + \left(\frac{dn}{dc}\right)_B w_B \qquad (2\text{-}9)$$

where $w_A + w_B = 1$. The temperature, solvent, and wavelength, as well as the refractive indices of both unipolymers A and B, must be known for this method of determination. Table 2-5 gives a summary of the results of various analytical methods for styrene–methyl methacrylate copolymers. The results from UV (ultraviolet) analysis differ markedly from those of other methods because the position of the UV bands is strongly dependent on the styrene sequence lengths.

Table 2-5. Results for a Styrene–Methyl Methacrylate Copolymer from Different Analytical Procedures[a]

Sample number	% Methyl methacrylate in the polymers				
	C,H,O	IR	UV	NMR	*dn/dc*
C12	74.4	74.0	78.5	73.5	72.8
C14	58.1	53.0	57.7	—	57.0
C16	42.2	41.0	48.5	40.2	41.5
C18	23.0	23.5	28.7	24.1	21.5

[a] Data from H.-G. Elias and U. Gruber.

The cloud-point titration method (Section 6.6.5) can also be applied to copolymers whose monomeric units are chemically not very different. In this method, solutions of various concentrations are titrated with nonsolvent to the first cloud point. By extrapolation to 100% polymer, a critical volume fraction ϕ_3^θ of the nonsolvent is obtained, which normally depends linearly on the copolymer composition.

Given certain assumptions, the cloud-point method allows a unipolymer to be distinguished from copolymers, and this enables the results of graft copolymerization to be monitored. All the other methods so far described do not allow any differentiation between copolymers and polymer mixtures (polymer blends). Polymer mixtures can also be characterized by ultracentrifugation in a density gradient (Section 10.7.5) and, in certain cases, also by fractionation (Section 6.6.4).

2.3.2.2. Constitutional Heterogeneity

A copolymer with two or more constitutionally different monomeric units is not sufficiently characterized by its average composition alone. A product with, for example, a 50% proportion of component A and 50% of component B can be a true copolymer with a composition that is constant for all the molecules present, a true copolymer with different A:B ratios among the component molecules, a polyblend from two unipolymers, or a corresponding mixture from two uni- or bipolymers.

Chemically uniform products can be distinguished quantitatively from those that are chemically heterogeneous by the dependence of the apparent molecular weight from light scattering measurements on the refractive index of the solvent (Section 9.5.3).

There are two principal methods that are suitable for quantitative determination of the constitutional heterogeneity: fractionation and equilibrium sedimentation (equilibrium centrifugation) in a density gradient. In fractionation, the different solubilities of the polymers of various compositions are utilized by adding a nonsolvent to, or altering the temperature of, the polymer solution. If a nonsolvent is added gradually to a solution of the polymer product, then varying fractions can be removed for different solvent–nonsolvent compositions. However, since the solubility depends not only on the chemical composition, but also on the molecular weight, the order of the fractions recovered does not always correspond to an increase or decrease in the content of a given monomeric unit. Table 2-6 shows the results of a fractionation of this kind on a copolymer of vinyl acetate and vinyl chloride. The proportion of vinyl chloride in the fractions tabulated in the third column does not conform to the fraction number (order in which the fractions were removed) reproduced in column one. The success of fractionations of this kind depends largely on correct choice of solvent and nonsolvent, as unsuitable solvent–nonsolvent pairs can, in some cases, simulate the behavior of homogeneous products in fractionation.

For equilibrium centrifugation in a density gradient, a mixture of two solvents is prepared, one of which has a lower density than the other, and the density of any component in the polymer product lies between the densities of the two solvents. At equilibrium, a density gradient forms at

Table 2-6. Results from the Fractionation of a Copolymer of Vinyl Acetate and Vinyl Chloride[a]

Fraction number	Mass, mg	Mass fraction, %	E_{vc}	$\sum w_i^*$
2	41.0	5.32	0.363	2.660
1	56.0	7.27	0.364	8.955
5	78.5	10.19	0.412	17.683
3	43.5	5.65	0.414	25.600
4	61.5	7.98	0.510	32.414
6	64.5	8.37	0.577	40.591
15	26.5	3.44	0.587	46.495
11	38.0	4.93	0.595q	50.681
13	38.0	4.93	0.595	55.613
7	72.5	9.41	0.625	62.784
9	51.0	6.62	0.636	70.798
8	63.5	8.24	0.638	78.228
10	32.0	4.15	0.642	84.425
14	56.0	7.27	0.665	90.136
12	48.0	6.23	0.676	96.885
	$\sum W_i = 770.5$	$\sum w_i = 100$	$\bar{E}_{vc} = 0.550$ $= (\bar{E}_w)_{vc} = \bar{E}_w$	

[a] The results are given in terms of increasing content of vinyl chloride monomeric unit E_{vc} in the fractions (from H.-J. Cantow and O. Fuchs).

a certain gravity field. If one of the polymer components is introduced into such a gradient, then at centrifugal equilibrium the component remains at that point in the density gradient where its buoyancy equals the gravitational pull due to centrifugation. Under suitable conditions, the distribution of the constitutional composition can be studied quantitatively by this technique (cf. also Section 9.7.5).

The constitutional composition distribution can be represented graphically or mathematically by various parameters. In practice, the mass distribution of the constitutional heterogeneity, i.e., the mass contribution of the individual components to the total mass, is the most important. The corresponding mean values thus represent mass averages, usually called weight averages.

The distribution in bipolymers can be displayed on a two-axis graph, in which the ordinate gives the sum of all the fractions $\sum w_i$ and the abscissa gives the properties E_i, in this case the compositions. A cumulative (integral) compositional distribution of this kind can be obtained from fractionation data after one or two calculations, as shown, for example, for the data given in Table 2-6.

It must be realized that the property E_i measured for a fraction i is a mean value in a distribution. Since it is impossible to fractionate a synthetic polymer into individual components, fractions obtained in practice contain a distribution of component sizes. To a first approximation, it can be assumed that the material in the fraction is distributed symmetrically about the mean property value. For fraction 2 of Table 2-6, with $E_{vc} = 0.363$, a value of $w_i^* = 2.66 = (5.32/2)\%$ has to be used, and not the value $w_i = 5.32$. Assuming for fraction 2 that half the material has a molecular weight greater than the mean value for the fraction, then the weight contribution of all the components up to this mean value is half the total weight of the material in fraction 2. In the next fraction (number 1), by the same reasoning, the value to be plotted on the ordinate is the whole of fraction 2 and half of fraction 1 (7.27/2 for fraction 1 and 5.32 for fraction 2) i.e., $\sum w_i^* = 8.95\%$. This calculation process is also used for fraction number 5, the next fraction:

$$\sum w_i^* = 5.32 + 7.27 + (10.19/2) = 17.685\%, \quad \text{etc.}$$

Plotting $\sum w_i^*$ against E_{vc} shows that the cumulative weight distribution of the vinyl chloride unit content in the polymer is by no means "continuous," i.e., it shows more than one point of inflection. There are marked gradations, which means, of course, that a second fractionation is needed to show whether the gradations are real and not an experimental artifact. It can also be seen that the distribution curve is distinctly skewed since the average composition $(\overline{E}_w)_{vc}$ corresponds to a mass fraction w_i^* of 0.35 of the total polymer, and not 0.5, a value expected for a distribution symmetric

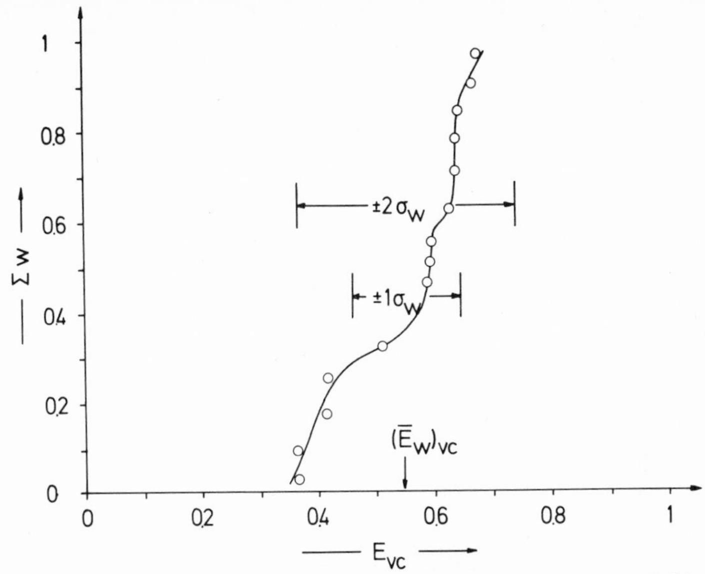

Figure 2-2. Integral mass distribution of the vinyl chloride content of a poly(vinyl chloride-co-vinyl acetate) with the weight-average composition $(\bar{E}_w)_{vc}$ and the weight average standard deviation σ_w of the distribution.

about the mean, i.e., nonskewed. Finally, the distribution does not extend over the whole of the spectrum, from $E_{vc} = 0$ to $E_{vc} = 1.0$, but from about 0.33 to 0.73, the lowest and the highest proportions of vinyl chloride units found in fractions of this polymer (Figure 2-2).

Three properties, average composition, width of the distribution, and the skew of the distribution, can be described by three statistical parameters: the weight average of the composition $\bar{E}_w$, the weight-average standard deviation σ_w, and the weight-average skew Ω_w. These three parameters are obtained by combining the moment μ about the origin of the component distribution,

$$\mu_w^{(q)}(E) \equiv \frac{\sum_i w_i E_i^q}{\sum_i w_i} \tag{2-10}$$

and the moment v about the mean of the distribution (the variance),

$$v_w^{(q)}(E) \equiv \frac{\sum_i w_i (E_i - \bar{E}_w)^q}{\sum_i w_i}, \qquad \bar{E}_w = \frac{\sum_i w_i E_i}{\sum_i w_i} \tag{2-11}$$

The standard deviation σ_w can be calculated from these moments:

$$\sigma_w \equiv [v_w^{(2)}]^{1/2} \equiv [\mu_w^{(2)} - (\mu_w^{(1)})^2]^{1/2} \tag{2-12}$$

The skewness of the distribution is determined from the second and third moments for the weight distribution,

$$\Omega_w \equiv \frac{\nu_w^{(3)}}{2(\nu_w^{(2)})^{3/2}} = \frac{\mu_w^{(3)} - 3\mu_w^{(1)}\mu_w^{(2)} + 2(\mu_w^{(1)})^3}{2[\mu_w^{(2)} - (\mu_w^{(1)})^2]^{3/2}} \tag{2-13}$$

Ω_w indicates whether the greater proportion of the properties lie above or below $\sum w_i = 0.5$. If the properties are plotted such that $\sum w_i = 0$ and $E = 0$, then Ω_w is negative if the larger part of the properties lie within the range $\sum w_i = 0$–0.5, and it is positive when the greater proportion lies above $\sum w_i = 0.5$. Here, the symbol (q) is the order of the moments. For a more detailed discussion of moments and averages, see Chapter 8.

For the examples given in Tables 2-6 and Figure 2-2, the following values are obtained:

$$\mu_w^{(1)} = \bar{E}_w = 0.550, \qquad \sigma_w = 0.106, \qquad \Omega_w = -0.300$$

Thus, on the average, the copolymer consists of 55.0% vinyl chloride. The standard deviation shows that the vinyl chloride units are not very widely distributed. If the standard deviation is formally calculated for a Gaussian distribution, then with this value of 1 σ_w, 68.3% will lie in the range $\bar{E}_w = 0.550 \pm 0.106$. In actual fact, however, only $\sum w_i^* = 86 - 28 = 58\%$ of all the components lie in this range (cf. Figure 2-2). Strictly speaking, such an interpretation of the standard deviation is only valid for a Gaussian distribution. However, the standard deviation can always be used as a qualitative measure of the width of a distribution, regardless of the kind of distribution of components in the polymer product.

Table 2-7 gives the various statistical parameters for a series of polymers (unipolymers, bipolymers, polyblends). Figure 2-3 shows the cumulative weight-distribution curves of some of the polymer products. One can infer from the table and figure that the standard deviation of the constitutionally homogeneous polymers 1, 2, and 3 is zero. In the heterogeneous polymers it increases from 6 through 5 and 4 to 7, as can also be readily seen from Figure 2-3.

Figure 2-3 and Table 2-7 also show that the skewness Ω_w is zero if the distributions are symmetric about $\sum w_i = 0.5$. In this case it does not matter whether it is a question of chemically uniform (numbers 1–3) or chemically different (numbers 4 and 5) fractions. The skewness Ω_w is relatively sensitive to impurities. With impurities of, for example, 10, 1, or 0.1% of a unipolymer with $E_1 = 0$ in a polymer of $E_2 = 1$, Ω_w becomes, respectively, -1.33, -4.92, and -15.7.

Furthermore, the individual moments can be used for the characterization of the fractions. If at any instant the fractions consists of chemically uniform substances (uni- or bipolymers), then $\nu_w^{(1)} = 0$ (numbers 1–7). If

Table 2-7. Various Moments $\mu_w^{(q)}$ and $v_w^{(q)}$, Standard Deviation σ_w and Skewness Ω_w of the Mass Distributions of Copolymers and Polyblends Composed of Fractions[a]

Polymer number and product	Mass fraction w_i		E		Component moments		Moments about the mean				
	w_{I}	w_{II}	$(E_A)_{\mathrm{I}}$	$(E_A)_{\mathrm{II}}$	$\mu_w^{(1)} = \bar{E}_w$	$\mu_w^{(2)}$	$10^3 v_w^{(1)}$	$10^3 v_w^{(2)}$	$10^3 v_w^{(3)}$	σ_w	Ω_w
1. Unipolymer	1	—	1	—	1	1	0	0	0	0	0
2. Constitutionally uniform/homogeneous bipolymer, e.g., azeotropic bipolymer	1	—	0.65	—	0.650	0.4225	0	0	0	0	0
3. Alternating bipolymer	1	—	0.50	—	0.500	0.2500	0	0	0	0	0
4. Bipolymer with linear increase in composition from $E_i = 0$ at $\sum w_i = 0$ to $E_i = 1$ at $\sum w_i = 1$	(see left-hand column)				0.500	0.3325	0	42.50	0	0.287	0
5. Mixture from a constitutionally homogeneous bipolymer and a unipolymer	0.5	0.5	0.65	1.0	0.825	0.7113	0	30.63	0	0.175	0
6. Same as 5	0.9	0.1	0.65	1.0	0.685	0.4803	0	10.94	+3.087	0.105	+1.349
7. Mixture of two unipolymers	0.35	0.65	0	1.0	0.650	0.6500	0	227.5	−68.25	0.477	−0.315

[a] The fractions w_i are always based on the component A with the composition $E_A = 1$ ($E_B = 0$).

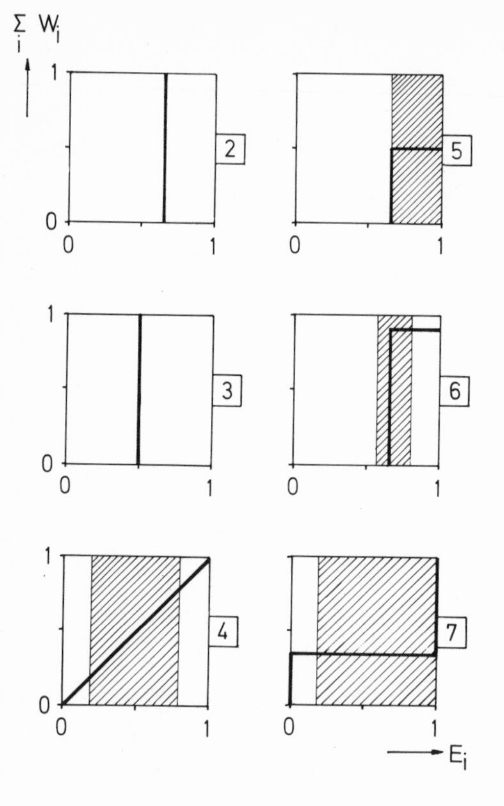

Figure 2-3. The integral mass distribution of various copolymers and polymer mixtures (polyblends). The numbers correspond to those listed in Table 2-7. The standard deviations $\pm\sigma_w$ are indicated by the shading.

the fractions are of unipolymers, then it follows that $\mu_w^{(1)} = \mu_w^{(2)}$ (numbers 1 and 7). It is possible to draw analogous inferences from the moments, the standard deviations, and the skews of multipolymers or polyblends of more than two components.

2.3.2.3. Sequence

The arrangement of the two types of monomeric units a and b of a bipolymer is determined by the polymerization mechanism, for example,

$$\text{a-b-a-a-b-b-a-b-a-a-b-a-b-b-b-a-a-a-b-b}$$

Thus there are sequences of 1, 2, 3, etc., monomeric units of the same kind in the chain. The sequences are written with capital letters to distinguish them from monomeric units and bonds, which are given lower case letters. The number average $(L_A)_n$ of the sequence length of monomeric unit a is, by definition,

$$(\bar{L}_A)_n = \frac{\sum_i (N_A)_i (L_A)_i}{\sum_i (N_A)_i} \tag{2-14}$$

$(N_A)_i$ is the number of sequences of length i and $(L_A)_i$ is their length. Consequently, for the example above, it follows that $(\bar{L}_A)_n = (3 \times 1 + 2 \times 2 + 1 \times 3)/(3 + 2 + 1) = 1.66$.

In radical polymerizations, the mean sequence length can be calculated, given certain assumptions (cf. Section 22.1.4). As a rule, it is not directly obtainable from experiments. However, in many copolymers, the fraction f_{ab} of all the ab groups (i.e., a bonded to b and b to a) can be determined by nuclear magnetic resonance spectroscopy; f_{ab} is defined by

$$f_{ab} = \frac{N_{ab}}{N_{aa} + N_{ab} + N_{bb}} = \frac{N_{ab}}{(\bar{X}_n - 1) N_{cop}} \tag{2-15}$$

The right-hand side of equation (2-15) stems from the fact that the number of bonds in *one* molecule is always one less than the degree of polymerization, and the total number of bonds, of course, still depends on the total number of copolymer molecules N_{cop}. The sum in the numerator of equation (2-14) is thus identical to the number $N_a$of all the monomeric units a, i.e., $\sum_i (N_A)_i (L_A)_i = N_a$.

In cyclic macromolecules, the sum of the sequences in the denominator is equal to half the number of all bonds N_{ab} between a and b monomeric units (i.e., a–b and b–a bonds). Thus it follows that $\sum_i (N_A)_i = 0.5 N_{ab}$. The same relationship, to a good approximation, is also applicable to noncyclic chains containing a suitably high number of sequences per macromolecule. With this relationship and equations (2-14) and (2-15), it follows that

$$(\bar{L}_A)_n = \frac{2N_a}{f_{ab}(\bar{X}_n - 1) N_{cop}} \tag{2-16}$$

The number of monomeric units is related to the mole number n_a, the mass m_a, and the formula molecular weight M_a of the monomeric units by $N_a = n_a N_L = m_a N_L / M_a$, where M_a is the formula molecular weight of the monomeric units and N_L is the Avogadro number. For the number N_{cop} of copolymer molecules one obtains, analogously, $N_{cop} = (m_a + m_b) N_L / \bar{M}_n$, with m_a and m_b the masses of the monomeric units a and b. For the number-average degree of polymerization $\bar{X}_n$ of the copolymer, we have

$$\bar{X}_n = \left(\frac{N_a + N_b}{N_{cop}} \right) = \bar{M}_n \left(\frac{w_a}{M_a} + \frac{w_b}{M_b} \right) \tag{2-17}$$

where the mass fraction of the a monomeric units in the copolymer is given by $w_a = m_a/(m_a + m_b)$. With this relationship, equation (2-16) becomes

$$\frac{1}{(\bar{L}_A)_n} = 0.5 f_{ab}\left[1 - \bar{M}_n^{-1}\left(\frac{M_a}{w_a}\right) + \frac{M_a w_b}{M_b w_a}\right] \tag{2-18}$$

In this way, the number-average sequence length of the A sequences can be calculated from equation (2-18). The value $(2 \times 10^2)\ [(\bar{L}_A)_n + (\bar{L}_B)_n]^{-1} = \bar{R}_n$ is designated the run number $\bar{R}_n$. Thus $\bar{R}_n$ is the total number of all blocks (A and B sequences) present per 100 monomeric units. For the example given above, $\bar{R}_n$ is 60.

A bipolymer can be described by the fractions f_{aa}, f_{ab}, and f_{bb} of the aa, ab, and bb bonds. A graphical representation on triangular graph paper* gives the values of f_{aa}, f_{ab}, and f_{bb} as coordinate values. In the triangular graph, the point aa corresponds to pure unipolymer a, ab to the alternating copolymer ab, and point bb to the pure unipolymer b. The point 0 shown in Figure 2-4 represents a copolymer with $f_{bb} = 45\%$, $f_{ab} = 25\%$, and $f_{aa} = 30\%$.

The width and skew of the sequence distribution can be calculated in exactly the same way as for chemical constitution distributions, i.e., via the corresponding moments (cf. Section 2.3.2.2). Since only the sequence length is determined experimentally, the number distribution (index n) is required here instead of the weight distribution.

The sequence length can be found from a series of physical and chemical methods. Generally, all the methods are strongly dependent on the polymer constitution; thus often they can only be applied to specific polymers.

The chemical methods depend almost exclusively on two principles—chain scission or neighboring side-group reactions. The methods that work on the chain scission principle use the fact that one of the two components of a bipolymer (or other copolymers) will be attacked by a specific reaction process, while the chain of the other component is stable. The copolymer of isobutylene with about 2% isoprene,

$$\text{+}CH_2\text{—}C(CH_3)_2\text{+}_i\text{+}CH_2\text{—}C(CH_3)\text{=}CH\text{—}CH_2\text{+}_j\text{+}CH_2\text{—}C(CH_3)_2]_k$$

which is used commercially, can be split at the double bond, for example,

* A triangular plot is read/constructed as follows: A graph is drawn with the point marked (Figure 2-4). The coordinates of this point are determined according to the rule that coordinate lines to the point for any given sequence are drawn parallel to the line opposite the corner of the triangle that represents 100% content of this sequence. To find the content f_{bb} of bb bonds or sequences, the corner in question is at the intersection f_{bb} and f_{ab} and the side is the f_{aa} side. One moves from zero on the f_{bb} axis along this axis until one meets the line that can be drawn from the point parallel to the f_{aa} axis. The point of intersection with the f_{bb} axis (at $f_{bb} = 45\%$) gives the f_{bb} content. Triangular graphs are particularly suitable for illustrating the dependence of the three monomeric unit fractions on the reaction conditions (time, temperature, etc.). In such a case, a curve is obtained instead of a point on the graph.

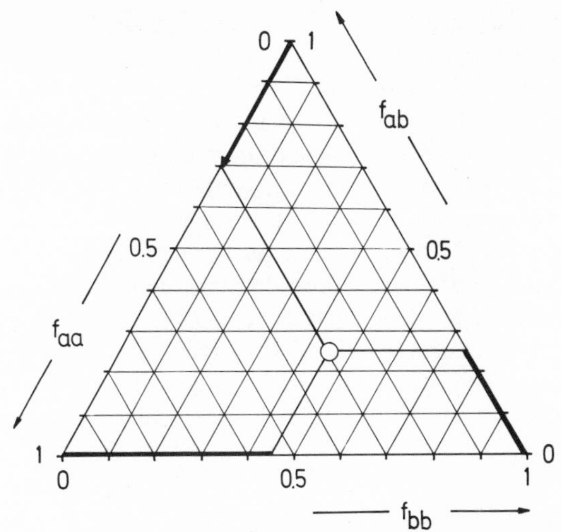

Figure 2-4. Description of the composition of a bipolymer by representation on a triangular graph.

by ozonolysis. The mean sequence length of the isobutylene sequences can be calculated from the molecular weight of the remaining isobutylene oligomer.

In reactions involving neighboring groups, the fact that in kinetically controlled reactions not all the groups can react is used (cf. Section 23.4.4). For example, the hydroxyl groups of poly(vinyl alcohol) cannot be completely acetylated with butyraldehyde, since isolated OH groups remain because of the random nature of the reaction. Of course, reactions must be carried out at a sufficiently high dilution so that possible intermolecular reactions do not occur. In addition, the solvent must be as "good" as possible, because then the polymer coil is greatly expanded and reaction between distant OH groups on the same polymer chain is reduced.

Two types of procedure can be similarly distinguished among the physical methods. One group deals with relatively short, the other with relatively long sequences. Nuclear magnetic resonance (NMR), UV, and IR spectroscopy belong to the first group, while X-ray analysis and differential thermal analysis belong to the second. In the IR spectrum, for example, the intensity of the $+CH_2+_n$ rocking frequency shifts from 815 ($n = 1$) to 752 ($n = 2$), to 733 ($n = 3$), to 726 ($n = 4$), and 722 cm^{-1} ($n \geq 5$), so that it is possible to assay short methylene sequences. In the far-IR, an isolated styrene unit ($n = 1$) in $+CH_2—CH(C_6H_5)+_n$ shows a broad band at 560 cm^{-1}, while a well-defined band is found at 540 cm^{-1} for $n \geq 6$. This band results from deformation of the aromatic ring, which stems from and is coupled with a deformation of the polymer chain; it can therefore

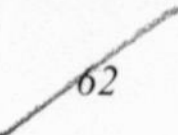

be used for the analysis of styrene–butadiene copolymers. In favorable circumstances, NMR enables the pentad sequences to be assayed, while UV studies are suitable up to triad sequences. X-Ray and differential thermal analysis can be employed for sequence analysis because longer stereoregular sequences can crystallize more readily than shorter ones, and longer atactic sequences show a distinctly different glass-transition temperature than shorter ones. In order to obtain anything conclusive from such a procedure, there must, in general, be a minimum of 15–20 units per sequence. Therefore this procedure does not distinguish between block copolymers and polymer blends. However, both X-ray analysis and differential thermal analysis are less direct than the other methods mentioned, since "false" groups can, in certain circumstances, be built into the chain without altering the ability to crystallize or shift the glass-transition temperature.

2.3.3. Substituents

Electrically neutral substituents in macromolecular chains exhibit no peculiarities with respect to low-molecular-weight compounds in terms of constitution, nomenclature, or, in general, modes of reaction. Polymers with substituents having ionizable bonds are called *polyelectrolytes*. Polyelectrolytes can dissociate to form a polyion and an oppositely charged gegenion. They can be *polyacids*, such as poly(acrylic acid)

$$\sim\!\sim CH_2\text{—}\underset{\displaystyle COOH}{\underset{|}{CH}}\text{—}CH_2\text{—}\underset{\displaystyle COOH}{\underset{|}{CH}}\text{—}CH_2\text{—}\underset{\displaystyle COOH}{\underset{|}{CH}}\sim\!\sim$$

The *polyion* is then a *polyanion*. A *polybase*, such as poly(vinyl amine),

$$\sim\!\sim CH_2\text{—}\underset{\displaystyle NH_2}{\underset{|}{CH}}\text{—}CH_2\text{—}\underset{\displaystyle NH_2}{\underset{|}{CH}}\text{—}CH_2\text{—}\underset{\displaystyle NH_2}{\underset{|}{CH}}\sim\!\sim$$

can accept protons and then becomes a *polycation*. The salts of polyacids are called *polysalts*. Macromolecular compounds that carry both positive and negative charges are called *polyampholytes*.

The gegenions can be mono-, bi-, or polyvalent, so that, theoretically, macrogegenions should also exist. However, if polyanions and polycations are mixed at a finite concentration, intermolecular salt bonds are formed before complete ionic dissociation can be achieved. A cross-linked network results, and the substance becomes insoluble. Conversely, it is to be expected that small concentrations of polyelectrolytes with monovalent gegenions are particularly soluble in water.

Polyions should be distinguished from *macroions*, which carry only one or two ionizable groups per chain. The chain carrier resulting in a cationic polymerization is, for example, a macromolecule with a positive charge at the growing end of the chain, and as such is a macrocation. Correspondingly, a *macroradical* is a macromolecule with a free electron. In radical graft reactions, it is possible in some circumstances to have *polyradicals* with several radical sites per molecule.

2.3.4. End Groups

The identification of the end groups of a macromolecule yields information about the synthetic mechanism, and also, in favorable circumstances, allows the determination of the molecular weight and/or the degree of branching. Since the proportion of end groups per unit mass of sample decreases with increasing molecular weight, the accuracy of end-group analysis (determination of the proportion of end groups present) decreases with increasing molecular weight in a homologous polymer series. Since, in addition, the proportion of end groups present depends on the number of molecules available, the chemical end group analysis yields a number-average molecular weight.

The number-average molecular weight can only be determined by end-group analysis if all the types of end groups present are known and can be quantitatively determined. If, for example, an unbranched polyamide has a total of $N_e = 2$ end groups per molecule, and the end groups are only either amino or carboxyl groups, then the molecular weight of the polymer $(\overline{M}_n)_{\text{end}}$ can be determined by a suitable titration. For amino group titration, V_{acid} cm^3 of acid is used; for the carboxyl acid groups, V_{base} cm^3 of base is required; the normalities of the titrants are, respectively, t_{acid} and t_{base} in equivalents/dm^3. Since the less end groups there are present, i.e., the higher the molecular weight $(\overline{M}_n)_{\text{end}}$, the less the titrant required, then, with the mass m (in g) of the polymer sample,

$$(\overline{M}_n)_{\text{end}} = \frac{N_e}{(t_{\text{acid}} V_{\text{acid}} + t_{\text{base}} V_{\text{base}})/10^3 m} \tag{2-19}$$

For any chosen number i of different end groups, with N_e end groups per molecule, the following is analogously obtained:

$$(\overline{M}_n)_{\text{end}} = N_e \sum_{i=1}^{i=i} \left(\frac{t_i V_i}{10^3 m} \right)^{-1} \tag{2-20}$$

According to equation (2-20), the arithmetic mean does not result from the individually calculated equivalent molecular weights. Expressions are analogously obtained from, e.g., a determination of the activity of radioactively labeled end groups, or from the determination on a mass basis of the color intensity of end groups bonded to dyes. Titrating allows molecular weights up to about 40,000 to be determined, microanalytical determination of iodine in end groups up to 100,000, radioactively labeled groups up to 200,000, and end groups bonded to intense dyes up to about one million.

Equation (2-20) shows that the determination of molecular weights by end-group analysis does not represent an absolute molecular-weight-determination method since certain assumptions (i.e., the chemical nature of the end groups and the number per molecule) need to be made about the constitution of the macromolecule. In contrast, with absolute methods (see Section 9.1) the molecular weight is only determined by the mass of the sample taken and the magnitude of the quantity measured, e.g., in osmosis the osmotic pressure. In many polymers, however, neither the number of end groups per macromolecule nor the chemical nature of the end groups is known with certainty. In melt polycondensation with dicarboxylic acids, for example, decarboxylation reactions can occur, causing a reduction in the number of end groups that can be titrated. If the proportion of decarboxylated end groups is not determined, then a number-average molecular weight which is too high is obtained from calculations that ignore these decarboxylated end groups. When iodine-containing initiators have been used in vinyl polymerization, transfer reactions involving the iodine residues have to be similarly taken into account. The azo groups of azo-containing initiators can participate in polymerization, etc.

On the other hand, in addition polymerizations various kinds of chain termination can occur, so that similar uncertainties arise in the assumptions required for end-group analysis calculations. In contrast, branching reactions increase the number of end groups per molecule. Since the end-group determination method depends on the equivalent proportion of end groups per molecule, it thus represents an equivalent, not an absolute, method. However, a known number-average molecular weight allows valuable inferences to be drawn about the constitution of the macromolecule.

2.4. Bonding in Individual Chains

2.4.1. Branching

The simplest kind of polymeric chain is the unbranched, or "one-dimensional" chain, e.g., the sulfur chain

—S—S—S—S—S—S—S—

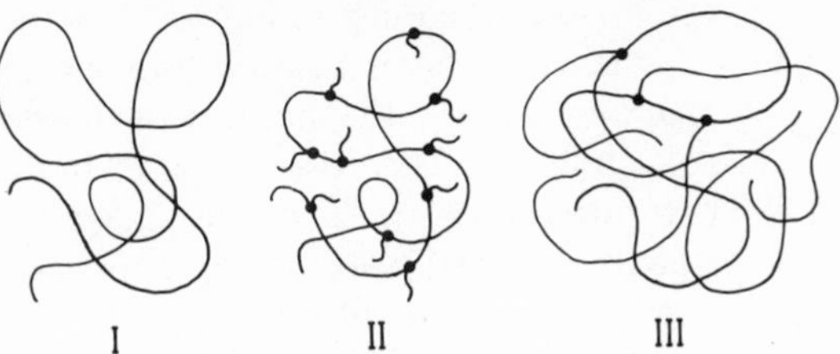

Figure 2-5. Schematic representation of "one-dimensional" chains. I: Unbranched; II: short-chain branching; III: long-chain branching. Branch points are represented by ●.

or the polymethylene chain

$$—CH_2—CH_2—(CH_2)_x—CH_2—CH_2—$$

These types of chains are also called linear chains, mainly on a historical basis, because it was originally thought that such a chain would occur fully extended in space. In fact, the random distribution of conformations dictates that, outside the crystalline state, an isolated chain of this type adopts a random coil shape (Figure 2-5).

Under certain reaction conditions, two or more growing polymer chains can unite irregularly in polyreactions that lead essentially to one-dimensional chains. The longest of the growing chains that so unite is called the main chain. The other chains joined to this are, according to length, called short- or long-chain branches. If the long-chain branches are themselves branched, then this is referred to as series branching. With very extensive series branching, the polymer has a kind of fir-tree-shaped structure.

In long- and short-chain branching, according to definition, each of the branches has the same chemical constitution as the main chain, i.e., as in polyethylene:

$$\sim CH_2—\underset{}{\overset{CH_3—(CH_2)_3—}{CH}}—(CH_2)_x—\underset{CH_2—CH_3}{CH}—(CH_2)_y\sim$$

short-chain branching ($x, y \gg 1$)

$$\sim CH_2—\overset{CH_3—(CH_2)_x—}{CH}—(CH_2)_y—\underset{(CH_2)_z—CH_3}{CH}\sim$$

long-chain branching ($x, y, z \gg 1$)

Side chains of different constitution which occur as substituents in the monomer are not classed as branching, as, for example, in poly(lauryl methacrylate):

$$\left[—CH_2—\underset{CO—O(CH_2)_{11}—CH_3}{\overset{CH_3}{C}}—\right]_x$$

Thus this polymer is not a branched chain, but an unbranched molecule. The residue —CO—O—$(CH_2)_{11}CH_3$ is called a "side group." With long-chain branching in macromolecules, the number of branch points (e.g., the $>$CH groups in polyethylene) is very low compared to the total number of chain links (—CH_2— and $>$CH—). Therefore, they cannot, in general, be determined by elemental analysis, spectroscopic methods, or chemical reaction. Determination of the number of branch points per molecule, the mean length of the branches, and the distribution in lengths, as well as the main-chain distance between branch points, thus presents a difficult analytical problem in macromolecular chemistry. The presence and extent of long-chain branching is mainly evaluated from the molecular dimensions of the branched molecule in solution. The branched-chain molecule must exhibit smaller dimensions than those of the unbranched chain of the same molecular weight, as can easily be visualized with regard to star-shaped branched macromolecules. With star-shaped macromolecules, several branches radiate out from one branch point. The designation of a main chain in such macromolecules is purely formal.

2.4.2. *Irregular Cross-Linked Structures*

Theoretically, branched molecules are soluble in some solvent, and are thus distinguishable from cross-linked networks. Conversely, however, not all insoluble polymers are cross-linked networks. Irregular cross-linked networks are the result of either certain uncontrolled, nonstereospecific reactions, or they are produced by the subsequent cross-linking of linear or branched molecules. For network formation to occur, it is essential that each macromolecule be cross-linked at two or more sites to two other polymer chains.

The cross-link can be of the same or of different atoms as occur in the main chain. An example of the first case is the product that results from the radical polymerization of butadiene to high conversions in the absence of transfer agent: A polymer with a large number of 1,4-*trans* links and a relatively high proportion of 1,2 double bonds is formed. Both the pendant double bonds and those in the main chain are less reactive than monomeric double bonds. The probability that the double bonds of the polymer will participate in the polymerization reaction is initially low and only becomes appreciable at higher conversions, when the concentration of the monomer is low and that of the polymer is large.

An example of different atoms forming the cross-link is to be found in the vulcanization of poly(butadiene) with sulfur, in which sulfur bridges are formed between polymer chains (Section 25.3.1.2). Another example is the cross-linking of unsaturated polyesters with styrene.

If the macromolecule is very small and/or there are only a few cross-link points per chain, then the macromolecules forming such a "cross-linked network" are still soluble. In the enzyme ribonuclease, for example, the only chain, the main chain, is joined to itself by four disulfide bridges (Figure 2-6), whereas in insulin, another protein, there are two chains, the A and B chains, which are connected by two disulfide bridges. The A and B chains differ in the nature and sequence of their α-amino acids.

True cross-link networks are only so designated when the molecule has so many cross-link points per primary chain that it is insoluble in all solvents. If the cross-linked molecule extends over the total volume of the reaction container, but the cross-linked product is still swollen by solvent, then this is known as a gel. Gels of very small size (i.e., between 300 and 1000 nm) are known as microgels. Such microgels behave as tightly packed spheres because of their high branch densities. Because of spherical shape and microdimensions, such microgels are in general "soluble," or more correctly, suspendable.

Gel formation occurs after a certain conversion, the gel point (see Section 17.3). Just past the gel point, some of the initial monomer charge is present as part of the cross-linked network which encompasses the whole of the reaction area, whereas the rest of the original monomer is in the form of branched, although still soluble, molecules. Such a product is considered to be partly cross-linked. Partly cross-linked structures are thus the complete reaction product. Cross-linked networks are considered to be single molecules.

Three-dimensional cross-linked networks are, according to definition, considered to be infinite in size. It is therefore pointless to consider their molecular weights. Such cross-link networks are classified according to the network chain lengths, branch type, and branch density.

The number of chain links between two branch points is the network

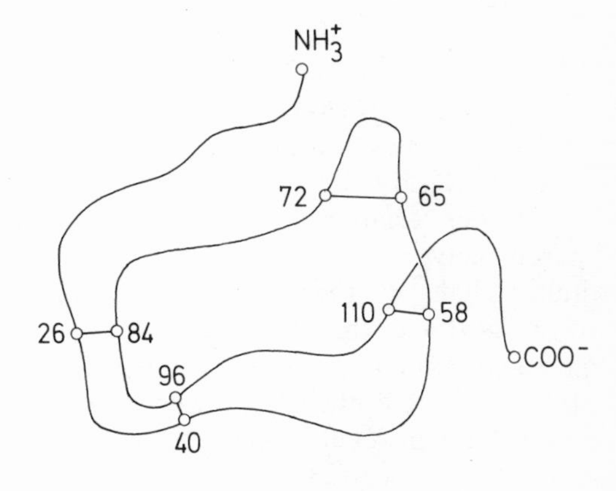

Figure 2-6. The enzyme ribonuclease, which consists of 124 amino acid units (—NH—CHR—CO—). The four intramolecular cross-linking points (26–84; 40–96; 58–110; 65–72) involve cystine units (cf. Chapter 30.1).

chain length. A branch point is defined as the point from which more than two chains, not necessarily of identical structure, radiate. The formula number-average molecular weight of such a network chain is defined as

$$(\overline{M}_c)_n = \overline{M}_u / x_c \tag{2-21}$$

where $\overline{M}_u$ is the average formula molecular weight of the monomeric unit and x_c is the degree of branching. x_c is also called the branch density and is given as the ratio of the moles of cross-linked monomeric units to the total moles of monomeric units present ($0 < x_c < 1$). Thus, it is the mole fraction of cross-linked monomeric units:

$$x_c = \frac{\text{moles of cross-linked monomeric units}}{\text{total moles of monomeric units present}} \tag{2-22}$$

The cross-link density or the cross-link index γ can be used to characterize network cross-link density. γ is the number of cross-linked monomeric units per primary chain, and is given by ($\overline{X}_n$ is the number-average degree of polymerization)

$$\gamma = \frac{(\overline{M}_n)_0}{(\overline{M}_c)_n} = \frac{(\overline{M}_n)_0}{\overline{M}_u} x_c = \overline{X}_n x_c \tag{2-23}$$

Here, $(\overline{M}_n)_0$ is the number-average molecular weight of the primary chain. A primary chain is the linear molecule before cross-linking.

All the quantities so far defined relate to ideal networks, i.e., continuous branched structures without free chain ends. In reality, the number of such free chain ends increases with decreasing primary chain molecular weight. The molar concentration $[M_c]_{\text{eff}}$ in mol/g of effective network chains can, according to P. J. Flory, be calculated from the molar concentration $[M_c]$ of all the chains present for $(M_n)_0 > (\overline{M}_c)_n$,

$$[M_c]_{\text{eff}} = [M_c]\left[1 - 2\frac{(\overline{M}_c)_n}{(\overline{M}_n)_0}\right] \tag{2-24}$$

$(\overline{M}_c)_n$ is the number-average network chain length. With very extensive cross-linking this formula cannot be used, because in such a case the number of free ends is too high.

Under certain reaction conditions so-called macroreticular or macroporous networks are produced. In such cross-linked networks, the cross-linked chains of the polymeric substance are not completely randomly distributed over the whole of the volume occupied by the substance. They give a structure that is more or less porous (Figure 2-7). With equal branch densities, macroreticular networks are much more permeable to solute and solvent molecules, which leads to their use as ion-exchange and gel-

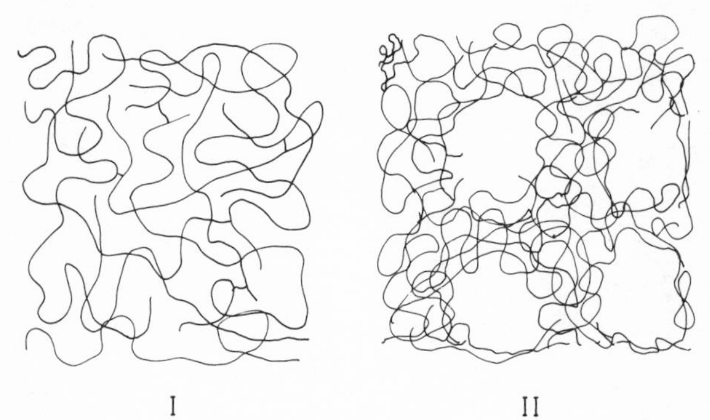

Figure 2-7. Schematic representation of a normal irregular (I) and a macroreticular (II) cross-linked network.

permeation chromatography supports (see Sections 23.4.3.2 and 9.8.2). Because of their more rigid structures, these materials swell less than the normal irregular cross-linked networks.

2.4.3. *Ordered Networks*

Regularly branched structures, in contrast to the irregular cross-linked networks, have only structurally equivalent units. They can be produced directly via stereospecific polymerization or polycondensation of rigid polyfunctional molecules in kinetically controlled reactions where the rate of cyclization is high. In reversible reactions it is usually necessary to work on the insoluble side of the macromolecule solubility equilibrium to obtain a high yield of regular branched structures. Regular branching is classified as 0, 1, 2, or 3 according to whether it extends a significant distance in 0, 1, 2, or 3 dimensions (Figure 2-8). There are generally several topological variations in each class.

The *0 types* form cage structures and are not macromolecular. Examples of this class are adamantane and bullvalene.

In *class 1*, it is possible to distinguish between bridge and spiro structures. Spiro structures occur frequently in inorganic macromolecules, e.g., in beryllium hydride (p. 4), dimethyl beryllium (p. 41), palladium chloride (p. 46), and silicon disulfide,

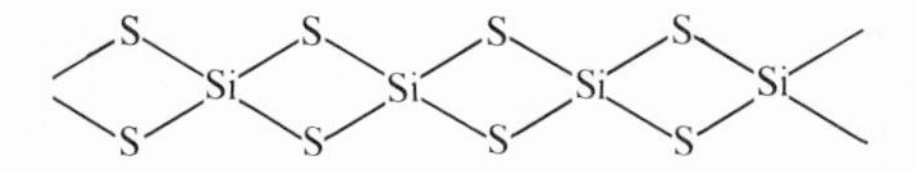

Theoretically, spiro polymers can also be considered as linear chains, the branch structure being part of the monomeric unit. Most inorganic spiro

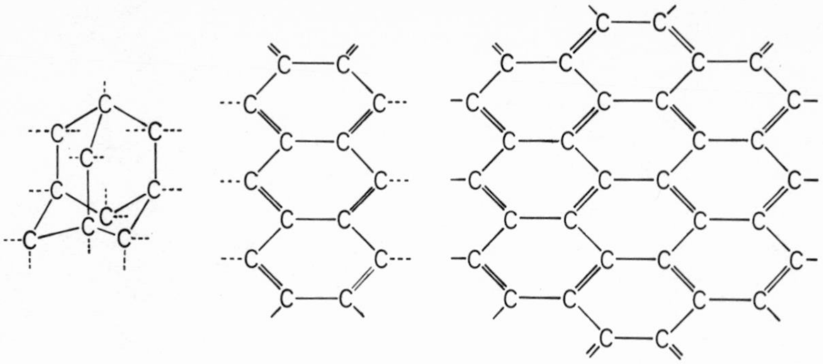

0 1 2

Figure 2-8. Ordered networks. 0 = adamantane as an example of a cage polymer (type 0); 1 = cyclized and dehydrogenated 1,2-poly(butadiene) as an example of a ladder or double-strand polymer (type 1); 2 = graphite as a layer or parquet polymer (type 2). (···) Carbon-to hydrogen bonds; (— and ═) carbon-to-carbon bonds.

macromolecules are highly crystalline and therefore brittle. When the spiro repeating unit is substituted with larger aliphatic groups, the ability to crystallize is lost. Consequently the substances are less brittle. For example, the "hybrid polymers" produced from zinc basic acetate (or a corresponding cobalt compound), phosphoric acid, and higher alcohols,

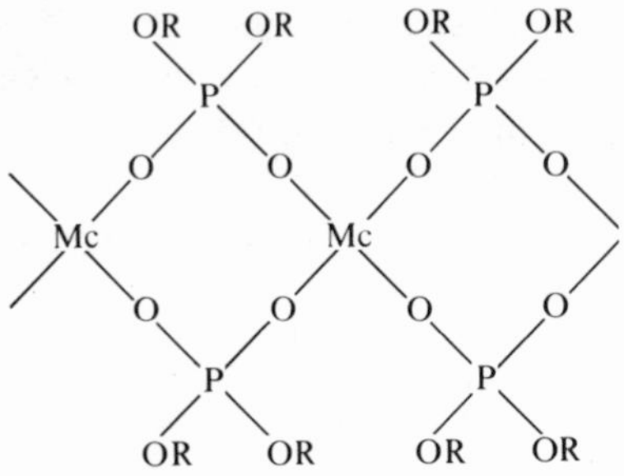

are still flexible at $-60°C$.

Bridged cross-links containing *type 1* polymers (double-strand polymers) occur extensively in inorganic polymer chemistry. In general, they are fibrous products such as valentinite (β-Sb_2O_3) or chrysotile asbestos,

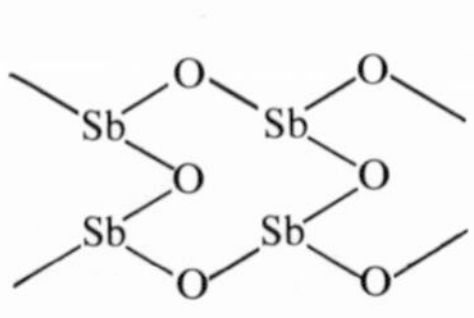

valentinite

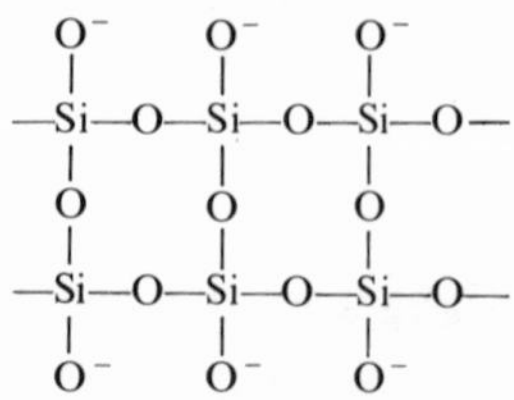

chrysotile asbestos

Because their structure is reminiscent of a ladder, type 1 polymers are also known as ladder polymers. Ladder polymer properties can also be improved by substitution with organic groups. One of the General Electric Company silicones, known as Phenyl-T, is a ladder polymer consisting of two syndiotactic siloxane chains with phenyl substituents. It has good electrical properties, double the tear strength of normal, single-chain polysiloxane, and only noticeably loses weight above 525°C.

Since about 1955, many ladder polymers have been synthesized from purely organic materials. For example, butadiene, under suitable polymerization conditions, can yield a syndiotactic 1,2-poly(butadiene) whose vinyl groups can form the bridge structure of a ladder polymer when subsequently reacted in a second polymerization step. On dehydrogenation of this, a black polymer with conjugated double bonds results (Figure 2-8). This polymer exhibits good thermal and electrical conductivity and can even be held in a flame (in the form of a cloth) without noticeable decomposition.

The good thermal stability of such polymers rests on the fact that simply breaking one bond does not lead to chain scission (degradation) and there is no molecular weight change. Statistically, it is highly improbable that two bonds in the same ring will be broken simultaneously. The electrical conductivity and the black color are a consequence of the double-bond conjugation.

Ladder polymers can be produced batchwise in a two-stage polymerization process if the two polymerizable groups are activated by different catalysts (mechanism). Vinyl isocyanate, for example, responds differently to radical and anionic catalysts and can be polymerized first by one and then by the other mechanism:

$$CH_2{=}CH(NCO) \xrightarrow{\text{radical}} -CH_2-CH(NCO)-CH_2-CH(NCO)- \xrightarrow{\text{anionic}} \text{ladder}$$

$$CH_2{=}CH(NCO) \xrightarrow{\text{anionic}} CH_2{=}CH\,CH_2{=}CH \;(\text{on }-N(-CO-)-\text{ chain}) \xrightarrow{\text{radical}} \text{ladder}$$

$$\text{ladder: } -CH_2-CH-CH_2-CH- \text{ backbone, each CH bonded to N of a } -N-CO-N-CO- \text{ chain} \qquad (2\text{-}25)$$

Paracyanogen is also included among ladder polymers of this type; it is a brown solid produced from cyanogen:

$$\text{fused rings: upper chain } -C{=}N-C{=}N-C{=}N- \text{, lower chain } -C{=}N-C{=}N-C{=}N- \text{, each upper C bonded to the lower C below it}$$

As a rule, ladder polymers are sparingly soluble or completely insoluble in all solvents. Paracyanogen, for example, is only soluble in concentrated sulfuric acid. The insolubility must, in most cases, result from some intermolecular cross-linking reaction occurring during the cyclization step. If two intermolecular cross-links are formed per 998 cyclization steps in an individual chain of degree of polymerization of 1000, then this is sufficient to lead to insolubility. The probability of intermolecular cross-linking can be decreased by working at higher dilutions (Ruggli–Ziegler dilution principle, cf. Section 16.1.2). In industrial processes, the necessary precautions to intermolecular cross-linking must be taken, since such cross-linked products can only be processed with difficulty, if at all. The cyclization reaction is therefore only allowed to proceed for as long as the polymer is still soluble; then the polymer product is cast as a film. Only then is the prepolymer intermolecularly cross-linked. Only a few cross-links are then necessary to produce a cross-linked network.

Theoretically, the various types of helices (cf. Chapter 4) can be considered to belong to type 1 polymers. Since their structure depends on a specific conformation, they will not be dealt with here.

Layer, parquet, or planar polymers of *type 2* are seen in graphite and its derivatives. Diamond is a network polymer of *type 3*. Network polymers exist exclusively, and parquet polymers almost exclusively, in the solid state. They are also known as monoaggregatable materials. Certain cell walls of bacteria consist of baglike macromolecules, which are a special case of parquet polymers.

In carbon compounds, the number of parquet and network polymers is limited by the tetravalence of carbon. In inorganic compounds, however, they exist in large numbers, for example, in quartz $(SiO_2)_x$, in black phosphorus $(P)_x$, etc. Theoretically, the same synthetic problems occurring with ladder polymers arise in the synthesis of parquet polymers. In order to obtain the desired arrangement in one dimension (type 1) or in two dimensions (type 2), it is necessary to avoid reaction paths leading to irregular structures. In the preparation of synthetic graphite, this is achieved by the meticulous exclusion of all centers of crystallization.

Literature

Section 2.1. The Concept of Structure

J. Haslam, H. A. Willis, and D. C. M. Squirrel, *Identification and Analysis of Plastics*, Iliffe, London, 1972.

Section 2.2. Atomic Structure and Chain Formation

M. F. Lappert and G. J. Leigh, *Developments in Inorganic Polymer Chemistry*, Elsevier, Amsterdam, 1962.

F. G. A. Stone and W. A. G. Graham, eds., *Inorganic Polymers,* Academic Press, New York, 1962.
F. G. R. Gimblett, *Inorganic Polymer Chemistry,* Butterworths, London, 1963.
J. Goubeau, Mehrfachbindungen in der anorganischen Chemie, *Angew. Chem.,* **69**, 77 (1957).
J. Goubeau, Force constants and bond orders of nitrogen compounds, *Angew. Chem., Int. Ed.* **5**, 567 (1966).
K. Andrianov, *Metallorganic Polymers,* Wiley–Interscience, New York, 1965.

Section 2.3. Monomeric Unit Bonding

S. Krimm, Infrared spectra of high polymers, *Fortschr. Hochpolym. Forschg.* **2**, 51 (1960/61).
J. C. Henniker, *Infrared Spectroscopy of Industrial Polymers,* Academic Press, New York, 1967.
G. Schnell, Ultrarotspektroskopische Untersuchungen an Copolymerisation, *Ber. Bunsenges.* **70**, 297 (1966).
U. Johnsen, Die Ermittlung der molekularen Struktur von sterischen und chemischen Copolymeren durch Kernspinresonanz, *Ber. Bunsenges,* **70**, 320. (1966).
A. Elliott, *Infrared Spectra and Structure of Organic Long-Chain Polymers,* E. Arnold, London, 1969.
I. C. Watt, Copolymers of naturally occurring macromolecules, *J. Macromol. Sci.* **C5**, 175 (1970).
J. L. Koenig, Raman scattering of synthetic polymers, *Rev. Appl. Spectrosc.* **4**, 233 (1971).
J. L. Koenig, Ramam spectroscopy of biological molecules: A review, *J. Polym. Sci., D (Macromol. Revs.)* **6**, 59 (1972).
Johannes Dechant, *Ultrarotspektroskopische Untersuchungen an Polymeren,* Akademie-Verlag, Berlin, 1972.
M. E. A. Cudby and H. A. Willis, Nuclear magnetic resonance spectra of polymers, *Ann. Rep. NMR (Nucl. Magn. Resonance) Spectrosc.* **4**, 363 (1971).
F. A. Bovey, The high resolution NMR spectroscopy of polymers, *Prog. Polym. Sci.* **3**, 1 (1971).
D. O. Hummel and F. Scholl, *Atlas der Kunststoff-Analyse,* C. Hanser, Munich 1968, 2 vols. (vol. 1 in two parts).
R. Zbinden, *Infrared Spectroscopy of High Polymers,* Academic Press, New York, 1964.
F. A. Bovey and G. V. D. Tiers, The high resolution nuclear magnetic resonance spectroscopy of polymers, *Fortsch. Hochpolym. Forsch.* **3**, 139 (1961–1964).
G. M. Estes, S. L. Cooper, and A. V. Tobolsky, Block copolymers and related heterophase elastomers, *J. Macromol. Sci. C (Rev. Macromol. Chem.)* **4**, 313 (1970).
S. R. Palit and B. M. Mandal, End group studies using dye techniques, *J. Macromol. Sci. C (Rev. Macromol. Chem.)* **2**, 225 (1968).
M. F. Hoover, Cationic quaternary polyelectrolytes—a literature review, *J. Macromol. Sci. A (Chem.)* **4**, 1327 (1970).
M. P. Stevens, *Characterization and Analysis of Polymers by Gas Chromatography,* M. Dekker, New York, 1969.
F. Oosawa, *Polyelectrolytes,* M. Dekker, New York, 1971.

Section 2.4. Bonding in Individual Chains

H.-G. Elias, Die Struktur vernetzter Polymerer, *Chimia* **22**, 101 (1968).
W. Funke, Ueber die Strukturaufklärung vernetzter Makromoleküle, insbesondere vernetzter Polyesterharze, mit chemischen Methoden, *Adv. Polym. Sci.* **4**, 157 (1965/67).
W. De Winter, Double strand polymers, *Rev. Macromol. Chem.* **1**, 329 (1966).
J. I. Jones, The synthesis of thermally stable polymers: A progress report, *J. Macromol. Sci. C (Rev. Macromol. Chem.)* **2**, 303 (1968).

W. Weidel and H. Pelzer, Bagshaped macromolecules—a new outlook on bacterial cell walls, *Adv. Enzymol.* **26**, 193 (1964).

C. G. Overberger and J. A. Moore, Ladder polymers, *Adv. Polym. Sci.* **7**, 113 (1970).

V. A. Grečanovskij, Branching in polymer chains, *Uspekhi Khim. (Russian Chem. Rev.)* **38**, 2194 (1969); *Rubber Chem. Technol.* **45**, 519 (1972).

G. Delzenne, Recent advances in photo-crosslinkable polymers, *Rev. Polym. Technol.* **1**, 185 (1972).

N. A. Plate and V. P. Shibaev, Comb-like polymers. Structure and properties, *J. Polym. Sci. D (Macromol. Rev.)* **8**, 117 (1974).

P. A. Small, Long-chain branching in polymers, *Adv. Polym. Sci.* **18**, 1 (1975).

Chapter 3
Configuration

3.1. Ideal Structures

3.1.1. Central Carbon Atom Asymmetry

Two possible nonsuperimposable arrangements exist in space for a carbon atom with four different substituents arranged tetrahedrally. One arrangement is the mirror image of the other. Carbon atoms arranged in this way are designated as asymmetric. The concept of asymmetry can, of course, be applied to all compounds with four substituents arranged tetrahedrally around a central atom (e.g., Si, P^+, N^+), as well as to central atoms with more than four substituents or ligands.

The two possible ways of arranging substituents about a central atom are distinguished from each other by the prefixes D and L or R and S. It is immaterial which arrangement is designated R and which is designated S as long as the consideration is confined to a definite molecule, CRR′R″R‴. When a comparison between the configurations of different molecules is being made, conventions are necessary. In the convention accepted in organic chemistry a seniority or priority is assigned to the substituents or ligands. Atoms of higher atomic number have priority over those of lower atomic number, i.e., atoms of higher atomic weight precede those of lower atomic weight. Thus, Cl comes before C. Then, the molecule is observed from a preferred position, that is, the position of the ligand with highest priority. If the priority of the other substituents decreases in a clockwise direction, then this chirality element is assigned the symbol R as a prefix. With an S configuration, the priority of the ligands decreases in a counterclockwise direction (chirality rule). Thus R and S are defined by convention.

A molecule with two central atoms can have four different configurative diads: RS—, RR—, SS—, and SR— (Figure 3-1). An example of such a case is 2,4-dichloropentane. Here, the central atoms each possess the ligands, in order of decreasing priority, Cl, $CH_2CHClCH_3$, CH_3, and H. Application of the chirality rules leads to assigning the configuration RS to the molecule I (Figure 3-1). The two central atoms in this case have, according to this rule, mirror-image configurations. Thus, molecule I is a *meso* compound. Finally, the configuration is assigned on the basis of each central atom being considered independent of the other central atoms: The observer always observes from the center of the asymmetric system being considered.

According to the convention outlined above, molecules I and IV (Figure 3-1) are *meso* compounds. They can be converted into each other by a 180° rotation of the molecules. Molecules II and III cannot be converted into each other; they are *racemic.*

The situation is similar with molecules having more than two central carbon atoms, i.e., in longer molecules such as the heptamers of propylene (Figure 3-2). Here, the molecules II and III are racemic mixtures in analogy to the case of structures II and III of 2,4-dichloropentane (Figure 3-1). Equally, structures I and IV in Figure 3-2 are *meso* compounds similar to structures I and IV of dichloropentane in Figure 3-1. The carbon atom marked in the middle of the oligopropylene chain is called a pseudoasymmetric carbon atom in organic chemistry. Pseudoasymmetric carbon atoms are so called because they are asymmetric when both end groups (in the sense of structural analysis and not process analysis) are different. In terms of organic chemistry configurational analysis, the configuration reverses about this pseudoasymmetric carbon atom.

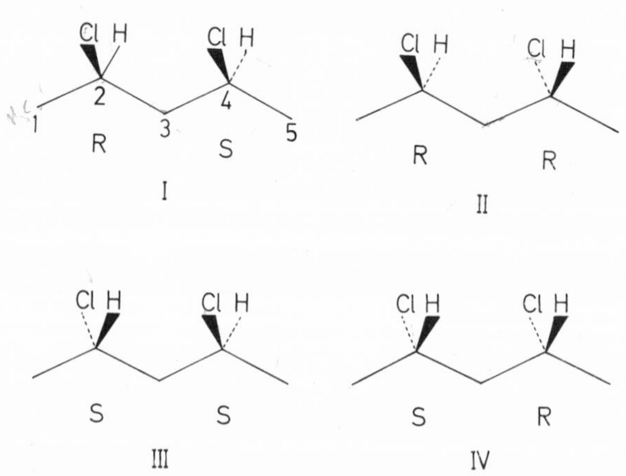

Figure 3-1. The different configurations of 2,4-dichloropentane.

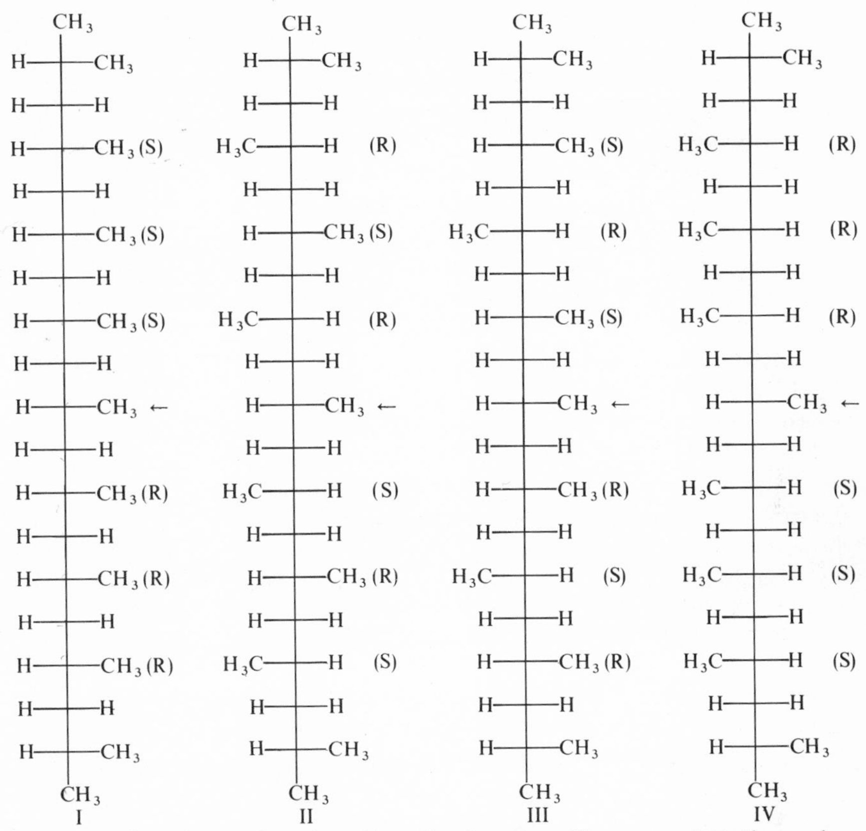

Figure 3-2. The various configurations of propylene heptamers. The arrows point to the pseudo-asymmetric carbon atoms. (I, IV) Isotactic ("meso"). (II, III) Syndiotactic ("racemic").

Such a "reversal of configuration" only occurs if the observer views the configuration from an observational position at each of the central carbon atoms in turn. If the observer views from a position outside the system, then a completely different picture is seen.

If the configuration of the central carbon atoms of the molecule I in Figure 3-1 is observed from the position of carbon atom C-1 in direction of carbon atoms C-2, C-3, etc., then the *nearest* substituents $H \rightarrow CH_2 \rightarrow Cl$ follow in a clockwise direction. With C-4 of molecule I in Figure 3-2, the sequence of substituents $H \rightarrow CH_2 \rightarrow CH_3$ is also in a clockwise direction. The sequences thus follow in the same direction. Such a diad is called an isotactic (it) diad. Consequently molecules IV (Figure 3-1) and I and IV (Figure 3-2) also contain only isotactic diads.

The central carbon atom C-2 of molecule II (Figure 3-1) has the clockwise substituent sequence $H \rightarrow CH_2 \rightarrow Cl$; the C-4 carbon atom of molecule II (Figure 3-2), however, has the clockwise sequence $CH_3 \rightarrow CH_2 \rightarrow H$.

In a more extended chain with the same configuration, the C-6 atom would have the substituent sequence $H \rightarrow CH_2 \rightarrow CH_3$ and the C-8 atom would have the sequence $CH_3 \rightarrow CH_2 \rightarrow H$, etc. Such sequences of configurative diads are called syndiotactic diads. Thus molecules II and III (Figure 3-2) are also syndiotactic.

If the observation is made from outside the system, i.e., observation of the relative configurations or diads, then, with isotactic molecules, all configurations about the central carbon atoms are equivalent. There is no configuration reversal about the pseudoasymmetric carbon atom as is observed when "absolute" configurations are being considered.

Ideal stereoregular polymers such as those shown in Figure 3-2 possess a translational symmetry in that the same configurations can be produced by shifting the central atoms along the chain. Molecules with a translational or rotational axis of symmetry but without mirror-image symmetry are called dissymmetric. Asymmetric molecules are molecules that do not have any axis of symmetry. Thus, it-polypropylene is dissymmetric but not asymmetric. Poly(L-alanine) or poly(D-alanine) $\text{+NH—CH(CH}_3\text{)—CO+}_n$, is, however, an asymmetric molecule.

The prefixes D and L, or R and S, are only assigned to molecules that have an asymmetric center.

3.1.2. Tacticity

There are two possible configurational observation points: the classical organic (observation form within the system) and the macromolecular (observation from outside the system). The macromolecular viewpoint is more suitable for long chains because it does not lead to an illusory configurational reversal in the middle of an isotactic chain. Detailed analysis shows that the configuration is only determined by the relative bonding positions between two main-chain central atoms. The smallest configurational unit of macromolecular chains is thus the configurational diad of two monomeric units. Such a diad must be either isotactic or syndiotactic. Isotactic diads are defined such that the relative configurations observed from one or the other main-chain central atoms are identical. With syndiotactic diads, the relative configuration seen from one of the pair of central atoms is the reverse of that seen from the other central atom.

These definitions can, with the aid of bonding priority rules, be easily extended from polymers with monomeric units of the type $\text{+CH}_2\text{—CHR+}_n$ to polymers with different base units, e.g., +CHR+_n or +X—CHR—Y+_n. A convention is again used to assign seniority or priority to the spatial arrangement of substituents about the central atom.

With a carbon atom as central atom, the three different substituents r, R, and ～ (chain), can be arranged, for example, in such a way that the size of the substituents relative to the bond — increases in a counterclockwise direction (Figure 3-3, left). The bond — leading to this central atom can be described as a (+) bond. The bond leading away from the central atom to the chain ～ is then, of necessity, a (−) bond. If, on the other hand, the substituents are arranged clockwise according to size, then the bond leading to the central atom will be a (−) bond, and that going away will be a (+) bond (Figure 3-3, right).

Two central atoms, or the monomeric units containing them, are configurationally identical if the corresponding bonds are characterized by the same (+) and (−) sign sequence. Polymers are defined as isotactic when all their central atoms have the same configuration. In the chain, therefore, (+) and (−) bonds alternate, i.e., (+) (−) (+) (−) (+) (−), etc. In syndiotactic polymers, on the other hand, every second central atom has the opposite configuration, and each central atom has one neighbor with the same, the other with the opposite, configuration, i.e., the bonds follow in the sequence (+) (−) (−) (+) (+) (−) (−) (+) (+), etc.

An isotactic carbon chain with the monomeric unit $+\!\!\!-CRr-\!\!\!+$ is considered as an example (Figure 3-4, top). Starting from bond 1, the three substituents at carbon atom I are arranged counterclockwise in relation to their size. Bond 1 is referred to as a (+), and bond 2 as a (−) bond in relation to the carbon atom. If one moves stepwise along the chain, then, according to definition, bond 2 with respect to carbon atom II of an isotactic polymer must be (+) and bond 3 (+). The three substituents around carbon atom III must likewise be arranged counterclockwise. This means that the substituent R in carbon atom II must lie below the plane of the paper. Thus, in an isotactic polymer with the base unit $+\!\!\!-CRr-\!\!\!+$, substituent R lies alternatively above and below the plane of the paper, whereas in a corresponding syndiotactic polymer all like substituents are found on one and the same side relative to the plane of the paper.

In an isotactic polymer of the type $+\!\!\!-CRr-\!\!\!+_n$, then, every individual bond is simultaneously (+) and (−) according to which of the central atoms at each end of the bond is being considered. Transferring to an isotactic polymer with the monomeric unit $+\!\!\!-CH_2-CRr-\!\!\!+$, every central

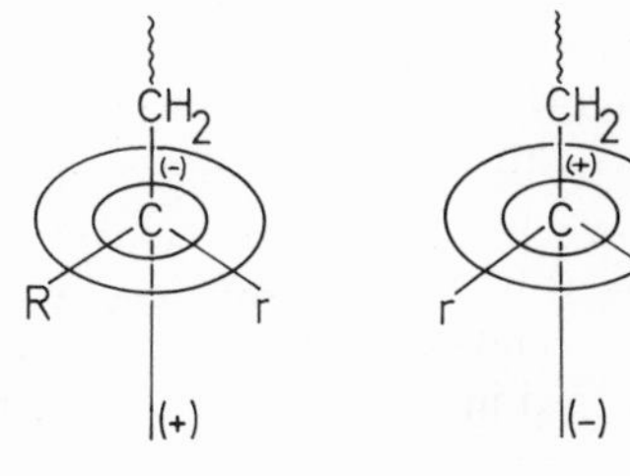

Figure 3-3. Definition of the (+) and (−) bonds around a central atom C, with the substituents r, R, and ～ (chain).

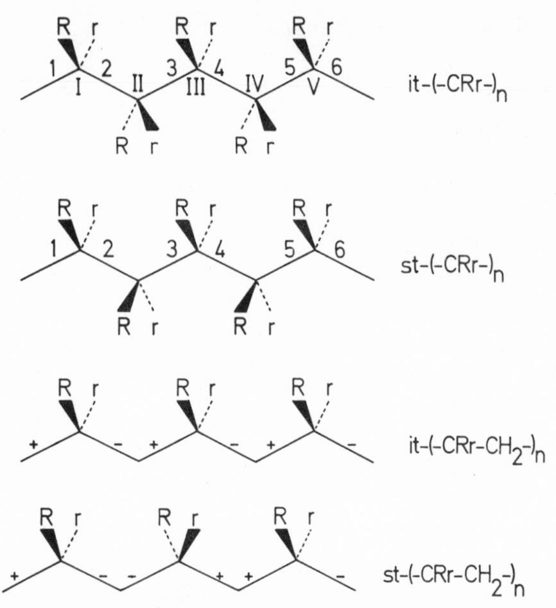

Figure 3-4. Representation of isotactic (it) and syndiotactic (st) polymers with the base units (—CRr—) and (—CH_2—CRr—), respectively.

atom has two CH_2 groups as neighbors. In a polymer of this type every bond is either (+) or (−). In projection, all the substituents project up from the plane of the paper (Figure 3-4, bottom). In a syndiotactic polymer with the monomer unit ⫞CH_2—CRr⫟, on the other hand, the substituents R lie alternatively above or below the paper surface.

3.1.3. Projections

The type of projection used in Figure 3-4 is only one of many types possible. In order to understand other types of projection, it is necessary to go briefly into the concept of conformation (see detailed discussion in Chapter 4).

In Section 3.1.1, the configuration was described as the arrangement of substituents around a central atom. This arrangement is not altered when the central atom is rotated around the bond to the neighboring central atom, although the relative position in space of those substituents that are not bonded to the central atom is changed. This position in space is referred to as conformation. Various types of projections result according to the chosen conformation. The most frequently used projections are compiled in Figure 3-5 for isotactic polymers with the base units ⫞CHR⫟,

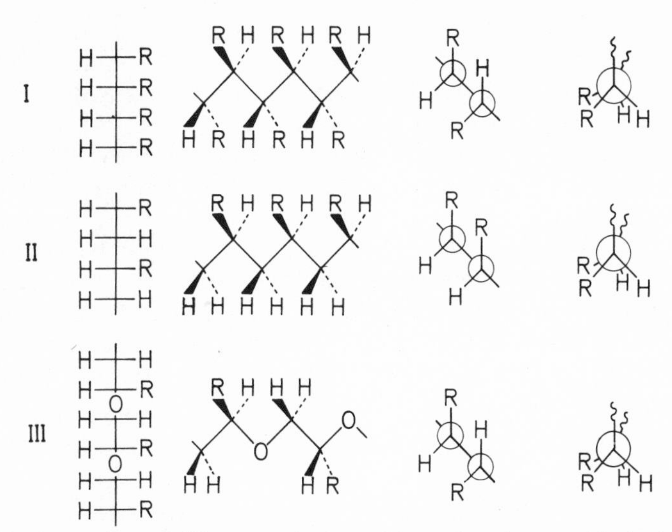

Figure 3-5. Comparison of different projections of isotactic polymers with the base units (—CHR—) (I), (—CH_2—CHR—) (II), and (—CH_2—CHR—O—) (III). From left to right: Fischer projection (*cis*-eclipsed chain conformation), Natta projection (*trans*-staggered), Newman projection (*trans*-staggered), and Newman projection (*cis*-eclipsed).

$+CH_2$—CHR$+$, and $+CH_2$—CHR—O$+$. Without going any further into individual types of conformations at this stage, it need only be mentioned that the Fischer projection corresponds to an eclipsed, and the Natta projection to a staggered conformation. Newman projections can represent both eclipsed and staggered conformations. As a rule, *trans*-staggered and *cis*-eclipsed conformations are used as the basis of such projections. (For the significance of these expressions see Section 4.5.)

Figure 3-5 shows that isotactic polymers in certain projections definitely do not have "their substituents always on the same side" as is frequently stated. Such an assumption is only correct for the two types of projections in the eclipsed conformation. In projections of *trans*-staggered conformations differences exist according to the monomeric unit.

3.1.4. *Monotactic Polymers*

Polymers with one stereoisomeric center per base unit are called monotactic. Examples of monotactic polymers are poly(ethylidene), $\text{+}CH(CH_3)\text{+}_n$, with one central asymmetric atom per chain link, poly-(propylene), $\text{+}CH_2$—$CH(CH_3)\text{+}_n$, with one asymmetric central atom per two chain links,. and poly(propylene oxide), $\text{+}CH_2$—$CH(CH_3)$—$O\text{+}_n$, with one asymmetric central atom per three chain links.

If all the bonds follow in the same sequence, i.e., (+) (−) (+) (−) (+) (−) or (+) (−) (−) (+) (+) (−) (−) (+), then one speaks of holotactic polymers. Holotactic, monotactic polymers thus show only isotactic or syndiotactic diads, whereas real polymers always possess a number of configurational defects, i.e., they contain iso- as well as syndiotactic diads (see also Section 3.2). The bonds of real polymers, therefore, do not always follow in the same sequence, i.e., they can be (+) (−) (+) (−) (−) (+) (+) (−), etc. The configurational defects make it necessary to bear in mind the randomness of the configuration when dealing with real structures.

Polymers with double bonds in the chains can occur in *cis*-tactic (ct) or *trans*-tactic (tt) configurations according to how the parts of the chain are arranged about the double bond. An example of this is 1,4-poly(butadiene), with the configurations

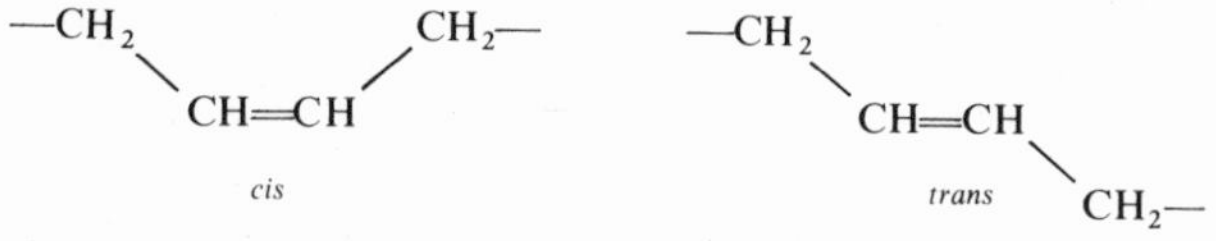

$$\begin{array}{ccc} -CH_2 & & CH_2- \\ & \diagdown\ \ \diagup & \\ & CH{=}CH & \end{array} \qquad \begin{array}{ccc} -CH_2 & & \\ & \diagdown & \\ & CH{=}CH & \\ & & \diagdown \\ & & CH_2- \end{array}$$

cis *trans*

3.1.5. Ditactic Polymers

Ditactic polymers possess two steroisomeric centers per constitutive base unit, and tritactic polymers possess three. Ditactic polymers may be formed by the polymerization of 1,2-disubstituted ethylene derivatives, as, for example, for pentene-2:

$$n\,\underset{CH_3}{\underset{|}{CH}}{=}\underset{C_2H_5}{\underset{|}{CH}} \longrightarrow \lfloor\!\!-\underset{CH_3}{\underset{|}{CH}}-\underset{C_2H_5}{\underset{|}{CH}}-\!\!\rfloor_n \qquad (3\text{-}1)$$

The resulting poly[(1-ethyl)-(2-methyl)ethylene] can theoretically occur in four different configurations, since two arrangements exist for each of the two asymmetric centers. However, the number of arrangements for the asymmetric centers is restricted by the fact that the centers in the base unit retain the configuration about the bond that joined them as monomer. Both centers are thus only isotactic or only syndiotactic, so that the polymer is either diisotactic or disyndiotactic. In analogy to the usual nomenclature of low-molecular-weight compounds, polymers with the same sequence of base units in the Fischer projection are referred to as erythropolymers, and those with alternating sequence as threopolymers (Figure 3-6).

In the erythro-diisotactic configuration (eit), the substituents R and R′ all lie on the same side in a Fischer projection, whereas in a Natta projection all the R substituents are found on one side, but the R′ substituents on either side, of the plane of the paper. In the Newman projection of the eclipsed

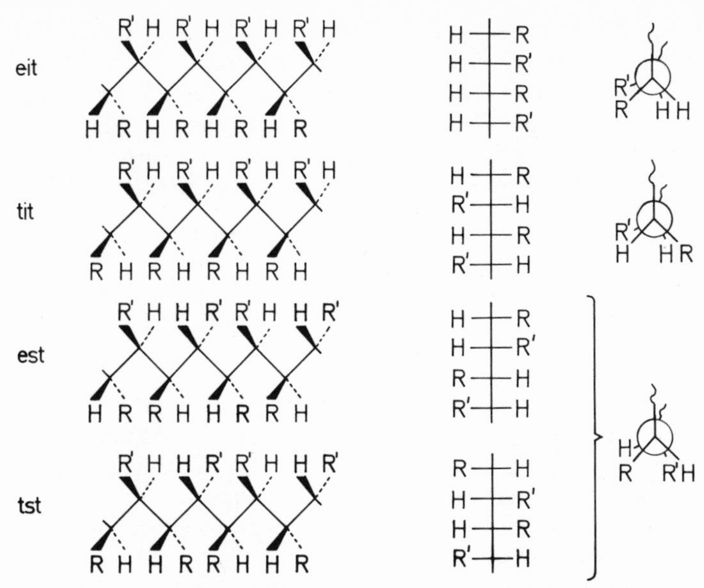

Figure 3-6. The four configurations of ditactic poly(2-pentene). eit = erythrodiisotactic, tit = threodiisotactic, est = erythrodisyndiotactic, tst = threodisyndiotactic; R = CH_3, R′ = C_2H_5 or vice versa.

conformation, R lies above R′ and H above H in the eit configuration. The characteristic features of the three other configurations can be seen from Figure 3-6.

By polymerizing the double bonds of unsaturated rings, it is possible

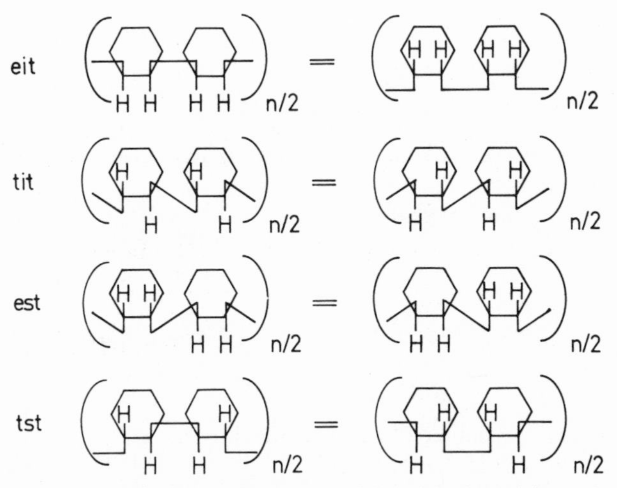

Figure 3-7. The four configurations of ditactic poly(cyclohexene).

to synthesize polymers with rings as stereoisomeric centers. The other ring atoms bonded directly onto the chain atoms of the ring should be treated as substituents. Poly(cyclohexene), therefore, forms four different configurations, just as poly(2-pentene) does (Figure 3-7). The only special case occurs for the bonds that represent the entry or exit points of the chain into the ring. They are *cis* in erythro and *trans* in threo configurations.

3.2. Real Structures

3.2.1. J-ads

The ideal, holotactic structures described in Section 3.1 possess a chain of infinite length and an absence of configurational defects. In real structures, however, the existence of end groups and an imperfect steric arrangement have to be taken into account, which can lead to a completely irregular arrangement of the configurational diads. The average arrangement and the sequence of the configurational diads must therefore be described by means of suitable statistical parameters in a similar way to the properties of a polymer.

Configurational *diads* must be either isotactic or syndiotactic. Thus, the sum of their mole fractions must be unity:

$$x_i + x_s \equiv 1 \tag{3-2}$$

Each configurational *triad* consists of a pair of diads. The pair of diads can be isotactic–isotactic or syndiotactic–syndiotactic or the triad can consist of a syndiotactic and an isotactic diad. The sum of the mole fractions of these three kinds of triads must also be equal to unity,

$$x_{ii} + x_{ss} + x_{is} \equiv 1 \tag{3-3}$$

where x_{is} is the mole fraction of what are called heterotactic triads. No directional distinction is made with heterotactic triads. Both is and si triads are included in x_{is}.

In an analogous manner, six different *tetrads* can be distinguished. The sum of their mole fractions must also equal unity:

$$x_{iii} + x_{iis} + x_{isi} + x_{iss} + x_{sis} + x_{sss} \equiv 1 \tag{3-4}$$

The number N_j of possible J-ad types (diads, triads, tetrads, etc.) is given by

$$N_j = 2^{j-2} + 2^{k-1} \tag{3-5}$$

When j is an even number (diads, tetrads, hexads, etc.), then $k = j/2$. If j is odd (triads, pentads, heptads, etc.), $k = (j - 1)/2$. Thus there are two kinds of diads, three kinds of triads, six kinds of tetrads, ten kinds of pentads, etc.

Relationships independent of polymerization mechanism must exist between the various *J*-ads since each of the *J*-ads from triads onwards consists of two or more diads. The relationships between the mole fractions of diads and triads are

$$x_i \equiv x_{ii} + 0.5x_{is}, \qquad x_s \equiv x_{ss} + 0.5x_{is} \tag{3-6}$$

Analogously for the tetrads,

$$\begin{aligned} x_{ii} &\equiv x_{ii} + 0.5x_{iis} \\ x_{is} &\equiv x_{isi} + x_{sis} + 0.5x_{iis} + 0.5x_{iss} \\ x_{ss} &\equiv x_{sss} + 0.5x_{iss} \end{aligned} \tag{3-7}$$

The macromolecular substance is better characterized when the experimentally quantified *J*-ad is largest. A block copolymer consisting of a block of isotactic diads and a block of syndiotactic diads

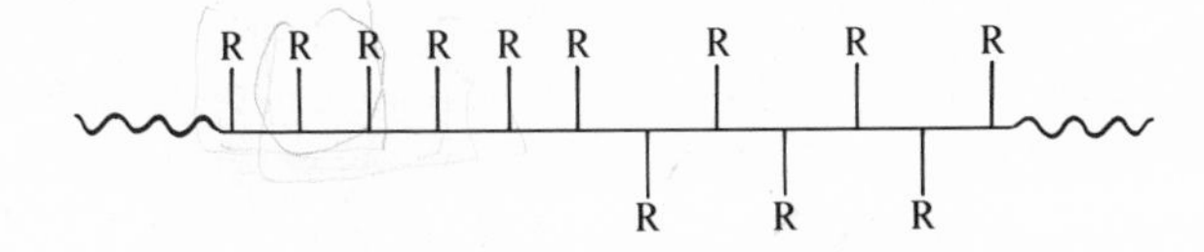

possesses, for example, the same number of isotactic and syndiotactic diad links as a polymer with alternating isotactic and syndiotactic diads

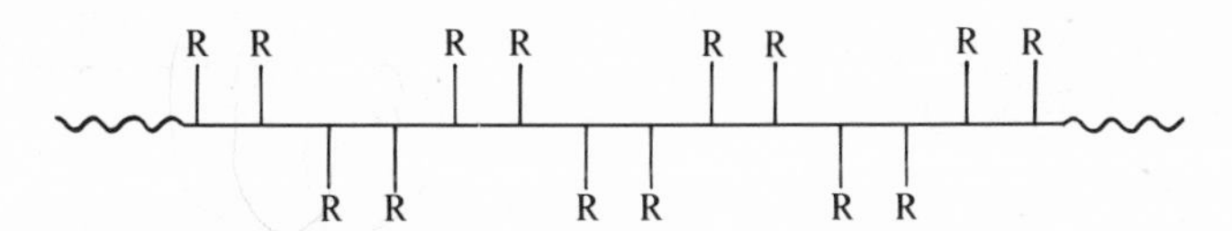

However, with the first polymer, the mole fraction of heterotactic triads is almost zero (only one heterotactic triad is to be found—in the middle of the polymer molecule). The second polymer is, on the other hand, 100% heterotactic ($x_{is} = 1$). Here, it must be remembered that any given diad is part of two triads, three tetrads, four pentads, etc.

Polymers are described as "atactic" in the literature when they do not completely or predominantly consist of only one type of *J*-ad. Strictly speaking, one must here distinguish between conventionally atactic and ideally atactic polymers. An ideally atactic polymer is described as one where, in the formation of each diad, a syndiotactic linkage and an isotactic linkage are equally probable. Thus, an isotactic diad can add on either an isotactic diad or a syndiotactic diad. Thus, $x_i = x_s = 0.5$; $x_{ii} = x_{ss} = 0.25$; $x_{is} = 0.5$ etc. Heterotactic triads can be formed in two ways (i → is and s → si), but isotactic and syndiotactic triads can only be formed in one way. Thus, an ideally atactic polymer contains twice as many heterotactic triads as isotactic or syndiotactic triads. Also, with ideally atactic polymers, the various diads, triads, etc., follow each other in random order. In contrast, conventionally atactic polymers can only be described as "not tactic" on the basis of results from any one measuring technique.

On transferring a constitutional uniform monomeric unit to copolymers with two or more constitutionally different monomeric units, the configuration statistics become very complicated. In a copolymer from two base units A and B, for example, monoisotactic, monosyndiotactic, and monoheterotactic diads are possible for each monomer, whereas there are four coisotactic, four cosyndiotactic, and eight coheterotactic diads. Admittedly, not all diads can be experimentally distinguished.

3.2.2. *Sequence Length*

A configurational or tactic sequence consists of at least one tactic linkage and, hence, at least two monomeric units. The number of successive linkages of a like nature is defined as the sequence length. The transition from an iso- and a syndiotactic sequence involves a heterotactic triad. The number-average sequence length $\bar{L}_n$ of all iso- and syndiotactic sequences of a polymer is thus given by the inverse of the mole fraction of all heterotactic triads:

$$\bar{L}_n = 1/x_{\text{is}} \tag{3-8}$$

The number-average sequence length $(\bar{L}_I)_n$ of isotactic sequences results from the definition of the number average of a property:

$$(\bar{L}_I)_n = \frac{\sum_j (N_I)_j (L_I)_j}{\sum_j (N_I)_j} \tag{3-9}$$

Here, $(N_I)_j$ is the number of sequences of a sequence length $(L_I)_j$. The index j can take values of 1, 2, 3, etc. The summation $\sum_j (N_I)_j (L_I)_j = N_{\text{id}}$ gives the number of isotactic diads present. The quantity x_{i}, on the other hand, is equal to the fraction of the total diads that are isotactic, i.e., $x_{\text{i}} = N_{\text{id}}/(N_{\text{id}} + N_{\text{sd}})$. The sum of the sequences present is equal to half the number of heterotactic triads, since every sequence of iso- or syndiotactic triads is sandwiched between two heterotactic triads. It thus follows that $\sum (N_I)_j = 0.5 N_{\text{ht}}$. The quantity x_{is} gives the fraction of the total triads that are heterotactic, i.e., $x_{\text{is}} = N_{\text{ht}}/(N_{\text{it}} + N_{\text{st}} + N_{\text{ht}})$. The sum of all the triads must be equal to the sum of the diads: $(N_{\text{it}} + N_{\text{st}} + N_{\text{ht}}) = (N_{\text{id}} + N_{\text{sd}})$. If one inserts all these relationships into equation (3-9), one obtains

$$(\bar{L}_I) = N_{id}/0.5N_{ht} = 2x_i/x_{is} \qquad (3\text{-}10)$$

Analogously, the number-average sequence length of the syndiotactic sequences is $(\bar{L}_S)_n = 2x_s/x_{is}$.

3.3. Experimental Methods

Methods for determining the configuration of polymers can be classified as relative or as absolute. Of these, the relative methods refer either directly to the configuration or indirectly to one of the properties that are dependent on the configuration.

3.3.1. X-Ray Crystallography

X-ray crystallography (Section 5.2.1) is an absolute metods. With it, it is possible to determine the distances between the atoms in crystalline regions, and then the configuration from the position and intensity of the diffractions. The method does not depend on knowledge of model compounds. However, it is only applicable to substances that crystallize well and have a high steric purity. X-ray crystallography is used in configurational studies to calibrate relative methods.

3.3.2. Nuclear Magnetic Resonance Spectroscopy

Nuclear magnetic resonance (NMR) spectroscopy of polymers in solution is a very important method of studying polymer configuration since noncrystalline as well as crystalline compounds can be studied. The method depends on the fact that the chemical shift of the signals of bonded hydrogen atoms (protons), ^{13}C and ^{19}F atoms, etc., in fixed chemical environments depends on the configuration of the main chain. In theory, the technique represents an absolute method, but, on technical grounds, it can often only be used as a relative method. An example of this is the analysis of the spectra of poly(methyl methacrylates) of various tacticities.

In poly(methyl methacrylate), $\text{-}[CH_2\text{—}C(CH_3)(COOCH_3)]\text{-}$, signals can be expected from the methylene protons CH_2, from the α-methyl protons CH_3, and from the methyl ester protons $COOCH_3$. The assignment of the signals from the three types of protons is made possible by comparison with the spectrum of methyl pivalate $(CH_3)_3C\text{—}COOCH_3$. In both poly(methyl methacrylate) and methyl pivalate, the α-methyl protons and the methyl ester protons appear at the same position in the NMR spectrum. Information about the tacticity of the polymer can be obtained as described below.

In st-poly(methyl methacrylate), the two methylene protons occur in a chemically equivalent environment, since every proton is flanked by an α-methyl group and a methyl ester group. In it-poly(methyl methacrylate), on the other hand, the two methylene protons are not chemically equivalent, since one proton is surrounded by two α-methyl groups, and the other by two methyl ester groups (Figure 3-5). It is immaterial whether the conformations represented in Figure 3-5 are really the only acceptable ones which occur or not, since only the average conformation is detected. The two equivalent methylene protons of st-PMMA thus lead to a single proton resonance signal, whereas the chemically nonequivalent methylene protons lead to an AB quartet (Figure 3-8).

Since the two hydrogen atoms of a methylene group in an isotactic diad are not NMR-chemically equivalent, they have been called "meso"

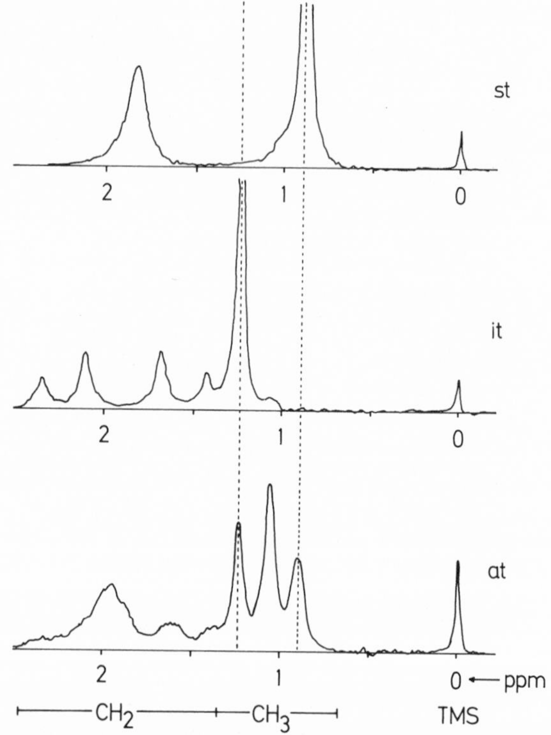

Figure 3-8. Section from the proton resonance spectra of isotactic (it), syndiotactic (st), and atactic (at) poly(methyl methacrylates). The signals of the methyl ester protons are not shown. TMS = reference signal of tetramethyl silane. (According to P. Goeldi and H.-G. Elias.)

(also heterosteric or diastereotopic) atoms. Consequently, the methylene group of a syndiotactic diad has analogously been called "racemic" (also homosteric or enantiotropic). For these reasons the mole fractions of isotactic and syndiotactic diads are often given in the literature as (*m*) or (*r*) instead of as x_i or x_s. The names racemic and meso are not equivalent to those used in classical organic chemistry and are therefore misleading. These terms, racemic and meso, are also superfluous, since the terms isotactic and syndiotactic are unambiguously defined in terms of configuration, and one should not base a structural definition on a phenomenon of a particular method of measurement.

The resonance signals of the α-methyl protons appear at various points in the spectrum according to tacticity. It is not possible to draw any conclusions about the configuration from the position of these signals alone, since it is only with difficulty that any inferences can be drawn about magnetic screening by neighboring groups. However, assignments can easily be made if the signals coming from the methylene protons are known.

Using the assignments thus obtained, it is possible to analyze the spectrum of nonholotactic PMMA. From the spectrum of a so-called atactic PMMA, it can readily be seen that evidence on the proportion of iso- and syndiotactic diads is only accessible with difficulty from the methylene proton signals. The signals of the singlets and the quartets are not very well defined. With the α-methyl proton signals, the situation is better. Here, three different signals are observed, one of which is in the position corresponding to a signal from the iso- and another from the syndiotactic polymer. The third signal lies between these two. One therefore concludes that the outer α-methyl proton signals correspond to the syndiotactic and isotactic triads and that the central signal arises from the heterotactic triads. The area beneath the signals is proportional to the proportion of the corresponding triads.

The methyl ester protons give rise to a single signal, which is independent of tacticity (not shown in Figure 3-8). Therefore, the methyl ester proton signal cannot be used for tacticity determinations. The chemical shift is not affected by the configuration of the main chain since they are too far removed from an asymmetric center.

Generally speaking, the signals obtained from polymer solutions are broader than those of low-molecular-weight model compounds. In low-molecular-weight compounds the higher the concentration and the lower the temperature, the broader are the signals. The broadening of the signals results from the strong magnetic interactions between different nuclei. If the concentration is lowered, the nuclei become less oriented and the signals become narrower. The same effect can be achieved by raising the temperature.

In polymers, however, the individual monomeric units are bonded together into a chain. Since the broadness of the signal depends on nearest-neighbor influence, dilution of the solution does not lead to well-resolved signals. The resolution of the signals in random coils is also largely independent of molecular weight. Sharper signals can therefore only be achieved by measurements at elevated temperatures. The splitting of the signals from nonholotactic polymers also depends to a certain extent on the nature of the solvent used. At the present time it remains to be shown whether this influence of the solvent results from a conformational shift or from a specific interaction between the solvent and the base units (solvation).

For polymers of the $\text{+}CH_2\text{—}CHR\text{+}$ type, spin–spin coupling of neighboring CH_2 and CH groups can lead to complex proton resonance spectra which are difficult to interpret. This difficulty can be overcome by the double-resonance technique and/or higher magnetic field strengths.

In general, 60-MHz proton resonance spectra can only yield data on diad and triad contents. Higher magnetic field strengths lead to greater chemical shifts, and so data on tetrads and pentads can then be obtained. such data are accessible using superconducting solenoids cooled with liquid helium. These magnets allow proton magnetic resonance spectra to be obtained at resonance frequencies up to 300 MHz. Higher J-ad contents can often be determined from ^{13}C spectra, since the chemical shift of ^{13}C is much greater than that of protons (up to 250 ppm, in contrast to up to 10 ppm).

3.3.3. *Infrared Spectroscopy*

Infrared spectroscopy is also frequently used in the quantitative determination of the proportion of diads. As a rule, the assignment of the different diad signals is made possible from results with polymers or oligomers of known configuration. In certain cases, the calculation of the absorption frequency for the individual types has already been performed. The CH and CH_2 deformational vibrations refer directly to various configurations. Since products of different stereoregularity crystallize to different extents, and since IR spectra are sensitive to crystallinity in the range 670–1000 cm^{-1}, then the diad content can also be determined by means of what are called crystallinity bands. However, the method is often unsuitable, since the crystallinity of the polymer depends on its previous thermal history (Chapter 5).

3.3.4. *Other Methods*

There is a series of other methods which likewise use the different crystallinities of polymers of various degrees of stereoregularity. However, none of these methods is completely unequivocal, for two reasons. On the one hand it is known that extensively "atactic" polymers, such as the poly(vinyl alcohol) obtained by saponification of radically polymerized vinyl acetate, can also crystallize relatively well. On the other hand, large substituents can impede or prevent the crystallization of stereoregular polymers. For example, isotactic poly(styrene), which can be crystallized, can, in a series of polymer analog reactions, be converted through noncrystallizable poly(*p*-iodostyrene) and poly(*p*-lithiumstyrene) back into crystallizable, isotactic poly(styrene) without any change in the configuration:

$$\underset{\displaystyle C_6H_5}{\underset{|}{\text{-(}CH_2\text{—}CH\text{)-}}} \xrightarrow{+I_2/HIO_3} \underset{\displaystyle C_6H_4I}{\underset{|}{\text{-(}CH_2\text{—}CH\text{)-}}} \xrightarrow{+Li} \underset{\displaystyle C_6H_4Li}{\underset{|}{\text{-(}CH_2\text{—}CH\text{)-}}} \xrightarrow{+H_2O} \underset{\displaystyle C_6H_5}{\underset{|}{\text{-(}CH_2\text{—}CH\text{)-}}} \qquad (3\text{–}11)$$

Stereoregular polymers can be separated from atactic ones in certain circumstances since crystalline polymers dissolve less readily than do amorphous ones. However, in addition to crystallinity, the solubility is also dependent on the degree of stereoregularity and the molecular weight. Thus, readily soluble fractions of high-molecular-weight, atactic polymers can also contain low-molecular-weight fractions of stereoregular material and vice versa. When samples have received identical thermal pretreatment, information on differing crystallinity and stereoregularity can be obtained from the values of the melt and glass transition temperatures (see Chapter 10).

Other methods used for stereoregularity determinations involve the use of dipole moments, streaming birefringence, rate of saponification, and cloud-point titration. However, all these methods are only applicable to special polymers and/or are only indirect methods, and so they have not found general application.

Literature

Section 3.1 –3.2. Ideal and Real Structures

M. Farina, M. Peraldo, and G. Natta, Cyclische Verbindungen als konfigurative Modelle sterisch regelmässiger Polymerer, *Angew. Chem.* **77**, 149 (1965).

M. L. Huggins, G. Natta, V. Desreux, and H. Mark, Report on nomenclature dealing with steric regularity in high polymers, *J. Polym. Sci.* **56**, 153 (1962).

G. Natta, Stereospezifische Katalysen und isotaktische Polymere, *Angew. Chem.* **68**, 393 (1956).

G. Natta, Von der stereospezifischen Polymerisation zur asymmetrischen autokatalytischen Synthese von Markomolekülen, *Angew. Chem.* **76**, 553 (1964).

G. Natta and F. Danusso, *Stereoregular Polymers and Stereospecific Polymerisations*, Pergamon Press, Oxford, 1967 (2 vols., original papers of the Natta school).

L. Dulog, Taktizität und Reaktivität, di- und tritaktische Polymere, *Fortschr. Chem. Forschg.* **6**, 427 (1966).

A. D. Ketley, ed., *The Stereochemistry of Macromolecules,* M. Dekker, New York, 1967/68, 3 vols.

R. S. Cahn, C. Ingold, and V. Prelog, Specification of molecular chirality, *Angew. Chem., Int. Ed. (Engl.)* **5**, 385 (1966).

F. A. Bovey, *Polymer Conformation and Configuration,* Academic Press, New York, 1969.

Section 3.3. Experimental Methods

F. A. Bovey and G. V. D. Tiers, The high resolution nuclear magnetic resonance spectroscopy of polymers, *Fortschr. Hochpolym. Forschg.* **3**, 139 (1961/64).

U. Johnsen, Die Ermittlung der molekularen Struktur von sterischen und chemischen Copolymeren durch Kernspinresonanz, *Ber. Bunsenges.* **70**, 320 (1966).

F. A. Bovey, The sterochemical configuration of vinyl polymers and its observation by nuclear magnetic resonance, *Acc. Chem. Res.* **1**, 175 (1968).

Hung Yu Chen, Application of high resolution NMR spectroscopy to elastomers in solution, *Rubber Chem. Technol.* **41**, 47 (1968) (also contains data on thermoplasts, etc.).

P. R. Sewell, The nuclear magnetic resonance spectra of polymers, *Ann. Rev. NMR Spectrosc.* **1**, 165 (1968).

S. Krimm, Infrared spectra of high polymers, *Fortschr. Hochpolym. Forsch.* **2**, 51 (1960).

G. Schnell, Ultrarotspektroskopische Untersuchungen an Copolymerisation, *Ber. Bunsenges.* **70**, 297 (1966).

F. A. Bovey, *High Resolution NMR of Macromolecules,* Academic Press, New York, 1972.

M. E. A. Cudby and H. A. Willis, Nuclear magnetic resonance spectra of polymers, *Ann. Rep. NMR (Nucl. Magn. Resonance) Spectrosc.* **4**, 363 (1971).

I. D. Robb and G. J. T. Tiddy, Macromolecules, *Nucl. Mag. Reson.* **3**, 279 (1974).

Chapter 4
Conformation

The physical structure of macromolecular materials is related to the structure of the isolated macromolecule (microstructure) and the structure of many macromolecules together (macrostructure). The macrostructure is essentially determined by the microstructure, and, in turn, the microstructure is determined by the conformation about σ bonds.

In macromolecular conformation, two concepts have to be distinguished. The microconformation, or, simply, the conformation, is concerned with the conformation about a single bond. Since many such microconformations will be adopted by different parts of the macromolecule, the whole molecule will itself adopt an overall conformation or macroconformation. The macroconformation determines the molecular shape of the macromolecule.

4.1. *Conformation of Single Molecules*

4.1.1. *The Concept of Conformation*

The conformation (or constellation) of a molecule is defined as one of the distinct spatial positions adopted by atoms or groups through rotation around single bonds or change in the relative positions of a given unit (for example, a ring) joined by several single bonds. Molecules with nonsuperimposable conformations are called conformers, rotamers, or conformational isomers. Of the many theoretically possible conformations about a given bond, only some will be energetically favorable.

Two conformers are singled out to describe the conformation of ethane (Figure 4-1). In the three identical staggered conformations, the hydrogen atoms on one carbon atom are at the greatest possible distance from those on the second carbon atom. With the three identical eclipsed conformations, the hydrogen atoms on neighboring carbon atoms are directly opposite and at the closest possible distance to each other. The ethane conformers can be transformed into each other by a 60° rotation about the C—C bond. The existence of such conformers was first proposed in the 1930s to explain the differences between calculated and measured entropies.

It is convenient in a discussion on conformation to distinguish between directly bonded ("bonded") and not directly bonded ("nonbonded") atoms and groups. Of the six hydrogens in ethane, three are not joined to any given one of the carbon atoms, and so are "nonbonded" atoms. For the atoms of the central C—C bond in butane, $CH_3—CH_2—CH_2—CH_3$, both methyl groups and two methylene hydrogens are also "not bonded" to a given central carbon atom in this sense.

Detailed information on the existence and stability of conformers can be obtained from energy calculations. All calculations proceed from a separate determination of the attraction and of the repulsion as a function of conformation angle. A typical calculation begins with a determination of the total energy E_{tot} of the ethane molecule. E_{tot} is made up of five separate contributions (see also Figure 4-2):

I. The energy E_{nn} between the nuclei of nonbonded hydrogen atoms.
II. The energy E_b between the electrons of the carbon–carbon bond.
III. The energy E_{ee} between the electrons of the carbon–hydrogen bonds.
IV. The energy E_{ne} between the hydrogen nuclei and the electrons of the bonds to the other hydrogen atoms.
V. The kinetic energy E_{kin} of the electrons.

Figure 4-1. The eclipsed (left) and staggered (right) conformations of ethane.

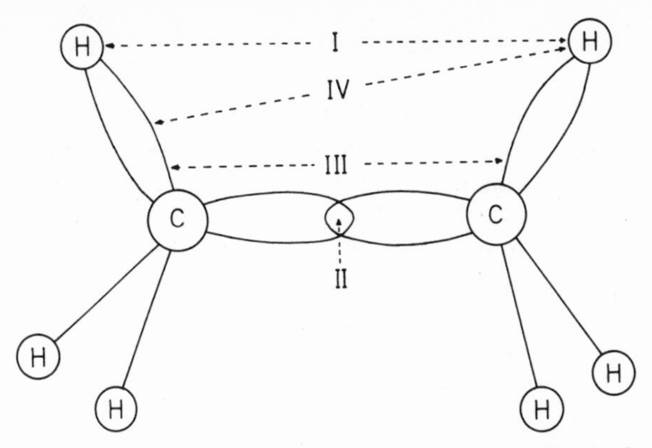

Figure 4-2. Schematic representation of the most important interactions in the ethane molecule (according to T. M. Birshtein and O. B. Ptitsyn). (For explanation of I–IV, see text.)

Only those interactions that are dependent on the angle of rotation are important in establishing conformations. Interactions between the electrons of the C—C bonds (case II) can therefore only contribute toward the conformation when there is no longer a prevalent cylindrical symmetry around the σ bonds. This symmetry would be broken if the $4f$ state took part in the C—C bond, since the corresponding regions of overlap of f bonds are not cylindrically symmetric about the C—C bond. Then, however, the eclipsed conformation would be more stable than the staggered, because of the greater overlapping of the electron clouds. In the staggered conformation of ethane, the H atoms are always staggered toward each other, whereas in the eclipsed conformation they are directly opposite to one another (Figure 4-2). Experimentally, however, exactly the reverse is found, i.e., the staggered conformation is more stable. Interactions between the f electrons of the C—C bond do not contribute, at least not to any measurable extent, to the conformation, i.e., $E_b \approx 0$.

The total attraction is given by E_{ne} and the total repulsion is given by $E_{nn} + E_{ee} + E_{\text{kin}}$. With the ethane molecule, three equal maxima are given in a curve describing either the attraction or repulsion as a function of angle of rotation. In this case, the phases of the attraction and repulsion energies are shifted by 120°. The energy difference between minimum and maximum for attraction is 82.5 kJ/mol (19.7 kcal/mol) and for repulsion is 93.8 kJ/mol (22.4 kcal/mol). The difference between these two energies is defined as the potential barrier. It is 11.3 kJ/mol (2.7 kcal/mol) for ethane and results from repulsion.

The ethane conformers are thus separated by a relatively low potential barrier, which can be overcome by thermal energy supplied by the collision of molecules. On the average, thermal energies of about $0.5RT$ per degree of freedom are transferred during such a collision. Because this is less than needed, the Maxwell–Boltzmann energy distribution dictates that a relatively small fraction of collisions will provide enough energy to cross the energy barrier. The majority of collisions will only deliver energy sufficient for oscillations of up to $\pm 20°$ about the potential minima. The majority of molecules thus remain in conformations associated with a minimum in the potential energy. They can be consequently treated as if they only exist in discrete rotational states. Fluctuations about the minima are not discounted; it is assumed that they compensate each other.

The potential barrier or potential energy barrier gives the energy difference between adjacent energy maxima and minima. It thus represents an activation energy $\Delta E^{\ddagger}$ for crossing the barrier associated with the energy maximum. The rate constant k_{conf} for this transfer can be calculated from $\Delta E^{\ddagger}$ via the equation

$$k_{conf} = (kT/h) \exp(-\Delta E^{\ddagger}/RT) \tag{4-1}$$

(see Table 4-1). The numerical values indicate that conformational transformations generally occur very fast. It is only in rare cases that k_{conf} has such a low value that the conformers can be preparatively separated (they are then called atropisomers). An example of an atropisomeric compound is 2,2′-dimethyl biphenyl.

Conformers can only be preparatively separated when no significant conformational change occurs during the time required for the separation. For conformers with very fast rates of transfer, it is necessary to use a technique that "sees" the molecules over a very short time interval. Electron spin resonance (ESR) spectroscopy is capable of observing states with lifetimes of only microseconds. This corresponds to an instantaneous picture of the conformer population. The conformers appear as distinctive species. On the ESR spectroscopy time scale, it is understandably difficult to distinguish between conformational isomers. For this reason and for other reasons, physicists often refer to conformations as configurations.

Table 4-1. Calculated Rate Constants for Conformational Transitions

$\Delta E^{\ddagger}$		k_{conf}, s^{-1}		
J/mol	kcal/mol	100 K	300 K	500 K
12,560	3	6×10^{5}	2×10^{10}	5×10^{11}
25,120	6	2×10^{-1}	3×10^{8}	2×10^{10}
41,870	10	3×10^{-10}	3×10^{5}	5×10^{8}

4.1.2. *Conformational Types*

The conformation about the C—C bond of ethane is determined by the three equivalent hydrogens as substituents on each of the carbon atoms. Thus, there are three equal energy maxima and three equal energy minima but only two different possible conformations: staggered and eclipsed.

With butane, $CH_3—CH_2—CH_2—CH_3$, on each side of the central C—C bond there are two hydrogen atoms and one methyl group attached to each carbon atom. Because of the *three* "nonbonded" substituents, there are three energy maxima and three energy minima. Only two of each are equivalent, however (Figure 4-3). Two *gauche* forms and one *trans* form of the staggered conformation can be distinguished. For the eclipsed form there are two skew forms and one *cis* form.*

With low-molecular-weight compounds, the *cis*-eclipsed position is given the rotational angle 0° in diagrams. This convention is unsuitable,

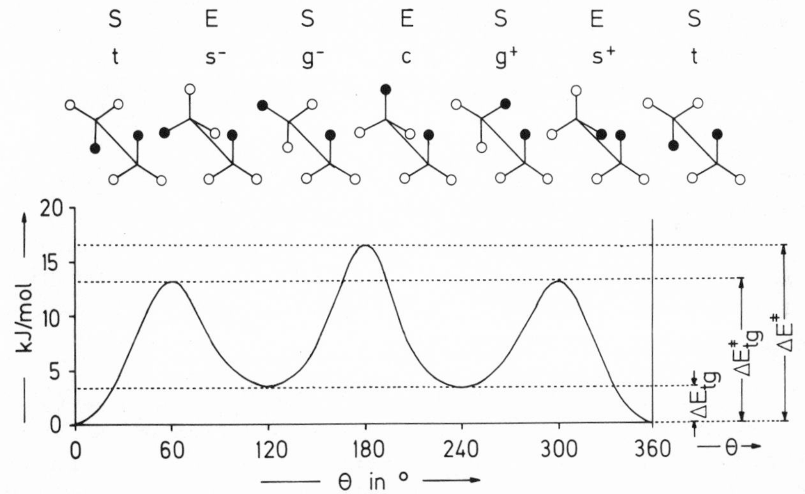

Figure 4-3. Conformations and rotational barrier potentials of the $CH_2—CH_2$ bond in butane, $CH_3CH_2CH_2CH_3$, as a function of the angle of rotation θ. ● = CH_3, ○ = H, S = staggered, E = eclipsed, t = *trans*, g = *gauche*, s = skew, c = *cis*.

---

* There are no uniformly accepted names for the various conformers. Other names used in the literature include: *trans*-staggered: *trans*, antiparallel, antiperiplanar (IUPAC), anti; *gauche*-staggered: *gauche*, skew-staggered, skew, synclinal (IUPAC); *cis*-eclipsed: *cis*, ecliptic, synperiplanar (IUPAC); *gauche*-eclipsed: anticlinal (IUPAC). The IUPAC terms, antiperiplanar (ap), synperiplanar (sp), anticlinal (ac), and synclinal (sc), have not become widely accepted in macromolecular chemistry.

since the *cis*-eclipsed position of polymers is sterically impossible. The *trans* position is given the 0° angle in this book because it often, but not always, represents the conformation of lowest energy.

Butane has two potential barriers: one between the lowest minimum (*trans*-staggered) and the highest maximum (*cis*-eclipsed) and one between the skew-eclipsed and the *trans*-staggered positions. Potential barriers determine the kinetic properties because they represent activation energies. Thermodynamic properties are governed by the conformational energies, i.e., the energy differences between energetically preferred conformations. Only one conformational energy difference is important with butane, and that is the conformational energy difference between the *trans* and the *gauche* conformations.

The actual influence of the chain on macromolecular conformation is only clearly apparent in *n*-pentane, since here, for the first time, there are two subsequent chain conformations to consider. Since one *trans* (t) and two *gauche* positions (g and g′) are possible in every chain bond of this kind, then for the two subsequent chain conformations there are four different physically distinct combinations or conformational diads (Figure 4-4). Of these, the diad tt possesses the lowest energy, and the combination gg′ the highest. Even with this simplified model, which does not take the eclipsed conformations into account, calculations of macroconformation are very difficult. For this reason, the model is further simplified by likewise excluding the combinations gg′ or g′g, and by taking the energy difference between g and t to be a constant (i.e., the energy difference should be independent of neighboring conformations).

The conformations of aliphatic hydrocarbons are characterized by threefold rotational potentials and a preferred *trans* conformation. These two characteristics do not always predominate. Twofold rotational potentials are produced by, for example, 1,4-phenylene groups and polymeric sulfur.

The repulsion forces dominating in aliphatic hydrocarbons lead to a *trans* conformation of the chain. The balance between attraction and repulsion forces can be changed for some of the possible conformations by the presence of neighboring atoms with lone electron pairs or by the presence of electronegative substituents. In these cases, the attractions

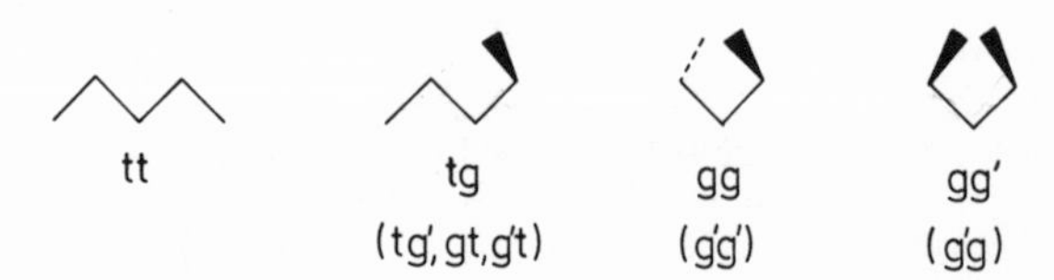

Figure 4-4. Conformational diads in pentane, $CH_3CH_2CH_2CH_2CH_3$.

between nuclei and electrons then become sufficiently large. When such atoms are present, the compounds tend to take up the conformations permitting the greatest number of *gauche* interactions between neighboring lone electron pairs and/or electronegative substituents (*gauche* effect). Thus, poly(oxymethylene), with the monomeric unit $\leftarrow O - CH_2 \rightarrow$, exists in the crystal in the all-*gauche* conformation, the lowest energy state.

4.1.3. Conformational Analysis

Conformational analysis attempts to determine the overall conformation of a molecule from the probability of the occurrence of the various microconformations. The energies of these conformations depend not only on their conformational states, but also on the conformations around neighboring bonds. The influence of conformations more than two bonds away from the bond being considered is however, usually ignored.

The conformational energy of the whole molecule is then calculated from the weighted sum of the contributions of the individual bonds. Here, a given reference conformation is given the energy zero and a statistical weight of unity. This conformation is, for example, the *trans* conformation for poly(ethylene) and the *gauche* conformation for poly(oxymethylene).

Several equations are used for the calculation of the conformational energy. They differ in the choice of the parameters and potentials. A typical equation is the following, in which the conformational energy E_r is determined from the contributions of the torsional energy [term (a)], the energy of interaction between nonbonded groups [term (b)], the electrostatic energy [term (c)], and the energy of hydrogen bonding [term (d)]:

$$E_r = \underbrace{\sum_i 0.5(E_i^{\ddagger})(1 - \cos N_i \vartheta_i)}_{\text{(a)}} + \underbrace{\sum_{i,j} 2\varepsilon_{i,j}\left(\frac{d_{i,j}}{r^{12}} - \frac{b_{i,j}}{r^6}\right)}_{\text{(b)}} + \underbrace{B \sum_{j,k} (e_j e_k / r)}_{\text{(c)}} + \underbrace{E_{\mathrm{H}}}_{\text{(d)}} \qquad (4\text{-}2)$$

$E_i^{\ddagger}$ is the potential barrier for rotation about the bond i, N_i is the symmetry of the rotation (usually 2, 3, or 6), ϑ_i is the angle of rotation, r is the distance between atomic nuclei, and e_j and e_k are partial charges derived from the dipole moments of the bonds. The factor B contains the Coulomb energy and the apparent relative permittivities (dielectric constants). $\varepsilon_{i,j}$, $d_{i,j}$, and $b_{i,j}$ are parameters describing the potential energy of the contributions from nonbonded atoms. A 9–6 potential is often used instead of the Lennard-Jones 12–6 potential shown in equation (4-2).

Although the various calculations can differ very strongly from each

Table 4-2. Individual Interaction Energy Contributions to the Stability of the Poly(L-alanine) α Helix

Source of contribution	E, J/mol	
	After Ooi, Scott, Van der Kooi, Scheraga	After Kosuge, Fujiwa, Isogai, Saitô
Rotation	2,050	2,430
Nonbonded atoms	−25,080	−29,940
Electrostatic interaction	−4,610	10,890
Hydrogen bonding	−7,290	−4,280
Total	−34,930	−20,890
Hydrogen-bonding contribution	20.8%	20.5%

other according to the choice of bond distance, interaction energy, potentials, etc., the general conclusions from given sets of calculations agree very well with each other. For example, two different groups of workers found that the hydrogen bonding in poly(L-alanine), which occurs in the form of an α helix with $\mathrm{-[NH-CH(CH_3)-CO]-}$ as the constitutional repeat unit, only contributes about 20% to the overall stability of the helix. This is in spite of the fact that the terms used by the different workers for the electrostatic contribution differ even in sign, one being positive, the other negative (Table 4-2).

4.1.4. Constitutional Influence

The height of the potential energy barrier, as would be expected, decreases with increasing bond distance (see ethane–methyl silane–disilane in Table 4-3). The potential barrier also decreases in going from trivalent CH_3— groups to monovalent OH— groups (see ethane–methanol).

The potential barrier increases with further substitution because of increasing steric hindrance, as can be seen from the series ethane, propane, isobutane, neopentane; methanol, dimethylether; and acetaldehyde, propylene, isobutylene.

In $\mathrm{-(CH_2-CO)-}$ and $\mathrm{-(CH_2-O)-}$ bonds, the potential barrier is considerably lower than in the $\mathrm{-(CH_2-CH_2)-}$ bonds. The transition from one conformation to another is therefore more readily possible: The macromolecule becomes kinetically more flexible. Since the flexibility of a macromolecule determines many of its properties, such as elasticity, melt viscosity, tendency to crystallize, etc., it is therefore an important technological property.

There are more energetically equivalent conformations for compounds of the type $+CH_2-CR_2+_n$ than there are for compounds of the type $+CH_2-CHR+_n$. Thus, the former example will probably have more conformations available than the latter one. Consequently, compounds with two like substituents tend to be more flexible than compounds with two unlike substituents.

Various factors are thus responsible for a high molecular flexibility: (1) a large bond length between chain atoms to give a low potential barrier, (2) many competing positions for like substituents, and (3) a low potential difference between *gauche* and *trans* positions because of the *gauche* effect. All three effects are present in poly(dimethylsiloxane), $+Si(CH_3)_2-O+$, i.e., the Si—O bond length of 0.164 nm is relatively long, the main chain is rotationally symmetric, and there are polar oxygen chain atoms. The high molecular flexibility is mainly responsible for the low glass-transition temperature (see Chapter 10). Linear poly(dimethylsiloxanes) are therefore highly viscous liquids up to molecular weights of millions.

Table 4-3. Potential Barriers for Rotation about 360° of Bonds Marked in Various Molecules

Compound	Potential barrier		Bond length, nm
	kcal/mol bond	kJ/mol bond	
$SiH_3\overset{\downarrow}{-}SiH_3$	1.0	4.2	0.234
$CH_3\overset{\downarrow}{-}SiH_3$	1.7	7.1	0.193
$CH_3\overset{\downarrow}{-}CH_3$	2.8	11.7	0.154
$CH_3\overset{\downarrow}{-}CH_2-CH_3$	3.3	13.8	0.154
$CH_3\overset{\downarrow}{-}CH(CH_3)_2$	3.9	16.3	0.154
$CH_3\overset{\downarrow}{-}C(CH_3)_3$	4.8	20.1	0.154
$CCl_3\overset{\downarrow}{-}CCl_3$	10	42	0.154
$CH_3\overset{\downarrow}{-}OH$	1.0	4.2	0.144
$CH_3\overset{\downarrow}{-}O-CH_3$	2.7	11.3	0.143
$CH_3\overset{\downarrow}{-}CHO$	1.0	4.2	0.154
$CH_3\overset{\downarrow}{-}CH{=}CH_2$	2.0	8.4	0.154
$CH_3\overset{\downarrow}{-}C(CH_3){=}CH_2$	2.4	10.0	0.154
$-CH_2\overset{\downarrow}{-}CO-CH_2\overset{\downarrow}{-}CH_2-$	2.3	9.6	0.154
$-CH_2\overset{\downarrow}{-}CO-CH_2-CH_2-$	0.8	3.35	0.154
$-CH_2\overset{\downarrow}{-}CO-O-CH_2-$	0.5	2.1	0.154
$-CH_2\overset{\downarrow}{-}CO-O\overset{\downarrow}{-}CH_2-$	1.2	5.0	0.143
$-CH_2\overset{\downarrow}{-}NH-CH_2-CH_2-$	3.3	13.8	0.147
$-CH_2\overset{\downarrow}{-}S-CH_2-CH_2-$	2.1	8.8	0.181
$CH_3\overset{\downarrow}{-}SH$	1.3	5.4	0.181

4.2. Conformation in the Crystal

The macroconformation of a crystalline macromolecule is determined by intra- and/or intermolecular factors. Intermolecular forces influence the mutual packing of components of the chain, which leads to varying densities. The maximum difference in density found in crystalline poly-(α-olefins) corresponds, however, to an energy difference of only 1200 J/mol monomeric unit. Intermolecular forces can therefore only influence the conformation of very flexible chains, since here the conformational energies are low. The successful calculation of the macroconformations from the intramolecular forces alone, without considering intermolecular effects, also indicates the limited influence of the latter on the conformation.

As a result, the conformation in the crystalline state is principally determined by intramolecularly active forces. They are governed by two principles. According to the equivalence principle, the chain conformation is governed by the sequence of like structural units. Usually, either a monomeric unit or a half-unit function as structural units. Successive structural units in a sequence occupy geometrically equivalent positions in relation to the crystallographic axis. According to the principle of the smallest intramolecular conformation energy, in a crystal, an isolated chain takes up the conformation with the smallest energy that the equivalence principle allows.

Thus, the conformation of polymers in the crystalline state can be evaluated with these two principles and the known van der Waals radii,* without it being necessary to know the detailed effect of individual interactions. The van der Waals radius of the H atom is 0.12 nm. It can be further calculated, from the C—C bond length of 0.154 nm and the C—C—C valence angle of 109.6°, that the H atoms bonded to different C atoms are 0.25 nm apart in the *trans* position. This distance is greater than the sum (0.24 nm) of the van der Waals radii. In the ideal crystalline state, therefore, poly(ethylene) shows an all-*trans* conformation (zigzag form), since the *trans* conformation corresponds to the lowest potential energy minimum. On transition to a *gauche* conformation, the distance between CH_2 groups on nonadjacent carbon atoms is increased, and the conformational energy increases to 33 kJ/mol bond. [See also Figure 4-3; the ● corresponds to a CH_2 group in poly(ethylene).]

The van der Waals radius of fluorine atoms is 0.14 nm. In an all-*trans* poly(tetrafluoroethylene) conformation, the separation of the fluorine atoms (0.25 nm) would thus be smaller than the sum of the van der Waals radii

* The van der Waals radii, and not the (in general smaller) atomic radii, are decisive in determining the conformation.

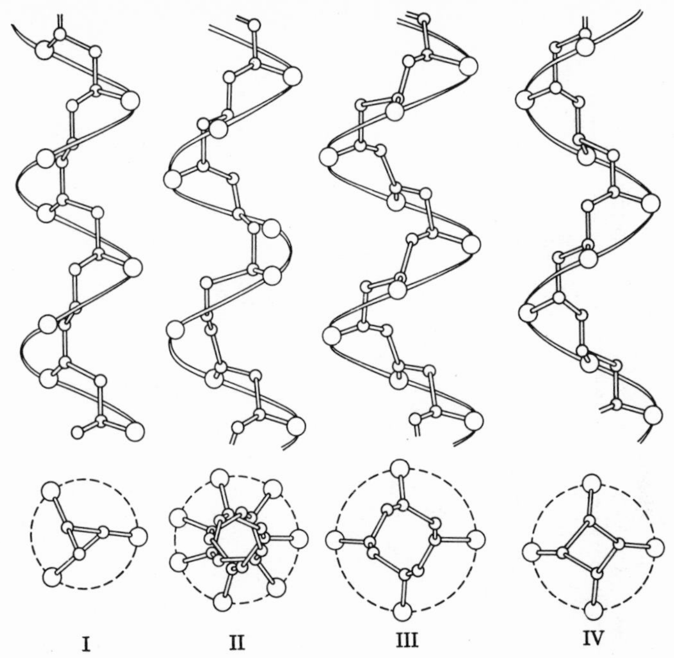

Figure 4-5. Schematic representation of the different kinds of helix occurring in various isotactic polymers $\text{+}CH_2\text{—}CHR\text{+}_n$. (I) 3_1; (II) 7_2; (III, IV) 4_1 (after G. Natta, P. Corradini, and I. W. Bassi).

0.28 nm). The chain atoms therefore deviate from the ideal all-*trans* position through a slight change in the angle of rotation from 0° to 16°, and form a 13_1 helix* (Table 4-4). In this conformation, the fluorine atoms attached to neighboring C atoms are now at a distance of 0.27 nm. The situation is similar in poly(isobutylene), $\text{+}CH_2\text{—}C(CH_3)_2\text{+}_n$: The large methyl groups force the chains in the crystal into an 8_3 helix. Since there are two possible *gauche* positions, left- and right-handed helices can exist, i.e., helices whose pitch is to the left or to the right.

Helix conformation occurs relatively often in macromolecules. Helices are distinguished by a number p_q, where p gives the number of monomeric units per q complete turns of the helix. The 3_1 helix of it-poly(propylene) thus exhibits three monomeric units per turn (Figure 4-5). In many macromolecules, two (e.g., DNA) or three (e.g., collagen) separate helices are

* The expression helix (spiral) originates from the name of the edible snail (*Helix pomatia*), which has a spiral shell.

Table 4-4. Important Conformational Forms of Macromolecules (after S.-L. Mizushima and T. Shimanouchi)

Conformation	Spatial representation: Perpendicular to the chain	Spatial representation: Along the chain	Monomeric units	Helix type[a]	Angle of rotation	Example
...ttt...		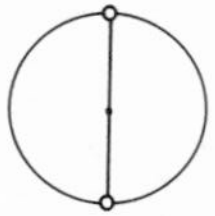	$-CH_2-CH_2-$	1_1	0/0	Poly(ethylene)
			$-CH_2-$	2_1	0/0	Poly(methylene)
			$-CH_2-CHCl-$	1_1	0/0	st-Poly(vinyl chloride)
		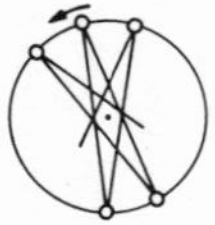	$-CF_2-CF_2-$	13_1	16/16	Poly(tetrafluoroethylene)
...tgtg...	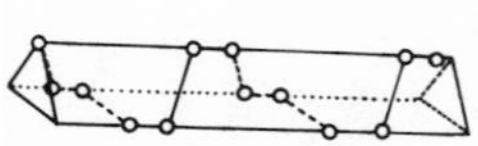	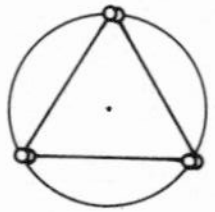	$-CH_2-CH(R)-$	3_1	0/120	it-Poly(propylene) ($R = CH_3$)
					0/120	it-Poly(styrene) ($R = C_6H_5$)
					0/120	it-Poly(5-methyl-1-heptene) ($R = CH_2-CH_2-CH(CH_3)-CH-CH_3$
		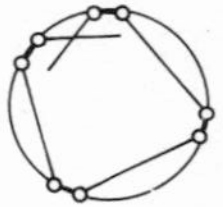	$-CH_2-CH(CH_2-CH(CH_3)-CH_3)-$	7_2	−13/110	it-Poly(4-methyl-1-pentene)
		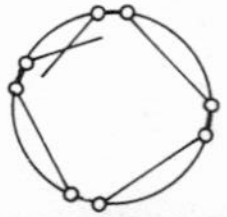	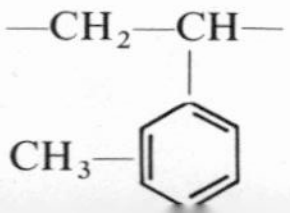 $-CH_2-CH(C_6H_4-CH_3)-$	11_3	−16/104	it-Poly(*m*-methyl-styrene)

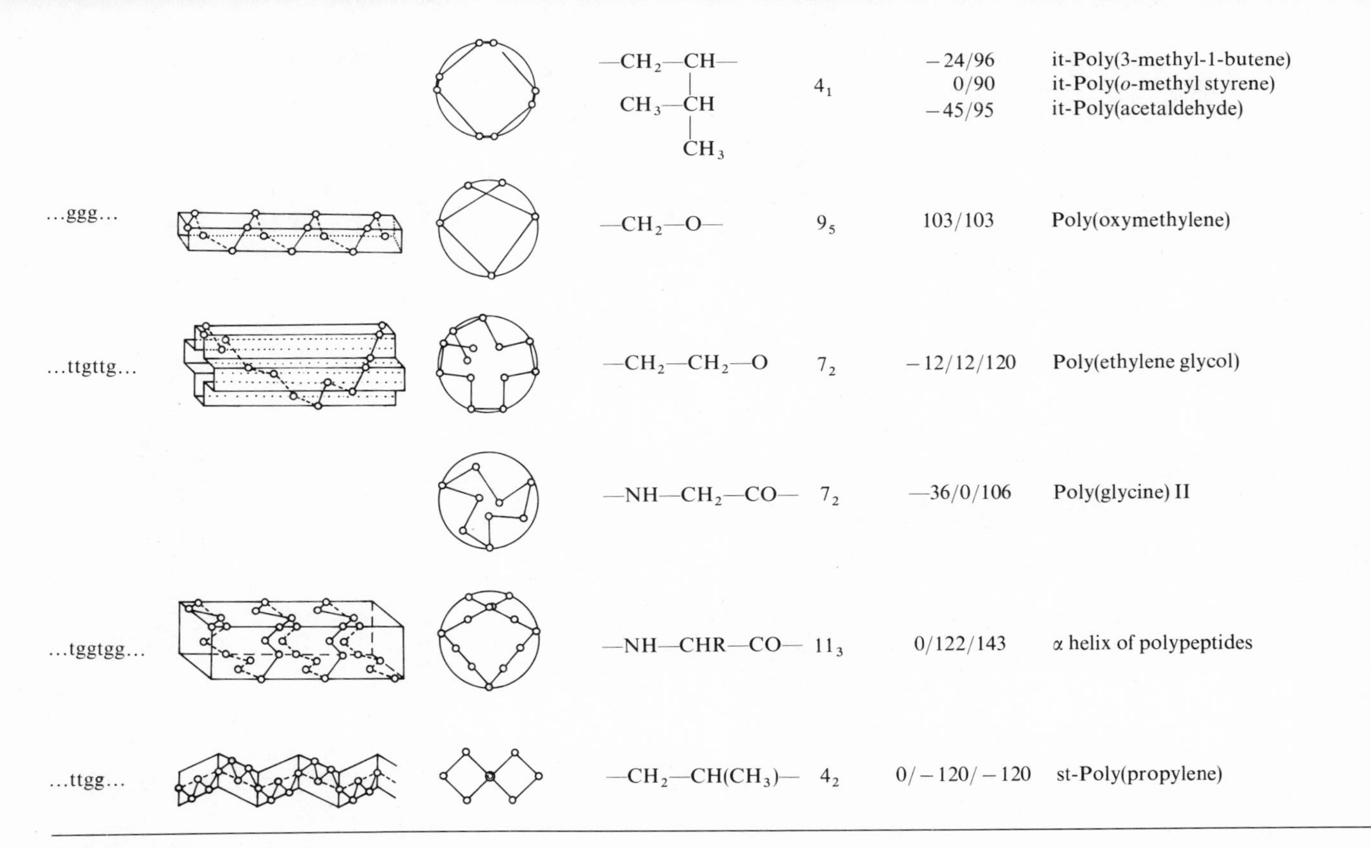

Conformation	Repeat unit	Helix[a]	Angles	Polymer
	$-CH_2-CH-$ ($CH_3-CH-CH_3$ side group)	4_1	−24/96 0/90 −45/95	it-Poly(3-methyl-1-butene) it-Poly(*o*-methyl styrene) it-Poly(acetaldehyde)
...ggg...	$-CH_2-O-$	9_5	103/103	Poly(oxymethylene)
...ttgttg...	$-CH_2-CH_2-O$	7_2	−12/12/120	Poly(ethylene glycol)
	$-NH-CH_2-CO-$	7_2	−36/0/106	Poly(glycine) II
...tggtgg...	$-NH-CHR-CO-$	11_3	0/122/143	α helix of polypeptides
...ttgg...	$-CH_2-CH(CH_3)-$	4_2	0/−120/−120	st-Poly(propylene)

[a]Monomeric units per number of turns.

intertwined. Such intertwined multiple helices are also referred to as superhelices (Figure 4-6).

In isotactic polyvinyl compounds, $\text{+}CH_2\text{—}CHR\text{+}_n$, the bulky substituents R on every second chain atom force the chain to change from the tttt conformation into the tgtg conformation. In it-poly(propylene), e.g., the ...tt... conformation energy is about 41.8 kJ/mol higher than that of the ...tg... conformation. it-Poly(propylene) and it-poly(styrene) exist in the crystalline state in the form of 3_1 helices with angles of rotation of 0° and 120°. The valence angle is also significantly distorted by increasing substituent size: 110° for poly(ethylene), 114° for it-poly(propylene), and 116° for it-poly(styrene). In going from it-poly(5-methyl-1-heptene) to it-poly(4-methyl-1-pentene), the methyl groups come closer to the main chain and the greater steric effect forces the chain atoms to deviate from the ideal *trans* and *gauche* positions (with rotational angles of 0° and +120°) and to adopt rotational angles of −13° and +110°. Poly(4-methyl-1-pentene) thus exhibits as a 7_2 helix. In poly(3-methyl-1-butene), the methyl groups are in the immediate vicinity of the chain, and the polymer forms a 4_1 helix.

Conformations that deviate by up to ±30° from the ideal position are mostly still considered to be ideal conformational positions. A conformation with an angle of rotation of −13° is thus called *trans*.

Isotactic polymers with two chain atoms per monomeric unit thus tend to occur in more or less ideal tg conformations. In addition, the low energy difference for slight deviations from the ideal rotational angle can lead to various helix types. Isotactic polymers therefore occasionally crystallize into various types of helices. Rapid crystallization of it-poly(butene-1) produces a 4_1 helix which, as a high energy form, changes into a 3_1 helix on annealing (see also Chapter 10.3.5).

The substituents of syndiotactic vinyl polymers in all-*trans* conformations are more widely spaced than in corresponding isotactic compounds. The ...tt... conformation is therefore generally the lowest energy conformation for syndiotactic polymers. 1,2-Poly(butadiene), poly(acrylonitrile), and poly(vinyl chloride) belong to this group. In a few cases a series of rotational angles of 0°, 0°, −120°, −120° is more advantageous, and therefore substances such as st-poly(propylene) generally take on a ttgg conformation, but can also crystallize in a ...tt... conformation since the energy differences are small.

Poly(vinyl alcohol), with the monomeric unit $\text{+}CH_2\text{—}CHOH\text{+}$, carries a hydroxyl group at every second chain atom. These OH groups can form intramolecular hydrogen bonds. Therefore, in contrast to it-poly-α-olefins), it-poly(vinyl alcohol) do not form a helix, but an all-*trans* conformation. For the same reason, st-poly(vinyl alcohol) does not exist as a zigzag chain, but as a helix.

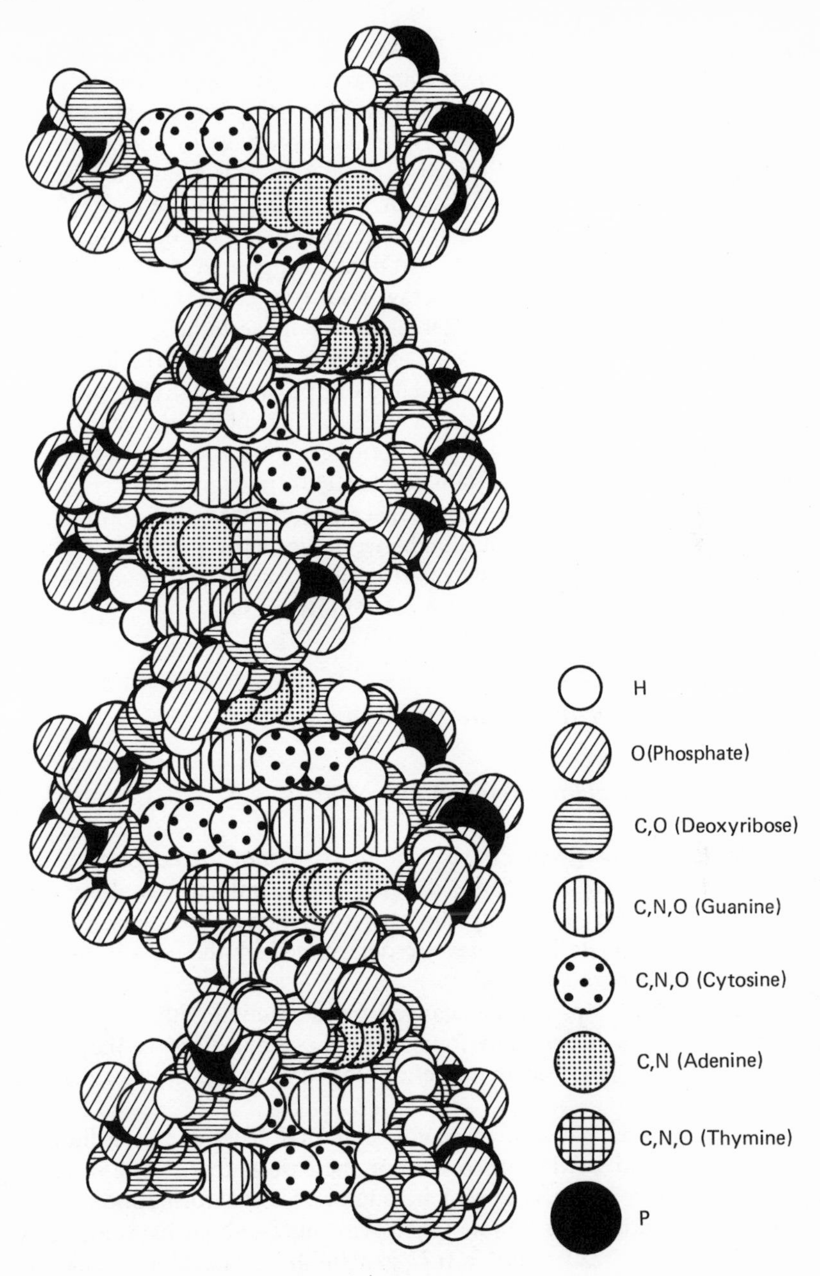

Figure 4-6. A part of the double helix of deoxyribonucleic acid. The frame formed by the heterocyclic bases, purine and pyrimidine, as well as the deoxyribose residues, are shown in monotone. Only the phosphate chain H atoms are shown (after M. Eigen).

The influence of interactions between the electron clouds of the bonds between the main chain atoms is much reduced in heterochain polymers. There are three bonds to be considered in the CH_2-group, but only one in the O-linkage. The potential barrier therefore falls to about one-third of the value in carbon chains (see also Table 4-3). This means, for example, that molecules with oxygen atoms in the main chain are more flexible than comparable ones with carbon chains. Because of the decreased atomic bond length of 0.144 nm for the C—O bond, as opposed to 0.154 nm in the C—C bond, methyl substituents draw relatively closer together, causing the diameter of the helix to be increased. it-Poly(acetaldehyde) thus exists as a 4_1 helix, but it-poly(propylene) exists as a 3_1 helix. If the influence of the methyl substituents is not present, as in poly(oxymethylene), then the effects of bond orientation make themselves particularly noticeable. Poly-(oxymethylene) thus exists as a ...ggg... conformation, whereas poly(ethylene glycol) is ...ttgttg.... Like poly(ethylene glycol), poly(glycin) II crystallizes into a 7_2 helix, but this is deformed because of hydrogen bonding. In it-poly(propylene oxide), the repulsion between the methyl groups is increased, and the bond orientation is decreased because of the methyl substituents: This polymer crystallizes in an all-*trans* conformation.

4.3. Microconformation in Solution

4.3.1. Low-Molecular-Weight Compounds

The discussions given above refer to molecules in the gaseous state or in the form of crystals. Since packing effects influence conformations in the lowest energy level only slightly in crystals, conformations in the gaseous and the crystalline states are determined by the constitution and the configuration of the molecules.

The conformations can be changed by interaction with like or unlike molecules. Thus, it is generally found that the proportions of the various conformers are different in the gaseous state and in the liquid state or in solution.

The differences between the conformational energies in the liquid or solution state and in the gaseous state increase with increasing *gauche* effect. The *gauche* effect is absent in butane. The conformational energy in the gaseous state (3350 J/mol = 0.80 kcal/mol) is about the same as that in the fluid state (3220 J/mol = 0.77 kcal/mol). 1,2-Dichloroethane does have a *gauche* effect, and its conformational energy in solution becomes more positive with increasing dielectric constant of its environment. In

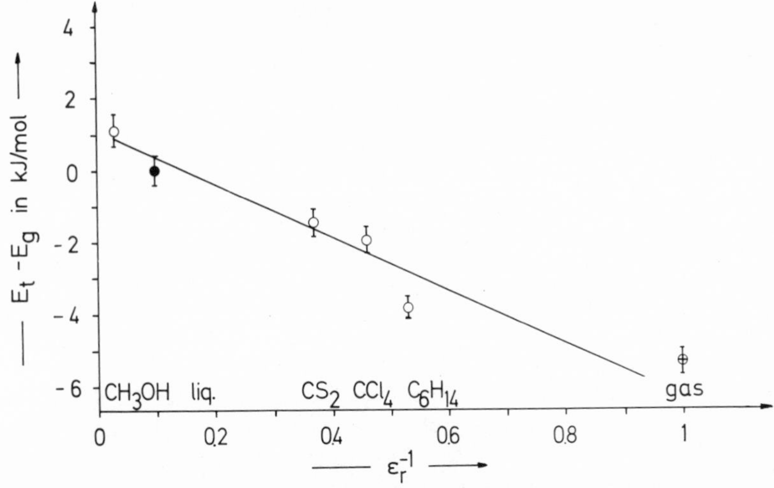

Figure 4-7. Conformational energy of 1,2-dichloroethane as a function of the relative permittivity (dielectric constant) in the liquid phase (●); in various solvents (○); and in the gas phase (⊕).

this case, the conformation energy $E_t - E_g$ is negative in most solvents, that is, the *trans* conformation is preferred. However, the *gauche* conformation dominates in methanol since the conformational energy is positive (Figure 4-7).

meso-2,4-Pentanediol also exhibits an increase in *gauche* conformations in polar solvents because of the *gauche* effect (Table 4-5). It is especially noticeable that the proportion of *gauche* diads with the same sign (g^+g^+ and g^-g^-) is zero for practically all solvents.

Table 4-5. Influence of Solvent on Conformation of meso-2,4-Pentanediol at 40°C

	Percentage of conformative diads, %			
Solvent	tt	tg^+ and g^-t	tg^- and g^+t	g^+g^+ and g^-g^-
CCl_4	70	10	10	10
CH_2Cl_2	90	10	0	0
Pyridine	45	48	7	0
DMSO	30	60	10	0
D_2O	5	70	25	0

4.3.2. Macromolecular Compounds

In low-molecular-weight compounds in solution, conformations are essentially determined by interactions between nearest neighbors and their interactions with the solvent. The conformation of high-molecular-weight compound is, however, determined by the whole chain, which can exclude certain conformations. A conformational change about one bond often initiates a series of conformational changes in neighboring bonds. Two limiting cases can be distinguished according to the strength of the interaction between the macromolecule and the solvent, as follows.

Strong interactions only occur between polar groups. They can, for example, lead to solvation of macromolecular groups. Alternatively, solvent molecules in the neighborhood of a macromolecular group can induce a *gauche* effect. In such cases, the conformational changes of the macromolecule in going from the crystalline state to the dissolved state are determined by group-specific effects, i.e., in the last analysis, by enthalpic effects. Only a few of the bonds retain the previous crystal conformation since practically every bond can adopt a new conformation. Thus, the sequence lengths of conformational diads of polar macromolecules are very short in very polar solvents.

In contrast, slight or no group-specific interactions occur in the dissolution of apolar polymers in apolar solvents. Neither solvation nor induced *gauche* effects are driving forces for conformational changes. Thus, conformational changes must be mostly due to entropic effects. For energetic reasons, only few conformations are transformed. Large sequences are retained in the original conformation.

The conformation of macromolecules in solution can also be influenced by the molecular order of the solvent itself. Benzene, for example, exists as money-roll-type aggregates, whereas CCl_4 exhibits no order. An ordered solvent can induce the macromolecular chain to adopt a certain series of conformational diads, at least over short chain segments.

The existence of ordered conformational sequences can be directly identified experimentally with some solutions of macromolecules. For example, the IR spectrum of crystalline st-poly(propylene) possesses a so-called crystalline band at 868 cm^{-1}, which disappears on melting. Theoretical calculations show that this band results almost exclusively from the helix structure of the polymer. The band does not occur with poly(propylene) solutions in CCl_4 but does occur with solutions in benzene. The intensity of the band decreases with increasing temperature of the solution: The helices "melt" (Figure 4-8).

The mean number of monomeric units occurring in helix segments, N_h, can be estimated from

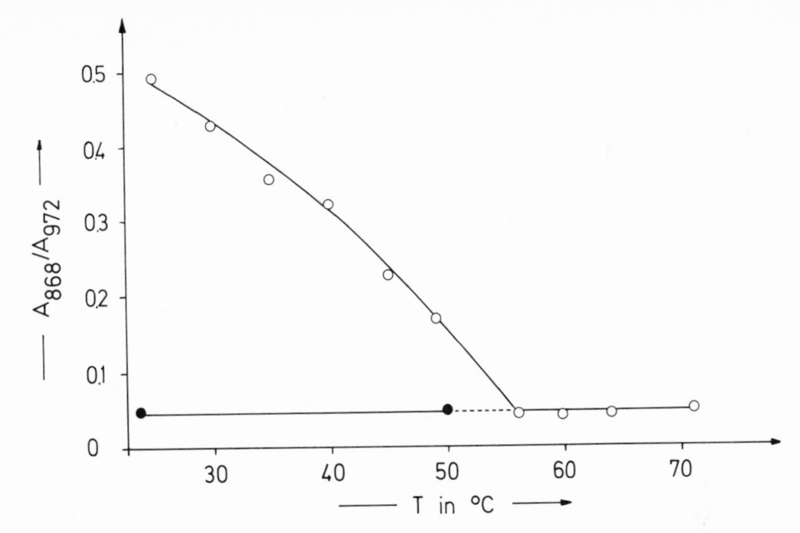

Figure 4-8. Change in intensity of the poly(propylene) IR crystalline band at 868 nm with temperature for solutions in benzene (○) and in carbon tetrachloride (●) (after B. Stofer and H.-G. Elias).

$$N_h = \frac{1 + \exp(-\Delta E/RT)}{\exp(-\Delta E/RT)} \tag{4-3}$$

The conformational energy of a bond ΔE is, however, equal to half of the Gibbs energy for the "reaction" of a left-handed with a right-handed conformational diad,

$$ll + dd \rightleftharpoons ld + dl \tag{4-4}$$

and so we have

$$\Delta E = 0.5\ \Delta G = -0.5RT \ln \frac{[ld]\,[dl]}{[ll]\,[dd]} = 0.5RT \ln \frac{g_{ll}g_{dd}}{g_{ld}g_{dl}} \tag{4-5}$$

The mole concentrations here can be replaced by the statistical weights. Each statistical weight is related to the conformational energy of a conformational diad E_{jk} by $E_{jk} = -RT \ln g_{jk}$. Thus from equation (4-5) we obtain

$$\Delta E = 0.5(E_{ll} + E_{dd}) - 0.5(E_{ld} + E_{dl}) \tag{4-6}$$

E_{ll} is the conformational energy of a left-handed monomer unit that follows another left-handed monomer unit; E_{dl} is the energy of a left-handed monomer unit that follows a right-handed monomer unit, etc. With completely dissymmetric chains, $E_{ll} = E_{dd}$ and $E_{ld} = E_{dl}$, but with chains with asymmetric main-chain atoms or asymmetric substituents, $E_{ld} \neq E_{dl}$. According to these calculations, about 6–10 monomer unit chain lengths are retained with optically inactive poly(α-olefins) in solution.

Similar evidence of order in solution can be obtained for other polymers with other experimental methods. Proton magnetic resonance studies on poly(oxyethylene) in benzene solution give a new signal (and a weaker one in CCl_4 solution) for heptamers and higher species which is obviously ascribable to another conformation. Alkanes, $CH_3(CH_2)_nCH_3$, in 1-chloronaphthalin give only one methylene proton signal for $n < 14$ in PMR measurements. When $n > 15$, two signals are obtained. This separation into two signals is not observed in CCl_4 solutions or with deuterated alkanes. The signal separation is attributed to an intramolecular chain folding, but could also be due to an intermolecular association of chain segments.

4.4. Ideal Coil Molecules in Solution

4.4.1. Phenomena

On dissolution of a constitutionally and configurationally uniform macromolecular chain, at least some of the conformations occurring in crystalline state will be transformed, and these new conformations will produce "bends" in the chain. Only a few of such bends are necessary to

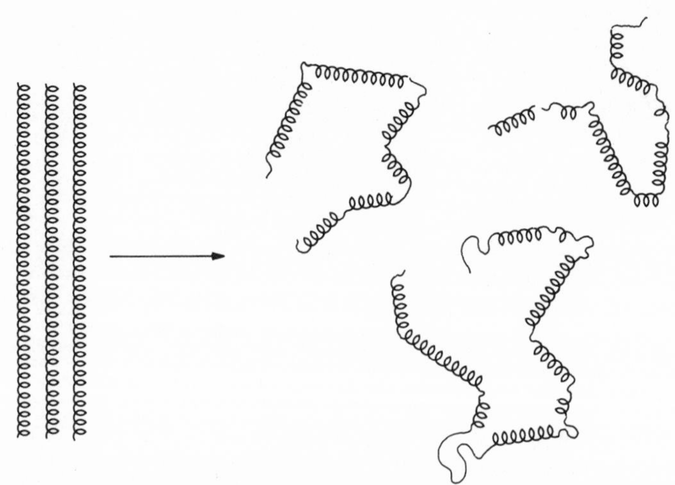

Figure 4-9. Dissolution of a macromolecule occurring in the crystalline state in the form of a helix. Only a few "kinks" are required to produce the macroconformation of a coil.

change the extended rodlike molecules of the crystal into a coil-like macroconformation in solution. Thus, it is very possible that a larger or smaller proportion of the original microconformation can be retained on dissolution (Figure 4-9).

If diameters are sufficiently large, the coil form of such macromolecules can be shown by electron microscopy. Since the molecules are always observed on a substrate, two-dimensional projections are obtained of the three-dimensional shapes occurring in solution. Figure 4-10 shows the

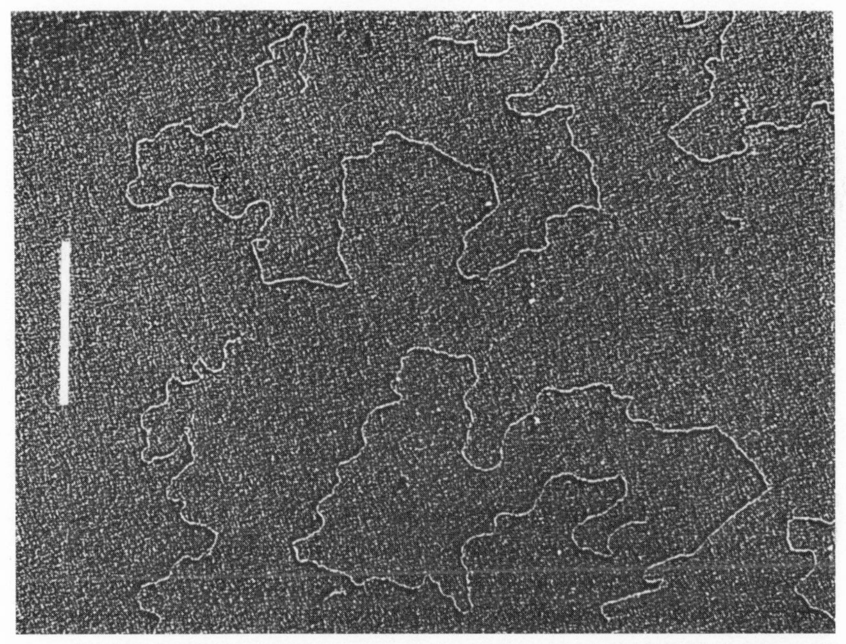

Figure 4-10. An electron microscope picture of the double chain of deoxyribonucleic acid (after D. Lang, H. Bujard, B. Wolff, and D. Russell).

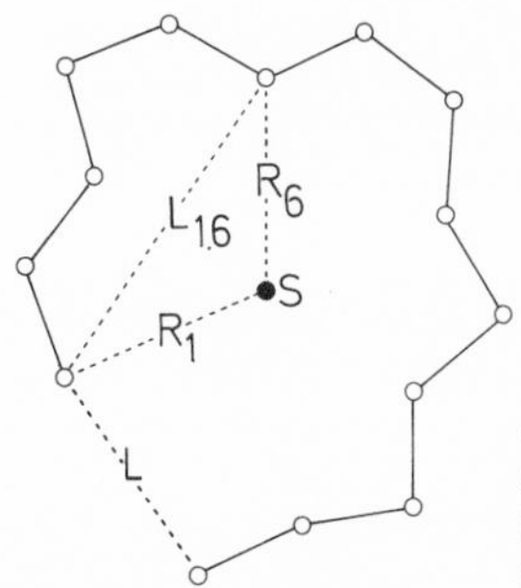

Figure 4-11. Schematic representation of a molecular coil consisting of 14 units of mass. The center of gravity of the molecule is at S. L is the chain end-to-end distance. The radii of gyration about S of the individual point masses are R.

coil form of the deoxyribonucleic acid molecule. The local conformation of the monomeric units is that of a double helix.

A coil-shaped molecule can be characterized by its end-to-end distance L or by its radius of gyration R_G (Figure 4-11). The relations between these two quantities and between each of them and the number and length of segments in the chain depend on the assumptions used in the calculation. These quantities will depend on the interaction between the coil and the solvent in which it is dissolved.

4.4.2. Chain End-to-End Distance and Radii of Gyration

4.4.2.1. The Random-Flight Model

In what is called the random-flight model, it is assumed that all segment lengths l are equal in length and infinitely thin. and that the angle ϑ between adjacent segments can take any value ("freely jointed"). Additionally, it is not specified what the physical interpretation of such a segment might be. The segment length is not restricted to being a bond length; it may also be a unit of many chain links.

The desired relationship between chain end-to-end distance L and N or l can be derived from vector analysis (see appendix to this chapter). The principles involved are more obvious on closer examination of the problem.

Figure 4-12 shows two segments of length l forming an angle ϑ. The distance L_{00} between the ends of the two segments is given by the cosine rule

$$L_{00}^2 = 2l^2 - 2l^2 \cos \vartheta \tag{4-7}$$

When there is a sufficiently large number of molecules or the observation time is sufficiently long, the number of angles ϑ between adjacent segments is also large. Then, the mean end-to-end chain distance $\langle L^2 \rangle_{00}$, replaces L_{00}^2 in equation (4-7) and $\cos \vartheta$ is replaced by the mean value $\langle \cos \vartheta \rangle$. Since all directions are equally probable, $\langle \cos \vartheta \rangle = 0$, and equation (4-7) becomes

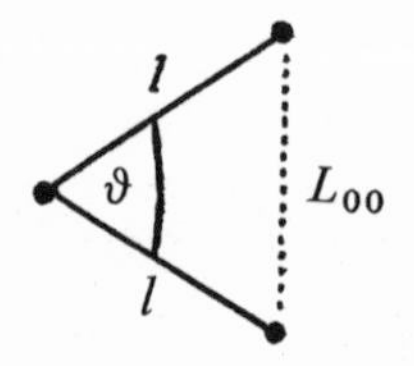

Figure 4-12. Definition of the end-to-end distance L_{00} of a chain with two segments.

$$\langle L^2 \rangle_{00} = 2l^2 \tag{4-8}$$

On going from two to N segments, one obtains

$$\langle L^2 \rangle_{00} = Nl^2 \tag{4-9}$$

4.4.2.2. Chains with Constant Valence Angle

The random-flight model given above allows any angle between adjacent segments. However, definite valence angles occur in real macromolecular chains. The relationship between the mean end-to-end distance L_{0f} of this valence angle chain (with implicit assumption of free rotation about single bonds) and the number N of bonds of length l in the chain is, for an infinitely large number N of constant angles ϑ,

$$\langle L^2 \rangle_{0f} = Nl^2(1 - \cos\vartheta)(1 + \cos\vartheta)^{-1} \tag{4-10}$$

Going from the freely jointed chain to one with fixed valence angles produces a coil expansion when the valence angle exceeds 90° [compare equations (4-9) and (4-10)].

The maximum possible chain length l_{max} is assigned to a constant valence angle chain in all-*trans* conformation. It can be calculated from simple geometric principles to be given by

$$l_{max} = Nl \sin(0.5\vartheta) \tag{4-11}$$

The contour length l_{cont},

$$l_{cont} = Nl \tag{4-12}$$

is, in contrast, the length of the fully stretched chain with a valence angle of 180°.

4.4.2.3. Constant Valence Angle Chains with Hindered Rotation

The constant valence angle chain with free rotation is also unreal, since it overlooks the existence of conformers. Each conformation can be assigned an angle of rotation θ (see Figure 4-3). By an analogous mathematical procedure to that used to determine the influence of the valence angle on the coil dimensions, the following is obtained for symmetrically constructed chains [i.e., $\text{+}CH_2\text{—}CR_2\text{+}_n$] of infinite molecular weight and for nonzero values of θ:

$$\langle L^2 \rangle_0 = Nl^2\left(\frac{1-\cos\vartheta}{1+\cos\vartheta}\right)\left(\frac{1+\cos\theta}{1-\cos\theta}\right) = Nl^2\left(\frac{1-\cos\vartheta}{1+\cos\vartheta}\right)\sigma_{sym}^2 \tag{4-13}$$

Since various microconformations are possible, a mean must be taken of all conformational influences. The term in equation (4-13) containing the rotational angle is often replaced by a new quantity, σ^2_{sym}, and is inserted as a squared term so that σ, as a measure for a change of dimensions, becomes comparable to the lengths L and l.

For a chain with completely free rotation, $\langle \cos\theta \rangle = 0$ and equation (4-13) becomes equation (4-10). An all-*trans* chain is a rigid chain with $\theta = 0$, which is, however, physically absurd with regard to equation (4-13). Thus, σ is a measure of the hindrance to rotation; it is often called the steric hindrance parameter. With tactic polymers, the relationship between θ and σ is more complicated. One can, however, always use an equation analogous to equation (4-13):

$$\langle L^2 \rangle_0 = Nl^2(1 - \cos\vartheta)(1 + \cos\vartheta)^{-1}\sigma^2 \tag{4-14}$$

The steric hindrance is a measure of the thermodynamic flexibility of a macromolecule. It can be calculated from the known molecular data (N, l, and ϑ) if $\langle L^2 \rangle_0$ is known. The mean end-to-end chain distance can be obtained from light scattering (Section 9.5), small-angle X-ray scattering, and neutron scattering measurements (see Section 9.6) or from viscosity data (Section 9.9.6).

When the mean chain end-to-end distance of a coiled macromolecule is only determined by the quantities N, l, ϑ, and σ as given by equation (4-14), the coil is said to be in its unperturbed state.

4.4.2.4. *Radius of Gyration*

The end-to-end chain distance is a readily conceivable but not a directly measurable quantity. Also, it loses all physical significance with branched chains, which possess more than two chain ends. A related quantity is the radius of gyration. It is directly measurable as the mean square radius of gyration, which is defined as the second moment of the mass distribution around the center of mass. The root mean square (rms) radius of gyration, often simply called radius of gyration, is thus

$$\langle R_G^2 \rangle^{1/2} = \left(\frac{\sum_i m_i R_i^2}{\sum_i m_i} \right)^{1/2} \tag{4-15}$$

A definite relationship exists between the end-to-end chain distance and the radius of gyration for linear (nonbranched) chains (with or without fixed valence angles or free rotation). The relationship is derived in the appendix of this chapter for the segment chain. It can be seen from equations (4-9), (4-10), and (4-13) that on transferring from the segment-chain model to either of the valence-angle-chain models, the end-to-end chain distance is increased for $\vartheta > 90°$. The radius of gyration likewise increases. Mathematical analysis shows that the same relationship between the chain end-to-end distance and the radius of gyration exists for all three chain models in the limiting case of infinitely high molecular weight:

$$\langle L^2 \rangle_0 = 6 \langle R_G^2 \rangle_0 \tag{4-16}$$

4.4.2.5. The Shape of Unperturbed Coils

The instantaneous shape of a coil-like chain molecule can be quantitatively characterized in the following fashion: The center of gravity of the molecule is placed at the origin of a Cartesian coordinate system. The molecule is oriented in this coordinate system so that the principal axes of inertia are along the coordinate axes. The vector radius $\mathbf{R}_i$ of each mass point of the chain molecule (see Figure 4-11) can be resolved into the three orthogonal components $(\mathbf{R}_i)_1$, $(\mathbf{R}_i)_2$, and $(\mathbf{R}_i)_3$, where

$$(\mathbf{R}_i)_1^2 + (\mathbf{R}_i)_2^2 + (\mathbf{R}_i)_3^2 = \mathbf{R}_i^2 \tag{4-17}$$

Likewise, the radius of gyration can also be resolved into three components, namely $R_{G,1}^2$, $R_{G,2}^2$, and $R_{G,3}^2$. Since these three components stand in special relationship to the three principal axes of inertia of the molecule, they are often called the main components of the radius of gyration of the chain.

For a fully spherically symmetric coil molecule, we have

$$R_{G,1}^2 = R_{G,2}^2 = R_{G,3}^2 = \tfrac{1}{3} R_G^2 \tag{4-18}$$

Calculations have shown, however, that the squares of the principal components are not equal in magnitude, but have about the relationship 11.8:2.7:1 to each other. The instantaneous shape of a coil molecule is not spherical; it is more in the form of a deformed ellipsoid. However, increased branching of the main chain will cause the molecule to become more spherical.

4.4.3. Steric Hindrance Parameter and Constitution

The steric hindrance parameter σ is a constant only in the case of apolar polymers in apolar solvents. A distinct dependence of the hindrance parameter on the type of solvent can be observed, however, for polar polymers and/or polar solvent combinations (see Table 4-6). Such effects are to be expected because of changes brought about in the *trans/gauche* ratios of conformers in the chain.

The hindrance parameter increases with the size of the substituents

Table 4-6. Hindrance Parameter of Various "Atactic" Polymers

Polymer	Solvent	Temperature, °C	σ
Poly(ethylene)	Tetralin	100	1.63
Poly(propylene)	Cyclohexanone	92	1.8
Poly(isobutylene)	Benzene	24	1.93
Poly(styrene)	Cyclohexanone	34	2.3
Poly(1-vinyl naphthalene)	Decalin/toluene	—	3.2
Poly(methyl methacrylate)	Benzene	30	2.1
	Toluene	30	2.12
	Benzene/cyclohexane	25	2.14
	Acetone	25	1.86
	Butanone	25	1.89
	Butyl chloride	25	1.87
Cellulose	Copper ethylene diamine	25	2.0
Hydroxyethyl cellulose	Methanol	25	1.9

(i.e., with increasing molecular weight M_u of the monomeric unit) for polymer–solvent pairs having approximately the same interaction. Table 4-7 shows this for the series polyethylene–poly(propylene)–poly(styrene)–poly (1-vinyl naphthalin), and Figure 4-13 shows it for the a series of poly-(methacrylic esters) and poly(itaconic esters).

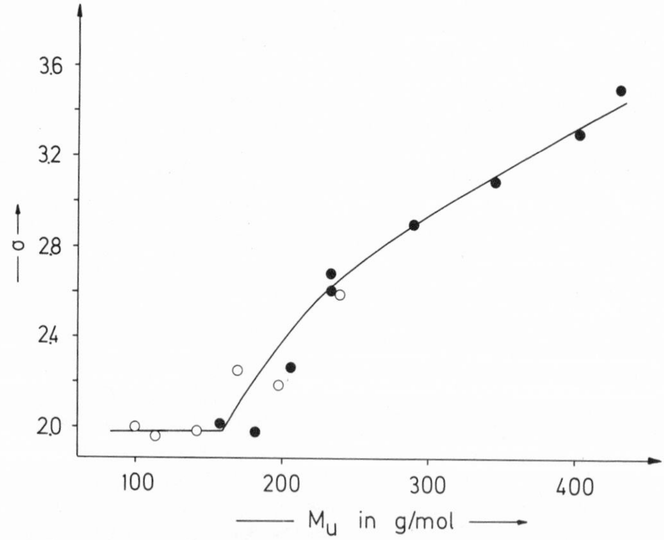

Figure 4-13. Dependence of the hindrance parameter σ on the formula molecular weight for a series of poly(methacrylic esters) $\text{+}CH_2\text{—}CCH_3(COOR)\text{+}_n$ (○), and poly(itaconic esters) $\text{+}CH_2\text{—}C(COOR)(CH_2COOR)\text{+}_n$ (●) in toluene solution at 25°C (after J. Veličkovič and S. Vasovič).

Cellulose and its derivatives have σ values of about 2, i.e., thermodynamically they are about as flexible as poly(isobutylene). Thus, cellulose chains are not extraordinarily stiff, although they are often assumed to be so on the basis of their high exponents in the intrinsic viscosity–molecular weight relationship (see Section 9.9.7). These high exponents are interpreted as arising from the particular (high) draining properties of the cellulose molecule.

4.4.4. *The Characteristic Ratio*

The number N of main-chain bonds, the bond length l, and the valence angle ϑ are assumed to be constants when calculating the hindrance parameter σ from the unperturbed mean square end-to-end chain distance. The assumption is correct for N and most probably also correct for the bond length since the bond energies of main-chain bonds are about 40–400 kJ/mol bond. The assumption of a constant valence angle is, however, critical. Deformation of a C—C—C valence angle by 5.6° only requires about 2 kJ/mol and a change of 10° only requires about 7 kJ/mol according to spectroscopic data and the heats of combustion of ring compounds.

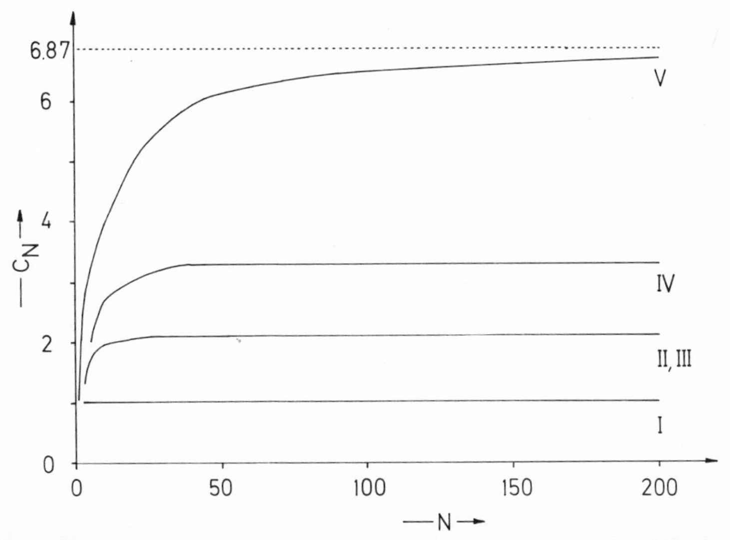

Figure 4-14. Variation of the characteristic ratio C_N with the number N of bonds in the chain molecule. (I) Segment chain; (II) valence angle chain with free rotation; (III) valence angle chain with hindered rotation and three rotational isomers of equal energy; (IV) valence angle chain with hindered rotation and a conformational energy of $E_g - E_t = 2090$ J/mol; and (V) valence angle chain with hindered rotation and neighboring group influence and $\Delta E_{g\pm} - \Delta E_{g\mp} = 8300$ J/mol (after P. J. Flory).

Since the conformational energies also lie in this range, the assumption of a constant valence angle can lead to problems.

Consequently, the valence angle component is often not separated from the hindrance parameter in calculations. Instead, a characteristic ratio C is defined as a measure of the expansion of a coil in the unperturbed state:

$$C \equiv \frac{\langle L^2 \rangle_0}{Nl^2} = (1 - \cos \vartheta)(1 + \cos \vartheta)^{-1} \sigma^2 \tag{4-17}$$

C increases asymptotically to a constant value with increasing number N of bonds (Figure 4-14).

4.4.5. Statistical Chain Element

Equation (4-13) can also be generalized as follows: In this equation, the chain end-to-end distance depends on the number of bonds, which is unaffected by constitution and configuration, and on quantities that are affected by them (bond length, valence angle, hindrance parameter, etc.). A given stiffening of the chain can be achieved, however, with a greater bond length or a higher valence angle, as well as by a larger hindrance parameter. Formally, all these factors can be included in one unit of several links, thus reducing the number of independent units requiring consideration. One can also write, therefore, in place of equation (4-9)

$$\langle L^2 \rangle_0 = N_s l_s^2 \tag{4-20}$$

where l_s is taken to be the statistical chain element, and there are N_s of these in the chain. The longer l_s, the stiffer will be the chain. l_s can therefore be used as a measure of flexibility in the same way as σ, but it has less physical significance than σ. However, since the calculation of σ is not entirely straightforward and that of l_s is free of these difficulties, l_s and σ can be considered to be of equal worth. Equation (4-20) corresponds to equation (4-9) for segment chains of unspecified segment length.

In this model (see Section 4.4.2.1) the contour length is, of necessity, given by the product of the length l_s and the number of statistical chain elements N_s,

$$l_{\text{cont}} = N_s l_s = Nl \tag{4-21}$$

so that equation (4-20) can also be written

$$\langle L^2 \rangle_0 = l_{\text{cont}} l_s = N_s l_s^2 \tag{4-22}$$

The length of the statistical chain element can thus be calculated from the contour length l_{cont} and the experimentally determined chain end-to-end distance.

4.4.6. Wormlike Chain

The angle between neighboring segments is not fixed in the random-flight model. If the bond length is taken as the segment, then the valence angle can be chosen as desired. The valence angle is, however, fixed in a real chain. Thus the segments following the first segment cannot occupy any desired position in space. The chain thus possesses persistence.

The rigidity resulting from this persistence can be described by a persistence length l_{pers}, which is the average projection of the end-to-end distance of an infinite chain in the direction of the first segment. It is defined as

$$l_{\text{pers}} \equiv \frac{l}{1 + \cos \vartheta} \tag{4-23}$$

The chain end-to-end distance of a finite chain with free rotation is, according to equation (A4-28) in the appendix to this chapter,

$$\langle L^2 \rangle_{0f} = Nl \left\{ l \left(\frac{1 - \cos \vartheta}{1 + \cos \vartheta} \right) + \frac{2l \cos \vartheta}{N} \left[\frac{1 - (-\cos \vartheta)^N}{(1 + \cos \vartheta)^2} \right] \right\} \tag{4.24}$$

Combining equations (4-21)–(4-24), we obtain

$$\langle L^2 \rangle_{0f} = l_{\text{cont}} l_{\text{pers}} (1 - \cos \vartheta) + 2 l_{\text{cont}} l_{\text{pers}} \frac{\cos \vartheta}{N} \left[\frac{1 - (-\cos \vartheta)^N}{1 + \cos \vartheta} \right] \tag{4-25}$$

A chain with an infinite number of segments of zero length and a valence angle approaching 180° is called a wormlike chain (note that the contour length remains constant). This limiting case cannot be derived directly from equation (4-25), because, although $\sigma \to \pi$ (and thus $\cos \vartheta \to -1$), N simultaneously tends to infinity. Thus, the second term in equation (4-25) is expressed in terms of the contour length and $(1 + \cos \vartheta)$ is given in terms of the persistent chain length:

$$\langle L^2 \rangle_{0f} = l_{\text{cont}} l_{\text{pers}} (1 - \cos \vartheta) + 2 l_{\text{pers}}^2 (\cos \vartheta) \left[1 - (-\cos \vartheta)^N \right] \tag{4-26}$$

Equation (4-26) contains no term including N except for the expression $(-\cos \vartheta)^N$. In the limiting case of $\vartheta \to \pi$, that is, $(1 - \cos \vartheta) \to 2$, equation (4-26) becomes

$$\langle L^2 \rangle_{0f} = 2 l_{\text{cont}} l_{\text{pers}} - 2 l_{\text{pers}}^2 + 2 l_{\text{pers}}^2 (-\cos \vartheta)^N \tag{4-27}$$

In the limiting case $(-\cos \vartheta)$ is only slightly smaller than 1, but is raised to the power of N. Transformation of the above equations can lead to a more easily handled expression. First, $\cos \vartheta$ is expressed in terms of equation (4-23). Then cross-multiplication leads to the introduction of the contour length:

$$\lim_{\substack{N\to\infty \\ \vartheta\to\pi}} (-\cos\vartheta)^N = \lim_{N\to\infty}\left(1-\frac{l}{l_{\text{pers}}}\right)^N = \lim_{N\to\infty}\left(1-\frac{l_{\text{cont}}}{Nl_{\text{pers}}}\right)^N \tag{4-28}$$

and since

$$\lim_{x\to\infty}\left(1-\frac{1}{x}\right)^x = e^{-1} \tag{4-29}$$

then equation (4-28) can be rearranged and solved like equation (4-26):

$$\lim_{N\to\infty}\left(1-\frac{l_{\text{cont}}}{Nl_{\text{pers}}}\right)^N = \left[\lim_{N\to\infty}\left(1-\frac{l_{\text{cont}}}{Nl_{\text{pers}}}\right)^{Nl_{\text{pers}}/l_{\text{cont}}}\right]^{l_{\text{cont}}/l_{\text{pers}}}$$

$$= \exp\left(\frac{-l_{\text{cont}}}{l_{\text{pers}}}\right) \tag{4-30}$$

Inserting equation (4-30) into equation (4-27), we obtain

$$\langle L^2\rangle_{0f} = 2l_{\text{pers}}^2[y-1+\exp(-y)]; \qquad y = l_{\text{cont}}/l_{\text{pers}} \tag{4-31}$$

For the radius of gyration, an analogous derivation gives

$$\langle R_G^2\rangle_{0f} = l_{\text{pers}}^2\left[\frac{2}{y^2}[y-1+\exp(-y)]-\left(1+\frac{y}{3}\right)\right] \tag{4-32}$$

For flexible chains, the contour length is much larger than the persistence length. That is, y becomes much larger than one and the expression $\exp(-y)$ tends to zero. Thus, equation (4-31) becomes

$$\lim_{y\to\infty}\langle L^2\rangle_{0f} = 2l_{\text{pers}}l_{\text{cont}} \tag{4-33}$$

and equation (4-32) gives

$$\lim_{y\to\infty}\langle R_G^2\rangle_{0f} = \lim_{y\to\infty}\frac{l_{\text{pers}}l_{\text{cont}}}{3}\left(1-\frac{3}{y}+\frac{6}{y^2}-\frac{6}{y^3}\right) = \frac{l_{\text{pers}}l_{\text{cont}}}{3} \tag{4-34}$$

A comparison of equations (4-22) and (4-33) shows that the persistence length is exactly half the segment length l_s. In this case, the radius of gyration of a wormlike chain bears the same relationship to the chain end-to-end distance as does the radius of gyration of a valence chain with or without free rotation [see equation (4-33), (4-34), and (4-15)].

On the other hand, for very rigid chains $y\to 0$, and $\exp(-y)$ can then be developed into a series $1-y+(y^2/2!)-(y^3/3!)+\cdots$, and the relationship obtained for the chain end-to-end distance and the radius of gyration is

$$\langle L^2 \rangle_{0f} = l_{\text{cont}}^2 \left(1 - \frac{y}{3} + \frac{y^2}{12} - \cdots \right) = l_{\text{cont}}^2 \tag{4-35}$$

$$\langle R_G^2 \rangle_{0f} = \frac{l_{\text{cont}}^2}{12} \left(1 - \frac{y}{5} + \frac{y^2}{30} - \cdots \right) = \frac{l_{\text{cont}}^2}{12} \tag{4-36}$$

Thus, a very rigid chain behaves like a rod, since the end-to-end chain distance equals the contour length and the radius of gyration is smaller than the chain end-to-end distance by a factor of $(12)^{1/2}$.

The persistence length model therefore describes the whole spectrum from the more rodlike oligomers (small y) to the well-developed coils (large y). However, the model ignores the finite thickness of the chain, and so it only holds strictly for unperturbed coils. The error due to finite thickness may be neglected when the persistence length is much greater than the chain thickness.

4.5. Excluded Volume

4.5.1. Rigid Particles

Two molecules cannot occupy the same space; each excludes the other from its space ("excluded volume"). The excluded volume of rigid macromolecules is relatively easily calculated, since only the excluded volume between molecules need be considered.

The volume of an unsolvated sphere is given by

$$V_{\text{sphere}} = (4\pi/3)\, r_{\text{sphere}}^3 = M_{\text{sphere}} v_{\text{sphere}} / N_{\text{L}} \tag{4-37}$$

where r is the radius, M the molecular weight, and v the specific volume of the sphere. The closest distance that one sphere can approach to another is $2r_{\text{sphere}}$. The excluded volume u_{sphere} is therefore

$$u_{\text{sphere}} = (4\pi/3)(2r_{\text{sphere}})^3 = 8M_{\text{sphere}} v_{\text{sphere}} / N_{\text{L}} \tag{4-38}$$

and is therefore eight times the volume of the sphere.

In calculating the excluded volume of rods, these are assumed to be cylindrical with the volume

$$V_{\text{rod}} = \pi r_{\text{rod}}^2 l_{\text{rod}} \tag{4-39}$$

with the length l_{rod} and the radius r_{rod}. The problem here is the calculation of the mutual orientation of the rigid rods in space, since not all orientations

Table 4-7. Excluded Volume u and Mean Square Radius of Gyration $\langle R_G^2 \rangle$ as a Function of Characteristic Dimensions of Various Particle Shapes (V = volume)

Particle	u	$\langle R_G^2 \rangle$
Infinitely thin spherical shell of radius r	$8V$	r^2
Spherical shell of interior radius $r_i = Cr_e$ and exterior radius r_e	$8V$	$\frac{3}{5}\left(C^2 + \frac{C+1}{C^2+C+1}\right) r_e^2$
Sphere of radius r	$8V$	$\frac{3}{5}r^2$
Very thin disk of radius r and thickness h	$\pi(r/h)V$	$0.5r^2$
Rotational ellipsoid		
a. Prolate with length l and radius r ($l \gg r$)	$(3/8)\pi(l/r)V$	$\frac{1}{5}(l^2 + 2r^2)$
b. Oblate with thickness h and radius r ($r \gg h$)	$(3/2)\pi(r/h)V$	$\frac{1}{5}(r^2 + 2h^2)$
Rod of length l and diameter $2r$	$(l/r)V$	$\frac{1}{12}l^2 + r^2$
Coil in the theta state with the chain end-to-end distance $\langle L^2 \rangle_0^{0.5}$	[a]	$\frac{1}{6}\langle L^2 \rangle_0$
Coil with the hydrodynamically equivalent radius r_h	[a]	$[8/(3\pi^{0.5})]^2 \langle r_h^2 \rangle$
Coil with the relationship $\langle L^2 \rangle$ = constant $\times (M^{1+\varepsilon})$	[a]	$\frac{1}{6}\langle L^2 \rangle (1 + \frac{5}{6}\varepsilon + \frac{1}{6}\varepsilon^2)^{-1}$

[a] See Section 4.6.2.

are allowed with rods separated by a distance of less than l_{rod}. The results of the calculation lead to a substitution of the factor 8 in the excluded volume of a sphere by the factor l_{rod}/r_{rod}:

$$u_{rod} = \frac{l_{rod}}{r_{rod}} \frac{M_{rod} v_{rod}}{N_L} \tag{4-40}$$

The excluded volumes of other rigid particles can be analogously calculated. Excluded volumes, radii of gyration, and characteristic ratios of various rigid particles are given in Table 4-7.

4.5.2. Unbranched Macromolecules

4.5.2.1. Basic Principles

Real macromolecules have an external and an internal excluded volume. The external excluded volume results from the volume excluded by two different molecules and disappears at infinite dilution.

Conversely, the internal excluded volume results from the finite thickness of the molecular chain since one part of the chain cannot simultaneously reside in the same place already occupied by another part of the chain. This positively excluded volume expands the coil. A negative excluded volume can result when attraction forces exist between two parts of the chain. In such a case, the volume of two parts of the chain in contact is smaller than the sum of their individual volumes. Neither the positive nor the negative excluded volume disappears at infinite dilution.

In special cases, positive and negative excluded volumes can compensate each other. The coil then behaves as if it consisted of an infinitely thin chain. Consequently, it adopts the unperturbed state and the dimensions in this state are the unperturbed dimensions.

Thus the mean chain end-to-end distance of an unperturbed coil approximates that of a valence chain with restricted rotation [equation (4-13)]. The restricted rotation results from the interaction forces between substituents on neighboring chain atoms. These forces are therefore referred to as "short range." The interactions resulting from the excluded volume effect are called "long range" since they occur between components of the chain separated from each other by many chain components. The terms "short range" and "long range" apply to the separation along the chain of the chain components involved, not to the effective range of the forces themselves.

Both repulsive and attractive forces exist in real random coils. The effect on the radius of gyration of the coil expansion caused by the interaction of these two opposing forces can be formally described by the use of an expansion factor α_R:

$$\langle R_G^2 \rangle = \alpha_R^2 \langle R_G^2 \rangle_0 \tag{4-41}$$

The expansion factor is unity for an unperturbed coil. An expansion factor based on the end-to-end chain distance can similarly be defined.

The larger α_R, the greater is the expansion and the "better" the solvent is for the polymer in question. With increasing coil expansion, the spatial requirements of the coil are greater, and the viscosity of the polymer solution is higher. (In paint technology, however, the term "good solvent" refers to one that gives a polymer solution of very low viscosity.)

In equation (4-41), the expansion factor gives directly the average linear expansion of the coil. The expansion factor calculated in this way is, of course, fictitious, since the coil is not spherical in shape, and will probably have various expansion factors for various directions in space. The existence of unequal expansion factors over all directions in space means that the distribution of molecular segments of real coils is no longer the same as for ideal coils.

4.5.2.2. Cluster Integrals

A potential function $\psi(\mathbf{r})$ between two segments $\mathbf{r}$ apart must be assumed before the excluded volume of a coiled molecule can be calculated. The segments in question can belong to the same or to different molecules. Since every molecule possesses many segments, the expressions involved are very complex, and the derivations can only be sketched here.

The excluded volume of a segment u_{seg} is given by what is called the cluster integral,

$$u_{seg} = 4\pi \int_0^\infty \left[1 - \exp\left(- \frac{\psi(\mathbf{r})}{kT} \right) \right] r^2 \, d\mathbf{r} \tag{4-42}$$

All relevant theories agree that u_{seg} can be directly related to the expansion factor α_R. To simplify the mathematics, a parameter z is defined:

$$z \equiv (4\pi)^{-3/2} \frac{u_{seg}}{M_{seg}^2} \frac{M^2}{\langle R_G^2 \rangle_0^{3/2}} \tag{4-43}$$

The term u_{seg}/M_{seg}^2 is a constant, independent of the molecular weight M of the chain, because it describes the excluded volume caused by a pair of segments.

When the following assumptions are made, an expression for the expansion factor can be derived from the cluster integral with the aid of z:

1. The probability distribution of bond vectors follows a Gaussian function.
2. The potential for interaction between segments is additive.
3. The pair potential is given by the expression

$$\exp\left(- \frac{\psi(\mathbf{r})}{kT} \right) = 1 - u_{seg}\delta(\mathbf{r}) \approx \exp\left[-u_{seg}\delta(\mathbf{r}) \right]$$

where $\mathbf{r}$ is the vector distance between two segments and $\delta(\mathbf{r})$ is the three-dimensional delta function.

Using these assumptions, we obtain for the relationship between the expansion factor α_R (based on the radius of gyration) and z

$$\alpha_R^2 = 1 + (134/105)\, z - 2.082 z^2 + \cdots \tag{4-44}$$

and correspondingly, for the expansion factor α_L (based on the chain end-to-end distance):

$$\alpha_L^2 = 1 + (4/3)\, z - 2.075 z^2 + 6.459 z^3 - \cdots \tag{4-45}$$

These series expansions are exact for coils with pairwise additivity of interactions. But the series only converge very slowly and so are only

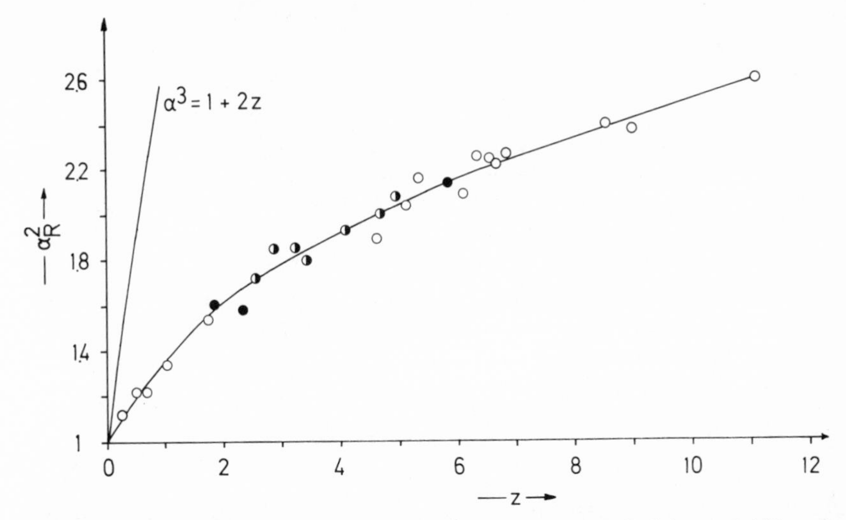

Figure 4-15. Variation of the expansion factor α_R^2 with z (see text). Poly(styrene) (○), poly-(vinyl acetate) (◐), and poly(ethylene) (●) in various solvents (after G. C. Berry and E. F. Casassa).

applicable for $z < 0.1$ (for α_R) or $z < 0.15$ (for α_L). The term α_R^2 should in any case only depend on z, that is, equation (4-45) should hold for all polymer–solvent–temperature systems (Figure 4-15).

A closed expression for the function $\alpha_R = f(z)$ has been sought without success and it is questionable if such a generally valid expression for $\alpha_R = f(z)$ exists. An expression often used is

$$\alpha_R^3 = 1 + 2z \tag{4-46}$$

This expression indeed shows the initial trend of the function correctly for $\alpha_R \approx 1$, but leads to strong deviations from experimental data for large expansion factors (Figure 4-15).

4.5.2.3. *Molecular-Weight Dependence of Coil Dimensions*

The excluded volume of a coil is characterized by the expansion factor α. The term α depends on z [equation (4-44)] and z depends on the molecular weight [equation (4-43)]. Thus with equation (4-44), α becomes a complicated function of the molecular weight.

If equation (4-14) is used for a chain of length l_e having N_e chain units, then with $N_e = M/M_e$ and equation (4-14), we have

$$\langle R_G^2 \rangle_0 = (l_e^2/6M_e)(1 - \cos\vartheta)(1 + \cos\vartheta)^{-1}\sigma^2 M = K_e M \tag{4-47}$$

Thus, with equations (4-47) and (4-41), for any desired good solvent

$$\langle R_G^2 \rangle = K_e \alpha_R^2 M \tag{4-48}$$

or, also empirically, since $\alpha_R = f(M)$, on including the expansion factor into an exponent of the molecular weight,

$$\langle R_G^2 \rangle = K_R M^{1+\varepsilon} = K_R M^{a_R} \qquad (\varepsilon \geq 0) \tag{4-49}$$

Such relationships between the radius of gyration and the molecular weight are often valid over an astonishingly wide molecular weight range (Figure 4-16).

Equation (4-49) describes the combined effects of long- and short-range interactions. These effects can be separated in the following manner: Combining equations (4-41), (4-43), and (4-46) and rearranging, we obtain

$$\left(\frac{\langle R_G^2 \rangle}{M}\right)^{3/2} = \left(\frac{\langle R_G^2 \rangle_0}{M}\right)^{3/2} + 2(4\pi)^{-3/2}\left(\frac{u_{\text{seg}}}{M_{\text{seg}}^2}\right) M^{1/2} \tag{4-50}$$

The first term on the right-hand side of equation (4-50) is a characteristic constant for the unperturbed coil, as can be seen on combining equations (4-14) and (4-16) with $N = M/M_u$,

$$\frac{\langle R_G^2 \rangle_0}{M} = \frac{l^2}{6M_u}(1 - \cos \vartheta)(1 + \cos \vartheta)^{-1} \sigma^2 = \text{constant} \tag{4-51}$$

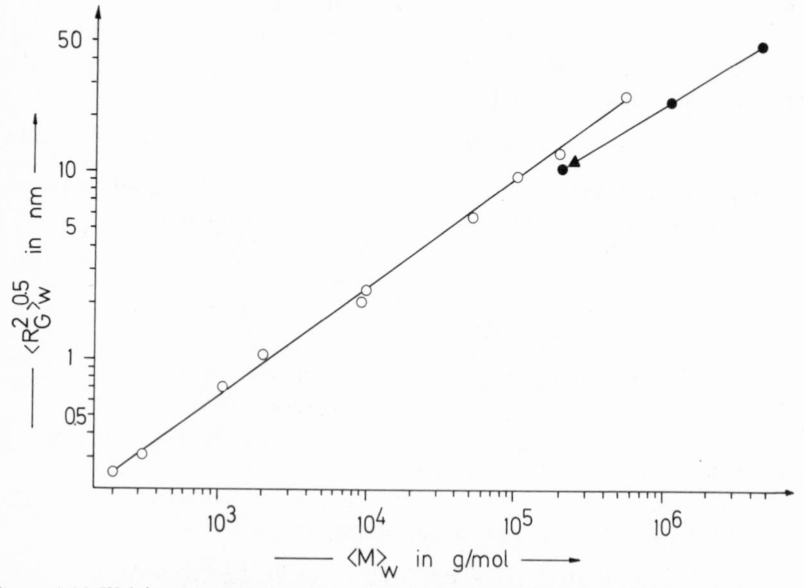

Figure 4-16. Weight-average radii of gyration of poly(methyl methacrylates) in acetone (○), butylchloride (●), and in the solid state (▲), as a function of the weight-average molecular weights. Measurements at 20 and 35.4°C (butylchloride). (After R. G. Kirste and W. Wunderlich.)

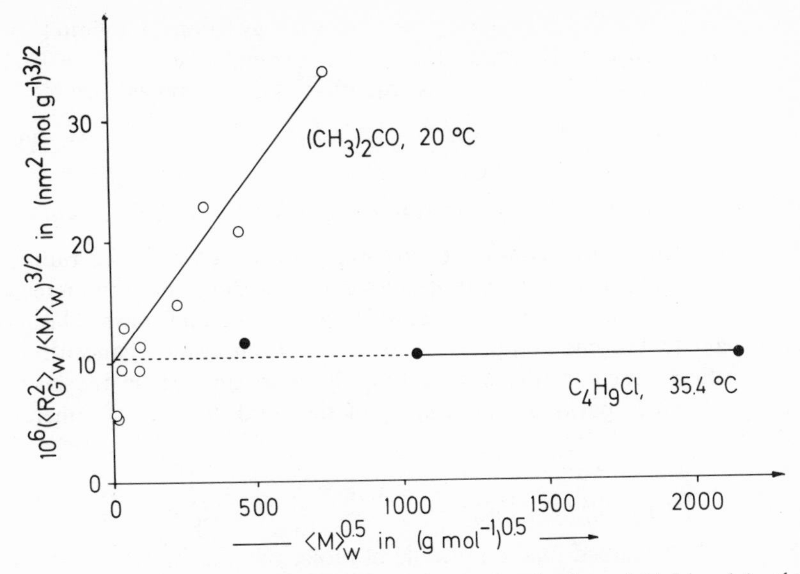

Figure 4-17. Reduced radii of gyration as a function of molecular weight for poly(methyl methacrylates) in acetone. Data from Figure 4-16.

Since $\langle R_G^2\rangle_0/M$ is a function of the hindrance parameter σ, it is a measure of the short-range interactions. Conversely, the slope contains the constant u_{seg}/M_{seg}^2, and so is a measure of the long-range interactions. Thus, if one plots $(\langle R_G^2\rangle/M)^{3/2}$ against $M^{1/2}$, long-range and short-range interactions can be separated from each other. The relationship is often valid over relatively wide molecular weight ranges (Figure 4-17), but should fail at high molecular weights, since the basic function $\alpha_R^3 = 1 + 2z$ then deviates strongly from experimental data.

It has been shown experimentally that iso- and syndiotactic polymers in good solvents (α high) have roughly the same dimensions, while in theta solvents ($\alpha = 0$) differences of up to 20% can occur. The effect is understandable since in theta solvents the dimensional properties are dominated by short-range interactions, i.e., microconformations. The microconformations, however, depend very much on the configurations (Section 4.2). In good solvents, on the other hand, long-range interactions, which are not affected by the microconformation, dominate.

A more complicated relationship for coil molecules in good solvents exists between the chain end-to-end distance and the radius of gyration for valence chains with restricted rotation:

$$\langle L^2\rangle = 6(1 + \tfrac{5}{6}\varepsilon + \tfrac{1}{6}\varepsilon^2)\langle R_G^2\rangle \qquad (4\text{-}52)$$

$\varepsilon = 0$ for coils with unperturbed dimensions and equation (4-52) reduces to equation (4-16). Conversely, the radius of gyration of infinitely thin rods is directly proportional to the molecular weight, i.e., $\langle R_G^2 \rangle = K_R M^2$. In this case, $\varepsilon = 1$. Thus, for rods, equation (4-52) becomes

$$\langle L^2 \rangle = 12 \langle R_G^2 \rangle \tag{4-53}$$

4.5.2.4. *Concentration and Temperature Dependence of Coil Dimensions*

All of the above considerations were concerned with the radius of gyration of unbranched chain molecules at infinite dilution. With increasing concentration the coils fill the available space more and more. The loose coils tend to become compressed above a certain critical concentration. This critical concentration can be roughly approximated on the basis of hexagonal close packing (about 75% of the total volume) of spheres of radius r:

$$c_{crit} = \frac{9}{16\pi(5/3)^{3/2}} \frac{M}{N_L \langle R_G^2 \rangle^{3/2}} = 1.38 \times 10^{-25} \frac{M}{\langle R_G^2 \rangle^{3/2}} \frac{\text{g}}{\text{cm}^3} \tag{4-54}$$

A spherically shaped macromolecule of molecular weight 1.3×10^6 g/mol and radius of gyration of $(\langle R_G^2 \rangle)^{1/2} = 100$ nm should then have a critical concentration of 1.8×10^{-4} g/cm³. Above this concentration, the coil will tend to be compressed and the radius of gyration will decrease (Figure 4-18). At still higher concentrations, the radius of gyration begins to increase again, which can indicate association.

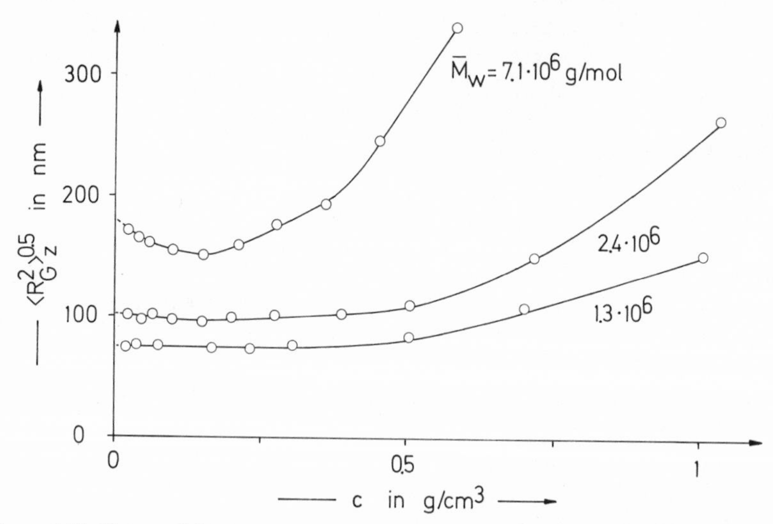

Figure 4-18. Change of the z average root mean square radius of gyration $\langle \overline{R_G^2} \rangle_z^{0.5}$ with concentration for poly(styrene) in benzene at 20°C (after H. Dautzenberg). The abscissa has to be multiplied by 0.01.

A generally valid relationship for the temperature dependence of the excluded volume has yet to be discovered. In the vicinity of the theta temperature (for definition, see Section 6.4.2), the following is a good approximation:

$$u_{\text{seg}} = \text{constant} \times \left(1 - \frac{\theta}{T}\right) \tag{4-55}$$

4.5.3. Branched Macromolecules

Branched macromolecules have a higher average segment density than unbranched macromolecules of the same molecular weight, and have a lower coil volume. This is easily seen by comparing a star-shaped branched molecule with a linear one. The influence of the branching on the dimensions at the theta temperature can be expressed by a g factor

$$\frac{\langle R^2 \rangle_{\text{branched}}^{\theta}}{\langle R^2 \rangle_{\text{linear}}^{\theta}} = g \tag{4-56}$$

The magnitude of g depends on the type of branching (star, comb) and on its regularity. The dimensions therefore decrease from linear macromolecules through regular comb branched, irregular comb branched, regular random branched, and irregular random branched to star-shaped branched molecules. The quantitative classification of the dimensions according to type, density, length, and regularity of the branches has not yet found a complete theoretical solution, since the effect also depends on the quality of the solvent. However, it does have some practical significance, since the determination of the dimensions is often the only possible way of detecting long-chain branching in polymers.

4.6. Compact Molecules

4.6.1. Helices

The helix has a high degree of internal order in comparison with the coil molecule. The chain structure of the helix is unambiguously defined in one direction: The molecule assumes the overall shape of a rod.

Each rod, however, possesses a certain flexibility per monomeric unit. The flexibility per *molecule* must consequently increase with increasing degree of polymerization. A macroscopic illustration of this effect is the flexibility of steel wires of constant thickness but with varying length. Even a perfect helix would thus exist as a random coil at sufficiently high molecular weight (see also Figure 4-10).

Thus, macromolecules in the form of a helix can often be well described by the wormlike model: At low molecular weights, the molecule is more rodlike, and at higher molecular weights the molecule is more coillike (see also Section 4.4.6).

4.6.2. Ellipsoids and Spheres

Proteins are copolymers with various α-amino acid units of the form ⟮NH—CHR—CO⟯. Certain sequences of these units adopt a helical form, whereas others do not. The tertiary structure of proteins (see Section 30.1 and Figure 4-19 for the tertiary structure of myoglobin) consists of spatial arrangement of helical and nonhelical elements. The tertiary structure essentially determines the overall conformation of the molecule. A molecule such as myoglobin has the external appearance of a sphere or an ellipsoid, which probably results from the fact that the hydrophilic groups try to take up positions on the "surface," and the hydrophobic groups try to reside within the sphere. For example, only two amino acid units having hydrophilic substituents reside within the myoglobin molecule.

Associations of two or more protein molecules lead to the so-called quaternary structures. At infinite dilution and with suitable solvents, quaternary structures dissociate into subunits. The thermodynamic stability of quaternary structures is often so high that no noticeable dissociation

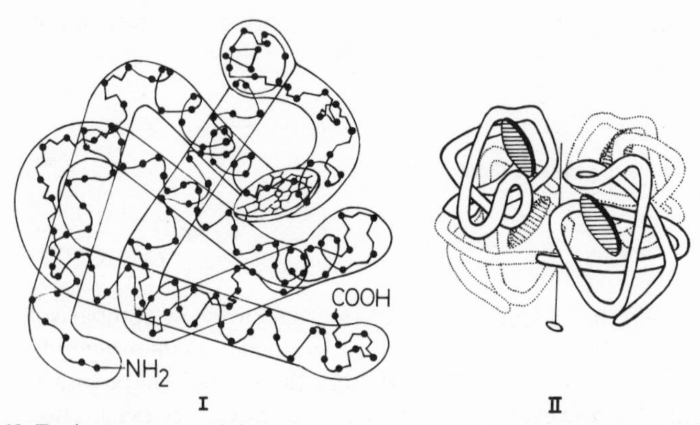

Figure 4-19. Tertiary structure of the protein myoglobin (I), and quaternary structure of the protein hemoglobin (II). The dots indicate amino acid residues. Hemoglobin consists of four myoglobin-type subunits. The heme planes are shown as cross-hatched areas (after M. F. Perutz). Myoglobin is shown on a larger scale.

occurs in measurable concentration regions. Consequently, quaternary structures appear to behave as molecules and not as associates.

Quaternary structure can have spherical or ellipsoidal external shape. Figure 4-18 shows the quaternary structure of the protein hemoglobin.

If, for example, the temperature of protein solutions is raised, then macroscopically perceptible changes often begin to occur in the solution. Solutions become, for example, turbid, and the precipitated products no longer possess, or no longer completely possess, the same biological activity. The process consists of two parts. First, through the action of heat, the conformation is changed, i.e., a "melting" of the helix usually results. This denaturing in the limited sense (thermal denaturing) is followed by a change in the particle size or particle weight, i.e., the thermal aggregation. In technology, the two effects are often not distinguished from one another, and the total effect is termed a denaturization. In addition to heat, denaturization can also be caused by chemical reagents, radiation, pressure, or shear forces. It is common to all denaturizations that a highly organized macromolecule changes from the ordered into a less ordered state.

4.7. Optical Activity

4.7.1. Basic Principles

If a linearly polarized electromagnetic wave meets a stereoisomeric center such as an asymmetric carbon atom, the plane of the polarized light is rotated. Since linearly polarized waves can be considered as the superposition of two circularly polarized waves of opposite rotation, the asymmetric electron configuration in the immediate vicinity of a stereoisomeric center causes the two propagation velocities of left-handed and right-handed polarized light to differ, hence resulting in a rotation of the plane of the polarized light.

The rotation of the plane of polarized light is measured as an optical rotation α. What is termed the "specific" optical rotation $[\alpha]$ in organic chemistry is a function of the mass fraction w_2 of the solute, the density of the solution ρ, and the length l of the sample cell, as well as the optical rotation:

$$[\alpha] = \frac{\alpha}{l w_2 \rho} \tag{4-57}$$

Traditionally, α is measured in degrees, l in dm, and ρ in g/cm^3.

The "molar" optical rotation $[\Phi]$,

$$[\Phi] = 10^{-2} [\alpha] M_u \tag{4-58}$$

relates the "specific" rotation to the formula molecular weight M_u of a monomeric unit or the molecular weight of low-molecular-weight compounds.

The "effective" molar rotation is also sometimes used. In this case, the refractive index is taken into account with a factor $(n^2 + 2)/3$:

$$[\Phi]_{\text{eff}} = \frac{3}{n^2 + 2}[\Phi] \tag{4-59}$$

$[\alpha]$, $[\Phi]_{\text{eff}}$, and $[\Phi]$ also depend on the temperature and the wavelength of the light used and also often on the concentration.

Generally, the wavelength dependence can be adequately represented by one of the following empirical equations:

$$[\Phi] = a_0\left(\frac{\lambda_0^2}{\lambda^2 - \lambda_0^2}\right) \quad \text{(one-term Drude equation)} \tag{4-60}$$

$$[\Phi] = a_0\left(\frac{\lambda_0^2}{\lambda^2 - \lambda_0^2}\right) + b_0\left(\frac{\lambda_0^2}{\lambda^2 - \lambda_0^2}\right)^2 \quad \text{(Moffit–Yang equation)} \tag{4-61}$$

a_0, b_0, and λ_0 are system-specific constants; λ_0 gives the wavelength of the subsequent absorption maximum. The dependence of optical rotation or quantities dependent on optical rotation on the wavelength is called optical rotatory dispersion (ORD). ORD derives from the differences in refractive index of right-hand and left-hand rotating components of polarized light.

Generally, the Drude equation describes the optical activity of coil molecules; the Moffit–Yang equation is more suitable for helices.

The left-hand and right-hand rotating components of polarized light are absorbed to different extents by optically active compounds. The dependence of the difference in absorption of left- and right-hand rotating components of polarized light on wavelength is called circular dichroism (CD).

Complex behavior is observed (Cotton effect) when measurements of the optical rotatory dispersion are carried out near an absorption band. On one side of the band it passes through a minimum (trough), on the other side a maximum (peak), and at the inflection point, the optical activity is zero. The Cotton effect is termed positive when the peak occurs at a higher wavelength than the trough. The inflection point of the curve $[\alpha] = f(\lambda)$ occurs at $[\alpha] = 0$ for nonoverlapping Cotton effects; it corresponds to the maximum in the UV absorption spectrum.

A Cotton effect always occurs when an absorbing group resides in an asymmetric environment. One component of the circularly polarized light is then absorbed more strongly than the other. The more weakly absorbed component has a greater velocity, and therefore a lower refractive index on the lower frequency side of the band. Since the Cotton effect is caused by the asymmetric environment of an absorbing group, its magnitude depends strongly on the helical content of the molecule.

4.7.2. *Structural Effects*

4.7.2.1. *General Considerations*

The molar rotation [Φ] caused by a given asymmetric carbon atom in compounds of the general type $R—{}^*CH(CH_3)—(CH_2)_y—CH_3$ is only slightly influenced by distant neighboring groups (Table 4-8). The measurable optical rotation depends on the sensitivity of the polarimeter and on any special experimental conditions used. For example, L-malic acid rotates to the left in dilute aqueous solution and to the right in concentrated solution. The optical rotation is zero at a certain concentration in water, although L-malic acid is chiral. All optically active systems are therefore chiral. Whether a chiral system is optically active or not depends on the conditions.

Optically active side groups, the configuration of the main-chain atoms themselves, as well as the conformation of the chain can all contribute to the optical activity of a polymer. The optical activity of the side groups corresponds exactly to that of low-molecular-weight compounds, so long

Table 4-8. Molar Optical Rotation $[\Phi]_{25}^{D}$ of Various Low-Molecular-Weight Compounds
$R—{}^*CH(CH_3)—(CH_2)_y—CH_3$
in the Liquid State at 589 nm and 25°C

	$[\Phi]_{25}^{D}$, 10^{-2} deg dm^{-1} cm^3 mol^{-1}				
R	$y = 1$	$y = 2$	$y = 3$	$y = 4$	$y = \infty$[a]
$(CH_2)_2H$	0	10	11.4	12.5	16.0
$(CH_2)_3H$	−10	0	1.5	2.4	6.0
$(CH_2)_4H$	−11.4	−1.7	0	0.8	5.0
$(CH_2)_5H$	−12.5	−2.4	−0.8	0	4.0
$(CH_2)_2Br$	−38.8	−21.3	−16.8	−14.7	−7.0
$(CH_2)_3Br$	−21.9	−14.5	−8.3	−6.2	−1.0
$(CH_2)_4Br$	−14.9	−7.8	−5.3	−4.0	−0.5
$(CH_2)_2OH$	−9.0	2.1	4.0	6.1	10.5
$(CH_2)_3OH$	−11.9	0	0.7	2.6	7.0
$(CH_2)_4OH$	−12.0	−1.7	0	0.8	5.5

[a]The data given for $y = \infty$ were obtained by extrapolating $[\Phi]_{25}^{D} = f(y^{-1})$ to $y^{-1} = 0$.

as the side groups are sufficiently far apart. Thus, only the influence of the chain configuration and conformation on the optical activity will be discussed here.

Stereoregular polymers of infinite molecular weight, dissymmetric main-chain atoms, and dissymmetric substituents show no optical activity resulting from the configuration, because of intramolecular compensation. Thus, isotactic poly(propylene) is not optically active. Stereoregular polymers with asymmetric main-chain atoms, such as $\text{+NH—CH(CH}_3\text{)—CO+}_n$, poly(L-alanine), and $\text{+O—CH(CH}_3\text{)—CH}_2\text{+}_n$, poly(L-propylene oxide), are optically active because of the configuration itself.

An additional and frequently dominating contribution to optical activity can result from the conformation in helix-forming polymers. A helix is either left- or right-handed. In the immediate vicinity of an asymmetry center, therefore, an additional asymmetry, and consequently an additional optical activity, is produced. For this reason, a helix should always be optically active. However, since a solution consists of many molecules, it is only possible for the helices to make a contribution to the optical activity if the helical rotation is the same in all the molecules, or if there is a larger proportion of one helix present.

Isotactic polymers possessing dissymmetric monomeric units, i.e., the poly(α-olefins), $\text{+CH}_2\text{—CHR+}_n$, with the dissymmetric substituent R, generally occur in the form of helices. The right- and left-handed helices of these polymers possess the same energy and thus are equally probable. The two helical conformers can only be separated when the potential barrier is very great. Since this is generally not the case, such polymers are optically inactive.

With polymers possessing an optically active main-chain atom, i.e., poly(α-amino acids), +NH—CHR—CO+_n, or with polymers having an asymmetric substituent, e.g., it-poly(3-methyl pentene-1), $\text{+CH}_2\text{—CHR+}_n$, where $R = CH(CH_3)(C_2H_5)$, one helix form is more energetically stable than the other. Such compounds are therefore optically more active than can be deduced from the optical activity of the monomeric units alone.

Since the contribution of the end groups to the molecular properties decreases with increasing molecular weight, end groups only influence the optical activity at low degrees of polymerization. Figure 4-20 shows that the specific rotation $[\alpha]$ of the polymer homologous series of poly-(γ-methyl-L-glutamate) in the hydrogen bond-destroying solvent dichloroacetic acid continuously decreases with rising degree of polymerization X, as the influence of the end groups on the optical activity becomes less and less. In the helicogenic (helix-producing) solvent dioxane, the optical activity falls from the monomer through the dimer, trimer, and tetramer, to rise sharply again in pentamers because of helix formation. At higher degrees of polymerization, the contribution per monomeric unit, which is

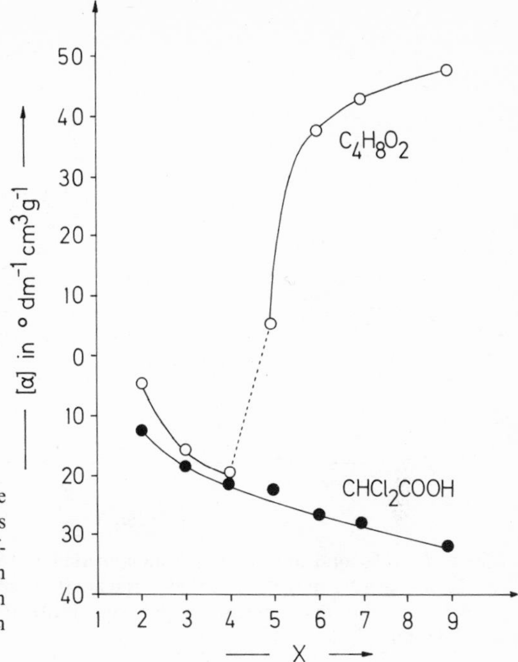

Figure 4-20. The dependence of the specific rotation $[\alpha]_\lambda$ of oligomers of poly(γ-methyl-L-glutamate) of different degrees of polymerization in dichloroacetic acid (coil) and in dioxane (helix) (after M. Goodman and E. E. Schmitt).

affected by helix formation, becomes less and less, until finally the optical activity becomes independent of the degree of polymerization. Practically constant values for the optical activity are already achieved by a degree of polymerization of ~10–15. The renewed increase in optical activity with degree of polymerization for the pentamers can be explained on the basis of the helical structure. Helices of poly(γ-methyl-L-glutamate) have 3.7 monomeric units per helix turn. Thus, before a helix can be formed, at least four amino acid residues must be joined together. But even with four amino acid residues, the helix is still not sufficiently stabilized by the contribution of nonbonded atomic interaction (see also Table 4-2).

4.7.2.2. *Poly(α-olefins)*

The molar optical rotation of optically active it-poly(α-olefins) depends not only on the wavelength and temperature, but also on the optical purity of the monomers (and thus, also, of the polymers) (Figure 4-21). The molar optical rotation of these polymers remains constant when the optical purity of the monomer is high.

The molar optical rotation of a polymer obtained by polymerizing a monomer of given optical purity under various conditions depends on the

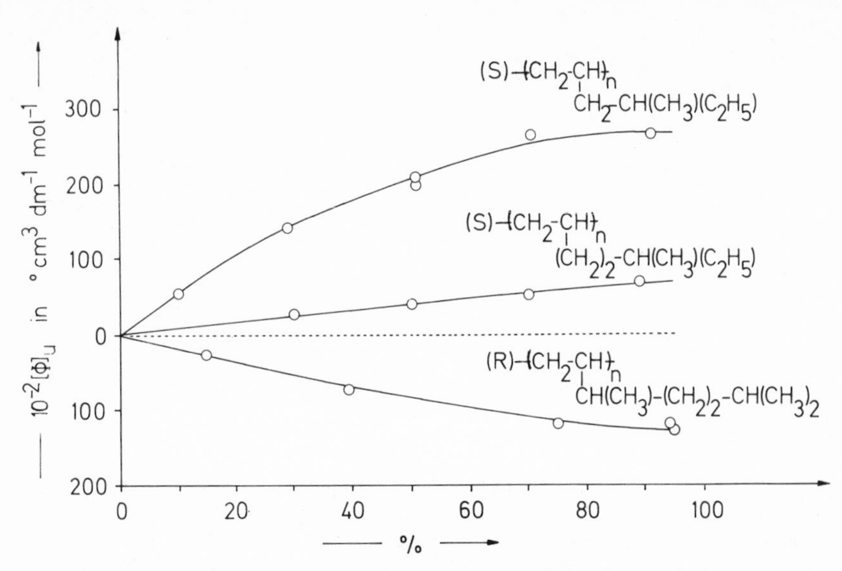

Figure 4-21. Molar optical rotation per monomeric unit $[\Phi]_u$ as a function of optical purity % of the monomer for poly[(S)-4-methyl-1-hexene] (top curve); poly[(S)-5-methyl-1-heptene], and poly[(R)-3,7-dimethyl-1-octene] (bottom curve) (after P. Pino, F. Ciardelli, G. Montagnoli, and O. Pieroni).

tacticity of the resulting polymer (Figure 4-22). Extrapolation of the reciprocal optical rotation to that of the holotactic polymer allows the molar optical rotation of the latter to be obtained.

A value of $\Phi = 292$ was found for poly[(S)-4-methyl hexene-1], whereas the value for the hydrogenated monomer chosen as model compound is only 9.9 (in each case, expressed in units of 10^{-2} deg dm^{-1} cm^3 mol^{-1}). The increased value for the polymer undoubtedly results from the contribution of the helical structure.

The dependence of the molar optical rotation of poly(α-olefins) on the wavelength of the polarized light used can be well represented by the one-term Drude equation. The constant λ_0 is of approximately the same magnitude for polymers and their hydrogenated monomers (Table 4-9). For polymers, however, the a_0 values are sometimes considerably larger than they are for the hydrogenated monomers. The values are only slightly influenced by the nature of the solvent used, i.e., the length of the helical segments is independent of solvent.

The molar optical rotation of poly(α-olefins) decreases with increasing temperature. This is interpreted as the "melting" of relatively long, left-handed helical segments. Calculations based on the same model say that the length of the relatively short, right-handed helical segments should not change very much with temperature.

Table 4-9. Constants a_0 and λ_0 of the One-Term Drude Equation for Various Synthetic Polymers and Their Hydrogenated Monomers as Model Low-Molecular-Weight Compounds[a]

	Monomer		λ_0, nm		a_0, 10^{-2} deg dm^{-1} cm^3 mol^{-1}	
	Name	Constitution	Model	Polymer	Model	Polymer
I	(S)-3-Methyl-1-pentene	$CH_2{=}CH{-}CH(CH_3)(C_2H_5)$	176	167	−113	1143
II	(S)-4-Methyl-1-hexene	$CH_2{=}CH{-}CH_2{-}CH(CH_3){-}C_2H_5$	170	165	3078	104
III	(IR, 3R, 4S)-1-Methyl-4-isopropyl-cyclohex-3-yl-vinyl ether	$CH_2{=}CH{-}O{-}$cyclohexyl ring with $CH(CH_3)_2$ and CH_3 substituents	155	165	−1144	−2169
IV	[(−)-*N*-Propyl-*N*-α-phenylethyl]-acrylamide	$CH_2{=}CH{-}CO{-}N(C_3H_7){-}CH(CH_3){-}C_6H_5$	280	272	−1518	−1188
V	[(1S, 2R, 4S)-1,7,7-Trimethyl-norborn-2-yl] 2-yl]-acrylate	$CH_2{=}CH{-}CO{-}O{-}$norbornyl ring with bridge $C(CH_3)_2$ (H_3C, CH_3) and CH_3	190	191	−485	−401
VI	[(S)-Methylbutyl]-methacrylate	$CH_2{=}CCH_3{-}CO{-}O{-}CH_2{-}CH(CH_3){-}C_2H_5$	191	188	59	53

[a]The measurements refer to room temperature, and the same solvent was used for each polymer–model-compound pair. The monomers were polymerized with Ziegler catalysts (I, II), cationically (III), anionically (IV), and radically (V, VI).

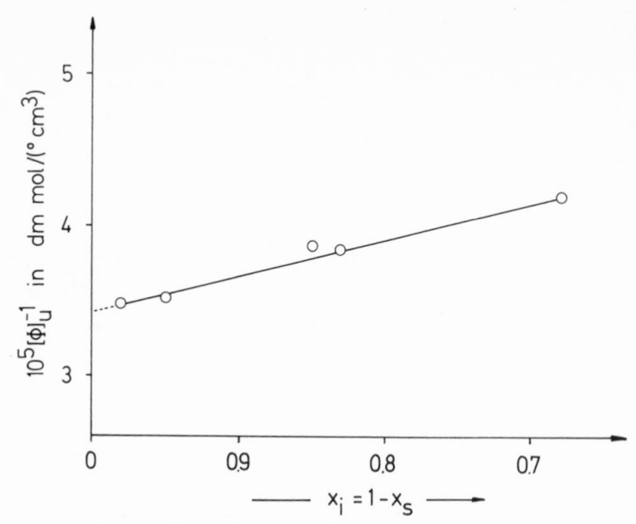

Figure 4-22. Variation of the reciprocal molar optical rotation $[\Phi]_u^{-1}$ with tacticity x_i and x_s for poly[(S)-4-methyl-1-hexene] polymerized under various conditions from a monomer of 93% optical purity. Tacticity was determined by IR. (After P. Pino *et al.*)

The molar optical rotation of configurational copolymers of (S) and (R) isomers of the same monomer is generally, in the case of poly(α-olefins), a hyperbolic and not a linear function of the optical purity of the monomers. Thus, the molar optical rotation of the copolymers is always greater than that obtained by additivity rules. Whether this is caused by long blocks in the polymers or by mixtures of configurative unipolymers has not been established yet.

4.7.2.3. *Poly(α-amino acids)*

Stereoregular poly(α-amino acids) form what are called α helices with 3.7 monomer units per turn or β structures (pleated-sheet structures) in the crystalline state (see Section 4.2) and sometimes after pretreatment (see Section 30.1). Generally, polymers with L-monomeric units form right-handed helices and polymers with D-monomeric units form left-handed helices. An exception is, for example, poly(β-benzyl-L-aspartate). In certain solvents, the helix structure is retained, while in others, e.g., dichloroacetic acid or hydrazine, it is destroyed and coils are formed.

The dependence of the molar optical activity of coils on the wavelength of polarized light can be described by a one-term Drude equation. In the case of helices, the Moffitt–Yang equation is better. The constants λ_0 are completely unaffected by the solvent, while a_0 and b_0 assume different values (Table 4-10). For a given polymer in different helicogenic solvents,

Table 4-10. Influence of Overall Conformation of Poly(γ-benzyl-L-glutamate) on Parameters λ_0, a_0, and b_0

Solvent	Overall conformation	Equation used in evaluation	λ_0	a_0	b_0
Dichloroacetic acid	Coil	Drude	190	—	—
Dichloroacetic acid	Coil	Moffitt–Yang	212	—	0
Hydrazine	Coil	Drude	212	—	0
Dimethylformamide	Helix	Moffitt–Yang	212	200	−660
Dioxane	Helix	Moffitt–Yang	212	220	−670
Dioxane	Helix	Moffitt–Yang	212	198	−682
Chloroform	Helix	Moffitt–Yang	212	250	−625
1,2-Dichloroethane	Helix	Moffitt–Yang	212	205	−635

b_0 is found to be more or less constant, while a_0 still depends on the solvent. For various poly(α-aminocarboxylic acids), b_0 has approximately the same value so long as these are in the helix conformation. b_0 is therefore a typical constant for the helix conformation of poly(α-aminocarboxylic acids), while a_0 contains contributions from the helix and the asymmetric carbon atoms.

4.7.2.4. Proteins

In proteins of natural origin, copolymers of α-aminocarboxylic acids of uniform sequence (see Chapter 30), the constant b_0 has different values according to the protein. Since proteins usually contain L-amino acids, the helix conformation does not depend a great deal on the size of the substituents, and proteins take up a very compact structure in aqueous solutions (see Section 4.6.2), the constant b_0 has been used as a measure of the helix content of proteins. $b_0 = -650$ has been fixed for a 100% helix conformation. The figures given in Table 4-11 were obtained for different proteins.

The evaluation of the helix content of proteins is important because

Table 4-11. Helix Content of Various Proteins

Protein	b_0	Helix content, %
Tropomyosin	−650	100
Serum albumin	−290	46
Ovalbumin	−195	31
Chymotrypsin	−95	15

it allows the influence of the solvent on the conformation to be estimated with regard to the conformation determined by X-ray measurements on the crystalline state. It presupposes a "two-phase" model, i.e., the independent and distinctly separate existence of helix and coil portions. This assumption is confirmed by observations of the helix/coil transition (see below). The determination of the helix content of proteins using b_0 is not completely straightforward, however, since helix portions that are too short do not make a full contribution to b_0 (see Figure 4-20), L-amino acids can occur in left-handed helices as well as right-handed helices with a change in sign for b_0, and, finally, there can be mixtures of left- and right-handed helices.

4.8. *Conformational Transitions*

4.8.1. *Thermodynamics*

According to type and extent of the conformational transition, it can be followed by group-specific (IR, UV, NMR, ORD, CD) or molecule-specific (radius of gyration, Staudinger index) methods. Changes in conformation can be brought about by varying solvent, temperature, pressure, etc.

The specific rotation of poly(γ-benzyl-L-glutamate) in a mixture of, for example, ethylene dichloride and dichloroacetic acid first increases slightly as more $CHCl_2COOH$ is added, then remains constant over a wide range of mixture compositions, and finally falls sharply to negative values at a ~75% $ChCl_2COOH$ content (Figure 4-23). Since ethylene dichloride is a helicogenic solvent, the initial increase is considered to be a change in the helix structure (an expansion?), but the decrease results from a helix/coil transition.

Conformational changes in macromolecules must be cooperative processes since, at least for regular conformational sequences, each conformation must be influenced by the conformation of neighboring bonds. An equilibrium constant can be defined for each individual conformational transition. The different conformations can be distinguished from each other by the symbols A and B. A and B can, for example, be *trans* or *gauche* conformations or even the *cis* and *trans* positions of the peptide groups in poly(proline), etc.

The process

$$\mathrm{AAB} \rightleftharpoons \mathrm{ABB} \tag{4-62}$$

(or BAA $\rightleftharpoons$ BBA) describes the *propagation* of already existing sequences of A or B conformations, and an equilibrium constant $K_p = K$ is assigned

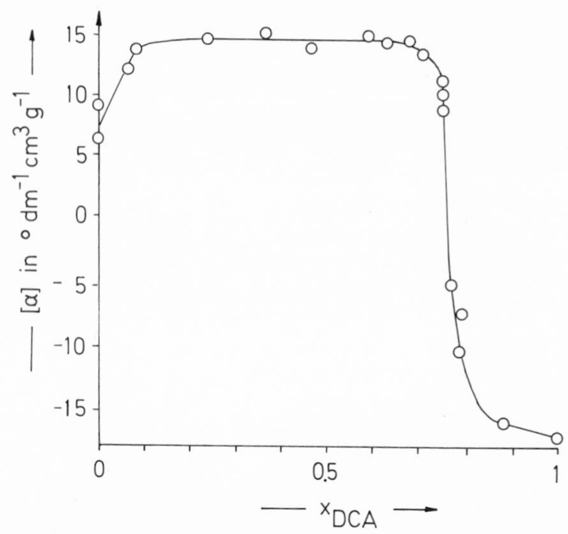

Figure 4-23. Specific rotation $[\alpha]_D$ of a poly(γ-benzyl-L-glutamate) ($\overline{M}_w = 350{,}000$) in ethylene dichloride–dichroroacetic acid mixtures at 20°C (after J. T. Yang).

to this process. In the process

$$AAA \rightleftharpoons ABA \tag{4-63}$$

however, a B sequence is begun or destroyed. An equilibrium constant $K_n = \sigma K_p = \sigma K$ can be assigned to this *nucleation* process. σ is a measure of the cooperativeness of the transition. The segments preferentially take on the conformation of neighboring groups for $\sigma < 1$. In this case, conformational diads AA or BB are preferred to AB or BA diads (positive cooperativity). When $\sigma = 1$, $K_p = K_n$; no cooperativity exists. There is no known case of a negative cooperativity or anticooperativity with $\sigma > 1$.

The nucleation process in the interior of a chain must be microscopically reversible. Thus, the equilibrium constant for the process

$$BBB \rightleftharpoons BAB \tag{4-64}$$

is $K^{-1}\sigma$. At each end of the chain, however, the conformations have only one neighboring conformation. The σ values for nucleation from the ends of the chain must therefore be different from those for the chain interior as well as depending on the type of conformation (A or B) of the chain ends. To a first approximation, however, the formation of a nucleus at the chain ends can also be described by σ when many transitions are involved.

In a chain consisting of $N = 4$ conformations, the all-A conformation can transform into the all-B conformation in four steps:

$$AAAA \rightleftharpoons BAAA \rightleftharpoons BBAA \rightleftharpoons BBBA \rightleftharpoons BBBB \tag{4-65}$$

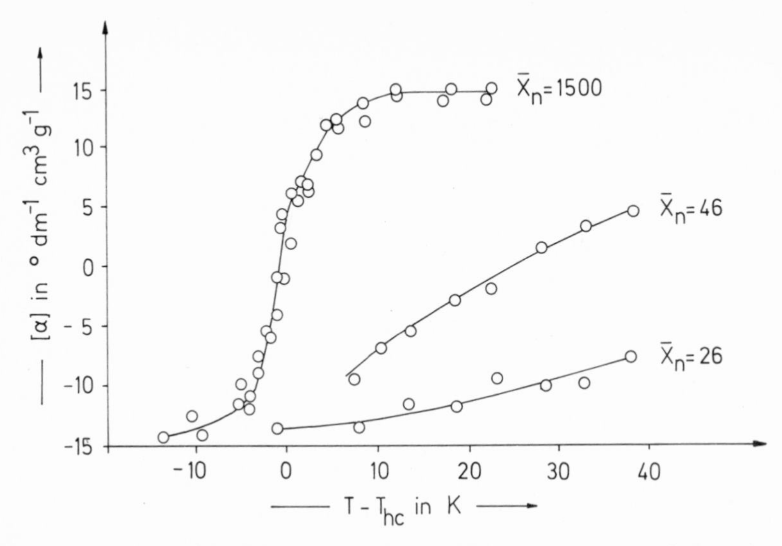

Figure 4-24. Specific rotation $[\alpha]_D$ of poly(γ-benzyl-L-glutamates) of various degrees of polymerization $\overline{X}_n$ as a function of the temperature difference $T - T_{hc}$ in 1,2-dichloroethane–dichloroacetic acid (20:80) mixture. T_{hc} is defined as the inflection point of the $[\alpha]_D = f(T)$ curve for the probe with the highest degree of polymerization. $T_{hc} = 30°C$. (After P. Doty and J. T. Yang, and B. H. Zimm, P. Doty, and K. Iso.)

Three propagation steps follow a nucleation step. The equilibrium concentration is thus given by

$$c_{\mathrm{BBBB}} = \sigma K \cdot K \cdot K \cdot K \cdot c_{\mathrm{AAAA}} = \sigma K^4 c_{\mathrm{AAAA}} \tag{4-66}$$

Thus, when $\sigma K^4 = 1$, $c_{\mathrm{BBBB}} = c_{\mathrm{AAAA}}$.

If, in this case, the equilibrium constant $K \gg 1$, then, $1/\sigma^{1/4} \gg 1$ since $\sigma K^4 = 1$. Thus, here, all intermediates must occur at low concentrations with respect to the extreme conformations (i.e., for $\sigma = 10^{-4}$: $c_{\mathrm{BBBB}} = 10c_{\mathrm{BBBA}} = 10^2 c_{\mathrm{BBAA}} = 10^3 c_{\mathrm{BAAA}}$). That is, the conformational transition occurs essentially completely or not at all.

The transition of a four-conformational chain is therefore described by the product σK^4. The expression for a chain with N conformations is consequently σK^N. The fraction f_{B} of B states formed is then

$$f_{\mathrm{B}} = \frac{\sigma K^N}{1 + \sigma K^N} \tag{4-67}$$

However, f_{B} can only be calculated from equation (4-67) when N is small. For an all-or-none process, $1/\sigma^{1/N} \ll 1$ or $\sigma^{1/N} \gg 1$ (see above). If the probabilities for the conformational transitions are the same for all conformations, then the probability per chain must increase with increasing N. Thus, for the chain transition probability, the product of N and $\sigma^{1/N}$ must be considered. The condition for equation (4-67) is thus

$$N\sigma^{1/N} \ll 1 \tag{4-68}$$

On the other hand, the chain transition is independent of N for large chain lengths. The degree of transition f_B in this case is calculated as

$$f_B = 0.5\left(1 + \frac{K - 1}{[(K - 1)^2 + 4\sigma K]^{0.5}}\right) \tag{4-69}$$

Thus, according to equation (4-69), K will always be unity no matter what the value of σ may be at the midpoint of the transition ($f_B = 0.5$). However, the sharpness of the transition increases with decreasing σ.

The expressions are more complex for transitions of chains of moderate length. In this region, the transitions depend significantly on the chain length (see also Figure 4-24).

4.8.2. *Kinetics*

With the exception of helix/coil transitions of polypeptides and polynucleotides, the kinetics of conformational transitions has only been investigated to a slight extent. Relatively large rate constants of 10^6–10^7 s^{-1} have been obtained for these helix/coil transitions. However, rate constants of 10^{-6} to 1 s^{-1} have been obtained for denaturing processes where helix/coil transitions also play a role (see also Chapter 4.6.2). The high rates of helix/coil transitions are undoubtedly due to the cooperativity of the process. The low rates for denaturing must therefore result from nonhelical regions.

A large part of the molecule must be set in motion in the rotation about a single bond in a long-chain molecule (Figure 4-25a). This should be very difficult in viscous media. Conversely, an increase in the activation energy

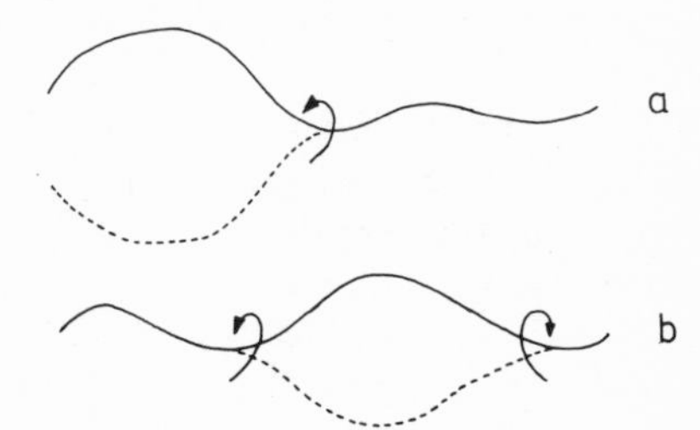

Figure 4-25. Schematic representation of the rotation about one single bond (a) and the coupled rotation about two single bonds (b) of a macromolecule.

for coupled rotation about two bonds (Figure 4-25b) with respect to similar low-molecular-weight compounds is to be expected.

This subject has been investigated for the case of piperazine polymers and diacoyl piperazines as model substances. The N—CO bond of these compounds possesses partial double-bond character. Consequently, the rotation about this bond is comparatively slow. Various absorption bands will be observed in the proton resonance spectrum according to whether neighboring groups are in the *cis* or the *trans* position about the N—CO bond. The Gibbs energy of activation $\Delta G^{\ddagger}$ can be determined from the temperature dependence of the band intensities.

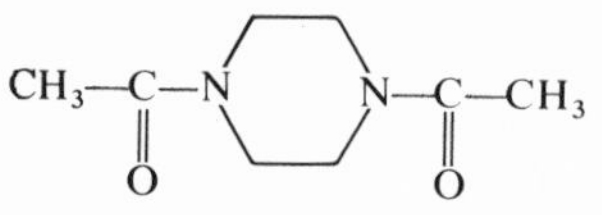

CH_3—C(=O)—N(piperazine)N—C(=O)—CH_3 (I)

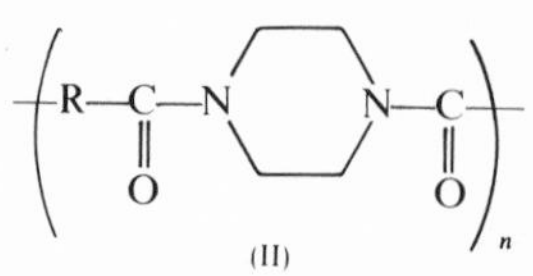

—[R—C(=O)—N(piperazine)N—C(=O)]$_n$— (II)

Experimentally it was found that the Gibbs activation energy for diacetyl piperazine (I) and poly(succinyl piperazine) [II, R = $(CH_2)_2$], poly(adipyl piperazine) [II, R = $(CH_2)_4$], and poly(sebacyl piperazine) [II, R = $(CH_2)_8$] are practically the same ($\Delta G^{\ddagger} = 7.6 \times 10^4$ J/mol). The reason for this is not clear. The activation energy can, for example, be stored in the polymer molecule and used for the rotation about another bond. Alternatively, the tension set up by rotation can be compensated by distortion of rotational and valence angles.

A4. Appendix to Chapter 4

A4.1. Calculation of the Chain End-to-End Distance

In the segment model the length and direction of each individual bond are given by a vector $\boldsymbol{l}_i$ (Figure 4-11). The vector distance $\mathbf{L}_{00}$ between both ends of the chain is then

$$\mathbf{L}_{00} = \boldsymbol{l}_1 + \boldsymbol{l}_2 + \cdots + \boldsymbol{l}_{n-1} = \sum_{i=i}^{i=n-1} \boldsymbol{l}_i \tag{A4-1}$$

For the mean square chain end-to-end distance $\overline{L_{00}^2}$ (average over all molecules and all molecular shapes in time of a molecule), we use the products of the vectors:

$$(\overline{L_{00}^2}) = \overline{\mathbf{L}\cdot\mathbf{L}} = (\overline{\sum_i \boldsymbol{l}_i \cdot \sum_j \boldsymbol{l}_j}) = \boldsymbol{l}_1\cdot\boldsymbol{l}_1 + \boldsymbol{l}_2\cdot\boldsymbol{l}_2 + \cdots + \boldsymbol{l}_{n-1}\cdot\boldsymbol{l}_{n-1} + 2\sum_i\sum_j \overline{\boldsymbol{l}_i\cdot\boldsymbol{l}_j} \tag{A4-2}$$

where the index j has the same significance as the index i and simply shows that every member of the first sum has to be multiplied by every member of the second sum. The scalar product $\boldsymbol{l}_i \cdot \boldsymbol{l}_{i+1}$ is equal to $l_i l_{i+1} \cos(180 - \vartheta)$, where $(180 - \vartheta)$ is the angle between two neighboring bonds and ϑ is the valence angle. In the chosen segment model, every angle ϑ has the same probability of occurring as the angle $(180 + \vartheta)$. Therefore, $\cos\vartheta = -\cos(180 + \vartheta)$. The double sum in equation (A4-2) vanishes, giving

$$\overline{(L_{00}^2)} = (N - 1)l^2 \approx Nl^2 \tag{A4-3}$$

If several different bonds of differing lengths l_q are present (for example, in polyamides), then equation (A4-3) is altered to the corresponding sum:

$$\overline{(L_{00}^2)} = \sum_q (N - 1)_q l_q^2 \tag{A4-4}$$

A4.2. Relationship between the Radius of Gyration and the Chain End-to-End Distance for the Segment Model

It is the radius of gyration $(\overline{R_G^2})^{0.5} \equiv \langle R \rangle$, and not the chain end-to-end distance $(\overline{L^2})^{0.5} \equiv \langle L \rangle$, that is experimentally accessible. In this model, nevertheless, the chain end-to-end distance is unambiguously related to the radius of gyration. As shown in Figure 4-8, the masses of the chain atoms can be concentrated into point masses connected together by bonds of length l. $\mathbf{r}_1$ is the vector from the center of gravity of the molecule to the first point mass, $\mathbf{r}_i$ the corresponding vector to the ith point mass, and $\mathbf{L}$ the vector between both point masses. Therefore, for each point mass

$$\mathbf{r}_i = \mathbf{r}_1 + \mathbf{L}_i \tag{A4-5}$$

The center of gravity is defined by requiring that the first moment of the segment distribution around the center of mass be zero

$$\sum_i m_i \mathbf{r}_i = 0 \tag{A4-6}$$

Since all point masses are identical, one can write for N point masses

$$\sum_{i=i}^{i=n} \mathbf{r} = N\mathbf{r}_i + \sum_{i=i}^{i=n} \mathbf{L}_i = 0 \tag{A4-7}$$

and therefore

$$\mathbf{r}_1 = -(1/N) \sum_i \mathbf{L}_i \tag{A4-8}$$

The radius of gyration $\langle R \rangle$ is now defined as the root mean square over all the radii $\mathbf{r}$ (see above), and this in turn as the second moment of the mass distribution

$$\mathbf{r}^2 = \frac{\sum_i m_i r_i^2}{\sum_i m_i} \tag{A4-9}$$

or for the mean over all squares (with, for this model, the index 00)

$$\overline{R_{00}^2} = \frac{\overline{\sum_i m_i r_i^2}}{\sum_i m_i} = \langle R^2 \rangle_{00} \tag{A4-10}$$

By definition, all the masses m_i are identical. Since it is possible to average over all sums first or alternatively over all products and then over all sums, one can also write for equation (A4-10)

$$\langle R^2 \rangle_{00} = \frac{m_i \sum_i \overline{r_i^2}}{\sum_i m_i} \tag{A4-11}$$

and, with $m = \sum_i m_i$ and N the number of chain links ($N = m/m_i$), from Equation (A4-11)

$$\langle R^2 \rangle_{00} = \sum_i \overline{r_i^2}/N \tag{A4-12}$$

With poly(methylene), the number of chain units N is identical to the degree of polymerization $\overline{X}_n$. However, for monomeric units of two units. e.g., poly(styrene), $N = 2\overline{X}_n$. The relation between the radius of gyration and the chain end-to-end distance is then

$$\langle R^2 \rangle_{00} = \frac{1}{N} \sum_{i=1}^{i=n} (\mathbf{r}_1 + \mathbf{L}_i) \cdot (\mathbf{r}_1 + \mathbf{L}_i) \tag{A4-13}$$

$$\langle R^2 \rangle_{00} = r_1^2 + \frac{1}{N} \sum_{i=1}^{i=n} L_i^2 + \frac{2}{N} \mathbf{r}_1 \cdot \sum_{i=1}^{i=n} \mathbf{L}_i$$

According to equation (A4-2), however,

$$r_1^2 = \sum_{i=1}^{i=n} \sum_{j=1}^{j=n} \mathbf{L}_i \cdot \mathbf{L}_j \tag{A4-2a}$$

and from equations (A4-8) and (A4-2)

$$\frac{2}{N} \mathbf{r}_1 \cdot \sum_{i-1}^{i=n} \mathbf{L}_i = -\frac{2}{N^2} \sum_{i=1}^{i=n} \sum_{j-1}^{j=n} \mathbf{L}_i \cdot \mathbf{L}_j \tag{A4-8a}$$

With equations (A4-2*a*) and (A4-8*a*), equation (A4-13) becomes

$$\langle R^2 \rangle_{00} = \frac{1}{N} \sum_{i=1}^{i=n} L_i^2 - \frac{1}{N^2} \sum_{i=1}^{i=n} \sum_{j=1}^{j=n} \mathbf{L}_i \cdot \mathbf{L}_j \tag{A4-14}$$

The vector product is solved with the cosine rule already used in equation (A4-2)

$$L_{ij}^2 = L_i^2 + L_j^2 - 2\mathbf{L}_i \cdot \mathbf{L}_j \tag{A4-15}$$

According to definition, the indices i and j have the same significance, so that the sums over the squares of the distances L_i^2 and L_j^2 are identical.

On inserting equation (A4-15) into (A4-14), one obtains

$$\langle R^2 \rangle_{00} = \frac{1}{2N^2} \sum_{i=1}^{i=n} \sum_{j=1}^{j=n} \overline{L_{ij}^2} \tag{A4-16}$$

According to equation (A4-2) or (A4-3), however, the average $\overline{L_{ij}^2}$ is the chain end-to-end distance of a chain of $j - i$ elements of length l:

$$\overline{L_{ij}^2} = |j - i| \, l^2 = (|j - i| \, \overline{L^2})/N \tag{A4-17}$$

The absolute difference corresponds to a product sum, where each sum can be evaluated individually. Summing over all j values, one obtains

$$\begin{aligned} \sum_{j=1}^{j=n} |j - i| &= \sum_{j=1}^{i} (i - j) + \sum_{j=i+1}^{n} (j - 1) \\ &= i^2 - 0.5i(i + 1) + 0.5(N - i)(N + i + 1) - i(N - i) \\ &= i^2 - iN + 0.5N^2 + 0.5N - i \end{aligned} \tag{A4-18}$$

and for the sum over all i values

$$\sum_{i=1}^{i=n} i^2 = 1^2 + 2^2 + \cdots + N^2 = \frac{N(N + 1)(2N + 1)}{N} \tag{A4-19}$$

thus giving

$$\sum_{i=1}^{i=n} \sum_{j=1}^{i=j} |j - i| = \tfrac{1}{3}(N^3 - N) \cong \tfrac{1}{3}N^3 \quad \text{for} \quad N \geq 1 \tag{A4-20}$$

From equation (A4-16) with equations (A4-17) and (A4-20), this gives

$$\langle R^2 \rangle_{00} = \frac{1}{2N^2} \frac{N^3}{3} \frac{\overline{L^2}}{N} = \frac{1}{6} \overline{L_{00}^2} = \overline{R_{00}^2} \tag{A4-21}$$

Equation (A4-21) allows the theoretically important end-to-end chain distance to be calculated from the experimentally accessible (e.g., by light scattering measurements) radius of gyration. This calculation remains valid for the linear valence angle chain and the linear valence angle chain with hindered rotation, but is not valid for polymers in good solvents.

A4.3. Calculation of the Chain End-to-End Distance for Valence Angle Chains

In the segment model, any given angle can occur between neighboring segments. Valence angles however, must be considered with real chains and, to a first approximation (see Section 4.4.2.3), they can be taken as constant. As with the segment model [see equation (A4-2)], all vectors for the N chain units (that is, $N - 1$ bonds) must be multiplied by each other:

$$
\begin{aligned}
(\overline{L_{0f}^2}) = \overline{\mathbf{L}} \cdot \overline{\mathbf{L}} &= (\boldsymbol{l}_1 \cdot \boldsymbol{l}_1 + \boldsymbol{l}_2 \cdot \boldsymbol{l}_2 + \cdots + \boldsymbol{l}_{n-1} \cdot \boldsymbol{l}_{n-1}) \\
&+ 2(\boldsymbol{l}_1 \cdot \boldsymbol{l}_2 + \boldsymbol{l}_2 \cdot \boldsymbol{l}_3 + \cdots + \boldsymbol{l}_{n-2} \cdot \boldsymbol{l}_{n-1}) \\
&+ 2(\boldsymbol{l}_1 \cdot \boldsymbol{l}_3 + \boldsymbol{l}_2 \cdot \boldsymbol{l}_4 + \cdots + \boldsymbol{l}_{n-3} \cdot \boldsymbol{l}_{n-1}) \\
&+ \cdots + 2(\boldsymbol{l}_1 \cdot \boldsymbol{l}_{n-2} + \boldsymbol{l}_2 \cdot \boldsymbol{l}_{n-1}) + 2(\boldsymbol{l}_1 \cdot \boldsymbol{l}_{n-1}) \qquad \text{(A4-22)}
\end{aligned}
$$

The scalar product of the vectors $\boldsymbol{l}_i$ and $\boldsymbol{l}_j$ that contain the angle $(180 - \vartheta)$ is defined as $\boldsymbol{l}_i \cdot \boldsymbol{l}_{i+1} = |\boldsymbol{l}_i||\boldsymbol{l}_{i+1}| \cos(180 - \vartheta)$. Therefore, from equation (A4-22),

$$
\begin{aligned}
(\overline{L_{0f}^2}) &= (N - 1)\, l^2 + 2(N - 2)\, l^2 \cos(180 - \vartheta) \\
&+ 2(N - 3)(\boldsymbol{l}_1 \cdot \boldsymbol{l}_3) + \cdots + 2(\boldsymbol{l}_1 \cdot \boldsymbol{l}_{n-1}) \qquad \text{(A4-23)}
\end{aligned}
$$

For the scalar product average we have

$$\boldsymbol{l}_1 \cdot \boldsymbol{l}_{j+1} = l^2 \cos^j (180 - \vartheta) \qquad \text{(A4-24)}$$

Equation (A4-23) then becomes

$$
\begin{aligned}
\overline{L_{0f}^2} &= (N - 1)\, l^2 + 2l^2 [(N - 2) \cos(180 - \vartheta) + (N - 3) \cos^2(180 - \vartheta) \\
&+ \cdots + \cos^{n-2}(180 - \vartheta)] \qquad \text{(A4-25)}
\end{aligned}
$$

It is possible to develop equation (A4-25) into a series. Writing $a = N - 1$ and $x = \cos(180 - \vartheta) = \cos\alpha$, and using the following relationships, valid for $x < 1$,

$$1 + x + x^2 + x^3 + \cdots = 1/(1 - x) \qquad \text{(A4-26)}$$

$$1 + 2x + 3x^2 + 4x^3 + \cdots = 1/(1 - x)^2 \qquad \text{(A4-27)}$$

one obtains from equation (A4-25)

$$\overline{L_{0f}^2} = (N - 1)\, l^2 \frac{1 + \cos\alpha}{1 - \cos\alpha} - 2l^2 (\cos\alpha) \frac{1 - \cos^{n-1}\alpha}{(1 - \cos\alpha)^2} \qquad \text{(A4-28)}$$

For many chain units, e.g., for $N - 1 \gg 1$, the last component is negligible compared to the first, and equation (A4-28) reduces to

$$\overline{L_{0f}^2} = N l^2 \frac{1 + \cos\alpha}{1 - \cos\alpha} = N l^2 \frac{1 - \cos\vartheta}{1 + \cos\vartheta} \qquad \text{(A4-29)}$$

One obtains, analogously, for the radius of gyration

$$6\overline{R_{0f}^2} = N l^2 \frac{1 + \cos\alpha}{1 - \cos\alpha} = N l^2 \frac{1 - \cos\vartheta}{1 + \cos\vartheta} \qquad \text{(A4-30)}$$

A4.4. Distribution of Chain End-to-End Distances

At any given instance, a number of chain molecules of equal length will have a random distribution of chain end-to-end distances. This information can be obtained from a derivation analogous to that used to derive the Maxwellian velocity distribution of molecules in an ideal gas.

The term $p(L_x)$ is the distribution function of the chain end-to-end distance along the x axis. Since space has reflective symmetry, $p(L_x) = p(-L_x)$, and, consequently, $p(L_x)$ must be an even-valued function of L_x and for this simplest case $p(L_x) = f(L_x^2)$.

The three distribution functions for the three possible directions in space must be interdependent when the number of bonds N is small. For $N = 1$, for example, $L_x^2 + L_y^2 + L_z^2 = l^2$ must hold, where l is the bond length. The three components of the overall distribution function become less dependent on each other with increasing number of bonds in the chain molecule. If N is very large, and L^2 is simultaneously much smaller than the square of the length of the fully stretched chain molecule, then the components can be considered to be independent of each other. The total probability is simply the product of the individual probabilities, i.e.,

$$p(L_x)\,p(L_y)\,p(L_z) = f(L_x^2)\,f(L_y^2)\,f(L_z^2) \tag{A4-31}$$

This probability cannot depend on the direction in space. It must be a function of the square of the end-to-end distance:

$$L^2 = L_x^2 + L_y^2 + L_z^2 \tag{A4-32}$$

and thus,

$$f(L_x^2)\,f(L_y^2)\,f(L_z^2) = F(L^2) = f(L_x^2 + L_y^2 + L_z^2) \tag{A4-33}$$

must also hold. The condition in equation (4-33) is only satisfied by one mathematical function, that is

$$p(L_x) = f(L_x)^2 = a\exp(-bL_x^2) \tag{A4-34}$$

The minus sign is used so that $p(L_x)$ tends to zero when L_x tends to infinity. The constants a and b can be determined as follows: The distribution can be normalized, i.e.,

$$p(L_x)\,dL_x = a\int_{-\infty}^{+\infty}\exp(-bL_x^2) = a(\pi/b)^{0.5} = 1 \tag{A4-35}$$

The second moment of the distribution function must give the mean square of the L components, i.e., $\langle L_x^2\rangle = \langle L^2\rangle/3 = Nl^2/3$, and thus

$$\langle L_x^2 \rangle = \int_{-\infty}^{+\infty} L_x^2 p(L_x)\,dL_x = a \int_{-\infty}^{+\infty} L_x^2 \exp(-bL_x^2)\,dL_x = \frac{a\pi^{0.5}}{2b^{3/2}} = \frac{Nl^2}{3} \tag{A4-36}$$

Dividing (A4-35) by (A4-36) and inserting into equation (4-34), we obtain

$$p(L_x) = \left(\frac{3}{2\pi Nl^2}\right)^{0.5} \exp\left(-\frac{3}{2Nl^2} L_x^2\right) \tag{A4-37}$$

Literature

Sections 4.1–4.3. Conformation

M. V. Volkenstein, *Configurational Statistics of Polymeric Chains,* USSR Academy of Science, Moskow, 1959; Interscience, New York, 1963.

T. M. Birshtein and O. B. Ptitsyn, *Conformations of Macromolecules,* Interscience, New York, 1966.

F. A. Bovey, *Polymer Conformation and Configuration,* Academic Press, New York, 1969.

P. J. Flory, *Statistical Mechanics of Chain Molecules,* Wiley–Interscience, New York, 1969.

G. G. Lowry, *Markov Chains and Monte Carlo Calculations in Polymer Science,* M. Dekker, New York, 1970.

A. J. Hopfinger, *Conformational Properties of Macromolecules,* Academic Press, New York, 1973.

Sections 4.4–4.5. Coils in Solution

H. Morawetz, *Macromolecules in Solution,* Wiley–Interscience, New York, second ed., 1975.

V. N. Tsvetkov, V. Ye. Eskin, and S. Ya. Frenkel, *Structure of Macromolecules in Solution,* Butterworths, London, 1970.

H. Yamakawa, *Modern Theory of Polymer Solutions,* Harper and Row, New York, 1971.

H. Yamakawa, Polymer statistical mechanics, *Ann. Rev. Phys. Chem.* **25**, 179 (1974).

Section 4.6. Compact Molecules

H. Sund and K. Weber. The quartenary structure of proteins, *Angew. Chem. Int. Ed.* **5**, 231 (1966).

G. N. Ramachandran, *Conformation of Biopolymers,* 2 vols., Academic Press, London, 1967.

R. E. Dickerson and I. Geis, *The Structure and Action of Proteins,* Harper and Row, New York, 1969.

Section 4.7. Optical Activity

B. Jirgensons, *Optical Rotatory Dispersion of Proteins and Other Macromolecules,* Springer, Berlin, 1969.

P. Pino, F. Ciardelli, and M. Zandomeneghi, Optical activity in stereoregular synthetic polymers, *Ann. Rev. Phys. Chem.* **21**, 561 (1970).

Section 4.8. Conformational Transitions

D. Poland and H. A. Scheraga, *Theory of Helix–Coil Transitions in Biopolymers—Statistical Mechanical Theory of Order–Disorder Transitions in Biological Macromolecules*, Academic Press, New York, 1970.

C. Sadron, *Dynamic Aspects of Conformation Changes in Biological Macromolecules*, Reidel, Dordrecht, 1973.

R. Cerf, Cooperative conformational kinetics of synthetic and biological chain molecules, *Adv. Chem. Phys.* **33**, 73 (1975).

Chapter 5
Supermolecular Structures

5.1. Phenomena

Macromolecular substances occur in an exceptionally large profusion of forms in the condensed state. Household utensils of poly(ethylene) are waxy to the touch and appear turbid. Films of the same material are transparent, but not as clear as films of poly(ethylene terephthalate). Poly(styrene) beakers are brittle, but beakers made from polyamides are not. Some products, such as leather, for example, are pliable, others, such as cured phenolic resins, are very rigid.

All these properties can be related to the physical structure of molecular aggregates, which is, in its turn, affected by the constitution, configuration, and conformation of the individual molecule. In addition, however, the physical structure of these aggregates of molecules is also strongly dependent on external conditions, as is shown, for example, by electron micrographs of morphological structures of a polyamide. If a solution of this polyamide in glycerine, heated to 260°C, is poured into glycerine at about 25°C, then globular structures occur (Figure 5-1a). If the same solution is cooled at 1–2 K/min, then fibrillar structures form (Figure 5-1b). At a rate of cooling of about 40 K/min, small lamellae occur (Figure 5-1c). From the evaporation of a solution of formic acid, on the other hand, dendritic or fibrillar structures are obtained (Figure 5-1d).

The globular structures show no degree of order that can be recognized electron-microscopically. A high degree of order may be presumed in

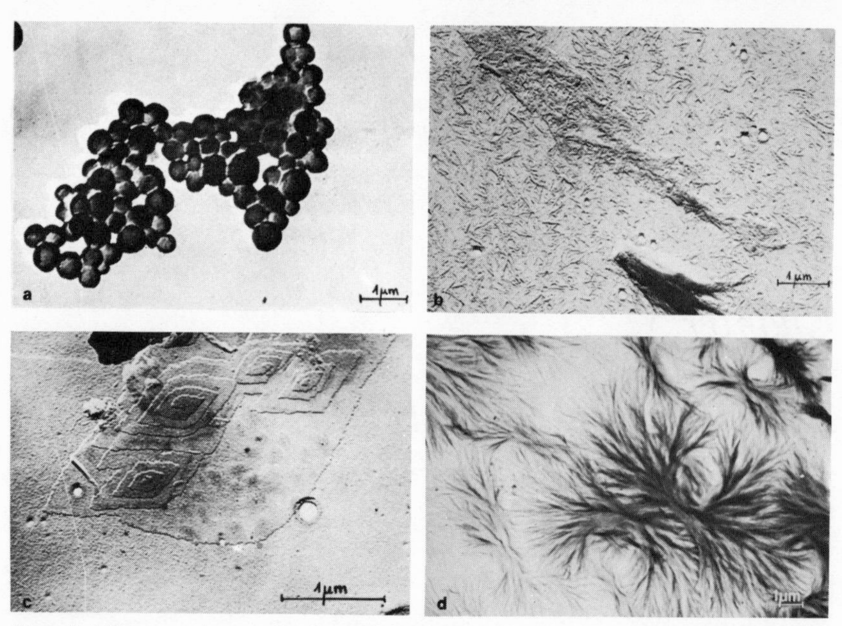

Figure 5-1. Electron micrographs of morphological features seen in poly(caprolactam) (nylon 6) specimens prepared under different conditions (after Ch. Ruscher and E. Schulz). (a) A hot solution (at 260°C) in glycerine has been poured into glycerine at room temperature. (b) A 260°C solution in glycerine has been cooled at about 1–2 K/min. (c) A 260°C solution in glycerine has been cooled at about 40 K/min. (d) A solution in formic acid has been diluted with formic acid at room temperature.

the lamellae; they are reminiscent of crystals. In the fibrillar and dendrite structures, likewise, there are undoubtably ordered structures. The nature of these states of order cannot be found from electron microscope photographs without additional methods.

In the extreme case, a substance can exist as 100% crystalline or 100% amorphous. Common salt forms good crystals; solid glycerine is normally completely amorphous. Both cases can be studied in low-molecular-weight chemistry by a wide selection of methods (X-ray crystallography, melting point, etc.) In macromolecular chemistry, however, the borderlines are less distinct.

In the middle of the 19th century, for example, a crystal was defined as a material with planar surfaces which intersected at definite angles. The electron micrograph in Figure 5-1c shows, for example, planar surfaces, but these are arranged spirally. The density, X-ray scattering, and melting point of this material do not correspond to a 100% crystallinity, however (see below). Toward the end of the 19th century, a crystal was redefined as a homogeneous, anisotropic, solid medium. "Homogeneous" means that the physical properties do not alter on transposing along the crystal axes. Crystals are anisotropic because the physical properties vary in different directions, i.e., with rotation. However, this definition also applies to drawn poly(styrene) produced by radical polymerization, which is certainly not crystalline according to all the experimental criteria.

At the beginning of the 20th century, crystals were redefined again on the molecular or atomic basis of the lattice concept. According to this view, highly ordered crystals must give sharp diffraction pictures on being irradiated with X-rays, because the wavelength of the X-ray beam is comparable with atomic distances. A sharp melting point also follows from the crystal lattice concept. Macromolecular substances, however, show ill-defined diffraction patterns in structures that had been regarded as ordered from electron microscopic evidence. These data can be interpreted as resulting from crystalline and amorphous regions lying next to one another (two-phase model), or from a crystal with imperfections (one-phase model). If, for example, the X-ray measurements on polyamides are analyzed according to the two-phase model, then the result is a crystallinity of less than 50%. Such a result, however, is barely consistent with the electron microscope photograph in Figure 5-1c. Density measurements lead to the same conclusion: The same density can be interpreted as the effect of a relatively large amorphous region (two-phase model), or a small number of imperfections (one-phase model) (Table 5-1).

It is debatable whether the two-phase model describes the relationships correctly. The concept of a phase has to be used with caution in the interpretation of the physical structure of macromolecular substances because equilibrium states do not usually exist, and the borderlines between "crystalline" and "amorphous" phases are not distinct. As the electron micrographs (particularly Figure 5-1d) show, different states of order can, in fact, occur alongside each other in a particular sample. The individual methods for determining crystallinity thus cover varying degrees of order, and will

Table 5-1. Comparison of the Crystallinity α_m of a Poly(ethylene) Sample Calculated from the Density ρ or the Specific Volume $v = 1/\rho$ Using a Two-Phase Model with the Percentage Number of Crystal Defects Calculated Using a One-Phase Model

Specification	Density ρ, g/cm^3	Specific volume v, cm^3/g	Crystallinity, %	Defects, %
100% crystalline	1.000	1.000	100	0
—	0.981	1.020	89	1.9
—	0.971	1.030	83	2.9
100% amorphous	0.852	1.174	0	—

Table 5-2. Comparison of Degree of Crystallinity as Determined by Various Methods[a]

	Degree of crystallinity, %		
		Poly(ethylene terephthalate)	
Method	Cellulose (cotton)	Undrawn	Drawn
Hydrolysis	93	—	—
Formylation	87	—	—
Infrared spectroscopy	—	61	59
X-ray analysis	80	29	2
Density	60	20	20
Deuterium exchange	56	—	—

[a] Calculated using the two-phase model. All values are for the same crystallization or draw conditions.

consequently lead to different degrees of crystallinity (Table 5-1). The concept of "crystallinity" in macromolecules is therefore as ambiguous as that of the molecular weight for polydisperse samples, as long as the method used to determine it is not specified. At the present time it is still not possible to specify to which degree of order the various methods of crystallinity determination refer. The crystallinity is therefore characterized by the method of determination used, and referred to as X-ray crystallinity, density crystallinity, infrared crystallinity, etc. Thus, according to the method used, different crystallinities may be obtained for the same polymer, as is shown in Table 5-2 for cotton and poly(ethylene terephthalate). In poly(ethylene) and crystalline 1,4-*cis*-poly(isoprene), on the other hand, the different methods yield consistent crystallinity values.

5.2. Crystallinity Determination

5.2.1. X-Ray Crystallography

When fast electrons impinge on matter, electrons are ejected from the inner shell of an atom of the target material, and the atom is ionized. Then electrons undergo transitions from the outer shells to the inner shells. Since the energy levels of the different shells are discrete (quantized), a coherent ray is emitted with energy equal to the difference in energy levels of the two shells. This coherent ray thus possesses a quite specific wavelength, e.g., 0.154 nm for the Cu $K\alpha$ radiation frequently used in X-ray measurements. Electron beams behave analogously, and also possess a specific wavelength, e.g., 0.0213 nm when the electrons are accelerated to 10,000 V.

Because the X-rays are of an electromagnetic nature, they must be diffracted at crystal planes if the distance between the planes is comparable to the wavelength of the X-ray. In crystals, which possess three-dimensional lattice order, the lattice plane arrangement causes diffraction to occur. The rays issuing from the different crystal planes interact systematically with one another, leading to discrete reflections. According to Bragg, the position of the reflection is given by the wavelength λ of the incident X-ray beam, the distance d_{Bragg} between parallel planes, and the angle θ between incident ray and lattice plane:

$$N\lambda = 2\, d_{\text{Bragg}} \sin\theta \tag{5-1}$$

where N is the order of the reflection. In polymers, the reflection with the strongest intensity is frequently found for $N = 1$. If the parallel planes of the different crystallites are randomly skewed relative to one another—as in crystal powders—then a monochromatic primary ray will find enough particles to yield all the reflection positions that fulfill Bragg's conditions (Figure 5-2). Since there are many small crystallites with many different crystal plane orientations, a system of coaxial cones of X-rays with a common apex at the center of the sample is obtained. A vertical section of this system of cones on a photographic plane gives a series of concentric circles, or, using a cylindrical film, sectors of a circle.

X-ray exposures of amorphous polymers on a photographic plate show weak rings upon a strong, diffuse background (Figure 5-3a). These

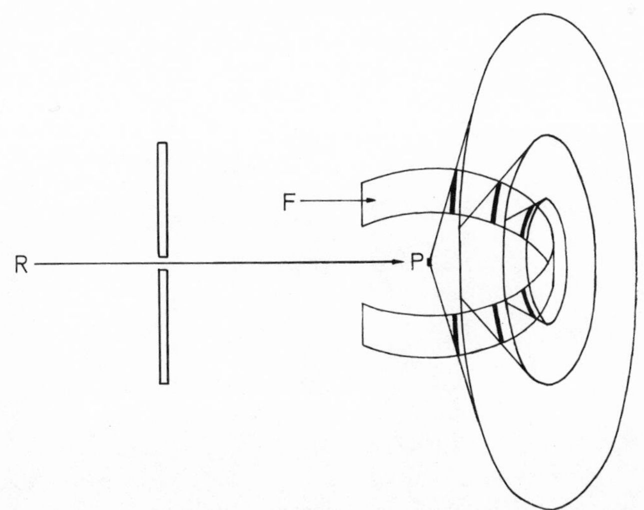

Figure 5-2. The Debye–Scherrer powder method. An X-ray R passes through a collimator and then meets a powder preparation P. The reflections caused by P lie on cones of reflection, which form crescents or arcs on a cylindrical film F.

weak maxima are also called halos, and come from the local order in amorphous polymers. Partly crystalline polymers also show these halos, but in addition, they show the relatively strong rings of crystalline reflections (Figure 5-3b). The fact that the diffuse background is always very strong in polymers comes mainly from the scattering by air, with small contributions from the thermal movement in the crystallites, and Compton scattering. Compton scattering is a quantized, incoherent scattering effect which occurs equally in all substances regardless of their physical state.

The intensity of reflections and halos is generally interpreted with regard to the two-phase model as the content of crystalline and amorphous phases. For this, the diffuse background is first separated (see Figure 5-4). To determine the amorphous content, one begins with the lowest scattering angle, since crystalline reflections are almost always absent here. In addition, crystalline diffraction is normally low in the minima between two maxima if the maxima are more than 3° apart. For further evaluation it is assumed that the diffracted intensity is proportional to the crystalline content and that the intensity of the amorphous halos is proportional to the amorphous content (at a specific angle or specific range of angles). The proportionality factors also depend on the observation angle and a specific function. These can be determined, for example, by comparison with completely amorphous or crystalline samples. Amorphous samples can be obtained, for example, by quenching (not always possible) or by a direct X-ray study of the polymer melt. Other methods are sometimes used, e.g., amorphous cellulose can be produced by grinding in a ball mill. The intensity of the reflections is thus a measure of the crystallinity.

The width of the reflections depends on both the size of crystallites and local lattice fluctuations (defects). The smaller the crystallites, the

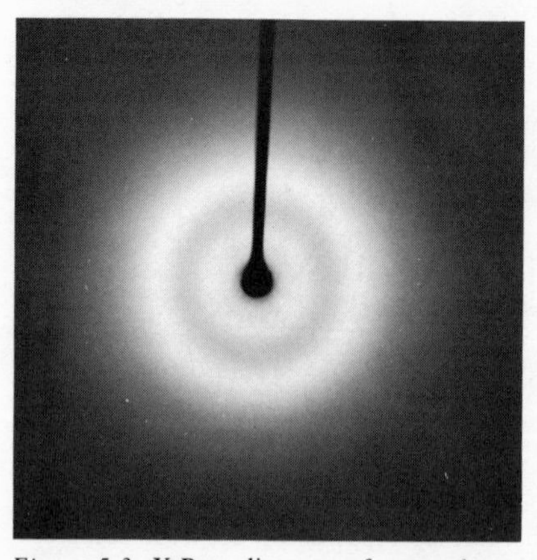

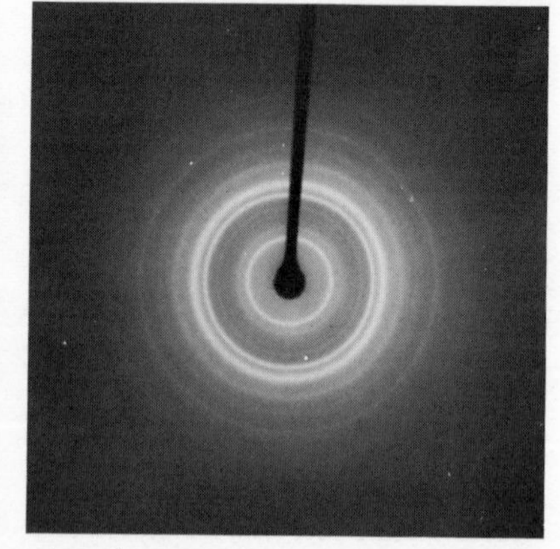

Figure 5-3. X-Ray diagram of amorphous atactic polystyrene (left) and drawn partially crystalline isotactic polystyrene (right).

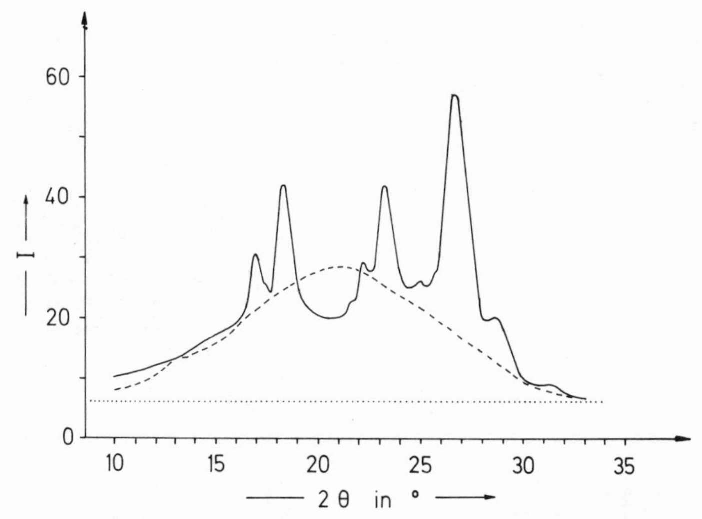

Figure 5-4. X-Ray intensity as a function of Bragg angle of amorphous (- - -) and crystalline (—) poly(ethylene terephthalate). The amorphous PETP was prepared by precipitation of the polymer from a solution in phenol–tetrachloroethane (1:1) with glycerol. Crystalline PETP was prepared by annealing (from A. Jeziorny and S. Kepka).

more the diffraction appears to be scattering. Very small crystallites are therefore extremely difficult to detect by X-rays. Also, the crystallites must be present above a certain minimum concentration, otherwise the corresponding reflection intensity is not obtained by this method. In this respect, the qualitative detection of crystallinity by polarization microscopy is more satisfactory in those cases where the crystallite size is greater than the wavelength of light. In order to obtain discrete X-ray diffractions, it is necessary to have a sufficiently high concentration of ordered, three-dimensional regions extending over distances of at least 2–3 nm. This means of calculating crystallite sizes from the width of the reflections is very suspect, however, in chainlike macromolecules. The positions of the monomeric units of the individual chains are, in fact, slightly displaced with regard to one another on crystallization, since, for kinetic reasons, parts of the chain are fixed in states of order before the whole chain attains its ideal lattice position. Local variations in the lattice constants are caused by this effect, and these similarly enlarge the width of the reflections.

Drawn fibers and films show X-ray diffraction pictures which are similar to the Bragg crystal rotation photographs (Figure 5-5). The crystal rotation method was originally carried out in order to orient favorably as many crystal planes as possible with regard to the X-ray beam. In drawn fibers and films, the molecular axes lie mainly in the direction of elongation (see Section 5.6). A ray that impinges normal to the draw direction will

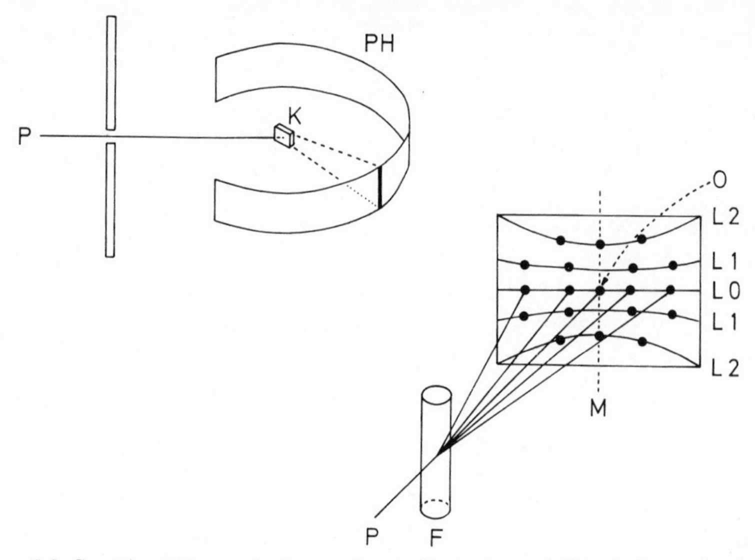

Figure 5-5. Crystal rotation method according to Bragg (upper left) and the production of a fiber diagram (lower right) by X-ray measurements. K, crystal; PH, photographic film; P, primary ray; F, fiber; M, meridian; 0, center; L0, L1, L2: zeroth, first, and second planes. L0, the equator.

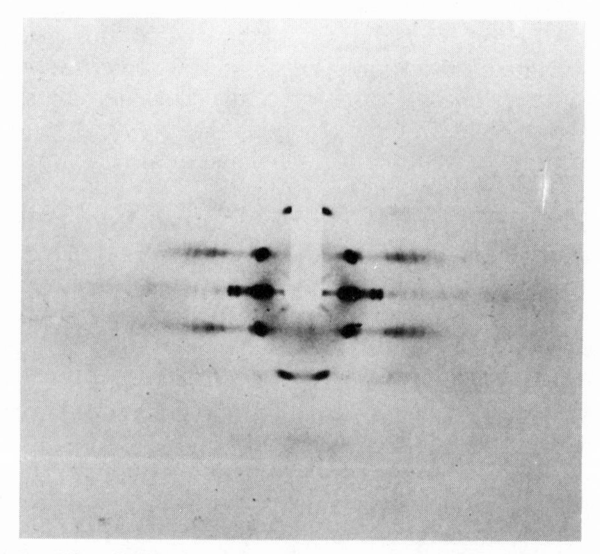

Figure 5-6. Fiber diagram of a drawn film of it-poly(propylene) (R. J. Samuels).

therefore produce reflections of varying sharpness on a photographic plate.

For historical reasons, such X-ray diffraction pictures are called fiber diagrams, although, of course, they are obtained also from drawn films. However, in contrast to crystal-rotation photographs, the fibers do not need to be rotated for fiber photographs, because many crystallites are already oriented. Reflections on the zero line are called "equatorial," and correspond to crystal planes lying parallel to the molecular axis (draw direction). Crystal planes that lie vertical ("normal" to the molecular axis produce what are called meridional reflections. Meridional reflections lie in a plane that bisects the equatorial line. When the crystallites are insufficiently oriented, the reflections (spots) degenerate into crescents (arcs) (see also Section 5.6). Thus, in shape, arcs lie between the spots of the fiber diagram with full crystallite orientation, and the circles produced by randomly oriented crystallites.

The conformation of helices can be elucidated from the observed number and spacing of layer lines in helix-forming macromolecules. In a 3_1 helix every fourth, seventh, etc., chain link is in the same position as the first. For this case, there should be three layer lines, as is shown in Figure 5-6 for a drawn film of it-poly(propylene).

5.2.2. *Density Measurements*

In general, molecules are more closely packed in the crystalline than in the amorphous state. The density of a crystalline polymer is correspondingly higher ($\rho_{cr} > \rho_{am}$) and its specific volume subsequently lower ($v_{cr} < v_{am}$). A degree of crystallinity α_m by mass can therefore be determined from the observed specific volumes v_{obs}, assuming the two-phase model and additivity for the specific volumes v_{cr} and v_{am}:

$$v_{obs} = \alpha_m v_{cr} + (1 - \alpha_m) v_{am} \tag{5-2}$$

or, solved for α_m,

$$\alpha_m = \frac{v_{am} - v_{obs}}{v_{am} - v_{cr}} \tag{5-3}$$

A degree of crystallinity α_v by volume can be defined analogously:

$$\alpha_v = \frac{\rho_{obs} - \rho_{am}}{\rho_{cr} - \rho_{am}} \tag{5-4}$$

The density ρ_{obs} or the specific volume v_{obs} is determined directly by experiment. The density-gradient column is suitable for this. A density-gradient column contains a liquid whose density increases continuously from the meniscus down to the base. Such liquids can consist, for example, of mixtures of organic solvents or of salt solutions. They should neither dissolve nor swell the macromolecular sample to be studied, but must wet it. With the appropriate mechanical apparatus, for example, it is possible to form density gradients in which the densities vary, the gradient becoming linear, concave, convex, etc., with the height of the column of liquid. A macromolecular sample then remains suspended at a particular level, according to its density.

The specific volume of an amorphous substance is obtained when the specific volume of the melt is extrapolated past the melting point to lower temperatures. An attempt may also be made to produce completely amorphous standard substances by quenching the melt, etc. The specific volume of a crystalline substance is calculated from the crystal structure. For this, one uses the volume V_e of the unit cell (see Section 5.3.1), containing the number N_i of atoms of type i having atomic weight A_i:

$$\frac{1}{v_{\mathrm{cr}}} = \rho_{\mathrm{cr}} = \sum \frac{N_i A_i}{V_e} \tag{5-5}$$

The densities of amorphous and crystalline polymers can differ by up to 15% (Table 5-3). Polymers with unsubstituted monomeric units, such as poly(ethylene) and nylon 6,6, for example, show the greatest difference in density. These chains crystallize in an all-*trans* conformation with particularly close packing of molecular chains. In helix-forming macromolecules with large substituents, such as it-poly(styrene), for example, the packing is, by contrast, less efficient.

Table 5-3. Polymer Densities in Completely Amorphous and Completely Crystalline States

Polymer	Density, g/cm³		
	Crystalline ρ_{cr}	Amorphous ρ_{am}	$\rho_{\mathrm{cr}} - \rho_{\mathrm{am}}$
Poly(ethylene)	1.00	0.852	0.148
it-Poly(propylene)	0.937	0.854	0.083
it-Poly(styrene)	1.111	1.054	0.057
Poly(vinyl alcohol)	1.345	1.269	0.076
Poly(ethylene terephthalate)	1.455	1.335	0.120
Bisphenol A-polycarbonate	1.30	1.20	0.10
Nylon 6,6 (α modification)	1.220	1.069	0.151
trans-1,4-Poly(butadiene)	1.020	0.926	0.094

5.2.3. Calorimetry

Differing states of order give rise to differing specific heats in crystalline and amorphous polymers. Presupposing a two-phase model, a degree of crystallinity can be calculated in partially crystalline polymers by means of the enthalpy, analogous to equation (5-3),

$$\alpha_{\text{cal}} = \frac{H_{\text{am}} - H_{\text{obs}}}{H_{\text{am}} - H_{\text{cr}}} = \frac{\Delta H_{\text{m}}}{\Delta H_m^{\circ}} \tag{5-6}$$

H_{am}, H_{cr}, and H_{obs} are the totally amorphous, totally crystalline, and test sample enthalpies, respectively. Correspondingly, ΔH_m is the heat of fusion of the sample and ΔH_m° is that of a perfectly crystalline material. ΔH_m° is very difficult to determine in macromolecules, since perfectly crystalline substances are never obtained (ΔH_m° is usually found from melting-point depression by diluent and occasionally from low-molecular-weight compounds, which naturally leads to some uncertainty). If the samples being studied have very small crystallites, then a quantity $\Delta H_m'$ is measured instead of ΔH_m, and this includes the surface energy γ at the end surfaces of the crystallites, as well as the length L of the crystallite:

$$\Delta H' = \Delta H_m - \frac{2\gamma}{\rho_{\text{cr}} L} \tag{5-7}$$

For this, it is assumed that the molecular axes lie parallel to the length L (see Section 10.4.1).

5.2.4. Infrared Spectroscopy

In the IR spectra of crystalline polymers, absorption bands often appear which are completely absent in amorphous polymers (Table 5-4). These bands lie mainly in the range 650–1500 cm^{-1}. Consequently, they originate mainly from bond angle deformations, which are, in turn, affected by the macromolecular conformation. These bands in the IR spectrum thus relate primarily to the conformation of the individual macromolecules, and not to intermolecular interactions. However, macromolecules can crystallize into varying conformations, thereby giving rise to differing crystal modifications (see Section 5.3.1). These modifications can, in turn, coexist in one sample. To determine the degree of crystallinity from IR measurements, it is first necessary to ascertain whether all the crystalline contributions are included in the chosen band.

A common method of determining the crystallinity from IR measure-

Table 5-4. IR Bands Used in Determining Crystallinity of Polyolefins

Polymer	IR bands, cm^{-1} Amorphous	Crystalline
Poly(ethylene)	1298	1894; 719
it-Poly(propylene)	4274	975; 894
it-Poly(butene-1) (orthorhombic modification)	4274	815; 922

ments relates the measured absorbance (previously: extinction) A_{cr} of a crystalline band to the absorbance A_{cr}° of a 100% crystalline sample via the degree of crystallinity α. For amorphous bands it follows analogously that $A_{am} = (1 - \alpha_{IR}) A_{am}^{\circ}$. This assessment is also based on a two-phase model. Both A_{am} and A_{cr} are naturally also proportional to the total quantity of the sample, but since only the ratio $D = A_{cr}/A_{am}$ is measured, the influence of the amount is eliminated. By inserting the expressions for A_{cr} and A_{am} into that for D, the following is obtained:

$$\alpha_{IR} = \frac{D}{D + (A_{cr}^{\circ}/A_{am}^{\circ})} \tag{5-8}$$

The absorbances A_{cr}° or A_{am}° of the totally crystalline and totally amorphous samples, respectively, must, in turn, be determined independently.

5.2.5. Indirect Methods

Indirect methods of determining the degree of crystallinity start from the fact that a given chemical or physical event proceeds differently in the crystalline phase and in the amorphous phase. Common physical experiments include, for example, the study of water vapor absorption of hydrophilic polymers or the diffusion of a dye into the polymer. Together with a series of chemical reactions (hydrolysis, reaction with HCHO, deuterium exchange), they are used in particular for determining the crystallinity of cellulose.

The degrees of crystallinity obtained by these indirect methods are not, however, very reliable. That is to say, swelling can occur when water and chemical reagents penetrate the solid polymer, but this results in a change in the accessibility of the separate regions, and the degree of crystallinity obtained consequently no longer relates to the original sample.

5.3. Crystal Structure

5.3.1. Molecular Crystals

Spherically shaped proteins readily crystallize out of aqueous solutions. The crystals so produced can contain up to 95% water of salt solution. The water or salt solution resides in the holes of the crystal lattice formed by the protein molecules. Since the holes adjoin one another, channels are produced. These holes and channels are so large that substrates can diffuse into the crystal and react, for example, enzymatically. Use is made of such diffusion (of heavy metals) in X-ray analysis, because the heavy metals allow the phases of the scattered ray distribution and thus the three-dimensional structure of the protein crystal to be determined.

Long-chain macromolecules only seldom form such large crystals. When they do, for example, with poly(2,6-diphenyl-1,4-phenylene ether) from tetrachloroethane, then they can also retain relatively large amounts of solvent (for this example, up to 35%) in the crystal.

5.3.2. Elementary and Unit Cells

While the diffracted X-ray intensity is a measure of the crystallinity of the sample, the positions of the diffracted rays give information about the crystal structure, i.e., the unit cell.

In low-molecular-weight substances, the unit cell contains at least one whole molecule. The unit cell therefore represents the lowest periodicity in an X-ray spectrum. The largest dimension of this unit cell corresponds, for example, in unbranched low-molecular-weight paraffins, to the length of the extended paraffin molecule in an all-*trans* conformation. Thus, in n-alkanes, $H(CH_2)_nH$, the length L of the unit cell is equal to the contour length up to a chain-link number of $n = 70$ (Figure 5-7). In other molecules, for example, in polyurethanes, the molecular axis is not perpendicular but diagonal to the base plane. In this case L is not equal, but only proportional, to the contour length. Since L increases with increasing n, the reflections resulting from the unit cell lengths are displaced to smaller and smaller angles. The two other dimensions of the unit cell are affected by the elongation of the molecular chain perpendicular to the molecular axis and the distances between the molecules.

Above $n \approx 80$, however, the length L of the alkane unit cell remains constant at room temperature, with $L \approx 10.5$ nm. Since the contour length continues to increase as n increases, the chain must consequently fold back on itself into the crystal. An analogous effect is observed in other polymers, for example, with polyurethanes.

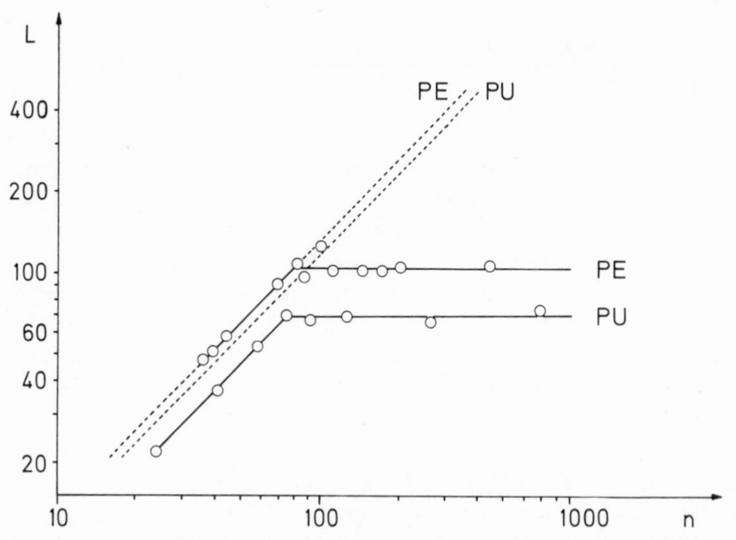

Figure 5-7. Dependence of the length L of the observed long period (in Å) on the number n of chain links in alkanes (PE) with the constitutional formula $H(CH_2)_nH$ (54 C. c direction), and in polyurethanes (PU) with the constitutional formula $HO{+}CH_2)_2{-}O{-}(CH_2)_2{-}[O{-}CO{-}NH{-}(CH_2)_6{-}NH{-}CO{-}O{-}(CH_2)_2{-}O{-}(CH_2)_2]_x{-}OH$ (room temperature). The long periods of the low-molecular-weight polyurethanes are considerably lower than those calculated for an all-*trans* conformation (---), the molecular axes must therefore be diagonal to the base plane. [Measurements on alkanes and poly(ethylenes) from various authors, and on urethanes from W. Kern, J. Davidovits, K. J. Rauterkus, and G. F. Schmidt.] (1 Å = 0.1 nm.)

In chain macromolecules, however, in addition to this large periodicity coming from the unit cell, a smaller periodicity is also observed. This results from the elementary cell—a subcell of the unit cell. In the unit cell of n-alkanes, the CH_2 links recur periodically. Since n-alkanes, and also poly(ethylene), crystallize into the all-*trans* conformation, every third, fifth, seventh, etc., CH_2 group has the same position as the first in the crystal lattice. As a result of the recurring methylene groups, an elementary cell is produced, which gives rise to short periodicity in the X-ray diagram. Subcell periodicity can be detected by X-ray scattering as strong reflections at relatively high angles. The arrangement of the molecular segments in the elementary cell can be deduced from the position of these reflections and their intensities. The chain structure of the macromolecule means that, in a crystal lattice, the atom intervals in the direction of the chain are different from those perpendicular to it. This anisotropy prevents the appearance of cubic lattices. The remaining six kinds of lattice (hexagonal, tetragonal, rhombohedral, orthorhombic, monoclinic, and triclinic) are observed, on the other hand, in long-chain macromolecules (Table 5-5). The direction of the molecular chain is usually termed the c direction. The

Table 5-5. Lattice Constants and Crystal Forms of Some Crystalline Polymers at 25°C (1 Å = 0.1 nm)

Polymer	Number of base units in the unit cell	Lattice constants, Å *a*	*b*	*c*	Helix	Crystal system
Poly(ethylene)	2	7.36	4.92	2.534	—	Orthorhombic
st-Poly(vinyl chloride)	4	10.40	5.30	5.10	—	Orthorhombic
Poly(isobutene)	16	6.94	11.96	18.63	8_5	Orthorhombic
it-Poly(propylene) (α form)	12	6.65	20.96	6.50	3_1	Monoclinic
it-Poly(propylene) (β form)	?	6.47	10.71	?	3_1	Pseudohexagonal
it-Poly(propylene) (γ form)	3	6.38	6.38	6.33	3_1	Triclinic
st-Poly(propylene)	8	14.50	5.81	7.3	4_1	Orthorhombic
it-Poly(styrene)	18	22.08	22.08	6.63	3_1	Triclinic
it-Poly(vinyl cyclohexane)	16	21.9	21.9	6.50	4_1	Tetragonal
it-Poly(*o*-methyl styrene)	16	19.01	19.01	8.10	4_1	Tetragonal
it-Poly(butene-1) (mod. 1)	18	17.69	17.69	6.50	3_1	Triclinic
it-Poly(butene-1) (mod. 2)	44	14.85	14.85	20.60	11_3	Tetragonal
it-Poly(butene-1) (mod. 3)	?	12.49	8.96	?	?	Orthorhombic

value of $c = 0.2534$ nm in poly(ethylene) is exactly what results from the carbon–carbon bond length of 0.154 nm, and the C—C—C bond angle of 112° for the distance between every second CH_2 group in the chain when poly(ethylene) crystallizes into an all-*trans* conformation (Figure 5-8). st-Poly(vinyl chloride) likewise crystallizes into an all-*trans* conformation, but only every second CHCl group is in the same position as the first, so that the lattice constant is doubled, becoming $c = 0.51$ nm. In poly(isobutene), on the other hand, the *c* value is not an integer multiple of 0.253 nm, so that the absence of an all-*trans* chain conformation can be concluded from this value alone. Indeed, poly(isobutene) adopts an 8_5 helical conformation in the crystal. The lattice constants are naturally very dependent on constitution and configuration, as can be seen from the four isomeric poly(butadienes) (Figure 5-9).

With increased temperatures, the measurements in the *c* direction, e.g., in poly(ethylene), remain constant, since the bond lengths and the

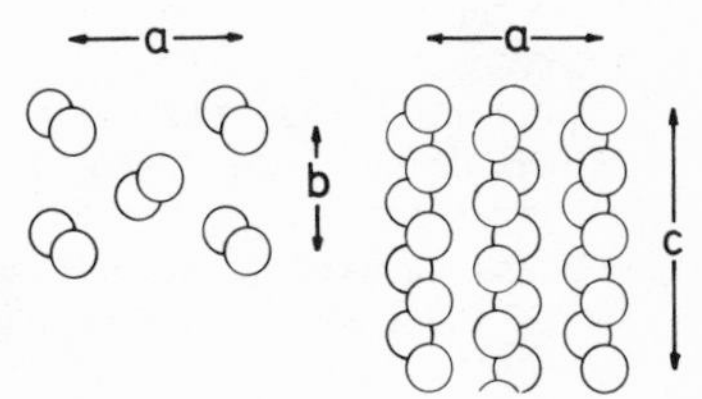

Figure 5-8. The arrangement of the CH_2 groups (shown as ○) in the crystal lattice of poly(ethylene). Because of chain folding, the chains proceed in an antiparallel direction. (After C. W. Bunn.)

valence angles of the chain remain essentially constant. However, since the forces between the molecules are affected by temperature, the a and b values must alter. In poly(ethylene), for example, for a temperature increase from $-196°$ to $+138°C$, the value of b increases by $\sim 7\%$.

In the packing of long-chain macromolecules already discussed, the lateral order plays a subordinate role. Effects of this kind are noticeable, however, if strong hydrogen bonds can be formed between the individual chains, as, e.g., in polyamides and proteins. A few representatives of this class of substance crystallize in the form of pleated-sheet structures, as is shown in Figure 5-10 for nylon 6 and nylon 6,6.

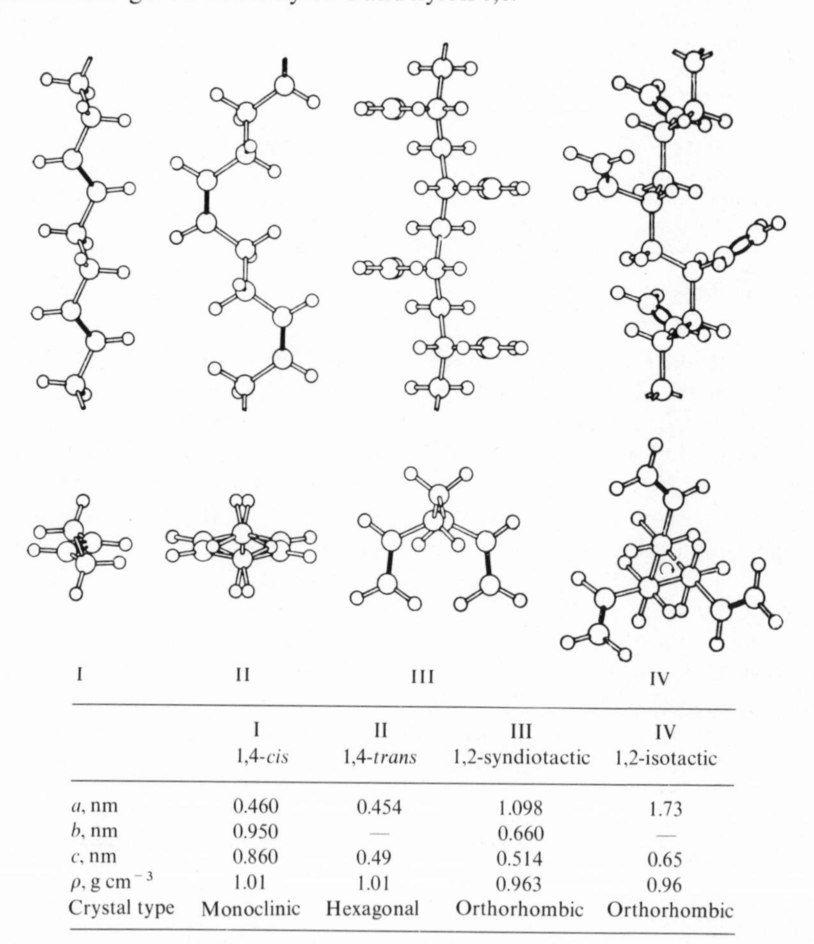

	I 1,4-*cis*	II 1,4-*trans*	III 1,2-syndiotactic	IV 1,2-isotactic
a, nm	0.460	0.454	1.098	1.73
b, nm	0.950	—	0.660	—
c, nm	0.860	0.49	0.514	0.65
ρ, g cm^{-3}	1.01	1.01	0.963	0.96
Crystal type	Monoclinic	Hexagonal	Orthorhombic	Orthorhombic

Figure 5-9. Conformations, lattice constants (a, b, and c), density ρ, and crystal forms of the four isomeric poly(butadienes) (after G. Natta and P. Corradini).

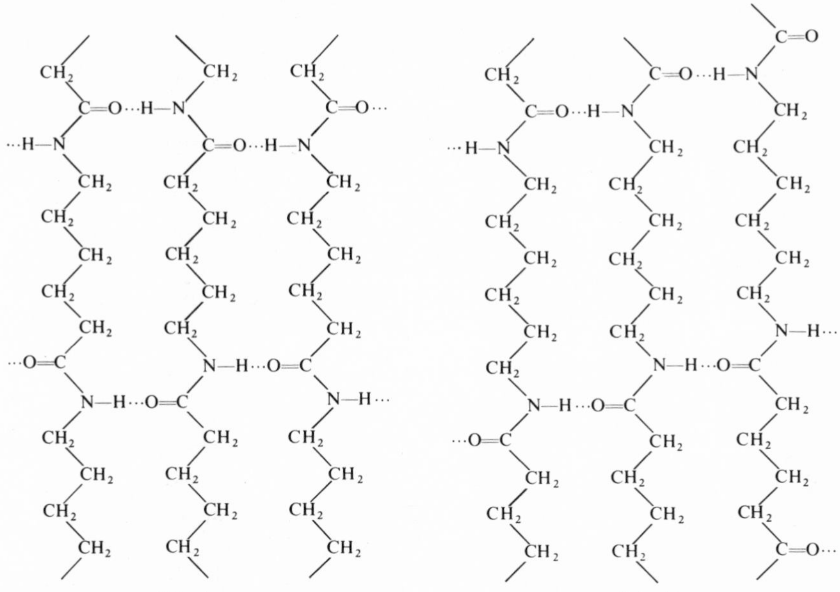

Nylon 6

Nylon 6,6

Figure 5-10. Pleated sheet structures of nylon 6 [poly(caprolactam)] and nylon 6,6 [poly-(hexamethylene adipamide)].

Certain proteins show both helix and coil sequences in their chain, and these can fold back on themselves in the low-energy state in such a way that compact spherical or ellipsoidal structures are produced (Chapter 4.6.2). In molecules of this kind, the unit cells can contain several molecules.

5.3.3. *Polymorphism*

Polymorphism is defined as the occurrence of different crystal modifications of the same molecule or polymer possessing the same monomeric unit. The various modifications are characterized by different lattice constants or lattice angles, and, consequently, different unit cells. The different unit cells result in microscopically perceptible differences in crystal shape, solubility, melting point, etc.

Polymorphism can result from conformational differences in a chain molecule or different packings of molecules with the same conformation. Such differences can be induced by slight alterations in the crystallization conditions, for example, by varying the crystallization temperature.

Polymorphism is observed relatively frequently in long-chain macromolecules, for which approximately isoenergetic structures exist. The stable crystal form of poly(ethylene), for example, possesses an orthorhombic lattice, but on elongation, triclinic and monoclinic modifications are observed. Three modifications are known in it-poly(propylene): α (monoclinic), β (pseudohexagonal), and γ (triclinic). Since the molecules are in a 3_1 helix conformation in all the modifications, differences in the packing of the chain must be responsible for this polymorphism. The three modifications appear at varying crystallization temperatures. In it-poly(butene-1), however, the various modifications correspond to different kinds of helix, so that variations in conformation must be important (see also Table 5-5).

5.3.4. Isomorphism

The phenomenon by which various monomer units can replace each other in the lattice is termed isomorphism. Isomorphism is possible in copolymers if the corresponding unipolymers show analogous crystal modifications, similar lattice constants, and the same helix type. For example, according to Table 5-5, the γ form of it-poly(propylene) and modification 1 of it-poly(butene-1) possess triclinic crystal form, similar lattice constants for the c dimension, and the same helix type. The copolymers of propylene and butene-1 therefore show isomorphism. Isomorphism occurs particularly readily in helix-forming macromolecules, since the helix conformations lead to "channels" in the crystal lattice, which can easily accommodate different substituents.

5.3.5. Lattice Defects

Crystal lattices can possess a number of different defects. Some of these are characteristic of all nonmetallic solids; others are specific to crystalline macromolecular substances. Phonons, electrons, holes, excitons, site defects, interstitial defects, and displacements are commonly occurring lattice defects. Typical crystalline macromolecular substance lattice defects result from end groups, kinks, jogs, Reneker defects, and chain displacements. Distortion of the whole crystal lattice can be conceived in terms of the paracrystal. The defects can be classified in terms of point, line, and network defects.

General Point Defects. Lattice atoms can oscillate thermally about their ideal positions. This oscillation can be conceived in terms of the oscillation of an elastic body with the energy $h\nu$. Such elastic bodies are called *phonons*. *Electrons* and *holes* are especially important with nonmetallic semiconducting solids. A semiconductor is considered to be perfect when it has an empty semiconducting band. An isolated electron in a perfect solid will, of course, produce a defect. Holes are quantum states in a normally filled semiconducting band. They behave in an electric field like a positive charge. Electrons and holes can be produced by thermal motion or the absorption of light. *Excitons* are electron/hole pairs. Excitons are produced when an electron takes up energy, but not a sufficient amount to escape the "hole" produced. Consequently, the electronic charge of an exciton is zero. The exciton can transport energy but cannot conduct an electric current. Empty lattice sites are called *site defects* or vacancies. Atoms residing in sites between lattice points are called *interstitial defects*.

Special Macromolecular Point Defects. *End groups* have a different chemical structure than that of the main-chain monomeric unit. They

Figure 5-11. Some lattice defects in poly(ethylene). From left to right: all-*trans* conformation (defect-free), Reneker defect, kink, and jog.

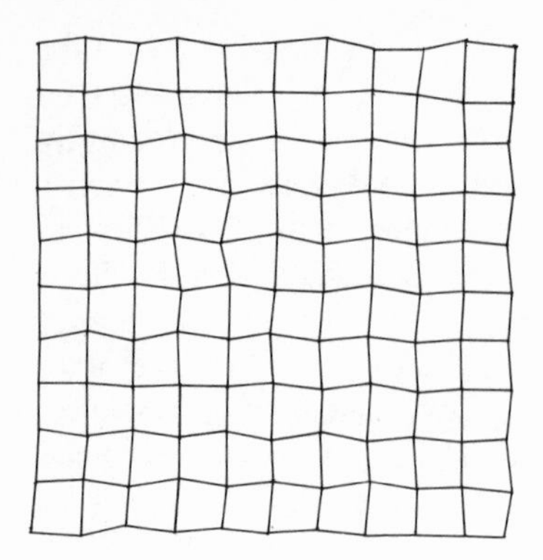

Figure 5-12. Paracrystal (schematic).

consequently produce a defect in the crystal lattice (see also Figure 5-18).

Kinks, jogs, and *Reneker defects* are conformational defects (see Figure 5-11). With kinks and jogs, a part of the chain is displaced perpendicular to the long axis by "false" conformations. This kind of defect is called a kink when the displacement is smaller than the interchain distance (example: ... $ttttg^+tg^-tttt$ ...). If, on the other hand, the displacement is larger than the interchain distance (e.g., ... $ttttg^+ttttg^-tttt$ ...), then the defect is called a jog. Kinks and jogs shorten planar chains and twist helices.

Reneker defects result from conformational defects as well as from changes in bond angles (see Figure 5-11). As with kinks and jogs, the chain is also shortened here. Reneker defects can pass along the polymer chain without causing any change in the relative position of the chain in the crystal aggregate. On the other hand, spacious chain movements would result if kinks and jogs were to move through the lattice system.

Network defects occur when the lattice atom positions are randomly displaced from their ideal lattice positions. Network defects can be conceived in terms of the paracrystal model (Figure 5-12).

5.4. Morphology of Crystalline Polymers

5.4.1. Fringed Micelles

In the early days of macromolecular physics, crystalline reflections were observed alongside amorphous halos in the X-ray photograph of gelatine (a degradation product of the protein collagen), and this was interpreted according to the two-phase model as the coexistence of perfect

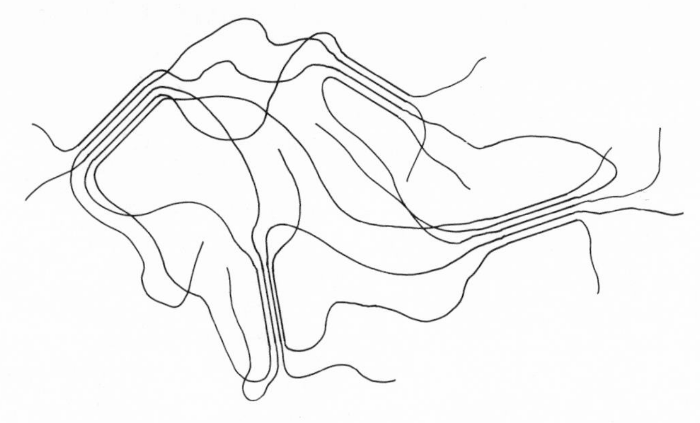

Figure 5-13. Fringed micelle model: One chain may run through several crystalline and amorphous regions.

crystalline regions and totally amorphous regions. Crystallite lengths of 10–80 nm were calculated from the line broadening of the reflections in these wide-angle X-ray exposures, and later from the positions of the reflections in small-angle X-ray scattering. The sizes of the crystallites were thus smaller than the maximum molecule length, which could be found from the molecular weight. In poly(oxymethylenes), it was also noticed that, with increasing molecular weight, the short periods resulting from the elementary cells remained, while the long periods vanished. This effect was ascribed to the absence of higher order, but higher order can only originate from lattices of high regularity. Since the then-known macromolecular substances could not be seen to crystallize either by direct observation or under the microscope, the alternative possibility—random lattice defects in crystals—seemed less probable. Thus, all these findings lead to the fringed-micelle model (Figure 5-13).

In this model, it is assumed that an individual molecular chain runs through several crystalline regions dispersed in an amorphous matrix. The model could explain the X-ray results and many other observations. Such observations and their explanations are: the smaller macroscopic density compared to the unit cell density, as a consequence of the amorphous regions; the appearance of arcs in the X-ray diagrams of drawn polymers as a result of crystallite orientation; the finite melting range as a result of crystallites of varying sizes; the optical birefringence of drawn polymers caused by the orientation of molecular chains in amorphous regions; and the heterogeneity in relation to chemical and physical reactions as a result of the greater accessibility of the amorphous phase compared to the crystalline.

5.4.2. Polymer Single Crystals

In 1957 it was found, however, that solutions of ~ 0.1 % of poly(ethylene) produced, on cooling, small, rhombohedral platelets, which were visible under the electron microscope (Figure 5-14). For a given solvent, the height of these platelets was always the same for a given crystallization temperature. Electron diffraction showed sharp, point reflections that suggested single crystals. The interpretation of the electron-diffraction diagram led to a model in which the direction of the molecular chain is vertical to the surface of the platelets. Since the height of these rhombohedral platelets is smaller than the contour length, it follows that the chain must fold back on itself (Figure 5-15). Polymer single crystals have been observed in many crystallizable macromolecular substances, for example, in poly(oxymethylene), poly(acrylonitrile), nylon 6 (see Figure 5-1), poly(acrylic acid), cellulose derivatives, and amylose.

Studies on tearing also indicate chain folding in single crystals. According to Figure 5-15, the molecular lamellae must change direction at the diagonals in a chain folding. Lamellar cracking (cleavage) should therefore

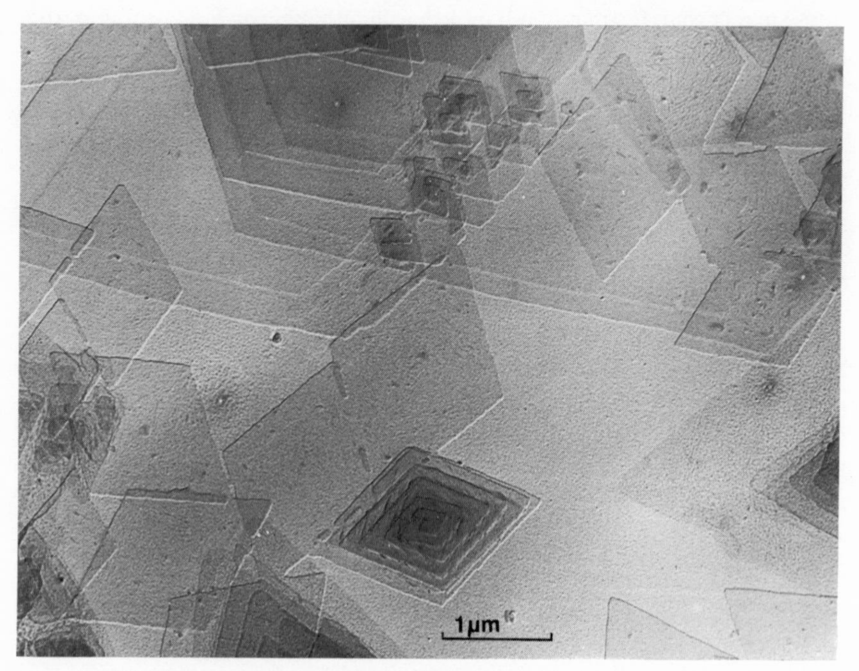

Figure 5-14. Single crystals of poly(ethylene) obtained from dilute solution. A spiral dislocation can be observed in lower center crystal. (After A. J. Pennings and A. M. Kiel.)

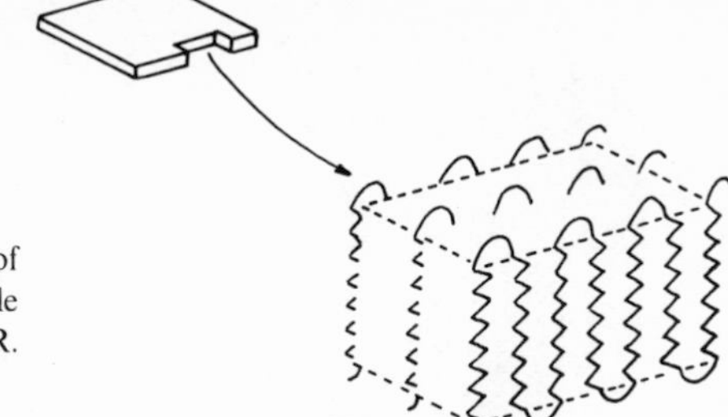

Figure 5-15. Schematic representation of chain folding in a poly(ethylene) single crystal (after W. D. Niegisch and P. R. Swan).

stop at the diagonals (Figure 5-16). Cracking causes the molecules to be drawn out perpendicular to the chain axis in the form of fibrils (Figure 5-17).

The exact nature of the fold surface of the polymer single crystal is not known. Back-folding certainly does not occur as regularly as shown in Figure 5-15. The single crystals are about 75–85% crystalline according to X-ray measurements. About 100% crystallinity is obtained if the surface layers are decomposed with fuming nitric acid. This leads to the conclusion that the surface layers are moderately disordered ("amorphous"). It is questionable whether reentry of the chain to the fold surface occurs in a neighboring position (as shown in Figure 5-18) or in a more distant position.

Crystallites of this type are also called chain-folded crystals since the

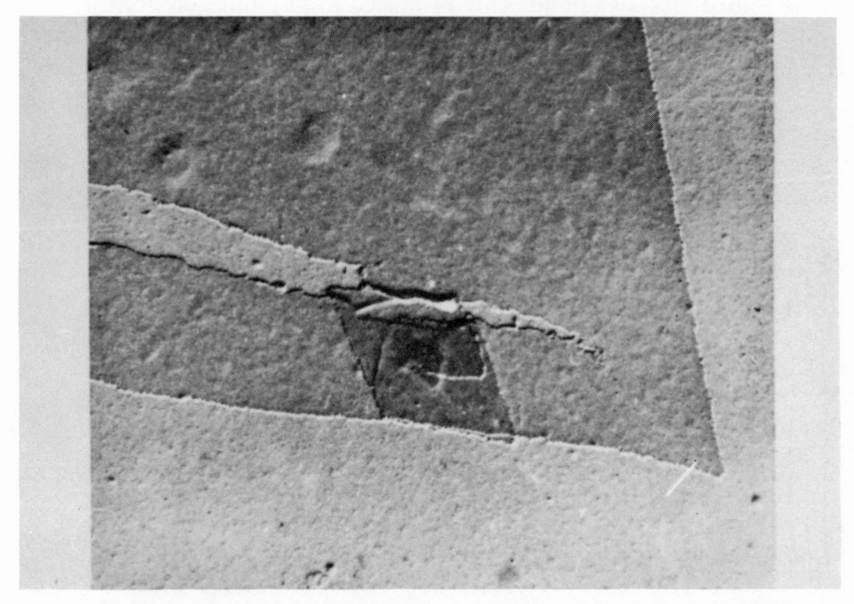

Figure 5-16. The termination of a dislocation in the direction of the plane of the chain at the diagonal of a polymer single crystal (after P. H. Lindenmeyer, V. F. Holland, and F. R. Anderson).

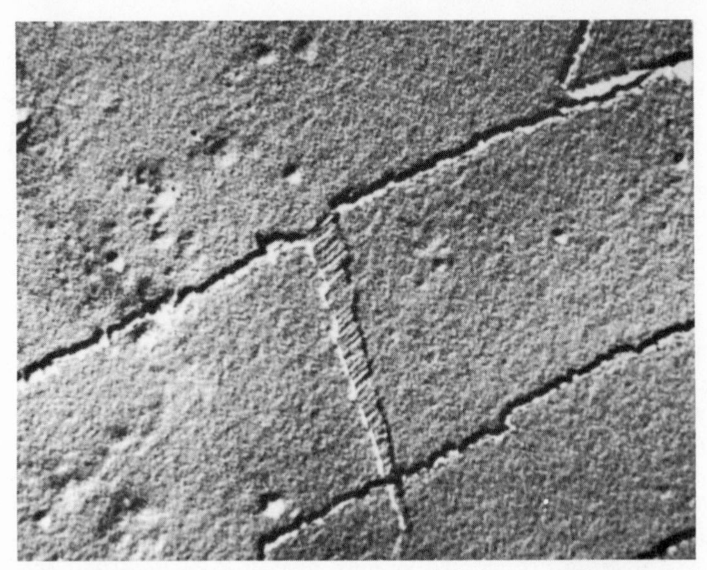

Figure 5-17. Tear perpendicular to the plane of the chain in a poly(ethylene) single crystal. The molecules are drawn out as fibrils at the point of tear. (After P. H. Lindenmeyer, V. F. Holland, and F. R. Anderson.)

macromolecules in polymer single crystals crystallize with chain folding. Chain-folded crystals are not only produced from dilute solutions (Figure 5-14), they also occur as lamellar structures on crystallizing from the melt (Figure 5-19). Lamellar height is strongly increased (i.e., the relative propor-

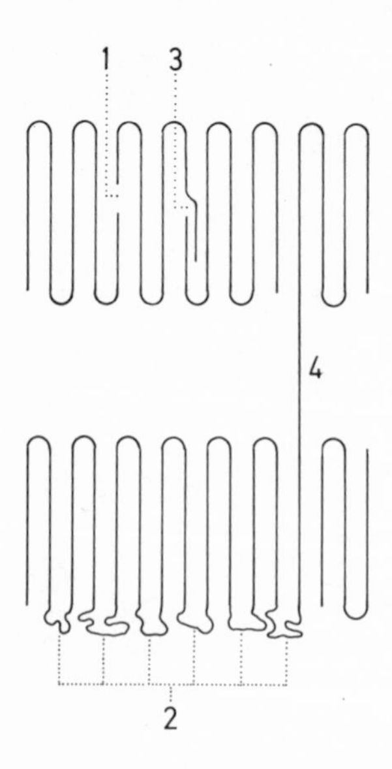

Figure 5-18. Some possible lattice defects with chain folding. 1: Chain ends; 2: disordered surface layer; 3: dislocations, 4: interlamellar linkages.

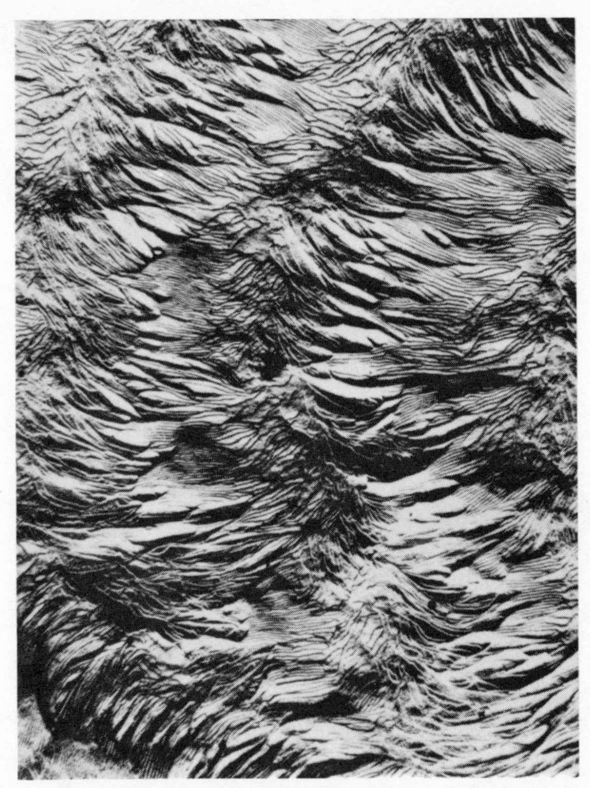

Figure 5-19. Lamellae of melt-crystallized poly(ethylene) (after P. H. Lindenmeyer, V. F. Holland, and F. R. Anderson).

tion of surface layer material is decreased) when crystallization from the melt occurs under pressure (Figure 5-20). Thus, such "extended chain crystals" correspond to the equilibrium state. Chain folding must therefore result from kinetic factors (see also Chapter 10).

The fold length or lamellar height can be increased by raising the crystallization temperature, as well as by increasing the pressure. The fold length for many polymers increases linearly with the reciprocal difference between the melting point and the crystallization temperature (Figure 5-21). This supercooling effect also suggests a kinetic mechanism for chain folding.

The fold length is independent of the supercooling in the case of polyamides, however. In this case, the fold length is determined by the number of hydrogens involved in hydrogen bonding. Nylon 3, nylon 6,6, and nylon 6,12 have 16 hydrogen bonds (i.e., four repeating units) per fold length. Nylon 10,10 and nylon 12,12 possess only 12 hydrogen bonds (e.g., only three repeating units) per fold length.

Figure 5-20. Extended chain crystals of poly(ethylene). Crystallization at 225°C and 4800 bar; $\overline{M}_w$ = 78.300 g/mol, $\overline{M}_n$ = 14,800 g/mol, 99% X-ray crystalline. (After B. Wunderlich and B. Prime),

Since the fold length increases sharply with increasing crystallization temperature, a polymer single crystal subsequently tempered at higher temperatures correspondingly increases in thickness. The material required to increase the fold length is taken from the interior of the crystal, thus producing holes there.

When crystallization is carried out from concentrated solution or from the melt, there is a high probability that parts of the same polymer molecule will crystallize into more than one lamella. Such interlamellar links (see Figure 5-18) were first shown to occur by jointly crystallizing poly(ethylene) with paraffin mixtures and subsequently dissolving the paraffins away (Figure 5-22). The number of interlamellar links increases with increasing molecular weight because the probability that parts of

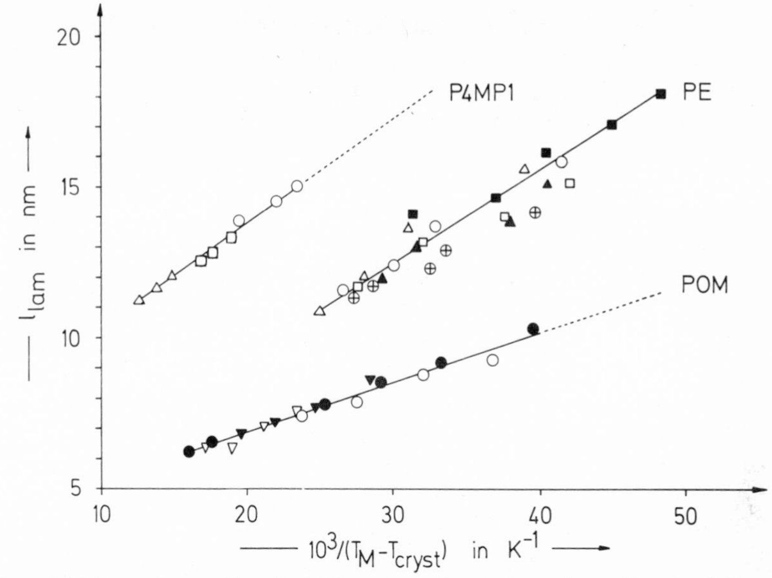

Figure 5-21. Dependence of lamella heights l_{lam} on the relative undercooling of poly(4-methylpentene-1) (P4MP1), poly(ethylene) (PE), and poly(oxymethylene) (POM) in tetralin, *p*-xylene, decalin, toluene, octane, hexadecane, *m*-cresol, furfuryl alcohol, and acetophenone (different symbols) (after A. Nakajima and F. Hameda).

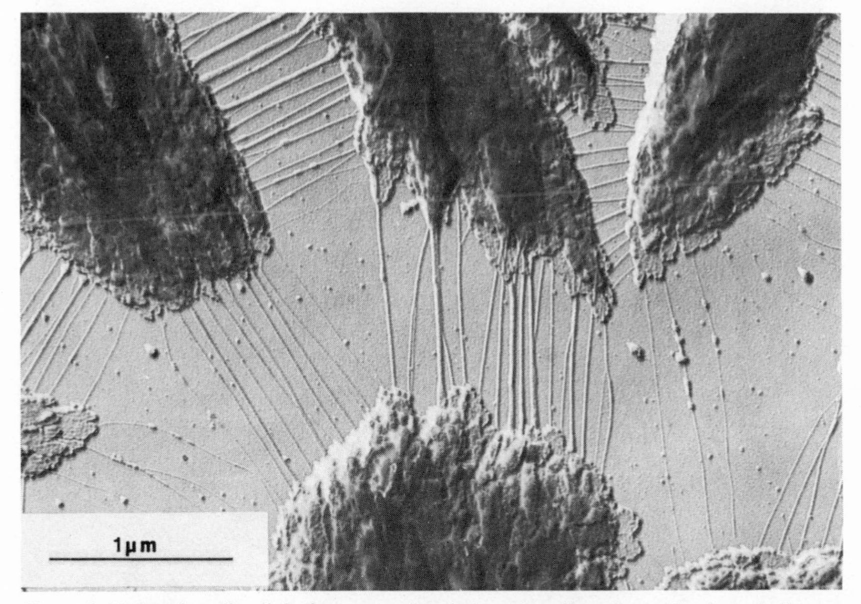

Figure 5-22. Interlamellar links between poly(ethylene) lamellae. Crystallization of mixtures of poly(ethylene) and paraffin wax (after H. D. Keith, F. J. Padden, and R. G. Vadimsky).

the same polymer molecule crystallize in more than one lamella increases with increasing molecular weight. Consequently, material crystallized from the melt always has a relatively high amorphous content. Interlamellar links are also called tie molecules.

5.4.3. Spherulites

In crystallization from the melt, polycrystalline regions sometimes occur which are called spherulites because of their spherical form and optical properties. Microtome sections show that their internal structure is radially symmetric. Circular structures of similar internal construction occur in the crystallization of thin films (Figure 5-23). They are therefore likewise termed spherulites, since they can be considered as cross sections of bulk-crystallized spherulites.

Spherulites with diameters between 5 μm and a few millimeters can be studied with an ordinary polarizing microscope, and those with diameters below 5 μm with an electron microscope or by small-angle light scattering. In polarized light, spherulites show the typical Maltese cross that is caused by birefringence effects (Figure 5-24). These effects occur because the speed of the light varies within the different spherulitic regions. The Maltese cross appears because the spherulites behave as crystals with radial optical symmetry, and in this case there are four extinction positions (see Section 7.4).

The differences in the speed of the light result from differences in the refractive index. If the highest refractive index is in the radial direction,

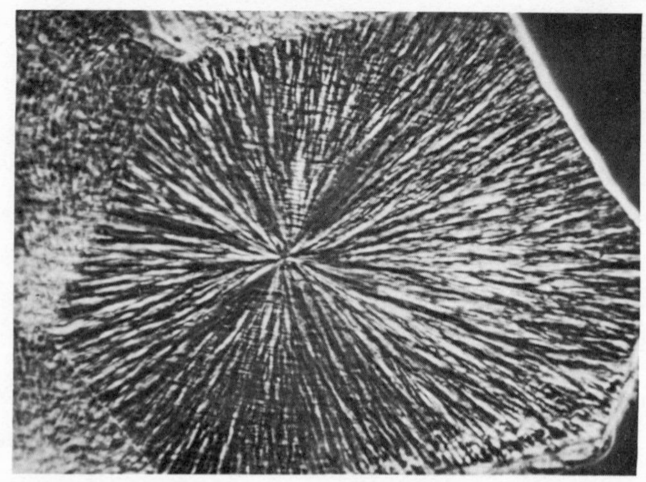

Figure 5-23. An it-poly(propylene) spherulite as seen under the phase-contrast microscope (after R. J. Samuels).

Figure 5-24. it-Poly(propylene) spherulites as seen under the polarizing microscope (after R. J. Samuels).

one talks of positive spherulites. Negative spherulites show the highest refractive index in the tangential direction. Thus, information about the microstructure of the spherulites can be gained from their optical properties.

To illustrate: In drawn poly(ethylene) fibers, the speed of light is less in the direction of the fibers than in the direction perpendicular to this. Here, light parallel to the fiber direction shows a higher refractive index. In drawn poly(ethylene) fibers, the molecular axes are largely parallel to the fiber axis. Since poly(ethylene) forms negative spherulites, the molecular axes must be at right angles to the spherulite radius.

In poly(vinylidene chloride), the refractive index is lower along, rather than perpendicular to, the molecular chain direction. Since the spherulites are positive, then here, too, the molecular axes must be arranged tangentially to the spherulite radius. This relationship occurs particularly in polymers with strongly polar groups, e.g., in polyesters and polyamides. In some cases, the same material can form both positive and negative spherulites, possibly even simultaneously. The negative spherulites of nylon 6,6, for example, have a higher melting point than do the positive spherulites.

Spherulites show an imperfect crystalline structure, since the melting point of the spherulite usually lies considerably below the thermodynamic melting point (see Chapter 10). Even then, a further increase in X-ray crystallinity can also be observed when the spherulites have filled the volume. Localized orientation of the crystalline region leads to the characteristic optical properties of spherulites. If spherulites are cross-linked by radiation, the identity of the individual spherulite is retained even after they have been heated above the melting point. The birefringence of oriented

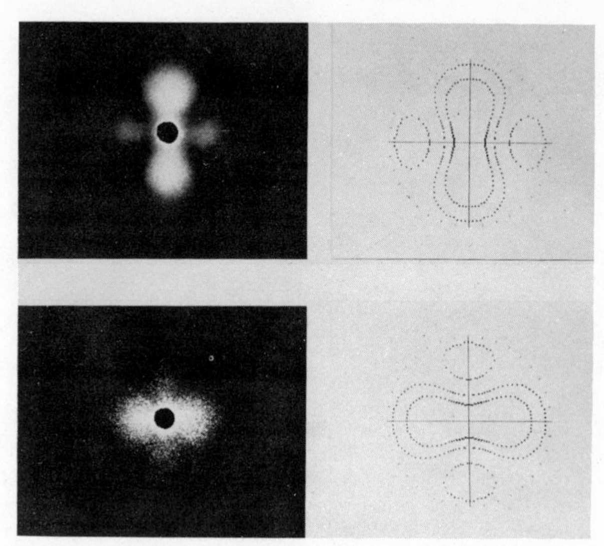

Figure 5-25. Narrow-angle light scattering of positive and negative spherulites. Photographs from incident and scattered light. Left: experimental; right: theoretical. Top: negative spherulites; bottom: positive spherulites. (After R. J. Samuels.)

sections of spherulites is lower, however, than in highly oriented fibers.

The orientation of the molecular axes in the spherulites can be followed particularly well through light scattering at very small angles. The scattering, for example, of vertically polarized incident light and the angular distribution of vertically polarized scattered light can be calculated. In this case, for example, positive and negative spherulites show different scattering diagrams (Figure 5-25).

Spherulites make films and foils opaque when their diameters are greater than half the wavelength of the light and when, in addition, inhomogeneities exist in relation to the density or to the refractive index. Spherulitic poly(ethylene), for example, is opaque, but spherulitic poly(4-methyl-pentene-1) is glass clear (at room temperature), even when the latter has the same number of spherulites with the same dimensions as poly(ethylene).

5.4.4. Dendrites and Epitaxial Growth

Spherulites are produced because the overall crystallization rate is the same for all directions in space. However, the growth rates of crystals in the spherulites may be directionally dependent. Conversely, if the overall crystallization rate is also direction dependent, then what are called dendrites are produced. Dendrites are structures that appear snowflake-like under

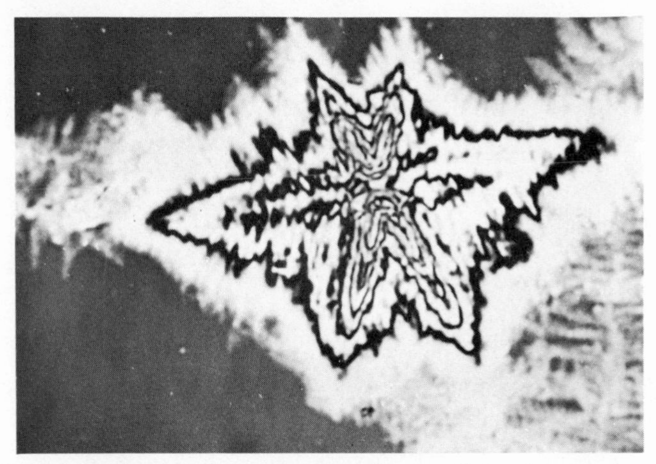

Figure 5-26. Poly(ethylene) dendrites crystallized from a dilute xylene solution at ~70°C (after B. Wunderlich).

the light or electron microscope (Figure 5-26). The amorphous material residing in the interior of the dendrites can be etched away by oxidation with nitric acid. The remaining crystalline component also has a lamellar structure of regular thickness.

Directionally dependent crystallization rates also lead to the "shish-kebab" (from the Armenian dish) structures (Figure 5-27). When a crystal-

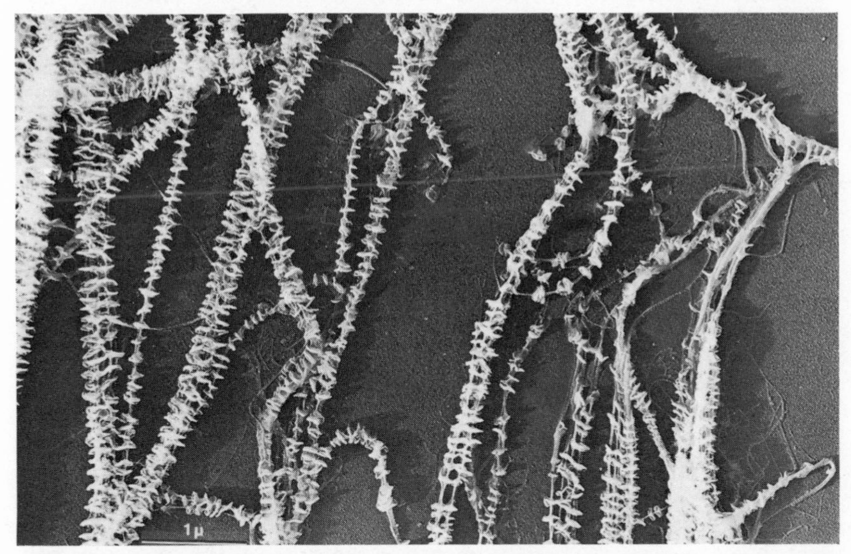

Figure 5-27. Shish-kebab structures of linear poly(ethylene) ($\overline{M}_w$ = 153,000, $\overline{M}_n$ = 12,000 g/mol). Crystallization from 5% xylene solutions with stirring at 102°C. (After A. J. Pennings and A. M. Kiel.)

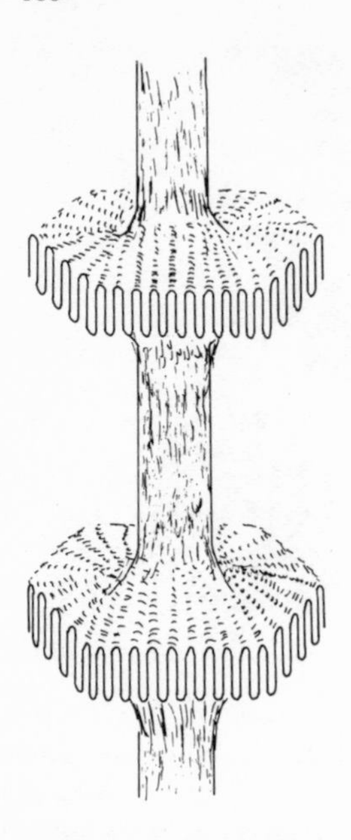

Figure 5-28. Schematic representation of arrangements of chains in shish-kebab structures (after A. J. Pennings, J. M. M. A. van der Mark, and A. M. Kiel).

lization from dilute solution occurs with strong stirring, shish-kebab structures are produced. The macromolecules are oriented along the direction of flow and settle out parallel to each other. X-ray scattering, electron diffraction, and birefringence measurements show that the chains lie parallel to the fiber axis. The fibrils produced order themselves into nucleation bundles, but the shear gradient is strongly reduced between these nucleation bundles. The remaining macromolecules in the solution between the fibrils crystallize out onto these in the form of folded-chain lamellae. The lamellae here lie vertical with respect to the fibrils (Figure 5-28).

Shish-kebab formation is a special case of epitaxial growth. The oriented growth of one crystalline substance on another is defined as epitaxy.

5.5. *The Amorphous State*

5.5.1. *Free Volumes*

By definition, no long-range order over extensive regions occurs in the amorphous state: For example, amorphous material is not X-ray crystalline. This definition, of course, does not consider the mutual arrangement in space of short chain segments or the order of the molecule itself.

A whole series of indications suggest that X-ray amorphous polymers possess a certain order. However, a definite number of vacant sites must be present. The density in the amorphous state is significantly different than that of the hypothetical liquid. The specific volume of polymer–monomer mixtures at first decreases linearly with increasing polymer concentration (Figure 5-29). At a certain polymer concentration, however, the viscosity of the mixture becomes so great that the polymer segments can no longer move freely. Because of this freezing-in process the specific volume of the amorphous polymer v_{am}° is larger than the specific volume of the liquid polymer v_l° would be at the same temperature. Conversely, the density of the liquid polymer is higher than the density of the solid: The solid polymer has vacant sites or what is called free volume. These vacant sites should be seen as having diameters of the order of those of atoms. The free-volume fraction f_{WLF} is calculated as

$$f_{WLF} = \frac{v_{am}^{\circ} - v_l^{\circ}}{v_{am}^{\circ}} \tag{5-9}$$

The same free volume also occurs in the Williams–Landel–Ferry dynamic glass-transition-temperature equation (see Section 10.5.2). The free-volume fraction is independent of the polymer type. It has a value of about 2.5% (Table 5-6).

In addition to the WLF free volume, a series of other free volumes can be defined and discussed. The vacant volume can be obtained from the specific volume of the amorphous polymer v_{am}° measured at the temperature

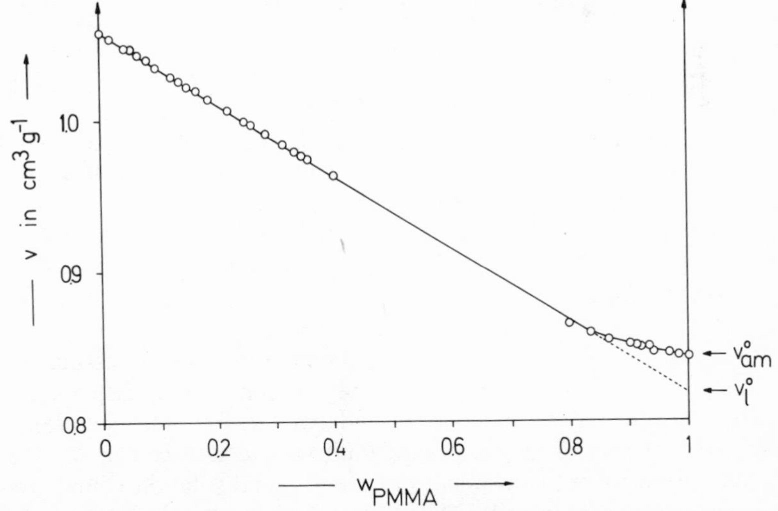

Figure 5-29. Specific volume of methyl methacrylate–poly(methyl methacrylate) mixtures as function of mass fraction of polymer at 25°C. $v_{am}^{\circ} = 0.842\ cm^3/g$, $v_l^{\circ} = 0.820\ cm^3/g$. (After D. Panke and W. Wunderlich.)

Table 5-6. Various Free-Volume Fractions of Amorphous Polymers Calculated from Crystalline Density at 0°C[a]

Polymer	Free-volume fraction			
	f_{vac}	f_{exp}	f_{WLF}	f_{fluc}
Poly(styrene)	0.375	0.127	0.025	0.0035
Poly(vinyl acetate)	0.348	0.14	0.028	0.0023
Poly(methyl methacrylate)	0.335	0.13	0.025	0.0015
Poly(butyl methacrylate)	0.335	0.13	0.026	0.0010
Poly(isobutylene)	0.320	0.125	0.026	0.0017

[a] Data from A. Bondi.

T and the specific volume v_{vdW}° calculated from the van der Waals radii. The vacant-volume fraction f_{vac} is

$$f_{vac} = \frac{v_{am}^{\circ} - v_{vdW}^{\circ}}{v_{am}^{\circ}} \tag{5-10}$$

As is true for simple liquids, the vacant-volume fraction here is considerable (Table 5-6). With macromolecules, the vacant volume is not completely available for thermal motion since not all vacant sites are accessible to monomeric units on conformational grounds. The volume available for thermal expansion f_{exp} can be calculated from the specific volumes of the amorphous and crystalline polymer at 0 K:

$$f_{exp} = \frac{(v_{am}^{\circ})_0 - (v_{cr}^{\circ})_0}{(v_{am}^{\circ})_0} \tag{5-11}$$

Finally, a fluctuation volume f_{fluc} can be determined from sound-velocity measurements, and this describes the motion of the center of gravity of a molecule as a result of thermal motion.

5.5.2. Morphology

It has been shown that polymers in the amorphous state assume the same unperturbed dimensions as in dilute solution. The evidence comes from the neutron scattering of deuterated polymers in solid solution in protonated polymers of the same type (and vice versa) (see Figure 4-16). These data seem to be independent of how the solid solution is prepared. In one case, the protonated polymer was dissolved in the deuterated monomer and the monomer subsequently polymerized. A kind of spaghetti structure was expected intuitively in this case. In another experiment, the protonated and deuterated polymer were mixed in dilute solution and the solution was concentrated. A balls-of-wool type packing was intuitively expected in this case. But similar structures were obtained in both of these experiments. Since the same unperturbed dimensions were found in both the solution and the amorphous state, the average interactions must be about the same in each case. This evidence does not, of course, exclude the presence of localized regions of order.

The spaghetti model predicts the density of the amorphous polymer to be about 65% of that of the crystalline polymer. A value of 85–95% is experimentally found. Thus, a certain degree of order must prevail in the amorphous state. Short lengths of chain or segments in parallel arrangement are conceivable.

X-ray studies also suggest the presence of order. Intensity maxima of 0.126, 0.223, 0.478, and 0.9–1 nm were obtained with amorphous polystyrene, for example. The value 0.126 nm corresponds to the normal distance between two neighboring carbon atoms in the chain. The value 0.223 nm is approximately the distance between one main-chain carbon atom and its next nearest neighbor. The value of 0.478 nm is also observed with monomeric styrene. It is displaced to higher values with increasing crosslinking and consequently must refer to an intermolecular distance.

The 0.478 nm and 0.9–1 nm values are, however, not true intermolecular distances l_{inter}, but are the distances between assumed Bragg network planes d_{Bragg}. When a certain scattering function is assumed, these quantities are related by the expression $l_{inter} = 1.22 d_{Bragg}$. As expected, the distance between two chains increases with increasing chain diameter, as can be seen in Table 5-7.

Spherical structures can also be observed electron-microscopically with the same polymers. These structures have a diameter of from about 2–4 nm [poly(styrene)] to about 8 nm [poly(ethylene terephthalate)]. It is

Table 5-7. Distances between Two Chain Segments at 20°C in Various Amorphous Polymers

Polymer	Distance l_{inter}, nm
Poly(ethylene)	0.55
cis-1,4-Poly(isoprene)	0.59
Poly(isobutylene)	0.78
Poly(methyl methacrylate)	0.81
Poly(styrene)	0.576 and 1.20
Poly(*p*-ethyl styrene)	0.585 and 1.44
Poly(*p*-*t*-butyl styrene)	0.620 and 1.55

still a matter of controversy as to what these structures signify. The question is whether they are real or are artifacts of the experimental procedure (due to insufficient care in focusing, surface effects produced by fracture or in the sample preparation, etc.).

5.5.3. Polymer Alloys

Mixtures of two or more different polymers are normally incompatible (for proof see Chapter 6). However, demixing is often incomplete, for kinetic reasons. Polymer mixtures in which the individual polymers to some extent lie separate from each other in submicroscopic regions are called polymer alloys. In the United States, block copolymers are also often called polymer alloys.

Polymer alloys play an important role in the industrial production of high-impact polymers (see Chapter 6). They can be produced by the mixing of solutions of appropriate polymers or by mixing of the melts. With fibers, a polymer fiber is swollen with an appropriate second monomer and the second monomer is polymerized. Conjugate fibers (see Chapter 12) have a partial polymer alloy character.

If the submicroscopic regions are amorphous and larger than about 5 nm, a polymer alloy exhibits two glass-transition temperatures. Under certain conditions, crystalline submicroscopic regions can be recognized by X-ray crystallography.

5.5.4. Block Copolymers

Block copolymers consist of blocks of two or more different monomeric units. According to the number of blocks, a distinction is made, for example (with binary block copolymers), between two-block copolymers A_nB_m, triblock copolymers $A_nB_mA_n$, and multiblock copolymers $(A_nB_m)_p$. Multiblock copolymers with short A_n and B_m blocks are also called segmented copolymers or segment copolymers.

In block copolymers, blocks of sufficient length are incompatible with each other. Thus, they tend to demix. Since the individual blocks are coupled to each other, this demixing is limited in extent. Consequently, blocks will form aggregates. The aggregate form is determined by the tendency to achieve the best possible packing. The following discussion will be confined to amorphous polymers.

Isolated chains of a noncrystalizable polymer tend to adopt the random-coil form (see Section 4.4). These random coils must pack as close as possible in the segregated regions. If the two blocks are of the same size, then the

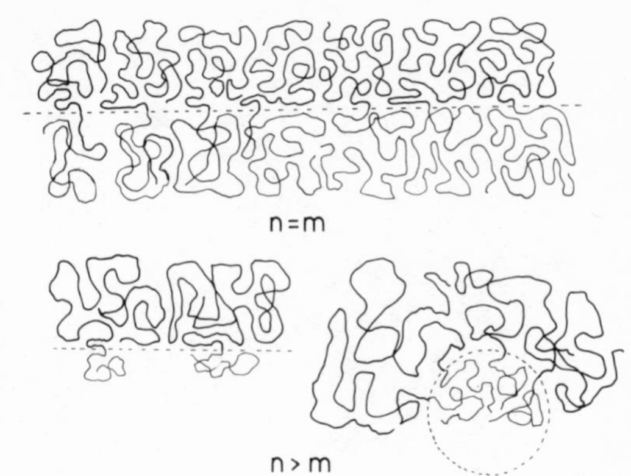

Figure 5-30. Schematic representation of volume requirements of blocks for two-block copolymers A_nB_m with different block ratios n/m. The block ratios n/m refer to dimensions, not to number of monomeric units.

space requirements are equal. The blocks can then adopt a lamellar structure (Figure 5-30). If, however, one block is very much larger than the other, it will require much more space. The smaller block trespasses on the principle of closest packing if it adopts a lamellar structure. It is much more favorable if the smaller blocks adopt spherical shape.

The transitional forms between spheres and lamellae are cylinders or rods. It can be expected that cylinders will be formed with block-size ratios intermediate between those favorable for the formation of spheres and those for the formation of lamellae. Forms other than spheres, cylinders, and lamellae are not to be expected since they transgress the principle of closest packing.

The three theoretically predicted forms and their dependence on the block-length ratio are actually experimentally found (Figure 5-31). The experimentally found block-length ratio of 40:60 for lamellae does not, of course, correspond to the theoretically required ratio of 50:50. This result, however, can be traced to the fact that the films were formed from evaporation of solutions. Solvents alter the chain dimensions and thus the morphology.

5.6. *Orientation*

5.6.1. *Definition*

When fibers or films are drawn, molecules and/or crystal regions can arrange themselves in the direction of the elongation, and thus orient themselves. Since the degree of orientation is difficult to measure and the distribution function of the orientation is so far practically impossible to

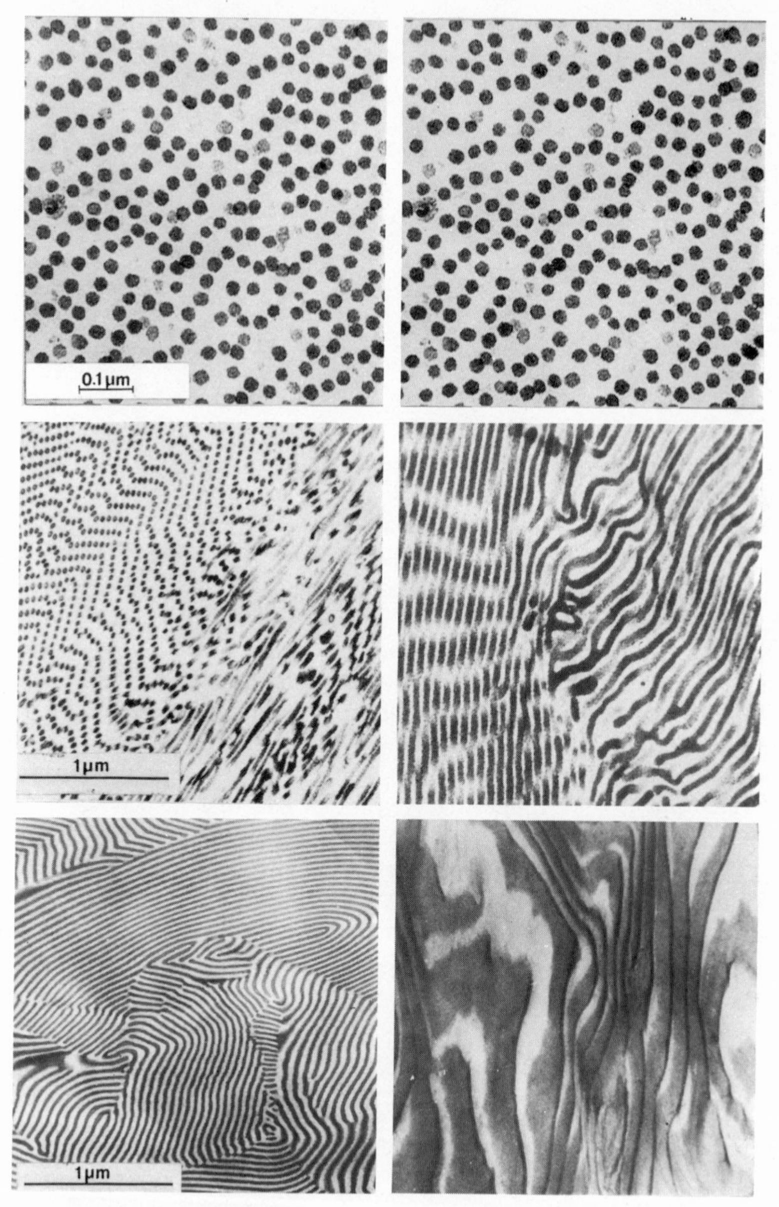

Figure 5-31. Electron micrographs of films of two-block and three-block copolymers of styrene and butadiene cut vertically (left) or parallel (right) to the film surface (M. Matsuo). Upper row: SBS polymer with S/B = 80/20 mol/mol. Spherical domains of poly(butadiene) segments embedded in a matrix of poly(styrene) segments. Center row: SB polymer with S/B = 60/40 mol/mol. Rods (cylinders) of poly(butadiene) segments in a matrix of poly(styrene) segments. Lower row: SBS polymer with S/B = 40/60 mol/mol. Lamellae of poly(butadiene) segments alternate with lamellae of poly(styrene) segments.

measure, the draw ratio is often taken as a measurement of the orientation. However, the draw ratio is not a good indication of the degree of orientation, since, in the extreme case, only viscous flow can result from drawing. The draw conditions therefore have a great influence on the degree of orientation reached. In addition. the degree of orientation reached at a given elongation naturally depends a great deal on the previous history of the material.

Methods used to characterize orientation are wide-angle X-ray scattering, IR spectroscopy, small-angle light scattering, birefringence measurements, polarized fluorescence, and sound-velocity measurements. They refer to effects resulting partly from chain orientation, partly from crystallite orientation, and partly from a combination of both types (combined effect).

5.6.2. X-Ray Diffraction

As the draw ratio increases, arcs first develop from the circular reflections at right angles to the draw direction and then point-shaped reflections in wide-angle X-ray pictures (Figure 5-32). The reciprocal length of an arc is a measure, therefore, of the extent of orientation of the crystallites, or, more precisely, the specific lattice planes. Arcs at various positions in the X-ray diagram correspond to the different lattice planes. Thus, an orientation factor f exists for each of the three spatial coordinates, and this is related to the angle of orientation β via

$$f = 0.5(3\,\overline{\cos^2\beta} - 1) \tag{5-12}$$

where β is defined as the angle between the elongation direction and main optical axis of the unit. f will be equal to 1 for complete orientation in the chain direction ($\beta = 0$), equal to -0.5 for complete orientation perpendicular to the chain direction ($\beta = 90°$), and equal to 0 for random orientation. When the optical axes of the crystallites are at right angles to one another, then $f_a + f_b + f_c = 0$. Uniaxial elongated polymers are charac-

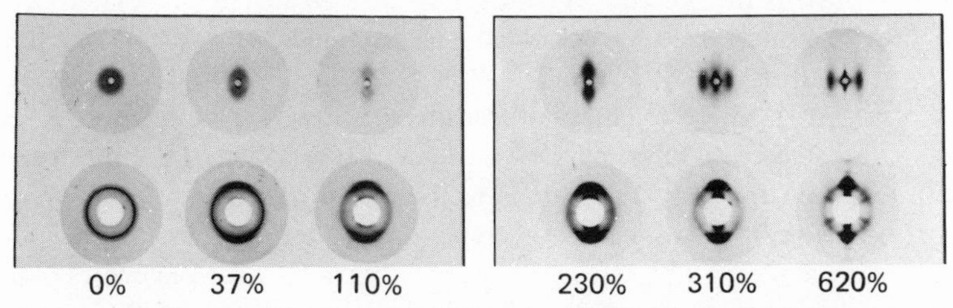

Figure 5-32. Small-angle (above) and wide-angle (below) X-ray interference by drawn poly(ethylene) (after Hendus).

terized by a single f value. The method is particularly suitable for low to medium degrees of orientation, since in some cases the crystallites will have become deformed at very high elongations.

5.6.3. Optical Birefringence

Every transparent material has three refractive indices n_x, n_y, and n_z along the three main axes. A material with $n_x = n_y = n_z$ is called isotropic. At least two of these refractive indices differ in anisotropic materials. The difference between each pair of these refractive indices is called the birefringence Δn.

Refractive indices vary according to the polarizability. For example, an alkane has greater polarizability along the chain than perpendicular to it because the electron mobility is greater along the chain.

Amorphous, nonoriented polymers are not optically birefringent, because their optically anisotropic monomeric units are randomly ordered with respect to one another. A birefringence first occurs when the chains are oriented or placed under strain. Generally, the following relation holds:

$$\Delta n = \sum \phi_i \Delta n_i + \Delta n_f + \Delta n_{st} \qquad (5\text{-}13)$$

Thus, each individual phase i contributes to the birefringence according to its volume fraction ϕ_i and birefringence Δn_i. These different phases can, for example, be the amorphous and crystalline phases of partially crystalline polymers, aggregates in block copolymers, fillers, or plasticized regions.

A form, or shape, birefringence Δn_f is produced when an electric field is distorted at the interface between two phases. In this case each phase must have dimensions of the order of the wavelength of light.

Amorphous polymers also become birefringent when placed under strain. The strain birefringence Δn_{st} depends on the strain applied and the anisotropy of the monomeric units. Strain birefringence is especially easy to see, even with unpolarized light, in the case of poly(styrene) with its strongly anisotropic phenyl groups. Strain birefringence is especially important in construction work with plastics since samples at the points of highest strain fail very readily.

Polarized light is generally required, however, for birefringence studies. The strongest interference colors is observed when the birefringent sample is at an angle of 45° to the oscillation direction of the polarizers. The order of the interference colors depends on the thickness of the sample and the difference in the refractive indices parallel and perpendicular to the strain direction. Refractive indices are determined by immersing the samples in inert liquids of known refractive index.

5.6.4. Infrared Dichroism

Light is absorbed when the direction of oscillation of its electrical vector has a component along the direction of oscillation of the absorbing group. The intensity of the absorption band of an oriented polymer thus depends on the direction of the electrical vector of the incident ray relative to the direction of orientation. Consequently, the absorption will be different according to the direction of oscillation of the incident polarized light. The degree of orientation is measured using the dichroic ratio R:

$$R = \frac{A_{\parallel}}{A_{\perp}} = \frac{\ln(I_0/I_{\parallel})}{\ln(I_0/I_{\perp})} \tag{5-14}$$

where I_0 is the intensity of the incident light and $I_{\parallel}$ and $I_{\perp}$ are the intensities of the transmitted light parallel and at right angles to the direction of elongation, respectively. The orientation factor f is derived from the dichroic ratio R and the corresponding value R^{∞} for complete orientation, analogously to the Lorentz–Lorenz formula, as

$$f = \frac{(R - 1)(R^{\infty} + 2)}{(R^{\infty} - 1)(R + 2)} \tag{5-15}$$

R^{∞} can be calculated if it is known that the dipole oscillation of a specific group in a uniaxially drawn polymer occurs at right angles to the chain axis, as is the case, for example, with hydrogen bonds between amide groups in polyamides. This method yields information about both "amorphous" and "crystalline" bands. First, however, every change in the IR bands produced by elongation must be shown not to have come from conformational changes in the molecule during drawing.

5.6.5. Polarized Fluorescence

Most organic polymers do not fluoresce. Consequently, about $10^{-4}\%$ (weight) of a fluorescing organic dyestuff is mixed in with the polymer. In the analysis of the results, it is assumed that the dyestuff added does not alter the morphology of the polymer and that the axes of the dyestuff molecules correspond to those of the polymer molecules. Also, the chromophore groups are not considered to rotate during the lifetime of the excited state, which is probably true because of the high viscosity. Since the dyestuff molecule generally does not have access to the crystal lattice, the method is only suitable for amorphous regions.

For measurement, plane-polarized light is allowed to fall on the fluorescing groups. The fluorescent light is also polarized.

When the polarizer and analyzer are parallel to the strain direction, the observed intensity varies with the fourth power of the cosine of the angle β between the strain direction and the molecular axis:

$$I_{\parallel} = \text{constant} \times \langle \cos^4 \beta \rangle \tag{5-16}$$

For fluorescent light observed with polarizers perpendicular to the strain direction and for uniaxial strain,

$$I_{\perp} = 0.5 \times \text{constant} \times (\langle \cos^2 \beta \rangle - \langle \cos^4 \beta \rangle) \tag{5-17}$$

5.6.6. *Sound Propagation*

The velocity of sound depends on the distances between the main-chain atoms and the intermolecular distances between the chains. In order to determine the orientation angle β from measurements of the sound velocity $\hat{c}$ in the fiber direction, it is necessary to know the sound velocities $\hat{c}_{\perp}$ and $\hat{c}_{\parallel}$ perpendicular and parallel to a polymer sample with completely oriented chains:

$$\frac{1}{\hat{c}^2} = \frac{1 - \langle \cos^2 \beta \rangle}{\hat{c}_{\perp}^2} + \frac{\langle \cos^2 \beta \rangle}{\hat{c}_{\parallel}^2} \tag{5-18}$$

In a completely unoriented sample, $f = 0$ according to equation (5-12) and thus $\cos^2 \beta = 1/3$. With these values, equation (5-18) gives

$$\hat{c}_{\perp}^2 = \frac{2\hat{c}_u^2 \hat{c}_{\parallel}^2}{3\hat{c}_{\parallel}^2 - \hat{c}_u^2} \tag{5-19}$$

Typical $\hat{c}_{\parallel}$ and $\hat{c}_{\perp}$ values are about 1.5 and 7–10 km/s, respectively.

The following procedure is adopted to determine the orientation angle. The velocity $\hat{c}_{\parallel}$ is either estimated or theoretically calculated. The velocity $\hat{c}_u$ is measured. Then $\hat{c}_{\perp}$ is calculated from equation (5-19). From all experience to date, the inequality $3\hat{c}_{\parallel}^2 \gg \hat{c}_u^2$ applies, so that the calculated value of $\hat{c}_{\perp}$ is fairly insensitive to the assumed value of $\hat{c}_{\parallel}$. Equation (5-19) can thus be given as

$$\hat{c}_{\perp}^2 = \tfrac{2}{3}\hat{c}_u^2 \tag{5-20}$$

and equation (5-18) as

$$\frac{1}{\hat{c}^2} = \frac{1 - \langle \cos^2 \beta \rangle}{\hat{c}_{\perp}^2} \tag{5-21}$$

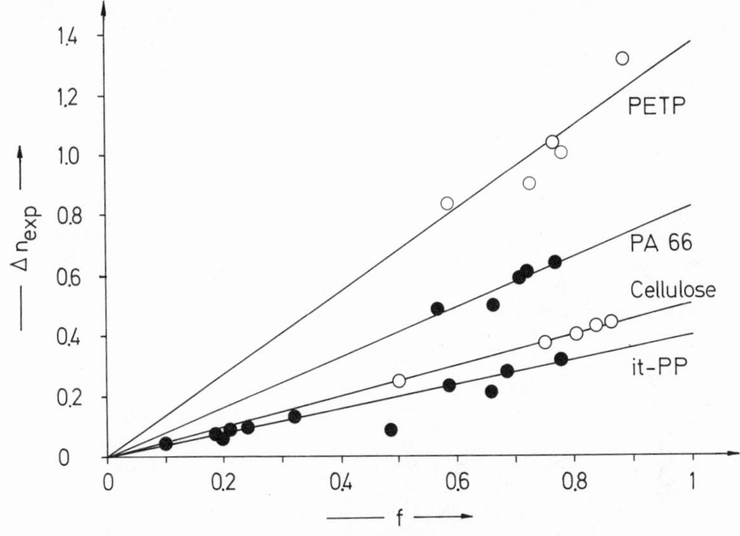

Figure 5-33. Optical birefringence Δn_{exp} as a function of orientation factor f from sound-velocity measurements of different polymers (after H. M. Morgan).

Combining equations (5-20) and (5-21), we obtain

$$\langle \cos^2 \beta \rangle = 1 - \frac{2\hat{c}_u^2}{3\hat{c}^2} \tag{5-22}$$

Consequently, an expression for the orientation factor f can be obtained from equations (5-22) and (5-12). f can be determined from the sound velocity in the fiber and in the unoriented sample:

$$f = 1 - \frac{\hat{c}_u^2}{\hat{c}^2} \tag{5-23}$$

The method allows the orientation factor to be determined for films and fibers during the actual straining or stretching process. A linear relationship between the orientation factor determined by sound-velocity measurement and that obtained by birefringence has been found experimentally (Figure 5-33).

Literature

General

E. P. Otocka, Physical properties of ionic polymers. *J. Macromol. Sci. C (Rev. Macromol. Chem.)* **5**, 275 (1971).

R. N. Haward, ed., *The Physics of Glassy Polymers,* Wiley, New York, 1973.

B. Wunderlich, *Macromolecular Physics,* Vol. 1, Academic Press, New York, 1973.

Journals

Journal of Materials Science
Journal of Non-crystalline Solids **1** (1968) ff.
XRS-X-Ray Spectrometry **1** (1973) ff.

Section 5.2. Crystallinity Determination

P. H. Hermans, Röntgenographische Kristallisationsbestimmungen bei Hochpolymeren, *Experientia* **19**, 553 (1963).

R. L. Miller, Crystallinity, *in: Encyclopedia of Science and Technology* (H. F. Mark, N. G. Gaylord, and N. M. Bikales, eds.), Wiley–Interscience, New York, 1964, Vol. 4, p. 449.

B. K. Vainshtein, *Diffraction of X-rays by Chain Molecules*, Elsevier, Amsterdam, 1966.

S. Kavesh and J. M. Smith, Meaning and measurement of crystallinity in polymers, a review, *Polym. Eng. Sci.* **9**, 331 (1969).

L. E. Alexander, *X-ray Diffraction Methods in Polymer Science*, Wiley, New York, 1969.

A. Elliot, *Infra-Red Spectra and Structure of Organic Long-Chain Polymers*, E. Arnold, London, 1969.

M. Kakudo and N. Kasai, *X-ray Diffraction by Polymers*, Kodansha, Tokyo, and Elsevier, Amsterdam, 1972.

V. J. McBrierty, N. M. R. of solid polymers: a review, *Polymer (London)* **15**, 503 (1974).

Section 5.3. Crystal Structure

C. W. Bunn, *Chemical Crystallography*, Clarendon Press, Oxford, 1946.

S. Krimm, Infrared spectra of high polymers, *Fortschr. Hochpolym. Forschg.* **2**, 51 (1960).

F. Danusso, Macromolecular polymorphism and stereoregular synthetic polymers, *Polymer (London)* **8**, 281 (1967).

G. Allegra and I. W. Bassi, Isomorphism in synthetic macromolecular systems, *Adv. Polym. Sci.* **6**, 549 (1969).

R. Hosemann, The paracrystalline state of synthetic polymers, *Crit. Rev. Macromol. Sci.* **1**, 351 (1972).

Section 5.4. Morphology of Crystalline Polymers

P. H. Geil, *Polymer Single Crystals*, Wiley, New York, 1963.

W. O. Statton, Small angle X-ray studies of polymers, *in: Newer Methods of Polymer Characterization* (B. Ke, ed.), Wiley–Interscience, New York, 1964.

E. W. Fischer, Electron diffraction, *in: Newer Methods of Polymer Characterization* (B. Ke, ed.), Wiley–Interscience, New York, 1964.

H. Brumberger, ed., *Small Angle X-ray Scattering*, Gordon and Breach, New York, 1967.

D. A. Blackadder, Ten years of polymer single crystals, *J. Macromol. Sci. C (Rev. Macromol. Chem.)* **1**, 297 (1967).

J. Willems, Oriented overgrowth (epitaxy) of macromolecular organic compounds, *Experientia* **23**, 409 (1967).

L. Mandelkern, Thermodynamics and physical properties of polymer crystals formed from dilute solution, *Prog. Polym. Sci.* **2**, 163 (1970).

R. A. Fava, Polyethylene crystals, *J. Polym. Sci. D***5**, 1 (1971).

R. H. Marchessault, B. Fisa, and H. D. Chanzy, Nascent morphology of polyolefins, *Crit. Rev. Macromol. Sci.* **1**, 315 (1972).

A. Keller, Morphology of lamellar polymer crystals, *in: Macromolecular Science* (Vol. 8 of Physical Chemistry Series 1) (C. E. H. Bawn, ed.), MTP International Review of Science, 1972.

R. D. B. Fraser and T. P. MacRae, *Conformation in Fibrous Proteins*, Academic Press, New York, 1973.

Section 5.5. The Amorphous State

R. N. Haward, Occupied volume of liquids and polymers, *J. Macromol. Sci. C (Rev. Macromol. Chem.)* **4**, 191 (1970).
T. G. F. Schoon, Microstructure in solid polymers, *Br. Polym. J.* **2**, 86 (1970).
G. S. Y. Yeh, Morphology of amorphous polymers, *Crit. Rev. Macromol. Sci.* **1**, 173 (1972).
R. E. Robertson, Molecular orientation of amorphous polymers, *Ann. Rev. Mater. Sci.* **5**, 73 (1975).

Section 5.6. Orientation

G. L. Wilkes, The measurement of molecular orientation in polymeric solids, *Adv. Polym. Sci.* **8**, 91 (1971).
C. R. Desper, Techniques for measuring orientation in polymers, *Crit. Rev. Macromol. Sci.* **1**, 501 (1973).
I. M. Ward, ed., *Structures and Properties of Oriented Polymers*, Halsted Press, New York, 1975.

Part II

Solution Properties

Chapter 6

Solution Thermodynamics

6.1. Basic Principles

According to the second law of thermodynamics, the Gibbs energy ("free energy") G is related to the enthalpy H, the entropy S, and the thermodynamic temperature T by

$$G = H - TS = U + pV - TS \tag{6-1}$$

where U is the internal energy, p is the pressure, and V is the volume. The Helmholtz energy A is given by

$$A = U - TS = G - pV \tag{6-2}$$

$\Delta G \approx \Delta A$ frequently holds for isobaric processes in condensed systems since the change in volume is often (but not always) negligibly small.

The change of Gibbs energy of a system caused by the addition of 1 mol of component i to an infinite system is called the partial molar Gibbs energy $\tilde{G}_i^m$, or the chemical potential μ_i:

$$\left(\frac{\partial G}{\partial n_i}\right)_{T,p,n_{j \neq i}} \equiv \tilde{G}_i^m \equiv \mu_i \tag{6-3}$$

Differentiation of the chemical potential of component i gives (see textbooks on chemical thermodynamics)

$$\partial \tilde{G}_i^m = \frac{\partial \tilde{G}_i^m}{\partial p} dp + \frac{\partial \tilde{G}_i^m}{\partial T} dT + \frac{\partial \tilde{G}_i^m}{\partial n_i} dn_i$$

$$= \tilde{V}_i^m dp - \tilde{S}_i^m dT + RT d(\ln a_i) \tag{6-4}$$

$\tilde{V}_i^m$ is the partial molar volume of component i of relative activity a_i. The complete differential form of G^m is

$$dG^m = \sum_i \tilde{G}_i^m dn_i + \sum_i n_i d\tilde{G}_i^m \tag{6-5}$$

The left-hand side of this equation must be identical to the first term of the right-hand side, according to equation (6-3). The so-called Gibbs–Duhem relationship thus states

$$\sum_i n_i d\tilde{G}_i^m = \sum_i n_i d\mu_i = 0 \tag{6-6}$$

With $dp = 0$ and $dT = 0$ for an isothermal–isobaric process, one obtains from equation (6-4) after integrating and converting to chemical potentials

$$\mu_i = \mu_i^\circ + RT \ln a_i = \mu_i^\circ + RT \ln x_i \gamma_i \tag{6-7}$$

The integration constant μ_i° is the chemical potential of the pure substance. The relative activity is often further separated into the mole fraction x_i and the activity coefficient γ_i. The contribution from the mole fraction is often called the ideal component or function, and that coming from the activity coefficient is the excess function:

$$\Delta\mu_i = \mu_i - \mu_i^\circ = RT \ln x_i + RT \ln \gamma_i = \Delta\mu_i^{\text{id}} + \Delta\mu_i^{\text{exc}} \tag{6-8}$$

Solutions or mixtures can be classified into four types according to the magnitude and sign of the excess functions: ideal, athermal, regular, and irregular (or real) solutions. In ideal solutions, the total contribution to the Gibbs energy of mixing comes solely from the ideal entropy of mixing (see Section 6.3.1). The enthalpy of mixing is also zero with athermal solutions, but the entropy of mixing is different from the ideal entropy of mixing. There is no excess entropy of mixing with regular solutions, but in this case the enthalpy of mixing is not zero. An enthalpy of mixing as well as an excess entropy of mixing are observed with irregular solutions.

The pseudoideal or theta solution is an important special case of irregular solutions in macromolecular science. The enthalpy of mixing and the excess entropy of mixing exactly compensate each other at a certain temperature with the dilute theta solution. Theta solutions at this theta temperature thus behave like ideal solutions. In contrast to ideal solutions, however, the enthalpy of mixing is not zero and the entropy of mixing differs considerably from the ideal entropy of mixing. Thus, an ideal solution exhibits ideal behavior at all temperatures, the pseudoideal solution only behaves ideally at the theta temperature. Consequently, the theta temperature corresponds to the Boyle temperature of real gases.

6.2. Solubility Parameter

6.2.1. Basic Principles

Thermodynamic analysis allows solutions to be classified *after* thermodynamic parameters have been determined. It cannot, however, predict the solubility or miscibility of two substances without the aid of additional assumptions. A prediction of this type is possible with the concept of the solubility parameter, which is based on the following considerations.

The transfer from the liquid to the gaseous state requires overcoming an interaction energy of $z\varepsilon_j/2$ per molecule and consequently $N_L z\varepsilon_j/2$ per mole. This is, however, exactly equal to the negative internal molar energy of vaporization $(\Delta E_{vap})_j$. The ε_j is the energy per bond. One molecule has z neighbors. The corresponding quantity related to the molar volume V^m is called the cohesive energy density:

$$e_j = \frac{\Delta E_{vap\,j}}{V_j^m} = -0.5\frac{N_L\varepsilon_j z}{V_j^m} \tag{6-9}$$

The solubility parameter is defined as the square root of the cohesive energy density:

$$\delta_j \equiv e_j^{0.5} \tag{6-10}$$

Interaction energies ε are related to each other in the following manner. Mixing solvent 1 and polymer 2 produces two solvent–polymer 1–2 bonds for every broken solvent–solvent 1–1 and polymer–polymer 2–2 bond. The change in interaction energy during the mixing process is consequently

$$\Delta\varepsilon = \varepsilon_{12} - 0.5(\varepsilon_{11} + \varepsilon_{22}) \tag{6-11}$$

$$-2\Delta\varepsilon = (\varepsilon_{11}^{0.5})^2 - 2\varepsilon_{12} + (\varepsilon_{22}^{0.5})^2 \tag{6-12}$$

From quantum mechanics it is known that the interaction energy of two different spherical molecules due to dispersion forces is equal to the geometric mean of the mutual interaction energies of the molecules themselves, i.e.,

$$\varepsilon_{12} = -(\varepsilon_{11}\varepsilon_{22})^{0.5} \tag{6-13}$$

The minus sign occurs in equation (6-13) because the interaction energy ε_{12} represents a geometric mean of two normally negative interaction energies ε_{11} and ε_{22}.

Inserting equation (6-13) into equation (6-12), we obtain

$$\Delta\varepsilon = -0.5(|\varepsilon_{11}|^{0.5} - |\varepsilon_{22}|^{0.5})^2 \tag{6-14}$$

Assuming equal molar volumes of solvent and polymer monomeric units, combination of equations (6-10) and (6-14) gives

$$\frac{0.5zN_{\mathrm{L}}\,\Delta\varepsilon}{V^m} = -0.5(\delta_1 - \delta_2)^2 \tag{6-15}$$

The difference in solubility parameters thus yields a measure of the interaction between solvent and solute with respect to the mutual interactions between like components. Now, if $\varepsilon_{11} \gg \varepsilon_{12}$ and/or $\varepsilon_{22} \gg \varepsilon_{12}$ then there will be practically no interaction between solute and solvent. The difference $|\delta_1 - \delta_2|$ will then be very large. With equal interactions between 1–1, 2–2, and 1–2, on the other hand, $\delta_1 - \delta_2 = 0$ and good solubility is obtained. There must therefore be a maximum permitted difference $|\delta_1 - \delta_2|$ at which it is still just possible to have mixing. The experimentally obtained maximum differences vary, according to the polarity of the solvent, between ± 0.8 and ± 3.4 (Table 6-1).

Solubility parameters are traditionally given without units, although, strictly speaking, the figures given in literature have the units $(\mathrm{cal/cm^3})^{1/2}$. Note that 1 $(\mathrm{cal/cm^3})^{1/2} = 2.05\,(\mathrm{J/cm^3})^{1/2}$.

The concept of solubility parameters is an attempt to quantify the old rule-of-thumb, "like dissolves like." It must of necessity fail when the interaction forces differ greatly in nature. Recently the solubility parameter has been separated into three component parameters in order to refine the quite crude process of trying to predict the possible mixability of a polymer and a solvent. The three components describe the interaction between dispersion, dipole, and hydrogen-bonding forces:

$$\delta^2 = \delta_d^2 + \delta_p^2 + \delta_h^2 \tag{6-16}$$

As expected, the contribution due to dispersion forces δ_d varies only very slightly from system to system. Consequently, solubility diagrams are constructed so that δ_h values are plotted against δ_p values. Each solvent

Table 6-1. δ-Regions for Polymers

Polymer	Solubility parameter δ of solvents that dissolve the polymer: Apolar solvents	Polar solvents[a]
Poly(styrene)	9.3 ± 1.3	9.0 ± 0.9
Poly(vinyl chloride-co-vinyl acetate)	10.2 ± 0.9	10.6 ± 2.8
Poly(vinyl acetate)	10.8 ± 1.9	11.6 ± 3.1
Poly(methyl methacrylate)	10.8 ± 1.2	10.9 ± 2.4
Cellulose trinitrate	11.9 ± 0.8	11.2 ± 3.4

[a] Alcohols, esters, ethers, ketones.

will have its own δ_d value. Points are then drawn on the plot representing the different solvents for the polymer in question (for example, drawn in color). Finally a contour is drawn with the aid of the δ_d values for all solvents that actually dissolve the polymer. Normally, the solubility increases with increasing δ_d for like values of δ_p and δ_h. Thus, if a substance has a δ_d value which is within the contour for this numerical value, then it is a solvent for the polymer.

6.2.2. Experimental Determination

Solubility parameters δ_1 can be determined directly via equation (6-9). One has simply to subtract the work against the external pressure from the experimental negative enthalpy of vaporization. All δ_1 values compiled in Table 6-2 were determined this way.

Table 6-2. The Solubility Parameters [in $(cal/cm^3)^{1/2}$]

Solvent	δ_1	δ_d	δ_p	δ_h
Heptane	7.4	7.4	0	0
Cyclohexane	8.18	8.18	0	0
Benzene	9.05	8.99	0.5	1.0
Carbon tetrachloride	8.65	8.65	0	0
Chloroform	9.33	8.75	1.65	2.8
Dichloromethane	9.73	8.72	3.1	3.0
1,2-Dichloroethane	9.42	8.85	2.6	2.0
Acetone	9.75	7.58	5.1	3.4
Butanone	9.30	7.77	4.45	2.5
Cyclohexanone	10.00	8.65	4.35	2.5
Ethyl acetate	9.08	7.44	2.6	4.5
Propyl acetate	8.74	7.61	2.2	3.7
Amyl acetate	8.49	7.66	2.1	3.3
Acetonitrile	11.95	7.50	8.8	3.0
Pyridine	10.60	9.25	4.3	2.9
Diethyl ether	7.61	7.05	1.4	2.5
Tetrahydrofuran	9.49	8.22	2.7	3.9
p-Dioxane	9.65	8.93	0.65	3.6
1-Pentanol	10.59	7.81	2.2	6.8
1-Propanol	11.85	7.75	3.25	8.35
Ethanol	12.90	7.73	4.3	9.4
Methanol	14.60	7.42	6.1	11.0
Cyclohexanol	10.69	7.75	3.8	6.3
m-Cresol	11.52	9.14	2.35	6.6
Nitrobenzene	11.25	9.17	6.2	2.0
Dimethyl formamide	12.14	8.5	6.7	5.5

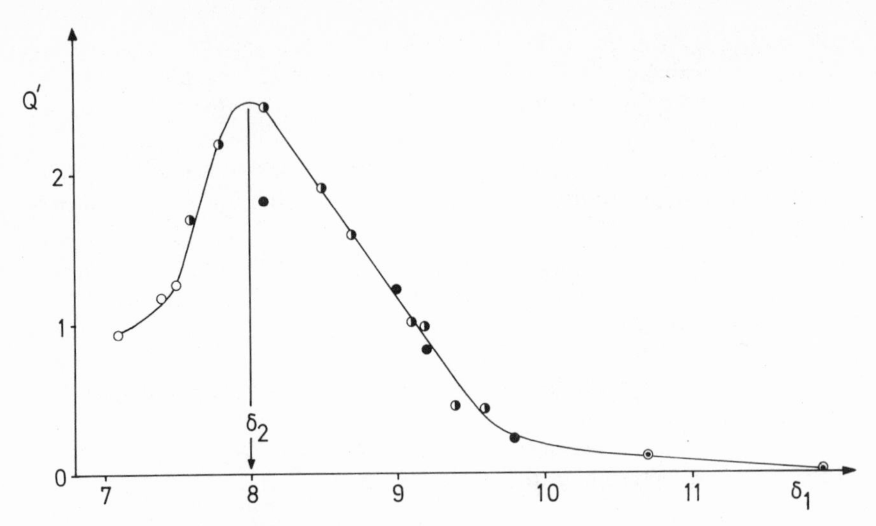

Figure 6-1. Degree of swelling Q' (= ml solvent/g polymer) of a cross-linked poly[(butadiene)$_{75}$-co-(styrene)$_{25}$] as a function of the solubility parameter δ_1 of the swelling solvent (according to G. Gee). ○ = hydrocarbons, ◐ = ethers and esters, ● = ketones, ⊙ = nitriles.

Macromolecules cannot be vaporized without degradation, because of the high cohesive energy density per molecule. Their solubility parameters are therefore often not determined directly, but are assumed to be equal to those of low-molecular-weight compounds. Alternatively, they can be estimated from the swelling of the cross-linked parent polymer. Cross-linked polymers swell more, the greater the interaction between polymer and solvent (see Section 6.6.7). A plot of the degrees of swelling against the solubility parameter δ_1 of the solvents thus gives a maximum at a certain δ_1, which corresponds to the δ_2 of the polymer (Figure 6-1).

Intrinsic viscosities $[\eta]$ can be measured for soluble polymers in various solvents. $[\eta]$ increases with increasing polymer–solvent interaction (Chapter 9.9.6). If the $[\eta]$ values are plotted against the solubility parameter of the solvents used, the maximum corresponds to the solubility parameter δ_2 of the polymer.

Both viscosity measurements and the swelling process yield quite unambiguous values for δ_2 so long as the polymers and solvents are not too polar. Both methods fail when forces other than dispersion forces predominate.

Theoretical equations are used to determine the individual solubility parameters δ_d, δ_p, and δ_h.

6.2.3. Applications

The solubility of a polymer can be estimated from its δ_2 value in many cases.

Apolar substances have low solubility parameters, whereas those of polar substances are high, since the heat of vaporization is higher for the latter. Apolar, noncrystalline polymers will therefore dissolve well in solvents with low δ_1 values. Predictions about solubility on the basis of the solubility parameter are still quite permissible for polar, noncrystalline polymers in polar solvents (see Table 6-3). It is more difficult in the case of crystalline polymers or apolar polymers in polar solvents, and vice versa, since equation (6-13), which was derived for pure dispersion forces, no longer applies in these cases.

Dilute solutions of poly(styrene) ($\delta_2 = 9.3$), for example, are readily obtained with butanone ($\delta_1 = 9.3$) and dimethylformamide ($\delta_1 = 12.1$), but not with acetone ($\delta_1 = 9.8$). That is, in liquid acetone, the acetone molecules form dimers through dipole–dipole interactions. In these dimers,

Table 6-3. The Solubilities and Solubility Parameters of Polymers

Solvent		Solubility of the polymer[a]			
Name	δ_1	Poly(isobutene) $\delta_2 = 7.9$	Poly(methyl methacrylate) $\delta_2 = 9.1$	Poly(vinyl acetate) $\delta_2 = 9.4$	Poly(hexamethylene adipamide) $\delta_2 = 13.6$
Decafluorobutane	5.2	–	–	–	–
Neopentane	6.25	+	–	–	–
Hexane	7.3	+	–	–	–
Diethyl ether	7.4	–	–	–	–
Cyclohexane	8.2	+	–	–	–
Carbon tetrachloride	8.62	+	+	–	–
Benzene	9.2	+	+	+	–
Chloroform	9.3	+	+	+	–
Butanone	9.3	–	+	+	–
Acetone	9.8	–	+	+	–
Carbon disulfide	10.0	–	–	–	–
Dioxane	10.0	–	+	+	–
Dimethyl formamide	12.1	–	+	+	(+)
m-Cresol	13.3	–	+	+	+
Formic acid	13.5	–	+	–	+
Methanol	14.5	–	–	–	–
Water	23.4	–	–	–	–

[a] + means soluble, – means insoluble, (+) means soluble at high temperatures.

Table 6-4. Solubility of Polymers in Mixtures of Nonsolvents

Polymer		Solutions possible with mixtures of			
Type	δ_2	Nonsolvent I	δ_I	Nonsolvent II	δ_{II}
at-Poly(styrene)	9.3	Acetone	9.8	Cyclohexane	8.2
at-Poly(vinyl chloride)	9.53	Acetone	9.8	Carbon disulfide	10.0
at-Poly(acrylonitrile)	12.8	Nitromethane	12.6	Water	23.4
Poly(chloroprene) (radically polymerized)	8.2	Diethyl ether	7.4	Ethyl acetate	9.1
Nitrocellulose	10.6	Ethanol	12.7	Diethyl ether	7.4

the keto groups are shielded by the methyl groups, and so are no longer able to solvate the phenyl groups of the poly(styrene). The addition of cyclohexane ($\delta_1 = 8.2$) decreases the tendency of the acetone molecules toward association, and thus frees keto groups for solvation. For the same reason, it is also possible to have 40% solutions of poly(styrene) in acetone. Butanone, on the other hand, is "internally diluted" by the additional CH_2 group, and is therefore a solvent over the whole concentration range.

Similar reasoning applies to mixtures of solvents. The combination of a nonsolvent with a lower, and a nonsolvent with a higher, solubility parameter than that of the polymer often gives a good solvent for the polymer (Table 6-4). Conversely, a mixture of two solvents can be a nonsolvent. Poly(acrylonitrile) ($\delta_2 = 12.8$), for example, dissolves in both dimethylformamide ($\delta_1 = 12.1$) and malodinitrile ($\delta_1 = 15.1$), but not in a mixture of the two.

In order to dissolve crystalline polymers, it is necessary to consider the Gibbs energy of fusion. This additional energy expenditure is not taken into account in the concept of the solubility parameter. Crystalline polymers therefore often dissolve only above their melting temperatures and in solvents with roughly the same solubility parameter. Unbranched, highly crystalline poly(ethylene) ($\delta_2 = 8.0$) only dissolves in decane ($\delta_1 = 7.8$) at temperatures close to the melting point of ~135°C.

The crystallinity of polymers is also responsible for the curious effect where a polymer at constant temperature first dissolves in a solvent and later, at the same temperature, precipitates out again. In these cases, the original polymer is of low crystallinity and therefore dissolves well. On dissolution, the chains become mobile. A crystalline polymer–solvent equilibrium is rapidly achieved with precipitation of polymer of higher crystallinity than the original material.

6.3. Statistical Thermodynamics

6.3.1. Entropy of Mixing

In ideal solutions it is assumed for the pair interactions that no energy is released on replacing a unit of group 1 by a unit of group 2, that is, $\Delta\varepsilon$ in equation (6-11) equals zero. Consequently, the enthalpy of mixing of an ideal solution is also equal to zero.

Since all energies are, by definition, equal in magnitude with ideal solutions, all environment-dependent entropy contributions can contribute nothing to the overall change in entropy. Consequently, the translational entropy, the internal rotation entropy, and the vibrational entropy are not changed by mixing. But the molecules of the components of the solution can be ordered relative to one another in many different ways. Thus, the many different arrangement combinations lead to an entropy contribution on mixing ΔS_{comb} (often called the "configurational entropy"). This entropy contribution can be calculated from the Ω possible arrangements of molecules (or monomeric units 1 and 2) with respect to each other, as long as the molar volumes are the same in each case (see textbooks of statistical thermodynamics):

$$\Delta S^{\text{id}} \approx \Delta S_{\text{comb}} = k \ln \Omega = k \ln \frac{(N_1 + N_2)!}{N_1!N_2!} \tag{6-17}$$

With the aid of the Stirling approximations,

$$N! \approx (2\pi N)^{0.5} N^N e^{-N} \tag{6-18}$$

$$N! \approx (N/e)^N \tag{6-19}$$

the factorials can be replaced by exponential expressions. The approximation (6-18) is in error by 6% for $N = 2$, 1.5% for $N = 3$, and 0.05% by $N = 10$. The approximation (6-19), on the other hand, is in error by 52% for $N = 10$, 4.5% for $N = 10^{10}$, and 2% by $N = 10^{23}$. Inserting equation (6-19) into equation (6-17), we obtain

$$\Delta S_{\text{comb}} = -k(N_1 \ln x_1 + N_2 \ln x_2) = \Delta S \tag{6-20}$$

or, on converting to mole fractions,

$$\Delta S^m_{\text{comb}} = -R(x_1 \ln x_1 + x_2 \ln x_2) \tag{6-21}$$

Volume fractions instead of mole fractions have to be used (and will be used in further derivations) if the molar volumes of the components are not identical.

6.3.2. Enthalpy of Mixing

In calculating the enthalpy of mixing, it is assumed that the distribution of molecules or monomeric units is not influenced by the enthalpy of mixing. This assumption allows entropies and enthalpies of mixing to be calculated independently of each other. The enthalpy of mixing ΔH is given by the difference between the enthalpies of solution H_{12} and the enthalpies H_{11} and H_{22} of the pure components

$$\Delta H = H_{12} - (H_{11} + H_{22}) \tag{6-22}$$

The enthalpies H_{12}, H_{11}, and H_{22} are calculated as follows: An interaction energy ε_{ij} exists between every two monomeric units. Each unit thus contributes $0.5\varepsilon_{ij}$. Furthermore, every unit is surrounded by z neighbors. Generally, a molecule consists of X monomeric units. Consequently, for the X_1N_1 monomeric units of all of the N_1 solvent molecules, H_{11} is given by the following, with the definition $\phi_1 = N_1X_1/N_g$ for the volume fraction:

$$H_{11} = N_1X_1z(0.5\varepsilon_{11}) = z(0.5\varepsilon_{11})\,N_g\phi_1 \tag{6-23}$$

where N_g is the total number of lattice sites. The enthalpy of the pure polymer is obtained analogously:

$$H_{22} = N_2X_2z(0.5\varepsilon_{22}) = z(0.5\varepsilon_{22})\,N_g\phi_2 \tag{6-24}$$

In calculating the enthalpy of the solution it is necessary to take into account the interaction energies between each segment unit and its z neighbors. There are X_1N_1 solvent segment units, so a total of zX_1N_1 interactions have to be considered. A solvent molecule can be surrounded by other solvent units with the interaction energy $0.5\varepsilon_{11}$ per unit and/or by solute units with the interaction energy $0.5\varepsilon_{12}$ per unit. The relative contribution of these two possible interactions is given by the volume fraction of the two kinds of unit in the solution. A corresponding contribution also comes from the solute units. The enthalpy of the solution is therefore

$$H_{12} = X_1N_1z(0.5\varepsilon_{11}\phi_1 + 0.5\varepsilon_{12}\phi_2) + X_2N_2z(0.5\varepsilon_{22}\phi_2 + 0.5\varepsilon_{12}\phi_1) \tag{6-25}$$

With the definition of the mole fraction of monomeric units,

$$\phi_i \equiv x_i^u \equiv \frac{N_iX_i}{N_iX_i + N_jX_j} = \frac{N_iX_i}{N_g} \tag{6-26}$$

one gets from insertion of equation (6-23)–(6-25) into equation (6-22)

$$\Delta H = zN_g\phi_1\phi_2(\varepsilon_{12} - 0.5\varepsilon_{11} - 0.5\varepsilon_{22}) = zN_1X_1\phi_2\Delta\varepsilon \tag{6-27}$$

Next, an interaction parameter χ is defined as follows: $\Delta\varepsilon$ is the average energy gain per contact. Each solvent unit, however, is surrounded by z neighbors, and every solvent molecule possesses X_1 segment units. Thus, we have, with respect to the thermal energy kT,

$$\chi \equiv \frac{zX_1\,\Delta\varepsilon}{kT} \tag{6-28}$$

Thus what is known as the Flory–Huggins parameter χ is, by definition, a measure of the interaction energy $\Delta\varepsilon$. However, $\Delta\varepsilon$ is in reality a measure of the Gibbs energy and not of the enthalpy. Consequently, χ also contains an entropy contribution, which is often found to depend on the concentration. A linear dependence of the interaction parameter on the volume fraction ϕ_2 of the solute can be assumed as a first approximation:

$$\chi = \chi_0 + \sigma\phi_2 \tag{6-29}$$

With relationships (6-28) and (6-29), equation (6-27) becomes

$$\Delta H = kTN_1\phi_2(\chi_0 + \sigma\phi_2) \tag{6-30}$$

or with respect to moles, $n_1 = N_1/N_L$, or mole fractions, $x_1 = n_1/(n_1 + n_2)$, the molar enthalpy of mixing is given by

$$\Delta H^m = \frac{\Delta H}{n_1 + n_2} = RTx_1\phi_2(\chi + \sigma\phi_2) \tag{6-31}$$

6.3.3. Gibbs Energy of Mixing for Nonelectrolytes

Using the relationships $\phi_2 = N_2X_2/N_g$, $N_g \equiv n_gN_1$, $N_g = N_1 + N_2$, $k = R/N_L$, $\Delta G^m = \Delta G/n_g$, and $x_1 = \phi_1$, combination of equation (6-1) with (6-20) and (6-30) leads to

$$\frac{\Delta G^m}{RT} = X_1^{-1}[\phi_1\phi_2\chi_0 + \phi_1\phi_2\sigma + \phi_1 \ln \phi_1 + X_1X_2^{-1}\phi_2 \ln \phi_2] \tag{6-32}$$

$\Delta G^m/RT$ is plotted according to equation (6-32) as a function of the volume fraction ϕ_2 of monomeric units of solute (Figure 6-2). We may have $X_1 = X_2 = 1$ for mixtures of some low-molecular-weight compounds. The molar Gibbs energy is always negative for these interaction parameters (e.g., $\chi_0 = 0.5$) and has a minimum at $\phi_2 = 0.5$. Such mixtures can never demix.

The function becomes asymmetric when the degree of polymerization X_2 goes from 1 to 100 for the same interaction parameter $\chi_0 = 0.5$. This behavior is caused by the last term in equation (6-32), that is, by the entropy

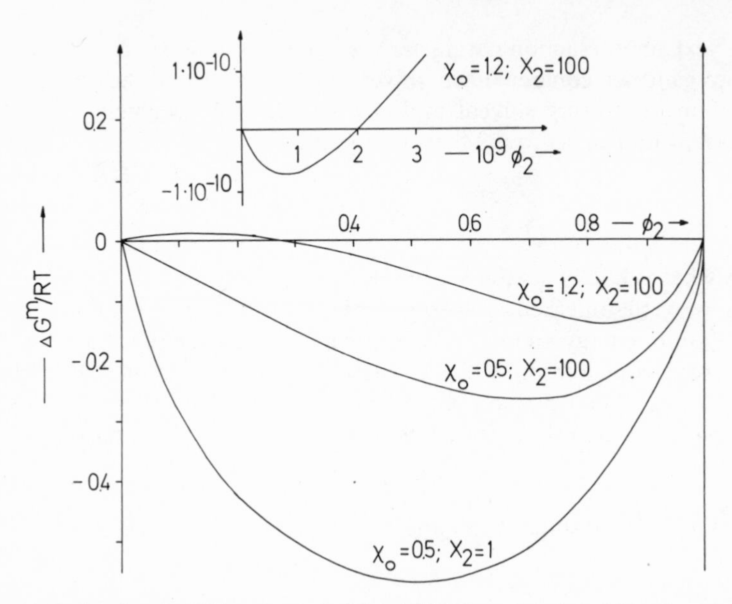

Figure 6-2. Reduced molar Gibbs energy of mixing $\Delta G^m/RT$ as a function of the volume fraction ϕ_2 of solute monomeric unit with different interaction parameters χ_0 and degrees of polymerization X_2 of the solute in low-molecular-weight solvents ($X_1 = 1$). Calculations according to equation (6-32) with $\sigma = 0$.

term. This behavior of high-molecular-weight polymer solutions deviates from the behavior of low-molecular-weight solution systems and is essentially caused by the difference in molecular size between the low-molecular-weight solvent and the high-molecular-weight solute.

If the value of the interaction parameter χ_0 increases from 0.5 to 1.2 at the same degree of polymerization $X_2 = 100$, then the molar Gibbs energy of mixing will even be positive between $\phi_2 = 2 \times 10^{-9}$ and $\phi_2 = 0.3$. Since there are two regions of concentration where the molar Gibbs energy of mixing is negative, however, a phase separation of the dilute starting solution into two solutions occurs. One of these solutions is very dilute, the other very concentrated in solute (see also Section 6.6).

The simple Flory–Huggins theory discussed above is based on a series of questionable assumptions: lattice sites of equal size for solvent segments and polymer monomeric units, uniform distribution of the monomeric units in the lattice, random distribution of the molecules, and the use of volume fractions instead of surface-area fractions in deriving the enthalpy of mixing. Proposed improvements, however, have led to more complicated equations or to worse agreement between theory and experiment. Obviously, various simplifications in the Flory–Huggins theory are self-compensating in character.

6.3.4. Gibbs Energy of Mixing for Polyelectrolytes

The molar Gibbs energy of mixing of polyelectrolytes ΔG_{el}^m is composed of the free energy of mixing of the uncharged polymer ΔG^m [see equation (6-32)], the contribution $\Delta G_{\mathrm{Coul}}^m$ for the Coulombic interaction between the polyion and the gegenion, and the contribution ΔG_{mm}^m for ionic interactions within the macromolecule itself:

$$\Delta G_{el}^m = \Delta G^m + \Delta G_{\mathrm{Coul}}^m + \Delta G_{\mathrm{mm}}^m \tag{6-33}$$

The magnitude of ΔG_{mm} is determined by the distribution of the ions within the macromolecule. This distribution is not yet experimentally determinable. A specific model is therefore assumed for the distribution of ions in the macromolecule. For example, the model of a rod is very suitable for true rod-forming macromolecules (viruses, nucleic acids) or for long-chain macromolecules at high degrees of ionization. In the case of the latter, the many like charges repel one another along the whole length of the chain, resulting in rigidity and rodlike behavior. For coiled molecules at low degrees of ionization or rigid spheres (e.g., globulin), on the other hand, spherical models are more suitable.

6.3.5. Chemical Potential of Concentrated Solutions

According to equation (6-3), the chemical potential of the solvent is defined as the derivative of the Gibbs energy of mixing with respect to moles of solvent. Consequently, with the condition $\phi_1 = 1 - \phi_2$, differentiation of equation (6-32) gives

$$\Delta\mu_1 = RT[(\chi_0 - \sigma + 2\sigma\phi_2)\phi_2^2 + \ln(1 - \phi_2) + (1 - X_1 X_2^{-1})\phi_2] \tag{6-34}$$

and analogously, for the chemical potential of the solute,

$$\Delta\mu_2 = RT[(\chi_0\phi_1 + 2\sigma\phi_2\phi_1 - 1)X_2 X_1^{-1}\phi_1 + \phi_1 + \ln\phi_2] \tag{6-35}$$

In a polymer–solvent system the chemical potential of the solvent decreases slowly at first and then more rapidly to negative values with increasing volume fraction of polymer if χ_0 is zero (Figure 6-3). The initial shape of the curve is increasingly flatter with increasing value of the interaction parameter. In the chosen example, there is a practically horizontal portion of the curve between $\phi_2 > 0.05$ and $\phi_2 < 0.14$ after an initial slow decrease and before the strong decrease of $\Delta\mu_1/RT$ to negative values for $\chi_0 = 0.605$. The value of $\Delta\mu_1/RT$ passes through a weak minimum (not recognizable in Figure 6-3) before passing through a strong maximum for still higher values of χ_0. Thus, $\chi_0 = 0.605$ is a critical interaction parameter

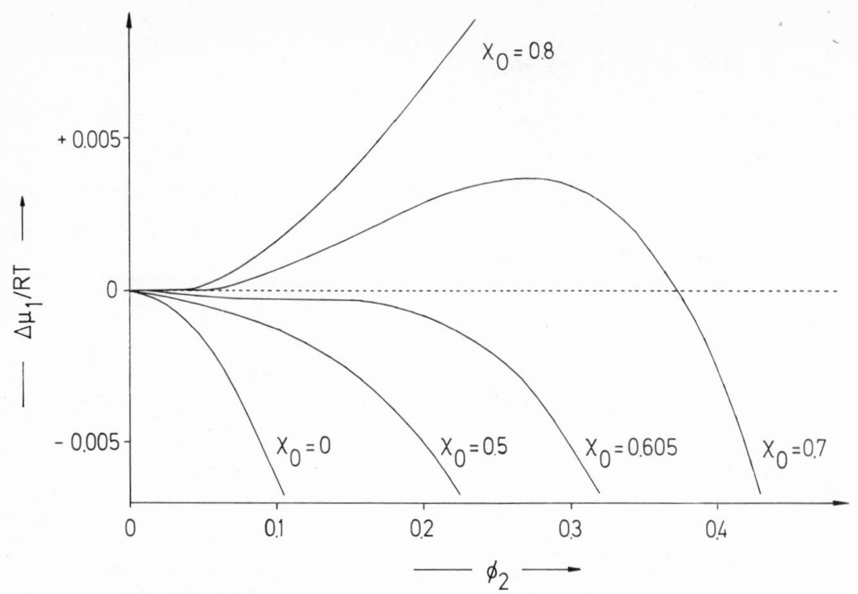

Figure 6-3. Reduced chemical potential $\Delta\mu_1/RT$ of the solvent (degree of polymerization $X_1 = 1$) as a function of the volume fraction ϕ_2 of monomeric units of a monodisperse solute with degree of polymerization $X_2 = 100$ and with different interaction parameters χ_0. Calculations according to equation (6–32) with $\sigma = 0$.

value for the chosen example. The critical concentration is defined as that volume fraction of the solute at which maximum, minimum, and inflection point coincide.

Chemical potentials for every volume fraction can be taken from a plot of $\Delta G^m = f(\phi_2)$ (see Figures 6-2 and 6-9). Equations (6-5) and (6-6) give, of course,

$$d(\Delta G^m) = \Delta\mu_1\, dn_1 + \Delta\mu_2\, dn_2 \tag{6-36}$$

With $\phi_i = N_i X_i / N_g$ and $n_i = N_i/N_L$, integration leads to

$$\Delta G^m = n_1\Delta\mu_1 + n_2\Delta\mu_2 = \frac{\phi_1 N_g X_1^{-1}\Delta\mu_1 + \phi_2 N_g X_2^{-1}\Delta\mu_2}{N_L} \tag{6-37}$$

$$N_L \Delta G^m = N_g X_1^{-1}\Delta\mu_1 + N_g\phi_2(X_2^{-1}\Delta\mu_2 - X_1^{-1}\Delta\mu_1) \tag{6-38}$$

The equation for the tangent to the curve of $\Delta G^m = f(\phi_2)$ at a point $\phi_2^§$ is

$$Y = A + B\phi_2^§ \tag{6-39}$$

The slope B is given by differentiation of equation (6-38)

$$B = \left(\frac{\partial \Delta G^m}{\partial \phi_2}\right)_{N_g}^§ = N_g(X_2^{-1}\Delta\mu_2^§ - X_1^{-1}\Delta\mu_1^§) \tag{6-40}$$

A is obtained from the argument that at the point $\phi_2^\S$ the value of ΔG^m is also given by equation (6-38), i.e., by

$$A = X_1^{-1} N_g \Delta\mu_1^\S \tag{6-41}$$

Consequently, for Y

$$Y = X_1^{-1} N_g \Delta\mu_1^\S + N_g (X_2^{-1} \Delta\mu_2^\S - X_1^{-1} \Delta\mu_1^\S)\, \phi_2^\S \tag{6-42}$$

Thus, for the limiting cases of $\phi_2 \to 0$ and $\phi_2 \to 1$

$$\lim_{\phi_2 \to 0} Y = X_1^{-1} N_g \Delta\mu_1^\S \qquad \text{and} \qquad \lim_{\phi_2 \to 1} Y = X_2^{-1} N_g \Delta\mu_2^\S \tag{6-43}$$

Thus, if a tangent is drawn to the $\Delta G^m = f(\phi_2)$ curve at a volume fraction of $\phi_2 = \phi_2^\S$, then the extrapolation of this tangent to the ΔG^m axis for values of $\phi_2 = 0$ and $\phi_2 = 1$ gives quantities from which the chemical potentials of solvent and solute, respectively, are obtained.

6.3.6. Chemical Potential of Dilute Solutions

Since the volume fraction of the solute ϕ_2 is very small in dilute macromolecular solutions, the expression $\ln(1 - \phi_2)$ can be expanded in a series

$$\ln(1 - \phi_2) = -\phi_2 - (\phi_2^2/2) - (\phi_2^3/3) - \cdots \tag{6-44}$$

and terminated after the second term. When this is inserted into equation (6-34), we obtain the following for high-molecular-weight solutes in low-molecular-weight solvents in a small volume element:

$$\frac{\Delta\mu_1^{\text{exc}}}{RT} = (\chi_0 - \sigma - 0.5)\, \phi_2^2 \tag{6-45}$$

Equation (6-45) contains the *excess* chemical potential instead of the chemical potential itself because the ideal term $X_1 X_2^{-1}$ has been omitted. This omission was necessary because the considerations have to be restricted to a small volume of solution where the segment distribution is uniform enough so that the lattice theory can be applied. Although in this equation the quantity $(\chi_0 - \sigma)$ is derived from the enthalpy of mixing and the factor 0.5 from the entropy of mixing, it is convenient to replace this combination of terms by another with a new enthalpy parameter κ and a new entropy parameter ψ:

$$\frac{\Delta\mu_1^{\text{exc}}}{RT} = (\kappa - \psi)\, \phi_2^2 \tag{6-46}$$

The chemical potential gives the partial molar Gibbs energy of dilution. The partial molar dilution enthalpy and the partial molar dilution entropy are given by

$$\Delta\tilde{H}_1^m = RT\kappa\phi_2^2, \qquad \Delta\tilde{S}_1^m = R\psi\phi_2^2 \tag{6-47}$$

In the theta state (see Section 6.1), $\Theta = \Delta\tilde{H}_1^m/\Delta\tilde{S}_1^m$, and, according to equation (6-47) with $T = \Theta$, we also have

$$\Theta = \kappa T/\psi \tag{6-48}$$

The combination of equations (6-34), (6-46), and (6-48) thus gives an expression for the temperature dependence of the Flory–Huggins interaction parameter:

$$(\chi_0 - \sigma) = (0.5 - \psi) + \psi\left(\frac{\Theta}{T}\right) \tag{6-49}$$

The expression $(\chi_0 - \sigma)$ takes on the value of 0.5 for $X_2/X_1 \gg 0$ in the theta state. It is smaller than 0.5 in good solvents, since then $T > \Theta$.

6.4. *Virial Coefficients*

6.4.1. *Definitions*

The chemical potential of solutions of nonelectrolytes can always be written in terms of a series of positive integral powers of the concentration

$$\Delta\mu_1 = -RT\tilde{V}_1^m(A_1c_2 + A_2c_2^2 + A_3c_2^3 + \cdots) \tag{6-50}$$

The proportionality coefficients of this series are called the first, second, third, ..., virial coefficients.*

On comparing with the expression for the osmotic pressure Π, we obtain the following (see textbooks of chemical thermodynamics):

$$\Delta\mu_1 = -\Pi\tilde{V}_1^m = -\frac{RTc_2\tilde{V}_1^m}{M_2} \tag{6-51}$$

showing that the first virial coefficient is equal to the reciprocal molecular weight of the solute. The second virial coefficient is a measure of the excluded volume (see Section 6.4.2).

* The name virial coefficient comes from the virial theorem, which was much used toward the end of the 19th century. This theorem states

$$\text{average of } (mv^2/2) = -\text{average of } 0.5(Xx + Yy + Zz)$$

Here m is the mass of the particles; v is their velocity; x, y, and z are their coordinates; and X, Y, and Z are the components of the forces which act upon them. The expression on the right-hand side was called "virial" because forces were considered [*vis* (Latin) = force]. The virial could be expanded into a series whose coefficients were consequently the virial coefficients.

It is important to pay attention to definitions of the virial coefficients when comparing their values reported in the literature. Equation (6-50), when used with osmotic pressures, corresponds to the expression

$$\frac{\Pi}{c_2} = RTM_2^{-1} + RTA_2c_2 + RTA_3c_2^2 + \cdots \tag{6-52}$$

Instead of this, the following definition is often used:

$$\frac{\Pi}{c_2} = RTM_2^{-1} + A_2c_2 + A_3c_2^2 + \cdots \tag{6-53}$$

Equation (6-52) generally will be used here. An apparent molecular weight M_{app} can be defined as a molecular weight calculated from experimental data at finite concentrations from an equation applicable to infinite dilution only:

$$M_{\text{app}}^{-1} \equiv \frac{\Pi}{RTc_2} = M_2^{-1} + A_2c_2 + A_3c_2^2 + \cdots \tag{6-54}$$

Consequently, virial coefficients can be determined from the concentration dependence of the reciprocal apparent molecular weight. But, since the various methods for measuring the molecular weight yield various averages of it (see Chapters 8 and 9), the virial coefficients obtained will be average values which vary according to the method used to determine them. Virial coefficients obtained from osmotic-pressure measurements (and all other measurements based on colligative methods) will give the average

$$A_2^{\Pi} = \sum_i \sum_j w_i w_j A_{ij} \tag{6-55}$$

while light scattering measurements, for example, give the average

$$A_2^{\text{LS}} = \frac{\sum_i \sum_j w_i M_i w_j M_j A_{ij}}{(\sum_i w_i M_i)^2} \tag{6-56}$$

6.4.2. Excluded Volume

The second virial coefficient depends on the excluded volume u. The macromolecules arrange themselves with little mutual interference since the total excluded volume N_2u is much smaller than the total volume V. The total number of possible ways of arranging these N_2 macromolecules is calculated from the partition function Ω,

$$\Omega = \text{constant} \times \prod_{i=0}^{N_2-1} (V - iu) \tag{6-57}$$

Polymer–solvent interactions can be considered in terms of the concept of the effective excluded volume. Consequently, ΔH is equal to zero and the addition of solvent occurs "athermally":

$$\Delta G = -T\Delta S = -kT \ln \Omega = -kT \ln\left[\text{constant} \times \prod_{i=0}^{N_2-1} (V - iu)\right] \quad (6\text{-}58)$$

The volume available to the second molecule is $V - u$, and to the third molecule is $V - 2u$, etc. Solving for the logarithmic expression, we obtain a sum instead of a product:

$$\Delta G = -kT\left[N_2 \ln V + \sum_{i=0}^{N_2-1} \ln\left(1 - \frac{iu}{V}\right)\right] + \text{constant}' \quad (6\text{-}59)$$

In dilute solutions, $iu/V \ll 1$. The logarithm can be expanded into a series: $\ln(1 - y) = -y - \cdots$, and the following is obtained:

$$\Delta G = -kT\left(N_2 \ln V + \sum_{i=0}^{N_2-1} \frac{iu}{V}\right) + \text{constant}' \quad (6\text{-}60)$$

Since u/V is constant, the sum term yields, for $N_2 \to \infty$,

$$\sum_{i=0}^{N_2-1} i = N_2^2/2 \quad (6\text{-}61)$$

The osmotic pressure is given by equations (6-51) and (6-3):

$$\begin{aligned} \Pi = -(\tilde{V}_1^m)^{-1}\left(\frac{\partial G}{\partial n_1}\right)_{n_2,p,T} &= -\frac{N_\mathrm{L}}{\tilde{V}_1^m}\left(\frac{\partial G}{\partial N_1}\right)_{N_2,p,T} \\ &= -\frac{N_\mathrm{L}}{\tilde{V}_1^m}\left(\frac{\partial G}{\partial V}\right)_{N_2,p,T}\left(\frac{\partial V}{\partial N_1}\right)_{N_2,p,T} \\ &= -\left(\frac{\partial G}{\partial V}\right)_{N_2,p,T} \end{aligned} \quad (6\text{-}62)$$

On considering equation (6-62), inserting equation (6-61) into (6-60), and differentiating with respect to V with $N_2/V = c_2 N_\mathrm{L}/M_2$, we obtain the following:

$$\frac{\Pi}{c_2} = \frac{RT}{M_2} + \frac{RTN_\mathrm{L}u}{2M_2^2} c_2 + \cdots \quad (6\text{-}63)$$

Equating the coefficients with those in equation (6-52), we obtain for the second virial coefficient

$$A_2 = \frac{N_L u}{2M_2^2} \tag{6-64}$$

Expressions for the excluded volume of rigid particles have already been derived in Section 4.5.1. The second virial coefficient for unsolvated spheres is given by equations (6-64) and (4-41) as

$$A_2 = \frac{4v_2}{M_2} \tag{6-65}$$

where v_2 is the specific volume. Thus, the second virial coefficient for spheres is reciprocally proportional to the molecular weight and is zero for infinitely large molecular weights.

The second virial coefficient for unsolvated rods is obtained from equations (6-64) and (4-43) with $M_2 = (\pi R_2^2) LN_L/v_2$:

$$A_2 = \frac{v_2^2}{2\pi R_2^3 N_L} \tag{6-66}$$

where R is the radius of the rod. Thus, the second virial coefficient of rod-shaped molecules is independent of length and molecular weight.

The dependence of the second virial coefficient of coil-shaped molecules is difficult to calculate since the excluded volume is a complicated function of the molecular weight (see Section 4.5.2). The usual method is to replace the excluded volume of the molecule u in equation (6-64) by the excluded volume of the chain segment u_{seg}; the molecular weight M_2 of the molecule is also replaced by the formula molecular weight M_u of the chain segment. The molecular weight dependence of the excluded volume is expressed in terms of a function $h(z)$ whose coefficients have been evaluated theoretically:

$$A_2 = \frac{N_L u_{seg}}{2M_u^2}(1 - 2.865z + 14.278z^2 - \cdots) \tag{6-67}$$

When $u_{seg} = 0$ and $z = 0$, $A_2 = 0$ also, that is, A_2 is zero under theta conditions (see also Section 4.5.2.2). A_2 decreases with increasing molecular weight since z increases less rapidly with increasing molecular weight than the square of the molecular weight. The dependence of A_2 on the molecular weight can be given by a power expression:

$$A_2 = K_A M_2^{a_A} \tag{6-68}$$

where K_A and a_A are empirical constants determined for each polymer–solvent–temperature system.

6.5. Association

6.5.1. Basic Principles

Macromolecules in solution can combine with one another to form larger, but still soluble, groups of macromolecules under certain conditions. We will call this phenomenon multimerization. The multimerization equilibrium is called association.

Association can be investigated by group-specific or molecule-specific methods. Group-specific methods are concerned with the behavior of specific groups, i.e., hydrogen-bonding groups, etc. However, these methods are often not sufficiently sensitive. One associating group per molecule is, of course, sufficient to cause association of a macromolecule. This corresponds to a group concentration of only 0.1% for a degree of polymerization of 1000. But the accuracy of group-specific methods is often not better than ±1%.

Molecule-specific methods are concerned with the molecular or particle molecular weight. The particle molecular weight will double for a total dimerization, that is, there is a 100% change. Consequently, molecule-specific methods are much more sensitive than group-specific methods. However, molecule-specific methods only show changes when intermolecular association occurs, and they yield no information on the molecular causes of association.

In molecule-specific methods, the apparent molecular weight is measured as a function of the concentration (see Section 6.4.1). All the expressions so far given, however, are only suitable for the case where an increase in the weight concentration leads to an equally large increase in the molar concentration (double the weight concentration gives double the molar concentration). This assumption does not hold for multimerization, where an increase in the weight concentration leads to a relatively smaller concentration of kinetically independent particles. The concentration dependence of the apparent molecular weight is consequently given by two different terms. The association term describes the change in concentration of independent particles relative to the change in the weight concentration. This association term also occurs in the absence of any polymer–solvent interaction when a polymer associates, i.e., it occurs also under theta conditions. The virial coefficients, on the other hand, take all other interactions into account. The following applies instead of equation (6-54):

$$M_{\text{app}}^{-1} = (M_{\text{app}})_{\Theta}^{-1} + \left(\frac{1}{c^2} \sum_i \sum_j (A_2)_{ij} c_i c_j \right) c + \cdots \tag{6-69}$$

$(M_{\text{app}})_{\Theta}$, is, of course, concentration dependent. The exact form of the concentration dependence depends on the stoichiometry of the association and on the effective associating unit.

Two simple cases can be distinguished for the stoichiometry. *Open association* is the term given to a consecutive process:

$$\begin{aligned} M_I + M_I &\rightleftharpoons M_{II} \\ M_{II} + M_I &\rightleftharpoons M_{III} \\ M_{III} + M_I &\rightleftharpoons M_{IV}, \quad \text{etc.} \end{aligned} \tag{6-70}$$

Thus, all possible "multimers" are in equilibrium with the "unimers."

Closed association is concerned with an "all or nothing" process with only two types of particles being involved:

$$N M_I \rightleftharpoons M_n \tag{6-71}$$

The effective associating unit can be the molecule or a segment of the molecule. The number of associogenic groups is independent of the molecular size in *molecule-related* association. An example of this is the association of end groups. Linear molecules have only two end groups per molecule. Consequently each molecule has two associogenic groups. In this case, the equilibrium association constant must obviously be related to the molar concentration.

Segments of several monomeric units are responsible for association in *segment-related* association. Examples are, e.g., syndiotactic sequences of sufficient length in an "atactic" polymer. The number of these associogenic segments will increase with increasing molecular weight. In this case, the equilibrium association constant is related to the weight concentration.

Macromolecules generally have a molecular-weight distribution. When they associate, a particle-size distribution is produced, which will be different from the molecular-weight distribution and will vary with the nature of the effective unit. The relationships between the molecular-weight distributions and the particle-weight distributions can be derived by statistical methods. Only the results are given here.

The number-average molecular weight of the N-mer is exactly N times as large as the number-average of the unimer in *molecule-related* association

$$(\overline{M}_N)_n = N(\overline{M}_I)_n \tag{6-72}$$

The weight-average molecular weight of the N-mer is given, however, by

$$(\overline{M}_N)_w = (\overline{M}_I)_w + (N - 1)(\overline{M}_I)_n \tag{6-73}$$

For *segment-related* association

$$(\overline{M}_N)_n = (\overline{M}_I)_n + (N - 1)(\overline{M}_I)_w \tag{6-74}$$

Only for a Schulz–Flory distribution do we have

$$(\overline{M}_N)_w = N(\overline{M}_1)_w \tag{6-75}$$

Consequently, the polydispersity $(\overline{M}_N)_w/(\overline{M}_N)_n$ for segment-related association as well as for molecule-related association is always smaller than the polymolecularity $(\overline{M}_1)_w/(\overline{M}_1)_n$. A linear relationship between these two quantities exists for molecule-related, but not segment-related, association:

$$\frac{(\overline{M}_N)_w}{(\overline{M}_N)_n} - 1 = N^{-1}\left(\frac{(\overline{M}_1)_w}{(\overline{M}_1)_n} - 1\right) \tag{6-76}$$

The distribution becomes narrower because the variation in molecular size now takes place within the particle.

6.5.2. Open Association

A series of particles $M_I, M_{II}, M_{III}, \cdots$ is produced by open association. The total molar concentration is therefore

$$[M] = [M_I] + [M_{II}] + [M_{III}] + \cdots \tag{6-77}$$

The equilibrium constant for molecule-related open association is defined as

$$({}^nK_{N-1})_0 = [M_N]/[M_{N-1}][M_I] \tag{6-78}$$

If the association occurs, for example, via the end groups, then it can be assumed that the equilibrium constant defined above is independent of the degree of association N of the multimers produced:

$${}^nK_0 = ({}^nK_I)_0 = ({}^nK_{II})_0 = ({}^nK_{III})_0 = \cdots \tag{6-79}$$

Inserting equations (6-78) and (6-79) into (6-77), we obtain

$$[M] = [M_I][1 + {}^nK_0[M] + ({}^nK_0[M])^2 + \cdots] \tag{6-80}$$

Since, according to equation (6-78), ${}^nK_0[M_I] = [M_{II}]/[M_I]$ still holds, and since the molar concentration of dimers must be smaller than that of monomers, ${}^nK_0[M_I]$ is always smaller than 1. Consequently, equation (6-80) can, according to the rules valid for such series, be given as

$$[M] = [M_I](1 - {}^nK_0[M_I])^{-1} \tag{6-81}$$

The total molar concentration is given as

$$[M] = \frac{c}{(\overline{M}_n)_{\text{app},\Theta}} \tag{6-82}$$

Combining equations (6-81) and (6-82), we obtain

$$[\mathrm{M_I}]^{-1} = {}^{n}K_0 + (M_n)_{\mathrm{app},\Theta}\, c^{-1} \tag{6-83}$$

and the weight concentration is given by

$$c = c_{\mathrm{I}} + c_{\mathrm{II}} + c_{\mathrm{III}} + \cdots \tag{6-84}$$

With

$$[\mathrm{M}_i] = c_i(\overline{M}_i)_n^{-1} \tag{6-85}$$

and equations (6-78), (6-79), and (6-72), we have

$$c = [\mathrm{M_I}]\,(\overline{M}_\mathrm{I})_n\,\{1 + 2({}^{n}K_0[\mathrm{M_I}]) + 3[{}^{n}(K_0[\mathrm{M_I}])^2 + \cdots]\} \tag{6-86}$$

or, for ${}^{n}K_0[\mathrm{M_I}] < 1$

$$c = \frac{[\mathrm{M_I}]\,(\overline{M}_\mathrm{I})_n}{(1 - {}^{n}K_0[\mathrm{M_I}])^2} \tag{6-87}$$

Combination of equations (6-87) and (6-83) gives

$$(M_n)_{\mathrm{app},\Theta} = (\overline{M}_1)_n + {}^{n}K_0(M_1)_n \frac{c}{(M_n)_{\mathrm{app},\Theta}} \tag{6-88}$$

Analogous calculations give for the apparent weight-average molecular weight in the theta state

$$(M_w)_{\mathrm{app},\Theta} = (\overline{M}_1)_w + 2[{}^{n}K_0(\overline{M}_1)_n] \frac{c}{(M_n)_{\mathrm{app},\Theta}} \tag{6-89}$$

It is seen from equation (6-88) that if the apparent number-average molecular weight in the theta state is plotted against $c/(\overline{M}_n)_{\mathrm{app},\Theta}$, the true number-average molecular weight is obtained from the ordinate intercept and the equilibrium constant for the association can be calculated from the slope. Equation (6-89) shows, however, that neither the weight-average molecular weight nor the association equilibrium constant can be determined from data on the apparent weight-average molecular weight alone when dealing with molecule-related association; the corresponding apparent number-average molecular weight must also be known.

Figure 6-4 shows the concentration dependence of the normalized reciprocal apparent number-average molecular weight of some poly(oxyethylenes), $H{+}OCH_2CH_2{)}_nOH$, in benzene. The high-molecular-weight material, H 6000, describes the course expected from equation (6-54) (i.e., no association). Association is obvious with the lower molecular weight material. But it cannot be assumed from such plots that the association decreases with increasing molecular weight. In fact, a plot according to equation (6-88) shows that the initial slope of the function $(\overline{M}_n)_{\mathrm{app},\Theta}/(\overline{M}_1)_n =$

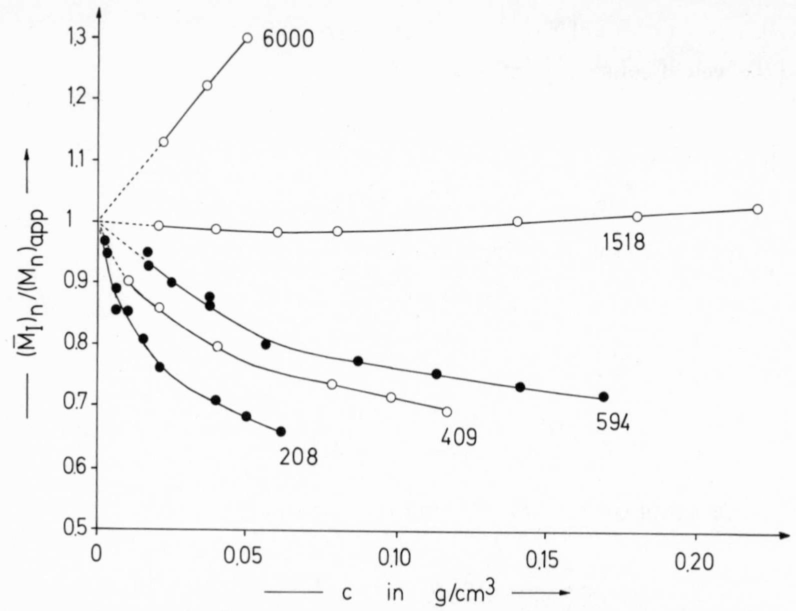

Figure 6-4. Concentration dependence of normalized inverse apparent number-average degrees of polymerization of α-hydro-ω-hydroxy-poly(oxyethylenes) in benzene at 25°C. Numbers indicate the number-average molecular weight of the unimers (H.-G. Elias and H. Lys).

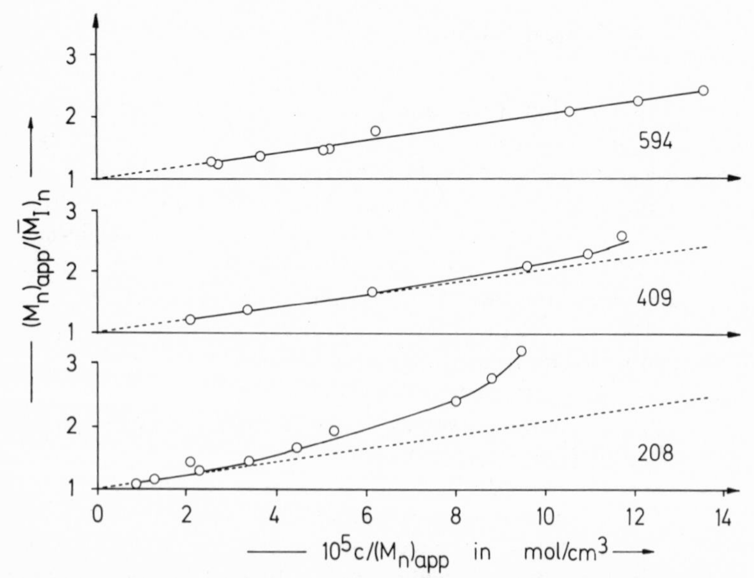

Figure 6-5. Plots of the data of Figure 6-4 under assumption of molecule-based open associations with particle-independent equilibrium constants of multimerization. Numbers indicate the number-average molecular weight.

$f(c/(\overline{M}_n)_{app,\Theta})$ has the same value independent of the molecular weight of the unimer. The equilibrium association constants must consequently also be equal in value. Increasing departure from linearity with decreasing molecular weight in Figure 6-5 indicates negative second virial coefficients.

6.5.3. Closed Association

Equilibrium constants for molecule-related closed association are defined by

$$^{n}K_c \equiv \frac{[M_N]}{[M_I]^N} \tag{6-90}$$

There is no closed expression for the concentration dependence of the

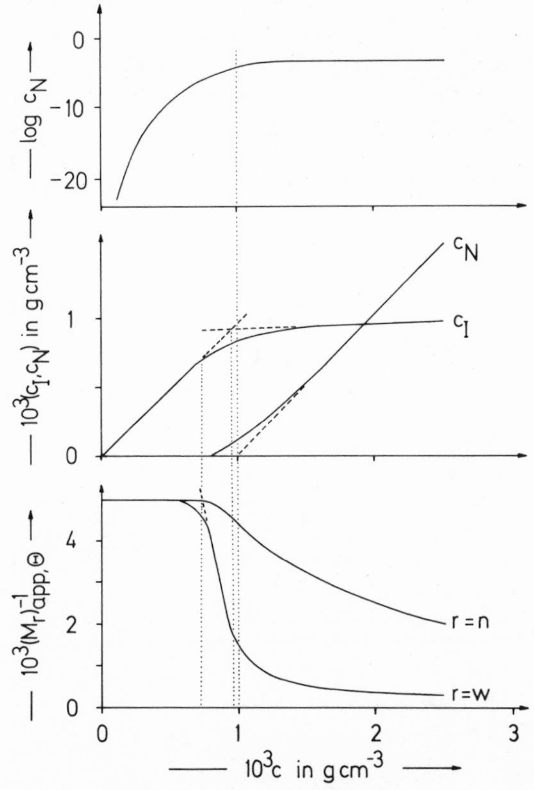

Figure 6-6. Calculated concentration dependences of unimer and multimer concentrations as well as of the inverse apparent number- and weight-average molecular weights for closed, molecule-based associations in the theta state (H.-G. Elias and J. Gerber). Calculations for $(\overline{M}_I)_n = (\overline{M}_I)_w = 200$ g/mol, $N = 21$, and $^{n}K_c = 10^{45}$ $(dm^3/mol)^{N-1}$. Critical micelle concentrations are determined by extrapolation of dotted lines.

apparent molecular weights. Generally, the molecular weights of the unimers, the equilibrium constants ${}^nK_c$, and the degree of association N are obtained by iteration.

The calculated dependence of the concentrations c_N of the multimer and the unimers c_1 on the total concentration c shows a more or less accentuated kink (Figure 6-6). This kink is generally known as the critical micelle concentration (cmc). As can be seen for the concentration dependence of the apparent molecular weights, such "critical micelle concentrations" are also observed there. The position of this critical micelle concentration depends on the measurement method used. The cmc is not a well-defined physical quantity, and it is certainly not the concentration at which associates first appear (see also Figure 6-6).

Solutions of detergents are especially prone to closed association. Closed association has also been observed for poly(γ-benzyl-L-glutamate) in various organic solvents (Figure 6-7). It has been shown in these cases that the Gibbs energy of association depends also on the reciprocal number-average molecular weight of the unimers. Thus, the association must occur here via the end groups. The apparent contradiction between this result and the occurrence of closed association is explained on the

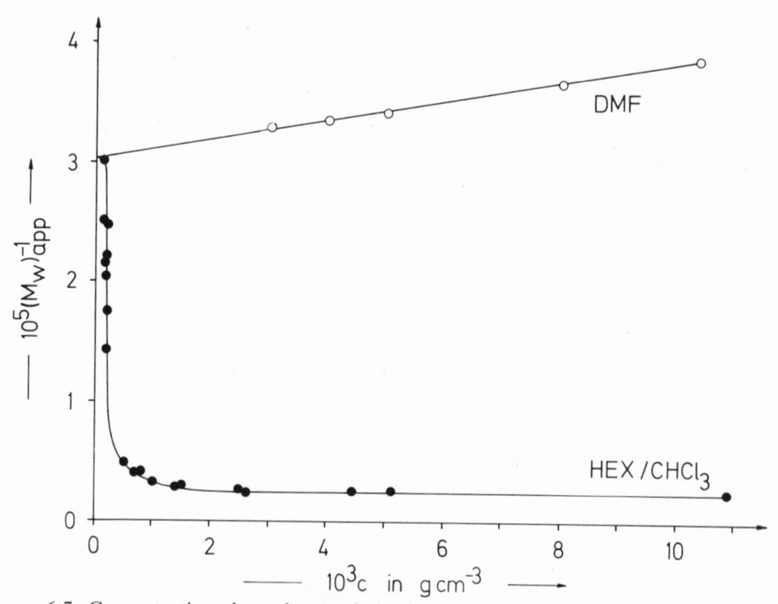

Figure 6-7. Concentration dependence of the inverse apparent weight-average molecular weight of a poly(γ-benzyl-L-glutamate) in dimethylformamide at 70°C and in hexane–chloroform (v/v = 47.7/52.3) at 25°C (H.-G. Elias and J. Gerber).

basis of the formation of ring-shaped associates. Since the molecules occur in the helix form and consequently are relatively rigid, a relationship must exist between the degree of association and the equilibrium association constant. This relationship has actually been observed (Figure 6-7).

6.5.4. *Concentrated Solutions and Melts*

In certain compounds, highly ordered association states (mesophases) occur in concentrated solutions and in melts. These states of order are distinguished from one another as smectic, nematic, and cholesteric. In the case of smectic systems the molecules arrange themselves in parallel layers (Figure 6-8) and the molecular axis is perpendicular to the plane of the layers. The molecules can be arranged regularly or irregularly with respect to each other within these layers. In the nematic state, the molecules are again parallel, but not in layers. The cholesteric state lies between the smectic and the nematic; the molecules are arranged in layers, but the molecular axes are parallel to the plane of the layers.

The layers of the smectic mesophase and the molecules in the nematic mesophase are easily displaced with regard to each other, so that melts of this type have a liquid character. Because of their ordered state, however, these melts are optically anisotropic and show characteristic colors. For this reason, such melts are also called "liquid crystals." Similar phenomena of order can also occur in solutions of rod-shaped macromolecules, so that one also speaks of tactoidal solutions.

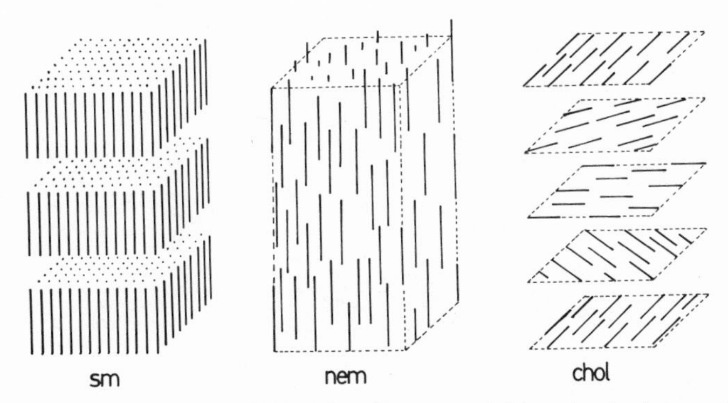

Figure 6-8. Schematic representation of the positioning of the molecules in smectic (sm), nematic (nm), and cholesteric (chol) mesophases.

6.6. Phase Separation

6.6.1. Basic Principles

If a phase separation occurs in a system, the chemical potential of each component in each phase must be the same at equilibrium. Thus, for a binary system of components 1 and 2,

$$\mu_1' = \mu_1'' \qquad \text{and} \qquad \mu_2' = \mu_2'' \tag{6-91}$$

and consequently

$$\begin{aligned} \Delta\mu_1' &= \mu_1' - \mu_1^0 = \mu_1'' - \mu_1^0 = \Delta\mu_1'' \\ \Delta\mu_2' &= \mu_2' - \mu_2^0 = \mu_2'' - \mu_2^0 = \Delta\mu_2'' \end{aligned} \tag{6-92}$$

The values of $\Delta\mu_1$ and $\Delta\mu_2$, however, are given as ordinate intercepts of the tangents to the $\Delta G^m = f(\phi_2)$ curve (see Section 6.3.5). The required equivalence of the chemical potentials in both phases can only be fulfilled if two points of the curve possess a common tangent (see Figure 6-9). But there is only one common tangent to a curve with two minima. The points of contact A and B of this tangent to the $\Delta G^m = f(\phi_2)$ curve determine the compositions ϕ' and ϕ'' of the two phases.

Only systems with $\phi_2 < \phi_2'$ and $\phi_2 > \phi_2''$ are stable. Every system with a composition of $\phi_2' < \phi_2 < \phi_2''$ will separate into two phases. A distinction is made between metastable and unstable regions in this composition range. The positions of the metastable and unstable regions are given by the positions of the inflection points C and D in the $\Delta G^m = f(\phi_2)$ curve.

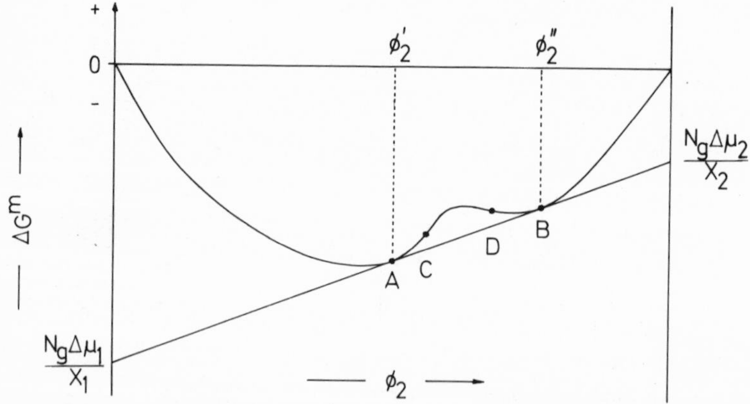

Figure 6-9. Schematic representation of the molar Gibbs energy as a function of the volume fraction of the solute for a partially miscible system.

Since $\partial^2 \Delta G^m/\partial \phi_2^2 > 0$ in the region AC and DB, the system remains stable with regard to a phase that differs by a vanishingly small difference in composition. The system is not stable, however, with respect to phases with the composition ϕ_2' or ϕ_2''. Such a system is called metastable.

The system is, however, unstable even with respect to a phase that differs by a vanishingly small difference in composition in the region CD, since $\partial^2 \Delta G^m/\partial \phi_2^2 < 0$ in this region. The inflection points C and D belong to what are called the spinodals. Spinodals are characterized by the condition $\partial^2 \Delta G^m/\partial \phi_2^2 = 0$.

The spinodals for the case of $X_1 = 1$ and $\sigma = 0$ are obtained by differentiation of equation (6-34):

$$\frac{\partial \Delta \mu_1}{\partial \phi_2} = RT[2\chi_0 \phi_2 - (1 - \phi_2)^{-1} + (1 - X_2^{-1})] = 0 \tag{6-93}$$

The volume fraction at which the maximum, minimum, and point of inflection of the curve $\Delta \mu_1 = f(\phi_2)$ coincide is defined as the critical point. Thus, differentiation of equation (6-93) gives

$$\frac{\partial^2 \Delta \mu_1}{\partial \phi_2^2} = RT[2\chi_0 - (1 - \phi_2)^{-2}] = 0 \tag{6-94}$$

Equations (6-93) and (6-94) are each solved for χ_0. Remembering that $(1 + X_2^{0.5})(1 - X_2^{0.5}) = 1 - X_2$ and taking the negative root of $[X_2/(1 - X_2)^2]^{0.5}$, we find that the critical point is given by

$$(\phi_2)_{\text{crit}} = (1 + X_2^{0.5})^{-1} \tag{6-95}$$

The higher the degree of polymerization of the solute, therefore, the lower will be the values of the critical volume fraction.

The critical value for the Flory–Huggins interaction parameter is obtained from a combination of equations (6-94) and (6-95):

$$(\chi_0)_{\text{crit}} = \frac{(1 + X_2^{0.5})^2}{2X_2} \approx 0.5 + X_2^{-0.5} \tag{6-96}$$

The critical interaction parameter has a value of 0.5 for infinitely high degrees of polymerization.

6.6.2. Upper and Lower Critical Solution Temperatures

To a good approximation, the temperature dependence of the Flory–Huggins interaction parameter can be given by

$$\chi_0 = \alpha + (\beta/T) \tag{6-97}$$

where α and β are system-dependent constants. The term β is generally positive (endothermal mixing). Consequently, χ_0 decreases with increasing temperature. Above a certain temperature ("upper critical temperature," UCST), complete solution occurs (Figure 6-10).

There are also systems that form one phase below a certain temperature ("lower critical solution temperature," LCST). They demix above the LCST. The terms UCST and LCST have nothing to do with the actual magnitude of the temperature at which demixing occurs. For some systems, the UCST, in fact, is at a higher temperature than the LCST (Figure 6-11) but for others, it is lower (Figure 6-10).

The case of UCST > LCST is observed with water-soluble polymers. Examples of these are poly(vinyl alcohol) (see Figure 6-11), poly(vinyl methyl ether), methyl cellulose, and poly(L-proline). The heating of aqueous

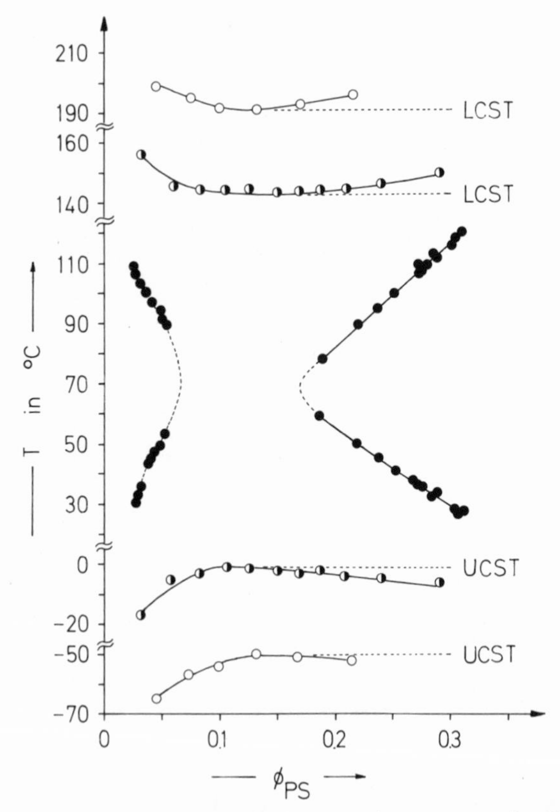

Figure 6-10. Precipitation temperatures of poly(styrene) in acetone as a function of the volume fraction of the solute. Measurement for molecular weights of 4800 (○), 10,300 (◐) and 19,800 (●). LCST is the lower critical solution temperature, UCST is the upper critical solution temperature (K. G. Siow, G. Delmas and D. Patterson).

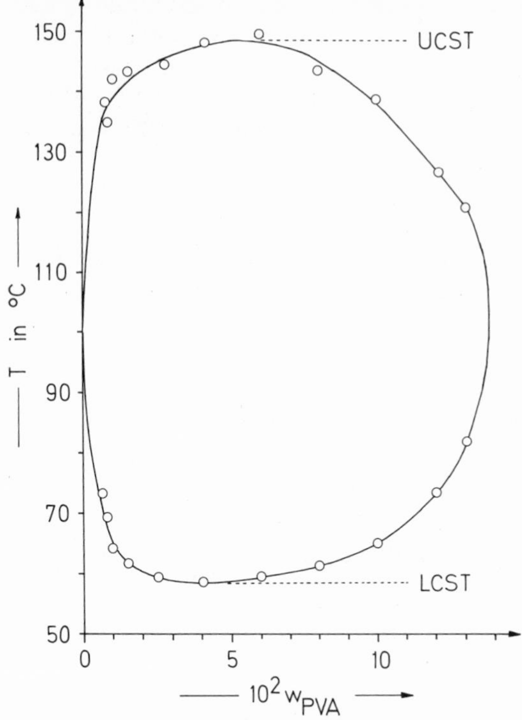

Figure 6-11. Precipitation temperatures of poly[(vinyl alcohol)$_{93}$-co-(vinyl acetate)$_7$] in water as a function of the mass fraction of the solute. $\overline{M}_n$ = 140,000 g/mol (G. Rehage).

solutions of these polymers causes a decreasing solvation of the polymer and thus a demixing. In some cases, closed miscibility loops can be observed.

The case of UCST < LCST occurs quite generally with solutions of macromolecules at temperatures above the boiling point of the solvent and at pressures of several bars. In these systems, a contraction occurs on mixing the dense polymer with the highly expanded solvent. This leads to negative entropies of mixing and, consequently, to lower critical solution temperatures. Since this effect is characteristic of all solutions of macromolecules, the "goodness" of the solvent must pass through a maximum between the upper and the lower critical solution temperature, that is, χ_0 must pass through a minimum with temperature.

6.6.3. *Quasibinary Systems*

All the points so far considered are concerned with genuine binary systems. A system is "binary" if both the solute and the solvent are monodisperse. But macromolecular substances generally possess a molecular-weight distribution: consequently they only form quasibinary systems with pure solvents.

The phase-separation behavior of quasibinary systems is different from that of binary systems. This phenomenon can be most simply described in

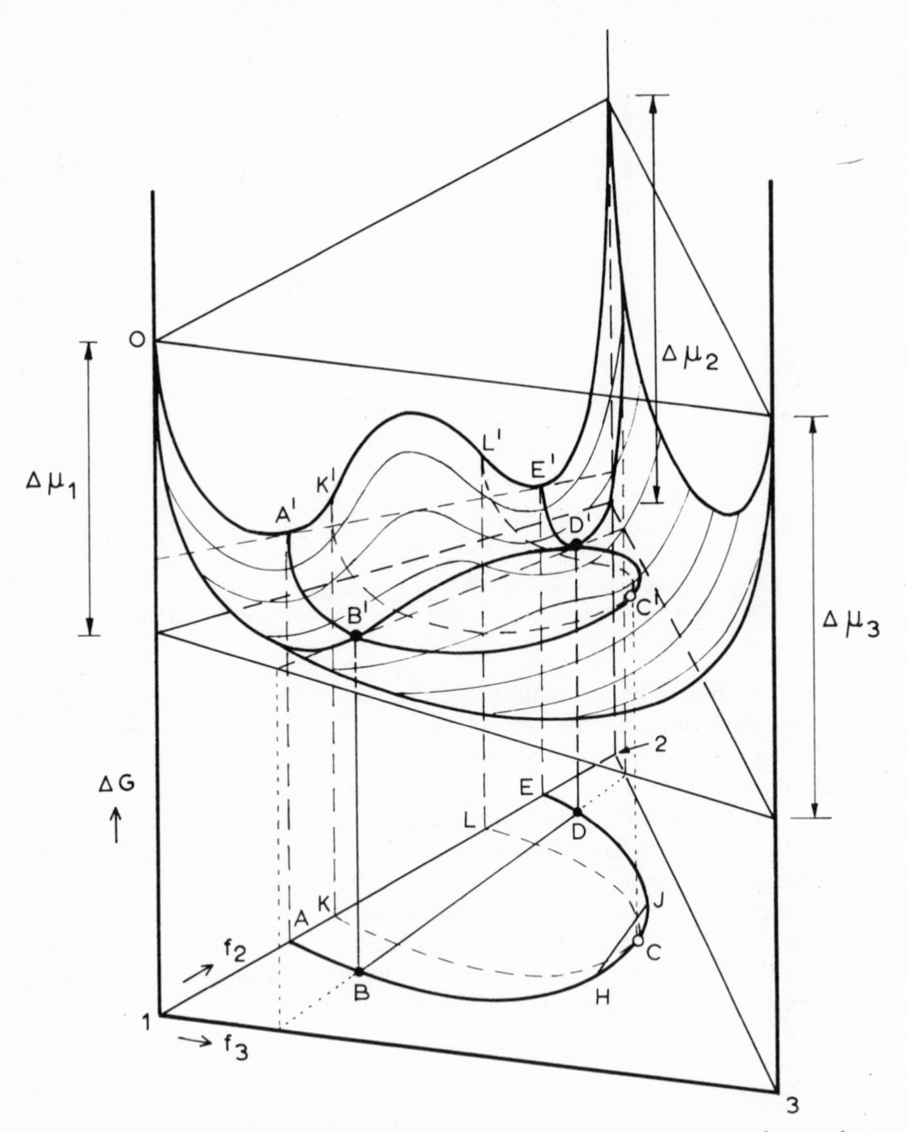

Figure 6-12. Surface of Gibbs energy for a partially miscible system of a solvent and two molecular homogeneous polymers. The binary system 1–2 exhibits partial miscibility, the systems 2–3 and 1–3 are completely miscible (R. Koningsveld).

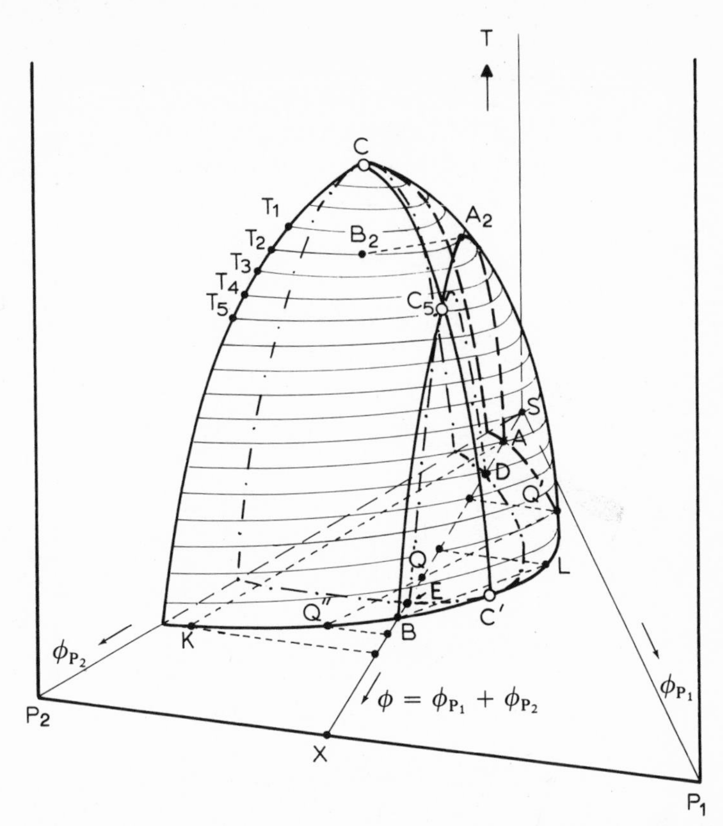

Figure 6-13. Surface of binodals of a ternary liquid system with a two-phase region. CC_5C' = connecting line for critical points, AA_2C_5B = quasibinary line (cloud curve) (R. Koningsveld).

terms of the cloud-point curves of ternary systems consisting of a solvent and two monodisperse solutes. The cloud-point curve corresponds to the special case of phase separation where the volume of one of the phases tends toward zero.

Such a ternary system has a Gibbs energy surface instead of the Gibbs energy curve associated with genuine binary systems (Figure 6-12). Also, a tangent plane occurs instead of a tangent. If a tangent plane is moved over the Gibbs energy surface of a ternary system of limited miscibility, two series of contact points (e.g., B' and D') are produced. The line $A'B'C'D'E'$ and its projection $ABCDE$ to the base surface are each called binodals. The binodals describe the boundary between stable and metastable mixtures and give the compositions of the coexisting phases. These compositions are related via the joining lines AE, BD, HJ, etc. The compositions of the coexisting phases are identical at the critical point C' (or C).

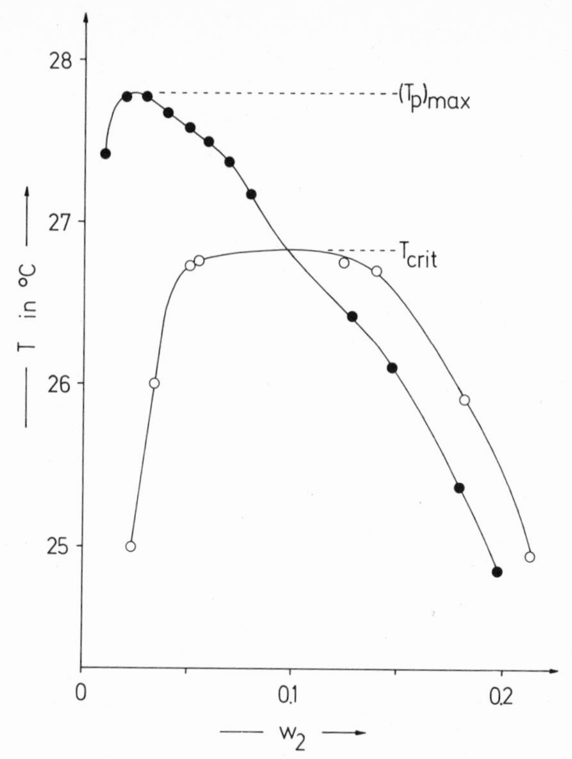

Figure 6-14. Dependence of the cloud point temperature T of a polydisperse polystyrene ($\overline{M}_z:\overline{M}_w:\overline{M}_n = 2.4:1.65:1$ and $\overline{M}_n = 210{,}000$ g/mol) on the mass fraction w_2 of the polymer (●–●). The curve (○–○) gives the curve of coexistence for a 6% initial solution at different temperatures T, i.e., the mass fractions w_2 at which the polymer is found in the two coexistent phases. (After G. Rehage and D. Moller.)

The Gibbs energy surface is altered by a change in temperature, and consequently, the positions of the binodals also change (see Figure 6-13). The critical points move along the joining line C–C_5–C'. The maximum of this joining line occurs at the critical point C for the pure polymer P_2. Consequently, the critical point can only be identical with the maximum in the cloud-point curve in the case of a genuine binary system. In contrast, the critical points of polymer mixtures lie lower than the cloud-point-curve maximum, that is, at larger polymer volume fractions. The point of intersection of the coexistence curve with the cloud-point curve gives the critical point (see Figure 6-14). Quantitative calculations give the critical volume fraction of a quasibinary system as [compare equation (6-95) for binary systems]

$$(\phi_2)_{\text{crit}} = (1 + \overline{X}_w \overline{X}_z^{-0.5})^{-1} \tag{6-98}$$

The critical interaction parameter is given as

$$(\chi_0)_{\text{crit}} = 0.5(1 + \overline{X}_z^{0.5}\overline{X}_w^{-1})(1 + \overline{X}_z^{-0.5}) \tag{6-99}$$

The difference between the volume fraction of the solute at the maximum of the cloud-point curve and the critical volume fraction can be used as a measure of the polymolecularity. The same holds for the difference between the maximum cloud-point temperature and the critical demixing temperature.

6.6.4. Fractionation and Microencapsulation

Fractionation of polymers according to molecular weight represents the most significant analytical application of phase-separation phenomena. The polymers of highest molecular weight separate out first on lowering the temperature of a quasibinary endothermic dilute solution system. Of course, this "precipitation" represents the formation of a highly concentrated "gel phase" and a dilute "sol phase." Successive decreases in temperature lead to further fractions, and the amounts and molecular weights of these are determined. The fractionation process should be so carried out that the fractions obtained reproduce the original molecular-weight distribution as closely as possible. Computer calculations on the basis of the Flory–Huggins theory show that this is best achieved by separating initially into e.g., five fractions and then separating each fraction into three subfractions (or vice versa). The fractions produced do not necessarily possess a much narrower molecular-weight distribution than the parent sample. The fractions may even have a much wider molecular-weight distribution than the original material.

Since precipitation temperatures may lie in experimentally unfavorable regions, precipitation fractionation is often carried out by addition of precipitant to the polymer solution at constant temperature. It is advantageous to use a 1% solution of the polymer in a poor solvent as initial solution and to add a weak nonsolvent as precipitant. To achieve a good fractionation, it is best to heat a precipitated system to redissolution and then cool the well-stirred solution to the original temperature. Further fractions are obtained by adding more precipitant.

Systems consisting of a polymer, a solvent, and a nonsolvent have become of commercial importance in what is known as the microencapsulation process. In this method, the substance to be encapsulated (e.g., carbon black) is dispersed in a polymer solution so that a two-phase system results. Through, for example, the addition of a precipitant, the dissolved polymer can then be precipitated as a new third phase. [A solution of a second polymer also acts as a precipitant (see next chapter).] In this third phase, the polymer is a highly concentrated solution in a mixture of solvent and precipitant. This highly concentrated solution envelops the substance that is to be encapsulated, coating it with a thick layer. Constant stirring ensures that the particles do not coagulate. This layer is then reinforced, for example, by cooling to below the glass-transition temperature, by cross-linking, etc. The result is microcapsules with diameters of between 0.005 and 5 mm containing the encapsulated material. This material can be released by pressure on the capsule, by dissolution, melting, or chemical degradation of the encapsulation layer. Microencapsulation is used in particular for certain carbon papers (the carbon black is encapsulated and so does not smudge the fingers on touching), in pharmaceuticals, and in the adhesives industry.

6.6.5. Determination of Theta States

The critical temperature is identical with the maximum of the cloud-point curve for binary systems (see Section 6.6.3). The dependence of the critical temperature on the degree of polymerization is given by combining equations (6-49) and (6-96):

$$\frac{1}{T_{\text{crit}}} = \frac{1}{\Theta} + \frac{1}{\Theta\psi}\left(\frac{1}{X_2^{0.5}} + \frac{1}{2X_2}\right) \tag{6-100}$$

Consequently, the critical temperature of mixing is identical with the theta temperature at infinitely high degrees of polymerization. The theta temperature is therefore the critical temperature of mixing of a polymer of infinitely high degree of polymerization.

The dependence of the critical temperature of mixing on the degree of polymerization required by equation (6-100) is found for quasibinary systems as well as for binary systems (Figure 6-15). The slope of line describing this dependence is determined by the entropy term ψ. If ψ is very small, the critical mixing temperature lies far from the theta temperature. For example, poly(chloroprene) has an entropy term of $\psi = 0.05$ and a theta temperature of 298.2 K. The critical temperature of mixing is -73°C for a molecular weight of 700,000 g/mol. Thus, under certain conditions, a quite wide temperature range must be studied to determine the critical temperature of mixing.

The method for determining the theta temperature based on equation (6-100) requires several samples of known degree of polymerization and is very time consuming. The theta temperature for a single sample of unknown molecular weight can be determined by another process. The method was first evolved for the determination of theta mixtures and was later used to determine the theta temperatures of binary and quasibinary systems.

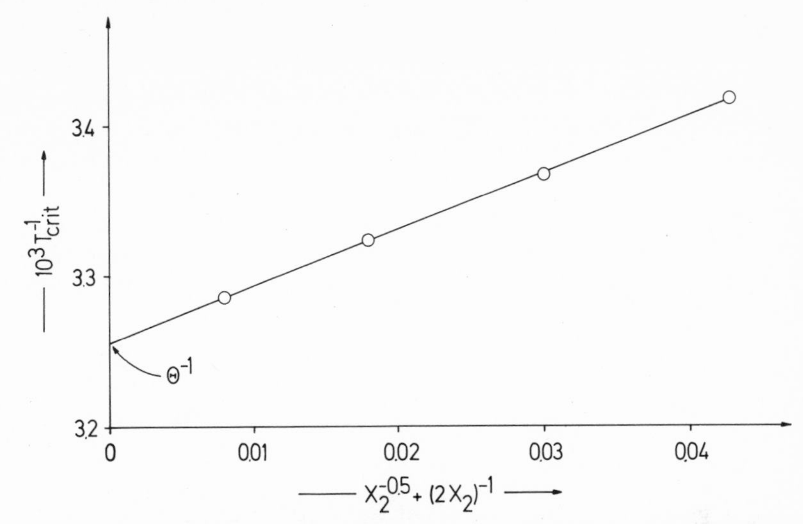

Figure 6-15. Determination of the theta temperature of poly(styrene) in cyclohexane from the critical temperature dependence on the degree of polymerization (after A. R. Shultz).

The process uses very dilute solutions, normally in the concentration range of ϕ_2 between 10^{-5} and 10^{-2}. In what is called the cloud-point titration method, dilute polymer solutions are titrated with nonsolvent at constant temperature to the first cloud point. The volume fraction ϕ_3 of nonsolvent to give the first cloud point is plotted against the logarithm of the volume fraction ϕ_2 of the polymer at the cloud point. The extrapolated straight line obtained for a homologous series at a point on the $\phi_2 = 1$ axis was found experimentally and theoretically to be equivalent to $(\phi_3)_\Theta$ (Figure 6-16). $(\phi_3)_\Theta$ corresponds to the solvent–precipitant theta mixture for the polymer at this temperature.

A first cloud point determined by lowering the temperature can also be obtained in an analogous manner to that produced by addition of precipitant. The dependence on concentration of the first cloud-point temperature is given by

$$T_p^{-1} = \Theta^{-1} + \text{constant} \times \log \phi_2 \tag{6-101}$$

Thus, the reciprocal of the theta temperature is obtained at $\phi_2 = 1$.

If the measurements are carried out with the same solvent–precipitant–temperature system on copolymers of different composition, then the values of $(\phi_3)_\Theta$ for the contents w_A of the various monomeric units of the copolymers lie on a straight line (Figure 6-17). It has been shown experimentally that the method yields data concerning the average composition of the copolymer, provided that copolymer compositions vary only slightly

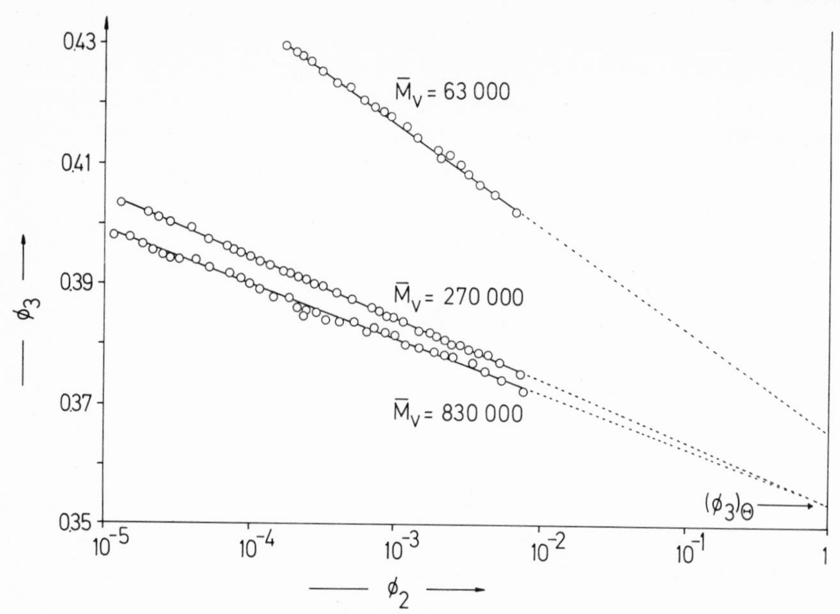

Figure 6-16. Dependence of the volume fraction ϕ_3 of the precipitant on the logarithm of the volume fraction ϕ_2 of the solute at the point of incipient turbidity for polystyrenes of different viscosity-average molecular weights in a benzene–isopropanol system at 25°C (after A. Stasko and H.-G. Elias).

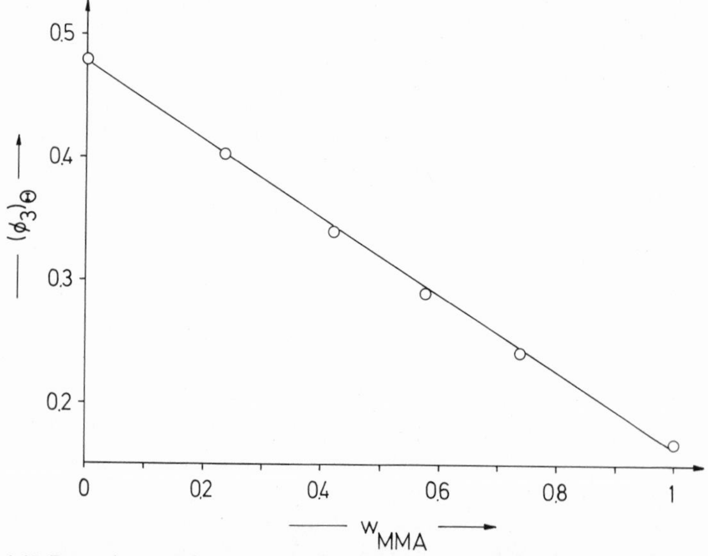

Figure 6-17. Dependence of the $(\phi_3)_\Theta$ values on the average mass fraction w_{MMA} of the methylmethacrylate monomeric units in poly(styrene-co-methyl methacrylates). Solvent: methylisopropyl ketone; precipitant: *n*-hexane; temperature: 25°C. (After H.-G. Elias and U. Gruber.)

from molecule to molecule. The values of $(\phi_3)_\Theta$ obtained are thus independent of whether the copolymers are alternating, random, branched, block, or graft copolymers. The method can therefore be used to elucidate or confirm the composition of copolymers. It is of particular interest in the study of graft copolymers, for example, since addition of a small amount of unipolymer with a lower $(\phi_3)_\Theta$ yields only this lower value and not that corresponding to the average composition of the mixture. If, on the other hand, another solvent–precipitant–temperature system is chosen in which this unipolymer now has a higher $(\phi_3)_\Theta$ value than the copolymer, then the copolymer composition can be determined independent of the amount of unipolymer present.

The molecular-weight distribution can, at least in principle, be determined by the cloud-point titration or turbidimetric titration method. In turbidimetric titration, a precipitant is added continually, with stirring, to a very dilute (~0.01 %) solution, and the increase in turbidity is observed as a function of the quantity of precipitant added. The turbidity curve which is obtained is a qualitative measure of the molecular weight distribution. These curves are difficult to evaluate quantitatively, however, since the turbidity continually changes because of the coagulation of the droplets during titration. The turbidity is therefore not entirely due to the molecular weight and concentration of the macromolecule.

6.6.6. *Incompatibility*

In a mixture of two different polymers, polymer 1 takes on the role of solvent for polymer 2. The degrees of polymerization X_1 and X_2 are of the same order of magnitude in a system polymer 1–polymer 2. Consequently, for $\sigma = 0$, equation (6-34) becomes

$$\frac{\Delta\mu_1}{RT} = \chi_0\phi_2^2 + \ln(1 - \phi_2) \qquad (6\text{-}102)$$

Thus, the total entropy contribution comes from the relatively small logarithmic term. Because of the high degree of polymerization X_1, however, the Flory–Huggins interaction parameter χ_0 will be very large [see equation (6-28)]. Numerical comparison of both terms shows that the chemical potential will already be positive for very small concentrations. Consequently, mixtures of two polymers are generally thermodynamically incompatible. Here, "incompatible" does not mean that the two polymers are not mixable over the whole of the concentration range. It only means that the systems show incompatibility in concentration ranges that are important in practice.

The theoretical prediction has been experimentally verified. Of a total of 281 pairs tested, 239 polymers were definitely incompatible. Even very similar polymer pairs such as poly(styrene)/poly(*p*-methyl styrene) are incompatible. On the other hand, nitrocellulose is compatible with quite a number of polymers.

In general the same principles apply to mixtures of two polymers in a common solvent. Theoretical calculations of the spinodals show that the incompatibility depends on the interaction parameter χ_{23} (polymer–solvent 3) at high polymer concentrations. Conversely, the difference between the interaction parameters χ_{12} and χ_{13} becomes important at low concentrations. If these interaction parameters differ greatly, a very strong solvent influence on the incompatibility will be observed in dilute solutions. The poly(styrene)/poly(vinyl methyl ether) system is, for example, compatible in toluene, benzene, or perchloroethylene, but not in chloroform or methylene chloride. Conversely, at high polymer concentrations, incompatibility in one solvent is normally accompanied by incompatibility in all other solvents.

Incompatible polymer mixtures can be recognized, in many cases, by purely optical means, such as the opaque appearance in the solid state. On the other hand, transparent samples are not a guarantee of compatibility of two polymers, since an opacity only occurs for sufficiently large refractive index differences between regions of sufficient size (see also Section 15.3). Consequently, incompatibility in transparent samples may require identification by electron microscopy. If the regions of pure polymer in the polymer mixture are of sufficient size, two different glass-transition temperatures can often be measured. In the case of incompatible polymer mixtures, such different glass-transition temperatures remain invariant with the composition of the mixture.

Incompatibility is not always undesirable in practice. It is even utilized in the case of block copolymers and high-impact-strength polymers (see Section 11.6.3).

6.6.7. *Swelling*

The equilibrium swelling of a chemically cross-linked macromolecule depends on the thermodynamic goodness of the solvent (Figure 6-18). Swelling occurs to an equilibrium value, since the solvent tries to fully dissolve the gel. However, elastic restraining forces resulting from the chemical cross-linking are effective. The following holds for equilibrium swelling:

$$\Delta G = \Delta G_{\mathrm{mix}} + \Delta G_{\mathrm{el}} = 0 \qquad (6\text{-}103)$$

where ΔG_{mix} is the Gibbs energy of mixing and ΔG_{el} is the Gibbs energy

Figure 6-18. Influence of the solvent power on the swelling of weakly cross-linked samples of polystyrene (cross-linked with divinylbenzene). From left to right: unswollen sample, swelling in the poor solvent cyclohexane (χ_0 high), swelling in the good solvent benzene (χ_0 low).

of elasticity. The chemical potential of the solvent in the gel is

$$\Delta\mu_1^{\text{gel}} = N_{\text{L}}\left(\frac{\partial \Delta G_{\text{mix}}}{\partial N_1}\right)_{p,T,N_2} + N_{\text{L}}\left(\frac{\partial \Delta G_{\text{el}}}{\partial N_1}\right)_{p,T,N_2} = 0 \qquad (6\text{-}104)$$

with

$$N_{\text{L}}\left(\frac{\partial \Delta G_{\text{el}}}{\partial \alpha}\right)_{N_2,p,T}\left(\frac{\partial \alpha}{\partial N_1}\right)_{N_2,p,T} = N_{\text{L}}\left(\frac{\partial \Delta G_{\text{el}}}{\partial N_1}\right)_{N_2,p,T} \qquad (6\text{-}105)$$

The three differentials in equations (6-104) and (6-105) can be evaluated as follows:

1. The chemical potential of the solvent

$$\Delta\mu_1 = \left(\frac{\partial \Delta G}{\partial n_1}\right)_{p,T,n_2} = N_{\text{L}}\left(\frac{\partial \Delta G}{\partial N_1}\right)_{p,T,N_2} \qquad (6\text{-}106)$$

is given by equation (6-34). For $\sigma = 0$ and with $R/N_{\text{L}} = k$, the following is obtained for a polymer of infinitely high degree of polymerization (i.e., cross-linked, $X_2 \to \infty$) and solvent of low molecular weight ($X_1 = 1$):

$$\left(\frac{\partial \Delta G_{\text{mix}}}{\partial N_1}\right)_{p,T,N_2} = kT[\chi_0\phi_2^2 + \ln(1 - \phi_2) + \phi_2] \qquad (6\text{-}107)$$

2. Following the derivation in Section 11.3.3, the Gibbs energy of elasticity depends on the effective molar chain concentration ν_e in the cross-linked system before cross-linking as well as on the expansion factor

$\alpha = \alpha_x = \alpha_y = \alpha_z$. From equation (11-31), the following is obtained with $\Delta G_{\text{el}} = -T\,\Delta S_{\text{el}}$:

$$\Delta G_{\text{el}} = 0.5kTv_e(3\alpha^2 - 3 - \ln \alpha^3) \tag{6-108}$$

or, after differentiation

$$\left(\frac{\partial \Delta G_{\text{el}}}{\partial \alpha}\right)_{p,T,N_2} = 0.5kTv_e(6\alpha - 3\alpha^{-1}) \tag{6-109}$$

3. A cross-linked polymer of volume V_0 in the unswollen state swells to the volume V. For an isotropic expansion $\alpha^3 = V/V_0 = \phi_2^{-1}$. Also $\phi_2 = V_0/(V_1 + V_0) = V_0(N_1 V_1^m N_{\text{L}}^{-1} + V_0)$ is valid when the volumes are additive. Inserting all these relationships into the expression for α and differentiating, we obtain

$$\left(\frac{\partial \alpha}{\partial N_1}\right)_{p,T,N_2} = \frac{V_1^m}{3\alpha^2 V_0 N_{\text{L}}} \tag{6-110}$$

Combining equations (6-104), (6-105), (6-107), (6-109), and (6-110) and transforming, we obtain

$$\chi_0\phi_2^2 + \ln(1 - \phi_2) + \phi_2 = -(v_e V_1^m V_0^{-1} N_{\text{L}}^{-1})(\phi_2^{1/3} - \tfrac{1}{3}\phi_2) \tag{6-111}$$

If the Flory–Huggins interaction parameter is known from other methods, the effective number of cross-linked network chains can be calculated from the observed volume fraction of polymer in the gel.

Equation (6-111) describes the behavior of weakly swollen, weakly cross-linked polymers quite well. The contributions of the polymer–solvent interaction and the dilution by solvent are negligibly small in comparison to the elasticity term. Highly cross-linked polymers swell to the same (small) extent in different good solvents.

6.6.8. *Crystalline Polymers*

All the derivations considered so far are only valid for demixing into liquid phases. A phase separation of a crystalline polymer, however, may result in the formation of a solid, crystalline phase and a liquid phase (as long as the phase separation occurs below the melting point of the crystalline polymer). According to equation (6-4), the change in chemical potential of the solute in the solid phase is

$$d\mu_i^{\text{sol}} = \tilde{V}_i^m\,dp - \tilde{S}_i^m\,dT + RT\,d(\ln a_i) \tag{6-112}$$

By definition, the activity of a pure crystalline substance is equal to one:

$$d\mu_i^{\text{cr}} = V_i^m\,dp - S_i^m\,dT \tag{6-113}$$

The chemical potentials in both phases are equal at solution equilibrium,

$$d(\mu_i^{\text{sol}} - \mu_i^{\text{cr}}) = (\tilde{V}_i^m - V_i^m)\,dp - (\tilde{S}_i^m - S_i^m)\,dT + RT\,d(\ln a_i) = 0 \quad (6\text{-}114)$$

$dp = 0$ and $dT = 0$, and, consequently, $d(\ln a_i) = 0$, i.e., $a_i =$ constant, for an isothermal–isobaric process. Crystalline substances cannot, therefore, exceed a certain saturation level at solution equilibrium. Consequently, in contrast to amorphous material, crystalline substances have only limited solubility. The saturation concentration can be raised by additives (salting in) or lowered by additives (salting out), since the saturation limit depends on the activity and the activity can be changed by additives.

The melting point T_{cr} of a monodisperse polymer–solvent system depends strongly on the concentration of the polymer. The chemical potential of the polymer in the crystalline phase is identical with the molar

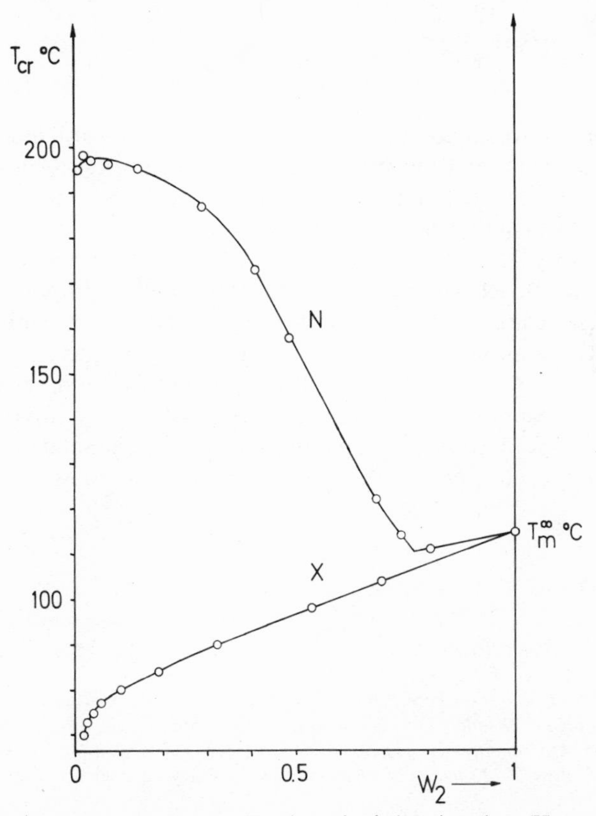

Figure 6-19. Crystallization temperature T_{cr} of a polyethylene in xylene (X) or nitrobenzene (N) as a function of the mass fraction w_2 of the polymer (after R. B. Richards).

Gibbs energy of fusion ΔG_M^m:

$$\mu_2^{cr} - \mu_2^0 = \Delta G_M^m = \Delta H_M^m - T_{cr}\Delta S_M^m = \Delta H_M^m\left(1 - \frac{T_{cr}\Delta S_M^m}{\Delta H_M^m}\right) \quad (6\text{-}115)$$

The equation $T_M^\infty = \Delta H_M^m/\Delta S_M^m$ holds for the melting point of the undiluted polymer, so equation (6-115) becomes

$$\mu_2^{cr} - \mu_2^0 = \Delta H_M^m\left(1 - \frac{T_{cr}}{T_M^\infty}\right) \quad (6\text{-}116)$$

Assuming the enthalpy and entropy of fusion to be temperature independent, and since at equilibrium

$$\mu_2^{cr} - \mu_2^0 = \mu_2^{sol} - \mu_2^0 \quad (6\text{-}117)$$

then inserting equations (6-116) and (6-35) into (6-117), we find

$$\frac{1}{T_{cr}} = \frac{1}{T_M^\infty} + \frac{R}{\Delta H_M^m}[X_2X_1^{-1}(\chi_0 + 2\sigma\phi_2)(1 - \phi_2)^2 + (1 - X_2X_1^{-1})(1 - \phi_2) + \ln\phi_2] \quad (6\text{-}118)$$

According to equation (6-117), the melting point should decrease with increasing solvent concentration when phase separation is into liquid and crystalline phases. This type of behavior is observed, for example, with solutions of poly(ethylene) in xylene (Figure 6-19).

The Flory–Huggins interaction parameter can, however, exceed its critical value above a certain temperature for solutions in poor solvents. In this case, separation into two liquid phases also occurs with crystalline polymers, as is shown, for example, by poly(ethylene) in nitrobenzene at mass fractions $w_2 < 0.75$ (Figure 6-19). On the other hand, separation into one crystalline and one liquid phase is observed at poly(ethylene) concentrations $w_2 > 0.75$. In xylene, separation into one liquid and one crystalline phase always occurs, no matter what the mass fraction is.

Literature

Section 6.1. Basic Principles

P. J. Flory, *Principles of Polymer Chemistry,* Cornell University Press, Ithaca, New York, 1953.

H. Tompa, *Polymer Solutions,* Butterworth, London, 1956.

H. Morawetz, *Macromolecules in Solution,* Wiley–Interscience, New York, 1965.

P. J. Flory, *Statistical Mechanics of Chain Molecules,* Wiley–Interscience, New York, second ed., 1975.

V. N. Tsvetkov, V. Ye. Eskin, and S. Ya. Frenkel, *Structure of Macromolecules in Solution,* Butterworth, London, 1970.

G. C. Berry and E. F. Casassa, Thermodynamic and hydrodynamic behavior of dilute polymer solutions, *Macromol. Rev.* **4,** 1 (1970).
H. Yamakawa, *Modern Theory of Polymer Solutions,* Harper and Row, New York, 1971.

Section 6.2. Solubility Parameter

J. L. Gardon, Cohesive-energy density, *in: Encyclopedia of Polymer Science and Technology* (H. F. Mark, N. G. Gaylord and N. M. Bikales, eds.), Wiley–Interscience, New York, 1966, Vol. 3, p. 833.
J. H. Meier zu Baxten, Löslichkeitsparameter und ihre Anwendung in der Praxis, *Farbe und Lack* **78,** 813 (1971).

Section 6.3. Statistical Mechanics

D. Patterson, Thermodynamics of non-dilute polymer solutions, *Rubber Chem. Technol.* **40,** 1 (1967).
H. Sotobayashi and J. Springer, Oligomere in verdünnten Lösungen, *Adv. Polym. Sci.* **6,** 473 (1969).
D. J. R. Laurence, Interactions of polymers with small ions and molecules, *in: Physical Methods in Macromolecular Chemistry* (B. Carroll, ed.), Dekker, New York, Vol. 2, 1972.
B. E. Conway, Solvation of synthetic and natural polyelectrolytes, *J. Macromol. Sci. C (Rev. Macromol. Chem.)* **6**, 113 (1972).
A. Katchalsky, Polyelectrolytes (IUPAC-International Symposium on Macromolecules, Leiden 1970), *Pure Appl. Chem.* **26**(3–4), 327 (1971).
F. Oosawa, *Polyelectrolytes,* Dekker, New York, 1971.
N. Ise, The mean activity coefficient of polyelectrolytes in aqueous solutions and its related properties, *Adv. Polym. Sci.* **7,** 536 (1971).

Section 6.5. Association

H.-G. Elias, Association and aggregation as studied via light scattering, *in: Light Scattering from Polymer Solutions* (M. B. Huglin, ed.), Academic Press, London, 1972.
H.-G. Elias, Association of synthetic polymers, *in: Order in Polymer Solutions* (Midland Macromol. Monographs, Vol. 2) (K. Solc, ed.), Gordon and Breach, New York, 1975.

Section 6.6. Phase Separation

M. J. R. Cantow, ed., *Polymer Fractionation,* Academic Press, New York, 1967.
R. Konigsveld, Preparative and analytical aspects of polymer fractionation, *Adv. Polym. Sci.* **7,** 1 (1970).
W. V. Smith, Fractionation of polymers, *Rubber Chem. Technol.* **45,** 667 (1972).
B. A. Wolf, Zur Thermodynamik der enthalpisch und entropisch bedingten Entmischung von Polymerlösungen, *Adv. Polym. Sci.* **10,** 109 (1972).
S. Krause, Polymer compatibility, *J. Macromol. Sci. C (Rev. Macromol. Chem.)* **7**, 251 (1972).
S. Krause, *Symposium on Microencapsulation* (Chicago 1973), Plenum Press, New York, 1974.
H.-G. Elias, Cloud point and turbidity titrations, *in: Fractionation of Synthetic Polymers* (L. H. Tung, ed.), M. Dekker, New York, in press.

Chapter 7

Transport Phenomena

Matter, energy, charge, momentum, etc., can be transported with observable effects. For example, the viscosity of gases is due to transfer of momentum, while the conduction of heat is the transport of thermal energy. The transport of matter can occur by diffusion, by sedimentation in a centrifugal field, by electrophoresis in an electric field, etc.

7.1. *Effective Quantities*

When matter is transported in solution, the effective mass and the effective volume are not identical to the mass and volume of the "dry" macromolecule. A transported protein molecule such as myoglobin (see Figure 4-14) drags associated solvent along with it, which contributes to the frictional resistance.

The effective mass m_h of a moving molecule consists of the mass m_2 of the "dry" macromolecule of molecular weight M_2 and the mass $m_1^{\square}$ of the solvent carried along with the macromolecule. If the mass of the solvent carried is expressed as a multiple $\Gamma_h = m_1^{\square}/m_2$ of the polymer mass, the following is then obtained for the effective mass of the molecule:

$$m_h = m_2 + m_1^{\square} = m_2(1 + \Gamma_h) = \frac{M_2(1 + \Gamma_h)}{N_L} \qquad (7\text{-}1)$$

The effective volume V_h is composed, analogously, of the volume of the dry macromolecule and that of the solvent moving with the macromolecule, where the volumes are replaceable by the specific volumes v_2 and v_1:

$$V_h = V_2 + V_1^{\square} = v_2 m_2 + v_1^{\square} m_1 = \frac{M_2(v_2 + \Gamma_h v_1^{\square})}{N_L} \tag{7-2}$$

The specific volume v_1 of the solvent in the macromolecule is different from the specific volume v_1 of the pure solvent, since part of the solvent which is in the macromolecule will undergo a specific interaction (solvation) with the macromolecule. The other part will be dragged along purely mechanically. The total volume V of the solution of a total mass m_1 of solvent and mass m_2 of dry macromolecule is thus given as

$$V = m_2 v_2 + (m_1 - m_1^{\square}) v_1 + m_1^{\square} v_1^{\square} \tag{7-3}$$

and, with $\Gamma_h = m_1^{\square}/m_2$, as

$$V = m_2 v_2 + m_1 v_1 + \Gamma_h m_2 (v_1^{\square} - v_1)$$

In very dilute solution, Γ_h is constant, and does not depend on concentration. The partial specific volume of the solute is obtained in this case by differentiation of equation (7-4),

$$\tilde{v}_2 = \left(\frac{\partial V}{\partial m_2}\right)_{p,T,m_1} = v_2 + \Gamma_h (v_1^{\square} - v_1) \tag{7-5}$$

Combining equations (7-2) and (7-5), we obtain

$$V_h = \frac{M_2}{N_L}(\tilde{v}_2 + \Gamma_h v_1) \tag{7-6}$$

The hydrodynamic volume therefore depends very much on the factor Γ_h. This factor measures the amount of solvent transported with the macromolecule by solvation and/or purely mechanically.

7.2. *Diffusion in Dilute Solution*

7.2.1. *Basic Principles*

Diffusion processes can be subdivided into various classes, e.g., translational, rotational, thermal, etc. Translational diffusion consists of the isothermal equilibration of matter between two phases of differing concentration. The rotation of molecules and particles around their own axis is termed rotational diffusion. Thermal diffusion is the equilibration of matter under the influence of a temperature gradient.

For the derivation of the elementary laws of translational diffusion it must be assumed that the particles or molecules are displaced randomly in time by distances L. In unit time, therefore, the particles move one-dimensionally by an absolute distance Δr in the positive or negative direction (Figure 7-1). This movement can be described by random flight statistics, and yields the mean square displacement $\overline{\Delta r^2}$ after N steps or after the corresponding time (cf. the analogous derivation for the relation between the end-to-end distance and the bond lengths in the appendix of Chapter 4):

$$\overline{\Delta r^2} = NL^2 \tag{7-7}$$

The time τ required to traverse the distance L is equal to the quotient of the time t necessary to move a distance Δr and the number of steps N, i.e., $\tau = t/N$. Equation (7-7) thus becomes

$$\frac{\overline{\Delta r^2}}{t} = \frac{L^2}{\tau} \tag{7-8}$$

According to experiments, $\overline{\Delta r^2}/t$ and, consequently, L^2/τ are constants for given systems. If a time interval of the size $dt = \tau$ is chosen, then consequently $\Delta r \approx dr = L$, and equation (7-8) can be written as

$$dr = \left(\frac{L^2}{\tau}dt\right)^{0.5} = L \tag{7-9}$$

Diffusion can also take place across a hypothetical interface separating two regions of mean concentrations c and $[c + (\partial c/\partial r)\,dr]$ (Figure 7-1). All the particles in the left region move away from their original positions because of Brownian motion. Half of the particles wander to the left, while

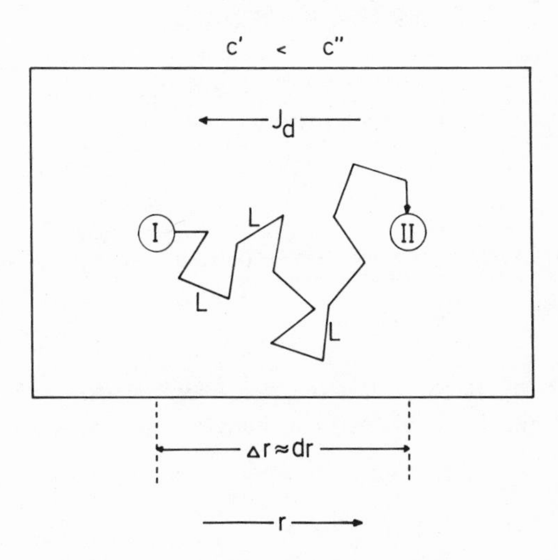

Figure 7-1. Schematic representation of the diffusion process. A particle moves from the chamber with the concentration c' into the chamber with the concentration c''. At the time considered, half of the particles have traveled out of the volume element I through the interface into volume element II, and vice versa. Because $c' < c''$, the flow J_d which is observed takes place from c'' to c'.

the other half wander to the right across the interface and enter the right region. The same diffusional motion occurs in the right half, and half of the particles from the right region cross the interface into the left region. The net transport of mass is therefore, with respect to interface A,

$$\frac{c\,dr}{2} - \frac{[c + (\partial c/\partial r)\,dr}{2} = -\frac{\partial c\,(dr)^2}{2\partial r} \equiv \frac{dm}{A} \tag{7-10}$$

In unit time, transport across the interface is

$$\frac{dm}{A\,dt} = -\left[\frac{\partial c\,(dr)^2}{2\,\partial r}\right]\Big/ dt \tag{7-11}$$

If the flow J_d is defined by

$$J_d \equiv \frac{dm}{A\,dt} \tag{7-12}$$

then from equation (7-9), we obtain Fick's first law of diffusion:

$$J_d = -\frac{L^2}{2\tau}\frac{\partial c}{\partial r} = -D\frac{\partial c}{\partial r} \tag{7-13}$$

in which the expression for the diffusion coefficient D

$$D = \frac{L^2}{2\tau} \tag{7-14}$$

is termed the Einstein–Smoluchowski equation.

In polymer solutions, it is usually not the displacement of the mass that is measured, and consequently not J_d, but the change in the concentration dc with time. During diffusion, the law of conservation of mass must be maintained, i.e., the hydrodynamic continuity condition established by inserting equation (7-12), bearing in mind the opposing signs of flow and change in concentration:

$$\frac{\partial c}{\partial t} = -\frac{\partial J_d}{\partial r} \tag{7-15}$$

Combining this equation with (7-13) gives Fick's second law of diffusion:

$$\frac{\partial c}{\partial t} = \frac{\partial\,[D(\partial c/\partial r)]}{\partial r} \tag{7-16}$$

If the diffusion coefficient D does not depend on concentration c, and thus also not on the path r, then equation (7-16) transforms to

$$\frac{\partial c}{\partial t} = D\frac{\partial^2 c}{\partial r^2} \tag{7-17}$$

The diffusion coefficient D is also related to the frictional coefficient f_D. The frictional force F_r increases with the velocity dr/dt of the particles; the proportionality constant is the frictional coefficient f_D:

$$F_r = f_D \frac{dr}{dt} \tag{7-18}$$

At the end of the time interval dt being considered, half of the particles have moved in one of the two directions by the average displacement dr. On the average, every particle retains its original kinetic energy E, i.e., the energy per particle, which, carried over by the diffusion process, is

$$E = 0.5 \frac{RT}{N_{\mathrm{L}}} \tag{7-19}$$

In the time interval considered, every particle takes up the energy E from another particle and passes it on again to a third particle. Since particles traveling in both directions take part in this process, there are on the average two times two energy units to consider. This energy is equal to that which is consumed in overcoming frictional forces over the distance dr, i.e., $4E = F_r\, dr$, or, with

$$f_D \frac{dr}{dt} dr = 4\left(0.5 \frac{RT}{N_{\mathrm{L}}}\right) \tag{7-20}$$

Combining equations (7-9), (7-14), and (7-20) leads to the Einstein–Sutherland equation, which relates the diffusion coefficient D to the frictional coefficient f_D,

$$D = \frac{RT}{f_D N_{\mathrm{L}}} \tag{7-21}$$

7.2.2. Experimental Methods

Fick's second law can be solved for different limiting conditions. A relatively simple solution is obtained for an initially infinite sharp boundary (or imaginary interface) ($dr = 0$ at $t = 0$) between two solutions with the concentrations c' and c'' or $\Delta c = c' - c''$. Furthermore, diffusion should only be observed over such a time period that the initial concentrations far from the interface are retained at the end of the diffusion process. The integration of equation (7-17) then yields

$$c(r, t) = \left(\frac{c' - c''}{2}\right)\left\{1 - \frac{2}{\pi^{0.5}} \int_{r_0}^{r} \exp\left[-\frac{(r - r_0)^2}{4Dt}\right] dr\right\} \tag{7-22}$$

The sharp boundary necessary for these limiting conditions is realized experimentally by several types of diffusion cells. In slide cells, a slide separates the lower chamber filled with the denser solution from the upper chamber with the less dense solution (mostly solvent). The slide is withdrawn at the beginning of the diffusion experiment and the diffusion followed by, e.g., the broadening of the concentration gradient curve. Another diffusion cell works on the lower layer principle. Here, the chamber is first half-filled with the less dense solution. The denser solution is then added in such a way that it forms a layer at the base of the chamber. This is done by adding the solution through a tube that reaches to the base of the chamber. Addition continues until the separating layer between the two solutions reaches the center of the observation window. In this position, slits are fixed at right angles to the cell window, through which the separating layer can be drawn by suction and thus sharpened.

The progress of the diffusion can be followed, e.g., by light absorption measurements (visible light, UV, IR) or by interference measurements. Both methods register a concentration-dependent quantity as a function of the position at a constant time (or vice versa), thus fulfilling the conditions of equation (7-22). The concentration gradients dc/dr, however, can be found as a function of the position by means of a suitable optical method, the Schlieren optics method. To monitor experiments by Schlieren optics, equation (7-22) has to be differentiated:

$$\frac{\partial c}{\partial r} = -\left[0.5\frac{c' - c''}{(\pi Dt)^{0.5}}\right]\exp\left[-\frac{(r - r_0)^2}{4Dt}\right] \tag{7-23}$$

The diffusion coefficients found using equation (7-22) or (7-23) usually depend also on the concentration of the initial solution. All the equations quoted so far, on the other hand, relate to infinitely dilute solutions. The dependence on the concentration of the diffusion coefficients in the case of dilute solutions can usually be given as

$$D_c = D(1 + k_D c) \tag{7-24}$$

The constant k_D contains a hydrodynamic and a thermodynamic term. It generally decreases with increasing molecular weight in a polymer homologous series. The diffusion coefficient D_c measured at a finite concentration c relates to the average between the two initial concentrations, i.e., to an average concentration of $c_0/2$, where c_0 is the initial concentration of one solution and the other solution is pure solvent. In polydisperse materials, D_c and, thus, D are obtained as average values which differ according to the evaluation method.

7.2.3. Molecular Quantities

The diffusion coefficient D (at $c \to 0$) is related to the frictional coefficient f_D [see equation (7-21). The coefficient f_D and, thus, D depend on a series of molecular quantities, as can be seen from the following reasoning. According to Stokes' law, the frictional coefficient f_{sphere} of an unsolvated sphere of homogeneous density is $f_{\text{sphere}} = 6\pi\eta_1 r_{\text{sphere}}$, where η_1 is the solvent viscosity. In a solvated sphere, the hydrodynamically effective radius r_h takes the place of the radius r_{sphere}. The deviation of the particle shape from that of an unsolvated sphere is described by an asymmetry factor $f_A = f_D/f_{\text{sphere}}$. Thus, the frictional coefficient f_D of a solvated particle of any given shape is, for $c \to 0$,

$$f_D = f_A(6\pi\eta_1 r_h) \tag{7-25}$$

The volume $V_h = (4\pi/3)\, r_h^3$ is expressed by equation (7-6), and the frictional coefficient f_D by equation (7-21). For the diffusion coefficient D, therefore, this gives

$$D = \left(\frac{RT}{6\pi\eta_1 N_L f_A}\right)\left[\frac{3M_2}{4\pi N_L}\left(\tilde{v}_2 + \Gamma_h v_1\right)\right]^{-1/3} \tag{7-26}$$

The diffusion coefficient D thus depends on the known or measurable quantities R, T, N_L, η_1, $\tilde{v}_2$, and v_1 and on three unknowns: the molecular weight M_2, the asymmetry factor f_A, and the parameter Γ_h. It cannot be interpreted molecularly without further assumptions. For unsolvated

Table 7-1. *Diffusion Coefficient D of Macromolecules in Dilute Solution*

Macromolecule[a]	Molecular weight, g/mol	Solvent	Temperature, °C	$10^7\,D$, cm^2/s
Ribonuclease	13,683	Water	20	11.9
Hemoglobin	68,000	Water	20	6.9
Collagen	345,000	Water	20	0.69
Myosin	493,000	Water	20	1.16
Deoxyribonucleic acid	6,000,000	Water	20	0.13
PMMA	34,100	Acetone	20	17.4
PMMA	280,000	Acetone	20	4.65
PMMA	580,000	Acetone	20	1.15
PMMA	935,000	Acetone	20	0.85
PMMA	200,000	Butyl chloride	35.6	7.18

[a] PMMA = poly(methyl methacrylate).

solid spheres, $f_A = 1$ and $\Gamma_h = 0$. Consequently, in a homologous series of such spherical molecules, the diffusion coefficient D decreases with $M^{-1/3}$. In a homologous series of molecules of different shapes, the molecular weight dependence of f_A and Γ_h also has to be taken into account. In general, it is found empirically for such a homologous series that

$$D = K_D M_2^{a_D} \tag{7-27}$$

K_D and a_D are shape- and solvation-dependent constants. a_D can be expressed through the exponents of the relationships between molecular weight and other hydrodynamic quantities [see equation (8-56)].

The diffusion coefficients D of macromolecules in dilute solutions have values of $\sim 10^{-7}$ cm^2/s (Table 7-1). The diffusion coefficients of the proteins ribonuclease, hemoglobin, collagen, and myosin, as well as deoxyribonuclease, were measured in dilute aqueous salt solutions. They were transformed into "pure water" diffusion coefficients by assuming that any differences are caused by differing viscosities only, and not by changes in the asymmetry or solvation coefficients. The diffusion coefficients of macromolecules in the melt are much lower, being $\sim 10^{-13}$–10^{-12} cm^2/s.

7.3. Permeation through Solids

7.3.1. Basic Principles

A gas on the higher pressure side of a membrane dissolves in the membrane and permeates through it to reach the side with the lower pressure. Permeation also takes place in the absence of macroscopic pores, in fact, by exchanging sites with the membrane material. If the gas is of sufficiently low solubility in the membrane, the difference in concentration Δc is, according to Henry's law, proportional to the difference in pressure Δp:

$$\Delta c = S\,\Delta p \tag{7-28}$$

The proportionality constant is the solubility coefficient S.

The rate-determining step for the permeation process is the diffusion that begins after the establishment of the solubility equilibrium. The diffusion in the steady state is determined by Fick's first law [equation 7-13)]. dc can be replaced by the concentration difference Δc, and dr by the thickness L_m of the membrane for concentration-independent diffusion coefficients:

$$-J_d = D\frac{\Delta c}{L_m} \tag{7-29}$$

Combining equations (7-28) and (7-29), we obtain

$$-J_d = DS\frac{\Delta p}{L_m} = P\frac{\Delta p}{L_m} \qquad (7\text{-}30)$$

The product DS is called the permeability P.

7.3.2. Experimental Methods

No noticeable gas permeation takes place up to a time t if a given pressure difference is produced at time $t = 0$ across both sides of a membrane with thickness L_m which was not previously in contact with the gas (Figure 7-2). A lengthy theoretical calculation shows that this time lag t_1 is related to the diffusion coefficient D of gases through membranes:

$$D = \frac{L_m^2}{6t_1} \qquad (7\text{-}31)$$

The transported mass of the gas, after passage through the membrane (or film), increases linearly with time t after the induction period t_1 [see also equation (7-12)]:

$$\Delta m = J_d A(t - t_1) \qquad (7\text{-}32)$$

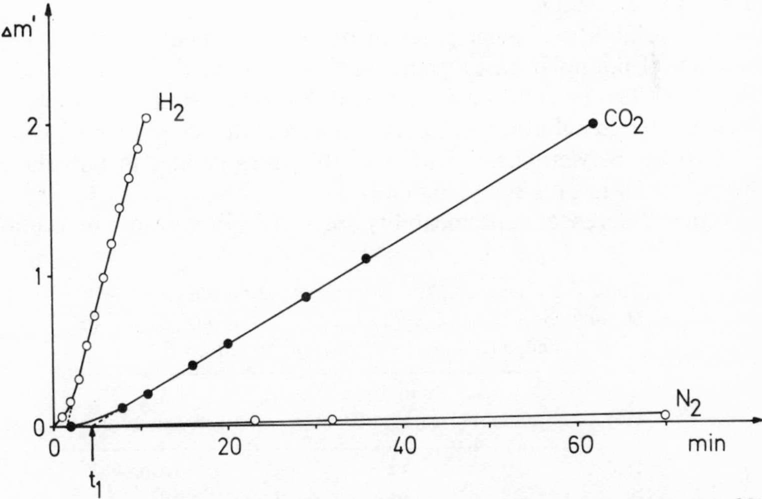

Figure 7-2. Time dependence of gas permeation through a film of a styrene copolymer at 25°C. With the chosen experimental conditions, the change $\Delta m'$ is proportional to the change in mass Δm. (After P. Goeldi and H.-G. Elias.)

Equations (7-30)–(7-32) give

$$\Delta m = \frac{DSA\,\Delta p}{L_m}\left(t - \frac{L_m^2}{6D}\right) \tag{7-33}$$

The diffusion coefficient D can therefore be determined from the induction period $t_1 = L_m^2/6D$, and the solubility coefficient S from the slope of the function $\Delta m = f(t)$ if the membrane surface area A, the membrane thickness L_m, and the pressure difference Δp are known.

7.3.3. Constitutional Influence

The permeation of gases through membranes or films is determined by both the diffusion coefficient D and the solubility coefficient S [see equation(7-33)]. The diffusion coefficient decreases in general with increasing molecular weight of the gas for a given membrane (Table 7-2). Since the solubility coefficient S depends on the interaction between the gas and the membrane material, however, there is no general relationship between the molecular weight of the gas and the permeability coefficient P.

The diffusion coefficient will be lower, the greater the distance the gas must travel in traversing the membrane. Thus bulky monomeric units, fillers, and crystalline regions lower diffusion coefficients (detour factor). The more flexible the chains of the membrane material, the less activation energy will be needed for the diffusion, and the greater also will be the diffusion coefficient.

The solubility of polar gases in polar membranes is generally higher than that of nonpolar gases. Individually, however, there are considerable differences. The permeability coefficient P for oxygen, for example, is 900 times greater in cellulose, 4.8×10^4 times greater in poly(vinyl chloride), 6.6×10^5 in poly(ethylene), and even 10^8 times greater in poly(dimethylsiloxane) than in poly (vinyl alcohol).

These differences in permeability are of great importance in technology.

Table 7-2. Permeability Coefficient P, Diffusion Coefficient D, and Solubility Coefficient S of Various Gases in Vulcanized cis-1,4-Poly(isoprene) at 25°C

Gas	M, g/mol	$10^7 P$, (cm²/s)/bar	$10^7 D$, cm²/s	S, (cm³/cm³)/bar
H_2	2	3.4	85	0.040
N_2	28	0.51	15	0.035
O_2	32	1.5	21	0.070
CO_2	44	10.0	11	0.90

[a]Data from R. M. Barrer.

For example, packaging films for groceries should let through very little oxygen. Conversely, a very high oxygen permeability is required for the use of membranes as artificial lungs.

The temperature dependence of a diffusion coefficient is determined by the activation energy $E_D^\ddagger$ of diffusion. The temperature dependence of the solubility coefficient is given by the enthalpy of solution ΔH. Hence, the temperature dependence of the permeability coefficient is given by

$$P = DS = (D_0S_0)\exp\left(-\frac{E_D^\ddagger + \Delta H}{RT}\right) \tag{7-34}$$

As the temperature increases, S usually falls, while D usually rises. According to the relative temperature dependence of S and D, therefore, P will become greater or smaller as the temperature increases.

Inert gases diffusing through amorphous polymeric membranes above the glass-transition temperature always show the ideal behavior described by equation (7-33). With vapors, and at temperatures below the glass-transition temperature of the membrane material, however, nonideal behavior is usually observed. Various effects may be responsible for this "anomalous" diffusion: a swelling of the membrane material, a change in crystallinity because of swelling, the condensation of vapor in the membrane, a transition from the glassy state to the "melt" state because of plasticization by the penetrating solvent, etc.

The permeation of liquids, particularly water, plays an important part in the weather resistance of plastics. It is also a problem in the drying of polymers. If polymer solutions are evaporated, it is frequently found that a considerable part of the solvent cannot be removed from the polymer even above the solvent boiling point. Such inclusion can amount, for example, to 20% for CCl_4 in poly(styrene), and up to 10% for dimethylformamide in poly(acrylonitrile). Allowances must be made for this inclusion in the interpretation of analytical data. Inclusion occurs because the permeation of liquids (solvents, monomers, etc.) through polymers is very low below their glass-transition temperature. It is best avoided by freeze-drying the polymer solution. In this case, $\sim$1–10% solutions of the polymer in solvents with sufficiently high melting points and which evaporate readily (sublime) as a solid (e.g., benzene, dioxane, water, formic acid) are completely frozen, and then, below the melting point, the solvent is sublimed out under vacuum. This method removes the solvent almost completely. Since the remaining polymer has a large surface area, however, and is therefore very prone to absorbing moisture, it is advisable to sinter the polymer carefully after the freeze-drying. Inclusion can also be avoided by adding certain nonsolvents to the solvent before evaporating; these nonsolvents should form azeotropes with the solvent. Another possibility is to dissolve the polymer in a poor solvent and precipitate it with a strong precipitant.

7.4. Rotational Diffusion and Streaming Birefringence

The longitudinal axes of nonspherical particles have a random distribution of angles with respect to each other in dilute solutions. This angular distribution of the longitudinal axes is disturbed by externally applied shear gradients or other fields. The longitudinal axes orient themselves parallel or perpendicular to the direction of the field, according to the type of field and its interaction with the particles. Orientation can be forced, for example, by an electric field (Kerr effect) or a magnetic field (Cotton–Mouton effect). When the field is no longer applied, the particles again move into their randomly oriented equilibrium positions by means of rotational diffusion. An analogous equation to that for the normal diffusion coefficient [equation (7-21)] holds for the rotational diffusion coefficient D_r:

$$D_r = \frac{RT}{f_r N_L} \tag{7-35}$$

where f_r is the rotational frictional coefficient.

For the rod-shaped particles of tobacco mosaic virus of length 280 nm at room temperature, for example, rotational diffusion coefficients of 550 s^{-1} were found. The rods thus require $1/550 = 0.0018$ s in order to return to their equilibrium (randomly oriented) positions.

The particles can also be oriented purely mechanically by a shear gradient, as well as by electric or magnetic fields. This type of flow with a linear shear gradient can be produced when the liquid to be studied is placed in a narrow space between two concentric cylinders (Figure 7-3).

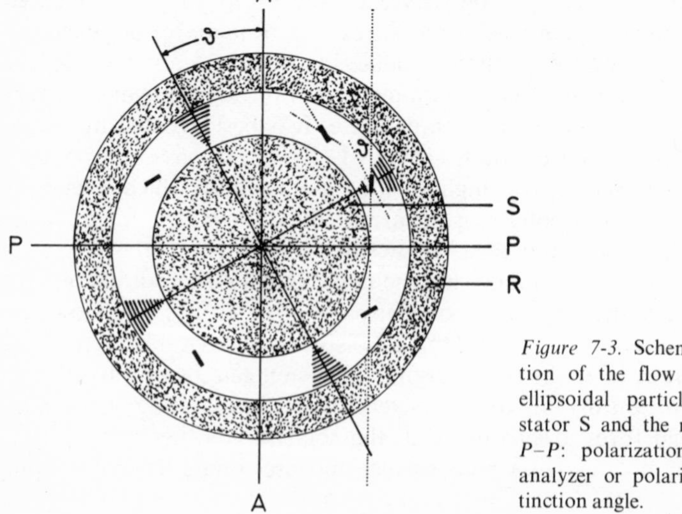

Figure 7-3. Schematic representation of the flow birefringence of ellipsoidal particles between the stator S and the rotor R. *A–A* or *P–P*: polarization planes of the analyzer or polarizer. ϑ is the extinction angle.

One of the two cylinders turns (rotor), while the other remains stationary (stator). Due to the partial orientation of the anisotropic molecules, the refractive indices n differ according to whether the long axis of the particles is parallel or at right angles to the direction of flow. The difference $n = n_{\perp} - n_{\parallel}$ of these two refractive indices is termed birefringence.

A dark cross against a light background is observed if the rotating solution is studied under crossed Nicol prisms. The effect comes about as follows: If horizontal polarized light enters an isotropic solution, complete extinction takes place. Under the same conditions, solutions with partially oriented anisotropic particles cause extinction only at those points where the optical axis of the anisotropic particles is parallel to the polarization plane of the polarizer or the analyzer (Figure 7-3). Extinction can therefore occur in four positions. As can be seen from Figure 7-3, all indicated particles form an angle ϑ to a tangent to the direction of flow. However, only at four regions in the circle, which form a cross, are they also parallel to the polarization planes A–A or P–P, as is shown for the particle in the upper right-hand quadrant. It is found experimentally that the cross lies below an angle at 45° to the two polarization planes at low shear gradients, and below an angle of 0° at very high shear gradients. The smaller of the angles between the polarization planes and the black cross is termed the extinction angle ϑ. Consequently, ϑ varies from 45° at low to 0° at high shear gradients. The extinction angle is therefore an indication of the orientation of the particles in the field of flow. The strength of the birefringence is an indication of the degree of orientation.

The orientation of the molecules counteracts the rotational diffusion. The smaller the molecules, the faster will be the rotational diffusion. For a given ratio of the molecular axes, therefore, there is a lower limit for which a particle length can be measured by birefringence. It is at ~20 nm. Shorter molecules than this require such high shear gradients that the flow becomes turbulent and the assumptions necessary for measurement and analysis no longer hold. Rigid, long molecules, on the other hand, require only low gradients. Thus, for example, for tobacco mosaic virus a strong effect is observed even at shear gradients of $\sim 5 s^{-1}$. With the flexible coils of poly(styrene), on the other hand, only a low streaming birefringence can be seen even at 10^4 s^{-1}. The method is therefore mainly applied in the case of rigid macromolecules, and then provides a means of measuring the particle lengths.

7.5. Electrophoresis

The movement of an electrically charged particle of mass m and charge Q under the influence of a uniform electric field of strength $\mathbf{E}$ is called electrophoresis. These particles may be biological cells, colloids, macromolecules, or low-molecular-weight substances. Even electrically neutral particles can be made electrophoretically mobile through formation of suitable complexes. An example is the formation of borate complexes by polysaccharides,

$$H_2BO_3^- + \begin{matrix} HO \\ HO \end{matrix} \!\!>\! R \sim\sim \rightleftharpoons \left[\begin{matrix} HO \\ HO \end{matrix} \!\!>\! B \!\begin{matrix} \cdots O \\ \cdots O \end{matrix}\!\!>\! R \sim\sim \right]^- + H_2O \tag{7-36}$$

The particles move in a solvent, typically aqueous salt solutions, in free electrophoresis or Tiselius electrophoresis. In carrier electrophoresis, the particles move in a swollen carrier [for example, paper, starch gels, cross-linked poly(acrylamide)].

The particle movement is induced by a force QE. The movement is resisted by the frictional force $f(dl/dt)$. Here, f is the frictional coefficient and dl/dt is the particle velocity. The resultant of these two forces is given by Newton's second law of motion as $m(d^2l/dt^2)$. Thus

$$m\frac{d^2l}{dt^2} = QE - f\frac{dl}{dt} \tag{7-37}$$

or, integrated once,

$$\frac{dl}{dt} = \frac{QE}{f}\left[1 - \exp\left(-\frac{f}{m}t\right)\right] \tag{7-38}$$

The quotient f/m is about $(10^{12}$–$10^{14})$ s^{-1} for molecular particles. Consequently, equation (7-38) reduces, for times greater than 10^{-11} s, to

$$\frac{dl}{dt} = \frac{QE}{f} \tag{7-39}$$

The electrophoretic mobility μ is defined as the velocity of movement induced by the effect of an electric field of 1 V/cm. On inserting into the Einstein–Sutherland equation (7-21), one obtains, with equation (7-39),

$$\mu = \frac{dl/dt}{E} = \frac{N_L QD}{RT} \tag{7-40}$$

In the laboratory, electrophoresis is used for analysis and to separate charged particles on the basis of their different electrophoretic mobilities. In the analysis of a mixture of proteins, the apparent protein content A, for example, depends on the ionic strength Γ as well as on the total protein concentration. Consequently, the apparent content of A is plotted against c/Γ and extrapolated to $c/\Gamma \to 0$.

In industry, electophoresis is used in the electro-dip-coating process or electrophoretic coating, especially in the automobile industry for painting car bodies. Here the metal component is connected as the anode in an electric circuit. On applying an electric field, the negatively charged particle (mostly latex particles) move to the anode and are discharged as a film there. An electroosmosis effect subsequently occurs and water molecules are expelled from the film. The solids content of the polymer film is increased up to 95%. A final electrophoresis can be applied to remove the last traces of water and dissolved ions. In contrast in other automated coating procedures, electro-dip-coating allows even relatively inaccessible corners and edges to be uniformly coated. Additionally, the solvent base of the paints used is water, so the costly installations required for the recovery of organic solvents from the vapor phase are not needed. Consequently, electro-dip-coating is being increasingly used for coating automobile bodies.

7.6. *Viscosity*

7.6.1. *Concepts*

Consider a ribbon of infinite length moving with the velocity v (cm/s) through a liquid contained between two infinitely long, parallel plates which are at a distance y (cm) apart. In the immediate vicinity of the plates the liquid is at rest, whereas in the immediate vicinity of the ribbon it will be moving at the velocity v. A velocity gradient $D = dv/dy$ with the unit s^{-1} thus exists between plate and ribbon. D is also called the shear rate.

Because of the problem of sealing at both ends, this so-called ribbon viscometer can only be realized for materials of extremely high viscosity. The properties of a ribbon viscometer are shown to a good approximation by a rotation viscometer of the Couette type (see Section 9.5.2). In Couette viscometers, a rotor revolves around a stator (or vice versa). The viscous liquid lies between the rotor and the stator. If the space between them is sufficiently narrow, the shear rate is constant across the whole distance between rotor and stator (Figure 7-4).

With the concentric lamellar flow of liquids in a capillary, however, dv/dy is not constant, but is a function of the distance y from the capillary wall. The shear gradient is the greatest near the capillary wall, while at the center of the capillary $D = dv/dy = 0$ (Figure 7.4).

The velocity gradient, measured perpendicular to the direction of flow, is the difference in velocity of two lamellae flowing past each other. Forces are active in the direction of flow at the contact surface between the two layers. They are called shear, thrust, or tangential forces. The ratio of the shear forces K to the contact surface A is called the shear stress σ_{ij}.

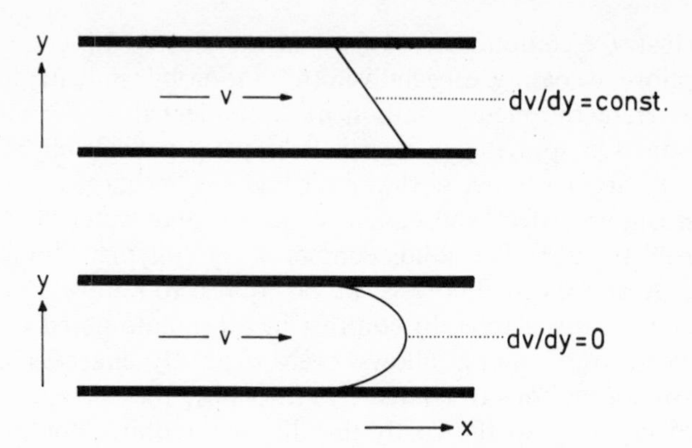

Figure 7-4. Definition of the shear gradient dv/dy in a current with velocity v between two parallel plates of a stator and a rotor (above), or in a capillary (below). With the plates, the shear gradient dv/dy = constant between the plates. With the capillary, $dv/dy = 0$ at the center of the capillary.

With liquids of low viscosity there is, according to Newton's law, a direct proportionality between the shear gradient D and the shear stress σ_{ij} for many liquids of low viscosity:

$$\sigma_{ij} = \eta D = \eta \frac{dv}{dy} \tag{7-41}$$

The proportionality factor η is called the viscosity and its inverse value $1/\eta$ is called fluidity.

Liquids that follow Newton's law are called Newtonian liquids. In non-Newtonian liquids, the quantity η, which can be calculated from the quotient, σ_{ij}/D, also changes with the velocity gradient, or with the shear stress. Newtonian behavior is usually observed for the limiting case $D \to 0$ or $\sigma_{ij} \to 0$. Melts and macromolecular solutions often exhibit non-Newtonian behavior. Non-Newtonian liquids are classified as dilatant, Bingham body, pseudoplastic, thixotropic, or rheopectic liquids.

In dilatant liquids, D increases less than proportionally with σ_{ij}, while in pseudoplastic liquids D increases more than proportionally (Figure 7-5). Expressed another way: In pseudoplastic liquids, the apparent viscosity $\eta_{app} = \sigma_{ij}$ falls as the shear stress increases (shear thinning), whereas in dilatant liquids it rises (shear thickening). The fluidities, on the other hand, increase with rising σ in pseudoplastic liquids and fall in dilatant liquids. At $D \to 0$, both dilatant and pseudoplastic liquids show a Newtonian behavior. Shear thickening is rare in melts and macromolecular solutions, but occurs often in dispersions.

Shear thinning is shown by flow-oriented, asymmetric, rigid particles and/or shear-deformed flexible coils. In the first case, the pseudoplastic viscosity should vary with the square of the concentration because of the interactions between two particles, and in the second case only with the concentration itself.

Plastic bodies are also called Bingham bodies. They exhibit a stress limit to flow (Figure 7-5). The limiting value is defined as the minimum value of σ_{ij} above which D begins to vary with σ_{ij}, i.e., above $(\sigma_{ij})_0$. It is also called yield value. Ideal plastic bodies show Newtonian behavior above the flow limit. Pseudoplastic bodies, on the other hand, show pseudoplastic behavior above $(\sigma_{ij})_0$. The plasticity or flow limit is interpreted as the breaking up of molecular associations. Plasticity is particularly desirable in paints.

When the shear stress changes in Newtonian, dilatant, or pseudoplastic liquids, as well as in Bingham bodies or fluids above the flow limit, the corresponding shear gradient or the corresponding viscosity is reached almost instantaneously. In some liquids, however, a noticeable induction time is necessary, i.e., the viscosity also depends on time. If, at a constant shear stress or constant shear gradient, the viscosity falls as the time increases, then the liquid is termed thixotropic. Liquids are termed rheopectic or antithixotropic, on the other hand, when the apparent viscosity increases with time. Thixotropy is interpreted as a time-dependent collapse of ordered structures. A clear molecular picture for rheopexy is not available.

If the rate of flow of low-molecular-weight liquids is increased very sharply, additional velocity components occur because of the surface

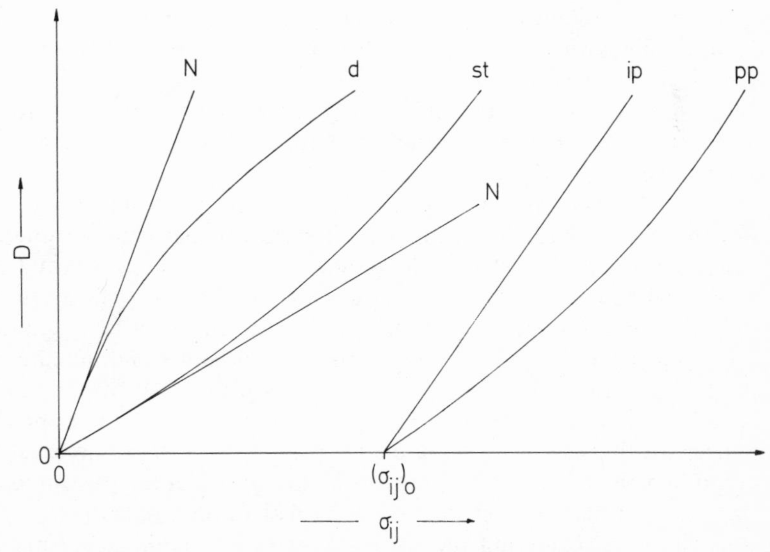

Figure 7-5. A plot of shear rate D as a function of shear stress σ_{ij} for Newtonian (N), dilatant (d), and pseudoplastic (st) liquids, and ideal plastic (ip) and pseudoplastic (pp) variants of Bingham bodies. $(\sigma_{ij})_0$ is the yield value.

roughness of the walls. As the flow rate increases, these disturbances of laminar flow finally become so large that they are no longer damped by the viscosity of the liquid. The individual liquid lamellae no longer flow in a parallel manner, and the flow becomes turbulent. The onset of turbulence is described by the Reynolds number, which gives the ratio of inertial forces to viscous forces.

The same effect is observed in melts of macromolecular substances. Since, in this case, the liquids are elastic, however, additional elastic oscillations of small liquid particles occur. The uneveness of the oscillations causes an elastic turbulence. This occurs at much lower flow velocities than the normal turbulence, i.e., at low Reynolds numbers. Elastic turbulence can also be recognized by the fact that the flow rate increases much more sharply with increasing pressure in the elastic turbulence region than in the laminar flow region; in normal liquids the increase in the flow rate is less in the turbulence region than in the laminar flow region. Elastic turbulence becomes apparent in the processing of plastics at what is called the melt break.

7.6.2. *Methods*

When making measurements on highly viscous solutions and melts, care must be taken to ensure that the systems are in thermal equilibrium. In many cases it is sufficient to anneal the systems for about a week at the measuring temperature. In a few studies, however, the equilibrium was only reached after half a year.

Viscosities or shear gradients and stresses can be measured with a range of instruments. The most important of these are rotation and capillary viscometers; both are available commercially.

Rotation Viscometers. In rotation viscometers, a rotor moves against a stator (see Figure 9-23). The Epprecht viscometer is particularly suitable for measurements on highly viscous solutions. Here, the angle of rotation of a torsion wire on which the stator is suspended is a measure of the torque produced by the rotating rotor on the liquid. Since all the other quantities (cylinder radius, width of the slit, gap between rotor and stator, number of rotations in unit time) are kept constant, the viscosity is easily calculated.

The Brookfield viscometer is of simpler construction than the Epprecht viscometer. In this case, a rotating metal plate or cylinder is immersed in the liquid and the frictional braking forces that act on it are measured.

In cone and plate viscometers, a cone revolves on a metal plate. The angle between the plate and the cone is kept as low as possible (smaller than 4°) so that the shear gradient remains uniform.

To calculate viscosity, shear stress, and shear gradient, it is usually necessary to make a series of corrections for the finite length of the cylinder, the variation of the shear gradient with distance, etc. These corrections vary from instrument to instrument.

Capillary Viscometers. Those used for measurements on concentrated solutions and melts usually consist solely of one capillary tube in a pressure chamber. Because of the high driving pressure, the capillaries are often of metal.

When a liquid flows through a capillary with radius R and length L under a pressure p, a force $\pi r^2 p$ is exerted on the liquid column at a radius r. It is counteracted by a frictional force $2\pi r L \sigma_{ij}$. At equilibrium, therefore,

$$\pi R^2 p = 2\pi r L \sigma_{ij} = 0 \quad (7\text{-}42)$$

or, solved for the shear stress σ_{ij},

$$\sigma_{ij} = \frac{pr}{2L} \quad (7\text{-}43)$$

The σ_{ij} calculated using equation (7-43) is equal for both Newtonian and non-Newtonian solutions because p, r, and L depend on the measuring apparatus only and not on the properties of the liquid being investigated. σ_{ij} is proportional to the radius and has a maximum value $\sigma_{ij}(\max) = pR/2L$ for capillary radius R.

If equation (7-43) is inserted into Newton's equation, equation (7-41), with $y = R$, the result is $dv = (pR/2\eta L)\,dR$. If η is Newtonian, then integration with the boundary condition $v_R = 0$ for the velocity v at distances $y \leq$ from the capillary wall yields

$$v = \frac{(R^2 - y^2)\,p}{4\eta L} \quad (7\text{-}44)$$

The flow in a capillary is envisaged as the telescoping movement of concentric hollow cylinders at different velocities. A volume q flows per unit time through such a hollow cylinder with radii y and $y + dy$:

$$q = 2\pi y v\,dy \quad (7\text{-}45)$$

The total flow volume is given by integration over the flow volumes of all the hollow cylinders,

$$Q = \int_{y=0}^{y=R} 2\pi y v\,dy \quad (7\text{-}46)$$

and, with equation (7-44),

$$Q = \int_{y=0}^{y=R} 2\pi y \frac{R^2 - y^2}{4\eta L} p\, dy$$

$$= \frac{\pi p}{2\eta L} \int_{y=0}^{y=R} (R^2 - y^2)\, y\, dy$$

$$= \frac{\pi p}{2\eta L} \left(\frac{R^2 y^2}{2} - \frac{y^4}{4} \right) \Bigg|_0^R = \frac{\pi p R^4}{8\eta L} \tag{7-47}$$

Equation (7-47) is the Hagen–Poiseuille law. By inserting equation (7-43) and (7-47) into (7-41) and solving for dv/dR, we obtain the following for the maximum shear gradient dv/dR at the capillary surface:

$$\frac{dv}{dR} = \frac{4Q}{\pi R^3} = D \tag{7-48}$$

The "viscosity" calculated from equation (7-47) for non-Newtonian fluids is an apparent viscosity only. The calculation of the correct viscosity requires a number of corrections. In non-Newtonian liquids, D is a more complicated function of σ_{ij} than is given by Newton's law, i.e., $dv/dR \neq D$. Equation (7-46), therefore, can only be partially integrated:

$$Q = 2\pi \left| \frac{y^2 v}{2} \right|_2^R - \int_0^R y^2 \frac{dv}{dy} dv \tag{7-49}$$

In both limiting cases, the first term will be zero. Considering that $D = \sigma_{ij}/\sigma_{\text{app}} = f(\sigma_{ij})$ and after introducing the shear stress $(\sigma_{ij})_R$ at the capillary surface, the second term gives

$$Q = \pi \int_0^R y^2 f(\sigma_{ij})\, dv = \frac{\pi R^3}{(\sigma_{ij})_R^3} \int_0^R \sigma_{ij}^2 f(\sigma_{ij})\, d\sigma_{ij} \tag{7-50}$$

According to equation (7-48), $dv/dR \neq 4Q/\pi R^3 = D$ is valid for non-Newtonian liquids. Equation (7-50) therefore becomes

$$D = \frac{4}{(\sigma_{ij})_R^3} \int_0^R \sigma_{ij}^2 f(\sigma_{ij})\, d\sigma_{ij} \tag{7-51}$$

Thus $dD/d\sigma_{ij}$ becomes

$$\frac{dD}{d\sigma_{ij}} = \frac{4}{(\sigma_{ij})_R^3} \sigma_{ij}^2 f(\sigma_{ij}) = \frac{4}{(\sigma_{ij})_R^3} \sigma_{ij}^2 D \tag{7-52}$$

from which one obtains

$$\frac{1}{4} \sigma_{ij} \frac{dD}{d\sigma_{ij}} = \frac{\sigma_{ij}^3}{(\sigma_{ij})_R^3} D \tag{7-53}$$

In analogy to the equation dv/dR for Newtonian liquids, $dv/dR = AD$ is assumed for non-Newtonian liquids. Since a factor $\sigma_{ij}^3/(\sigma_{ij})_R^3 \neq 1$ occurs in equation (7-53), A is separated according to $A = a + \sigma_{ij}^3/(\sigma_{ij})_R^3$. On inserting into equation (7-53), we find

$$\frac{dv}{dR} = aD + \frac{\sigma_{ij}^3}{(\sigma_{ij})_R^3} D = aD + \frac{1}{4}\sigma_{ij}\frac{dD}{d\sigma_{ij}} \tag{7-54}$$

Equation (7-54) must also apply to Newtonian liquids. Since it is true here that $D = \sigma_{ij}/\eta$ and $dD/d\sigma_{ij} = 1/\eta$, this also gives $dv/dR = a(\sigma_{ij}/\eta) + (1/4)\ (\sigma_{ij}/\eta)$ and, with $dv/dR = \sigma_{ij}/\eta$, it follows that $a = 3/4$. Equation (7-54) thus becomes the Weissenberg equation

$$\frac{dv}{dR} = \frac{3}{4}D + \frac{1}{4}\sigma_{ij}\frac{dD}{d\sigma_{ij}} \tag{7-55}$$

or, with $dv/dR = \sigma_{ij}/\eta$ and $D = \sigma_{ij}/\eta_{\text{app}}$

$$\frac{1}{\eta} = \frac{3}{4}\frac{1}{\eta_{\text{app}}} + \frac{1}{4}\frac{dD}{d\sigma_{ij}} \tag{7-56}$$

where $dD/d\sigma_{ij}$ is taken from a graph of $\sigma_{ij} = f(D)$.

On examining data from the literature, it should be noted that it is frequently not the shear gradient D but, according to Kroepelin, the average shear gradient G across the whole diameter of the capillary which is given [see equation (7-48)]:

$$G = \frac{8Q}{3\pi R^3} = \frac{2}{3}D \tag{7-57}$$

Technical Viscometers. These usually do not enable shear stresses and shear gradients to be calculated, since the conditions of measurement usually cannot be varied. Hence, the values calculated from these instruments are only relative values. Their advantage is simple construction and rapid measurement.

Calibrated vessels with a hole at the base are called Ford cups. The liquid runs out under its own pressure. The time taken for a standard quantity of liquid to flow out is an indication of the viscosity. Since the height of the liquid and, subsequently, the pressure change during the measurement, the shear stress also varies with time. Ford cups are particularly used in the paint industry. Apparatus for determining the melt index (also called the grade value or grade number) works on a similar principle. Here, the amount of melt which flows out in a given time under given conditions is measured. The melt index is thus proportional to the fluidity and not the viscosity.

With Höppler viscometers, the time required for a rolling ball to run along an inclined tube is measured. In Cochius tubes, the time taken for an air bubble to rise is a measure of the viscosity. Here, the true viscosities, shear stresses, and shear gradients are also difficult to determine.

7.6.3. Flow Curves

The viscosity itself frequently changes when the shear stress is varied by several orders of magnitude (Figure 7-6). In cases of this kind, it is more appropriate to plot log D against log σ_{ij} than D against σ_{ij}. The resulting curves are called the flow curves. At very small shear stresses, a flow curve has a slope of unity as long as it is not for a Bingham fluid with a yield stress. In this Newtonian area, it intercepts the ordinate at the value $1/\eta_0$ at log $\sigma_{ij} = 0 (\sigma_{ij} = 1)$. At very high shear stresses, a second "Newtonian" area is sometimes claimed to be observed with the viscosity η_∞. The existence of a true second Newtonian viscosity is, however, disputed by many rheologists.

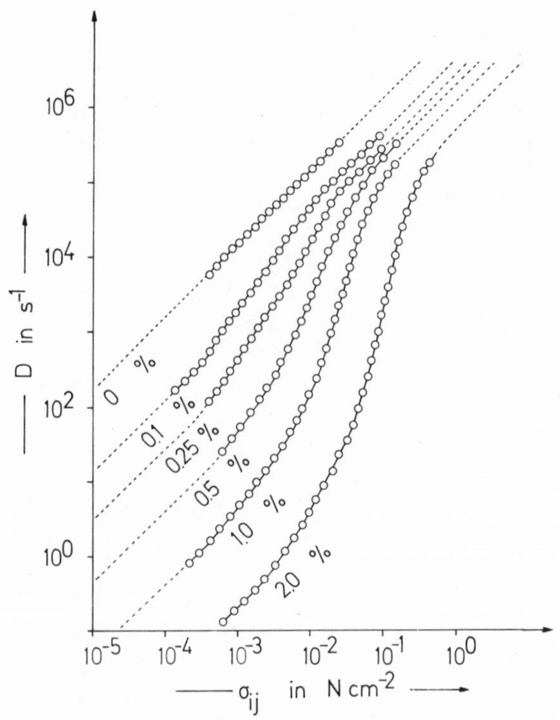

Figure 7-6. Flow curves of solution of different concentration of a cellulose nitrate (M = 294,000) in butyl acetate at 20°C (after K. Edelmann).

Between η_0 and η_∞ lies an area of non-Newtonian behavior. The relationship between D and σ_{ij} can often be reproduced by an empirical equation developed by Ostwald and de Waele:

$$D = \frac{1}{\eta_{\text{app}}}(\sigma_{ij})^n \qquad (7\text{-}58)$$

With plastic melts, the flow exponent n, as it is called, lies between ~ 2 and 3. Equation (7-58) can, of course, only apply to a limited range of shear stresses.

The Oswald–de Waele equation belongs to the many, so far unsuccessful, attempts to find a common flow law. For this reason it is also unsuitable to characterize the extent of the non-Newtonian behavior by a single point on the flow curve. The quotient η_∞/η_0 could be a good measure of the non-Newtonian behavior if there were a common flow law. Admittedly, η_∞ is often difficult to determine. The slope at the inflection point in the flow curve is often used as a means of classifying the non-Newtonian behavior.

7.6.4. *Viscosities of Melts*

At constant temperature, the melt viscosities of homologous series of substances increase with the weight-average molecular weight $\overline{M}_w$ (Figure 7-7). Below a given critical molecular weight M_c, the following relationship applies:

$$\eta = K\overline{M}_w \qquad (7\text{-}59)$$

and above M_c for $\sigma_{ij} \to 0$, the relationship

$$\eta = K'\overline{M}_w^{3.4} \qquad (7\text{-}60)$$

is found. The slope of the $\log \eta = f(\log \overline{M}_w)$ curve above M_c decreases with increasing σ_{ij}. The intersection of the $\log \eta_\sigma = f(\log \overline{M}_w)$ curve with the function $\log \eta_{\sigma \neq 0} = f(\log \overline{M}_w)$ often occurs at different molecular weights according to the value of σ_{ij} or D.

The beginning of pseudoplastic behavior at molecular weights $\overline{M}_w > M_c$ is interpreted as the effect of entanglements of parts of molecules. For this, of course, the chain must have a certain length or a sufficient number of chain links N_c. The number N_c is not a general constant; it depends on the constitution of the macromolecules (Table 7-3).

The tendency toward entanglement decreases with the rigidity of the chain. The parameter $\langle R^2 \rangle_\Theta/M$ can serve as an indication of the rigidity (see also: Section 4.4.2.4), where R is the radius of gyration and the subscript Θ indicates the theta state. For this, it is assumed that the radius of gyration

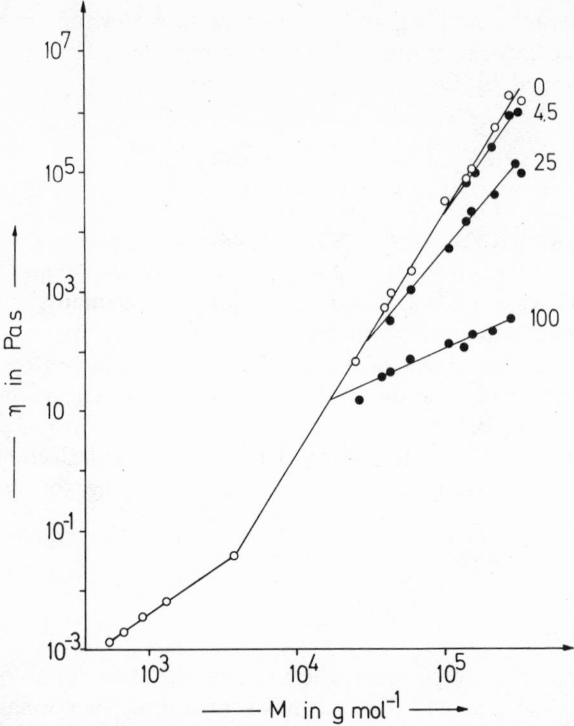

Figure 7-7. Melt viscosity η of unbranched poly(ethylenes) as a function of the molecular weight M at 190°C and shear stress $\sigma_{ij} = 0$ (○), or $\sigma_{ij} = 4.5$, 25, and 100 N/cm^2 (all ●) (after H. P. Schreiber, E. B. Bagley, and D. C. West); M_c occurs at $\log M = 3.6$.

of coil molecules in the melt is equal to the radius of gyration in the unperturbed state (see also Section 5.5.2). In order to compare the viscosities of melts and solutions, reference must also be made to the volume fraction ϕ_2 and specific volume v_2 of the polymers. In this way, it is possible to define

Table 7-3. Number N_c of Chain Links of Various Polymers at the Critical Molecular Weight M_c[a]

Polymer	N_c
Poly(ethylene)	286
1,4-*cis*-Poly(isoprene)	296
at-Poly(vinyl acetate)	570
Poly(isobutylene)	609
at-Poly(styrene)	730
Poly(dimethyl siloxane)	784

[a] Data from T. G. Fox.

a new quantity z_w, where N_c is the number of chain links:

$$z_w \equiv \frac{\langle R^2 \rangle_\Theta}{\overline{M}_w} \frac{N\phi_2}{v_2} \tag{7-61}$$

If $\log \eta$ is plotted against $\log z_w$, then the points of inflection all lie at roughly the same value of z_w (Figure 7-8).

In these relationships, the weight average is substituted for the molecular weight or the number of chain links since, according to experiments with polymers of differing molecular-weight distribution, the viscosity depends on the weight-average molecular weight. This finding applies only for $\sigma_{ij} \to 0$, however. At shear stresses $\sigma_{ij} > 0$,on the other hand, it has become evident that, at an equal $\overline{M}_w$, monodisperse polymers exhibit the

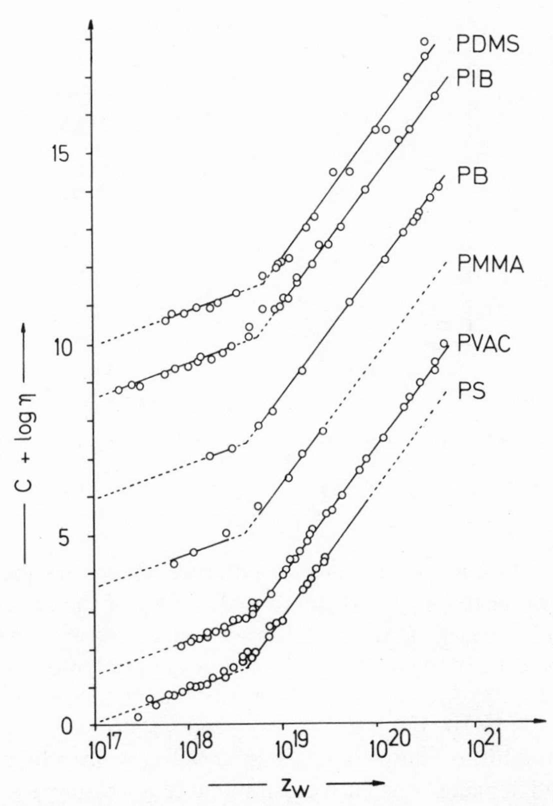

Figure 7-8. Dependence of the melt viscosity η of polymers on the parameter z_w (see text) at $\sigma_{ij} \to 0$. For easier comparison, the η values of the different types of polymer have all been multiplied by a constant factor of C. PDMS = poly(dimethyl siloxane), PIB = poly(isobutylene), PB = poly(butadiene), PMMA = poly(methyl methacrylate), PVAC = poly(vinyl acetate), PS = poly(styrene) (after T. G. Fox).

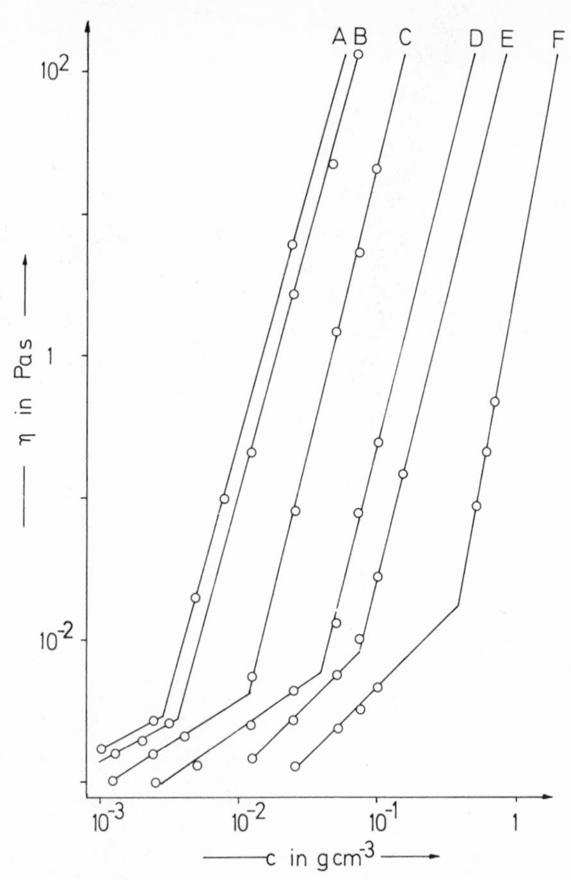

Figure 7-9. Viscosity η of solutions of poly(isobutylene) of different molecular weights in toluene at 25°C as a function of the concentration c. $\overline{M}_w = 7{,}270{,}000$ (A); 3,550,000 (B); 1,250,000 (C); 328,000 (D); 139,000 (E); 40,600 (F). (After J. Schurz and H. Hochberger.)

larger viscosity. Because the higher molecular weight species show a greater drop in viscosity at the higher shear rate, the effect of the lower molecular weight species has been occasionally described as "lubricating."

The temperature dependence of the viscosity of polymer melts typically does not follow an Arrhenius relationship unless temperatures are far above T_G ($T > T_G + 100$ K). If a viscosity ratio η_R (not to be confused with the ratio η/η_1 in dilute solutions, cf. Chapter 9.9) is defined by means of the viscosities η and densities ρ at the measurement temperature T and reference temperature T_l,

$$\eta_R = \frac{\eta \rho T}{\eta_1 \rho_1 T_1} \tag{7-62}$$

then the overall temperature dependence can be expressed by the semi-empirical Williams–Landel–Ferry equation (WLF equation)

$$\log \eta_R = \frac{-B(T - T_I)}{C + (T - T_I)} \tag{7-63}$$

(see also the derivation in Section 10.5.2). Here, B and C are constants specific to the material. Equation (7-63) is applicable for temperatures between T_G and $T_G + 100\,\mathrm{K}$.

7.6.5. Viscosity of Concentrated Solutions

If the viscosity η of solutions of high-molecular-weight, long-chain molecules is plotted against the concentration c in the log–log plot, then, in a similar way as for the dependence on molecular weight of the viscosities of melts, a curve is obtained which can be approximated by two straight lines with different slopes (Figure 7-9). The straight lines intersect at a concentration c_{crit}. Admittedly, this point of intersection is not as strongly marked as in the case of melts. c_{crit} is considered as that concentration at which effects due to entanglements first become apparent. c_{crit} also depends on the solvent power, which supports the hypothesis of entanglement. The slopes of the straight lines in this plot are, depending on the system, $\sim$2–4 below c_{crit} and $\sim$5–6 above c_{crit}.

Literature

Section 7.3. Permeation through Solids

J. Crank and G. S. Park, eds., *Diffusion in Polymers*, Academic Press, London, 1968.

H. J. Bixler and O. J. Sweeting, Barrier properties of polymer films, *in: The Science and Technology of Polymer Films*, Vol. II (O. J. Sweeting, ed.), Wiley–Interscience, New York, 1971.

C. E. Rogers and D. Machin, The concentration dependence of diffusion coefficients in polymer-penetrant systems, *Crit. Rev. Macromol. Sci.* **1**, 245 (1972).

V. Stannett, H. B. Hopfenberg, and J. H. Petropoulos, Diffusion in polymers *in: Macromolecular Science* (Vol. 8 of Physical Chemistry Series 1) (C. E. H. Bawn, ed.), MTP International Review of Science, 1972.

Section 7.4. Rotational Diffusion and Streaming Birefringence

V. N. Tsvetkov, Flow birefringence, *in: Newer Methods of Polymer Characterization* (B. Ke. ed.), Wiley–Interscience, New York, 1964.

H. Janeschitz-Kriegl. Flow birefringence of elastico-viscous polymer systems, *Adv. Poly. Sci.* **6**, 170 (1969).

Section 7.5. Electrophoresis

R. L. Yeates, *Electropainting,* Draper, Teddington, 1966.

K. Weigel, *Electrophorese-Lacke,* Wiss. Verlagsges., Stuttgart, 1967.

J. R. Cann and W. B. Goad, *Interacting Macromolecules, The Theory and Practice of Their Electrophoresis, Ultracentrifugation and Chromatography,* Academic Press, New York, 1970.

Section 7.6. Viscosity

W. Philippoff, *Viskosität der Kolloide,* D. Steinkopff, Dresden, 1942.

M. Reiner, *Deformation and Flow,* K. H. Lewis & Co., London, 1949.

F. R. Eirich, ed., *Rheology, Theory and Applications,* 5 vols., Academic Press, New York, 1956–1969.

S. Peter, Zur Methodik der Viscositätsmessung, *Chem.-Ing.-Techn.,* **32** 437 (1960).

J. R. Van Wazer, J. W. Lyons, K. Y. Kim, and R. E. Colwell, *Viscosity and Flow Measurement,* Interscience, New York, 1963.

S. Middleman, *The Flow of High Polymers,* Wiley–Interscience, New York, 1968.

G. W. Scott-Blair, *Elementary Rheology,* Academic Press, London, 1969.

G. C. Berry and T. G. Fox, The viscosity of polymers and their concentrated solutions, *Adv. Polym. Sci.* **5**, 261 (1968).

J. D. Ferry, *Viscoelastic Properties of Polymers,* Wiley, Chichester, 1970.

J. A. Brydson, *Flow Properties of Polymer Melts,* Iliffe, London, 1970.

A. Peterlin, Non-Newtonian viscosity and the macromolecule, *Adv. Macromol. Chem.* **1**, 225 (1968).

W. W. Graessley, The entanglement concept in polymer rheology, *Adv. Polym. Sci.* **16,** 1 (1974).

K. Walters, *Rheometry,* Halsted, New York, 1975.

Journals

Rheologica Acta **1** (1958) ff., D. Steinkopff, Darmstadt.

Transactions of the Society of Rheology **1** (1957) ff.

Journal of Non-Newtonian Fluid Mechanics **1** (1976) ff.

Chapter 8

Molecular Weights and Molecular-Weight Distributions

8.1. Introduction

In synthesizing macromolecules *in vitro* and *in vivo*, monodisperse macromolecules (i.e., those in which every molecule has the same molecular weight) only occur under specific mechanistic conditions. The overwhelming majority of reactions proceed more or less randomly, and the resulting macromolecular substance shows a more or less broad molecular-weight distribution.

An example of such a random reaction is the polycondensation of α,ω-dicarboxylic acids, HOOC—R′—COOH (S), with glycols, HO—R—OH (G). In the first reaction step, the monoester GS forms from G and S, with the elimination of water, and this has a degree of polymerization of 2. In the second step, this semiester can add on either another glycol molecule or another acid molecule, etc., as follows:

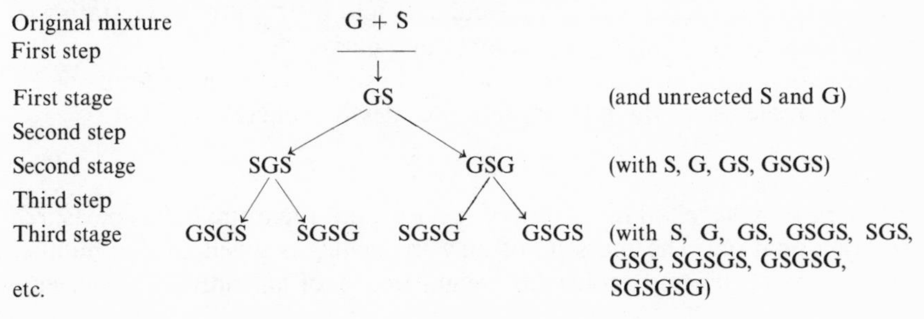

In these equilibrium reactions, not all glycol or acid molecules are converted into monoesters in the first step. If the reactivity of the individual hydroxyl and carboxyl groups is independent of the size of the molecules to which they are attached (the principle of equal chemical reactivity), then in the second step, in addition to molecules of degree of polymerization of 3 (SGS and GSG), new ones with degree of polymerization of 2 are simultaneously formed. In the third step, tetramers form from trimers and monomers or from two dimers, trimers from monomers and dimers, and dimers from two monomers, as well as, by this time, pentamers from dimers and trimers or monomers and tetramers, and hexamers from two trimers, etc., together with the original reactants formed by the reverse reactions.

Thus, in a polycondensation of this type various macromolecules occur according to extent of reaction or time: The substances produced possess a distribution of the degrees of polymerization. The kind of distribution depends on the reaction mechanism. For every kind of distribution, a characteristic relationship exists between the mole fraction x_i of species i and its degree of polymerization X_i. Inferences about the polymerization mechanism can be drawn if the nature of the distribution function is known.

It is considered sufficient in many cases to measure various moments of the distribution, or various degree-of-polymerization averages, instead of measuring the complete distribution. The relationships between the various moments or averages are also characteristic of the distribution type.

The distribution function, the moments, and the averages can be given in terms of the molecular weights rather than degrees of polymerization. Theoretically, a description in terms of the degrees of polymerization is more desirable since there is a more direct relationship between the formation mechanism and the degree of polymerization. However, it is the molecular weight and not the degree of polymerization that is measured experimentally.

8.2. Statistical Weights

Various statistical weights can be assigned to individual species i of a degree of polymerization or molecular weight distribution. The statistical weight g, can, for example, be a number or a mass.

From a mechanistic point of view, all events are given in terms of the reacting moles n_i or the number of molecules N_i, where

$$n_i = N_i / N_{\mathrm{L}} \tag{8-1}$$

The mass m_i of all molecules of species i, not their number, is measured by fractionation. The mass m_i of all i molecules is given by the number N_i of i molecules and molecular weight $(m_{\mathrm{mol}})_i$ of an individual molecule

$$m_i = N_i(m_{\mathrm{mol}})_i \tag{8-2}$$

With the definition of molecular weight (molar mass of IUPAC), we have

$$M_i = (m_{\mathrm{mol}})_i N_{\mathrm{L}} \tag{8-3}$$

Consequently, with equations (8-2) and (8-3)

$$M_i = \frac{m_i}{N_i} N_{\mathrm{L}} = \frac{m_i}{n_i} \tag{8-4}$$

Mass and mole are consequently related via the molecular weight:

$$m_i = n_i M_i \tag{8-5}$$

Equation (8-5) is valid for monodisperse fractions only. If the fractions are polydisperse, the number-average molecular weight (for definition, see Chapter 1 and Section 8.5.1) must be used instead of the molecular weight M_i:

$$m_i = n_i(\overline{M}_n)_i \tag{8-6}$$

Statistical weights of higher rank, for example, the z statistical weight, can be defined in analogy to the relationships (8-5) and (8-6):

$$z_i \equiv m_i(\overline{M}_w)_i = n_i(\overline{M}_n)_i (\overline{M}_w)_i \tag{8-7}$$

or a $(z + 1)$ statistical weight

$$(z + 1)_i \equiv z_i(\overline{M}_z)_i = m_i(\overline{M}_w)_i (\overline{M}_z)_i = n_i(\overline{M}_n)_i (\overline{M}_w)_i (\overline{M}_z)_i \tag{8-8}$$

Statistical weights of lower rank can also be defined:

$$(n - 1)_i \equiv \frac{n_i}{(\overline{M})_{n-1}} \tag{8-9}$$

Many measurable averages and moments correspond to z and $(z + 1)$ distributions (see below).

It can be readily shown by numerical examples that the average molecular weight corresponding to the statistical weight must always be used in multiplication with the statistical weight in question. Thus, $(z + 1)$ and $(n - 1)$ are not operators in the calculation—they are symbols. Of course, $(n + 1)_i$ can be used formally instead of m_i, and $(m + 1)_i$ or $(n + 2)_i$ instead of z_i, but this mode of description has not been adopted in the literature.

Mole fraction x_i, mass fraction w_i, z fraction Z_i, etc., can also be used instead of mole, mass, etc. Consequently, from their definitions and the equations above, we have

$$x_i = \frac{n_i}{\sum_i n_i} \tag{8-10}$$

$$w_i = \frac{m_i}{\sum_i m_i} = x_i \frac{(\overline{M}_n)_i}{\overline{M}_n} \tag{8-11}$$

$$Z_i = \frac{z_i}{\sum_i z_i} = w_i \frac{(\overline{M}_w)_i}{\overline{M}_w} = x_i \frac{(\overline{M}_n)_i (\overline{M}_w)_i}{\overline{M}_n \overline{M}_w} \tag{8-12}$$

Here, $(\overline{M}_n)_i$, $(\overline{M}_w)_i$, $(\overline{M}_z)_i$, etc., are number-, weight-, and z-averages of the individual i fractions, whereas $\overline{M}_n$, $\overline{M}_w$, $\overline{M}_z$, etc., are the corresponding averages of the complete sample. Note that the following is valid for the sum of masses:

$$\sum_i m_i = \sum_i [n_i(\overline{M}_n)_i] = n\overline{M}_n = m \tag{8-13}$$

and this, with equation (8-6), was used in the derivations leading to equations (8-10)–(8-12). We have, analogously, with equation (8-7),

$$\sum_i z_i = \sum_i [m_i(\overline{M}_w)_i] = m\overline{M}_w = z \tag{8-14}$$

for the sum of all z.

8.3. *Molecular-Weight Distributions*

8.3.1. *Representation of the Distribution Functions*

Distribution functions can be classified as discontinuous or continuous. Discontinuous distribution functions are subdivided into frequency distributions and cumulative distributions. Continuous distribution functions are further classified as differential and integral distribution functions.

In a *discontinuous* distribution function, the frequency distribution gives the distribution of the statistical weights (or their fractions) of the components i as a function of the property E. For example, E may be the degree of polymerization. Typical property distributions are

$$x_i(E_i), \qquad w_i(E_i), \qquad Z_i(E_i), \quad \text{etc.}$$

The *cumulative* distribution gives the summation over all statistical weights up to a certain value i. Typical cumulative distributions are therefore

$$\sum_{k=1}^{k=i} n_k(E_k), \qquad \sum_{k=1}^{k=i} w_k(E_k), \qquad \sum_{k=1}^{k=i} Z_k(E_k), \quad \text{etc.}$$

Frequency and cumulative distributions are stepwise distributions (see Figure 8-1). They can be transformed into continuous distributions by the process described in Section 2.3.2.2.

Discontinuous distribution functions can, of course, be transformed into continuous distribution functions when the difference between two neighboring properties is very small compared with the whole range of property values. Frequency distributions convert to the corresponding differential distributions and cumulative distributions convert to integral distributions,

$$X(E), \qquad W(E), \qquad Z(E), \quad \text{etc.}$$

and cumulative distributions convert to integral distributions,

$$\int_0^E x(E')\,dE', \qquad \int_0^E w(E')\,dE', \qquad \int_0^E Z(E')\,dE', \quad \text{etc.}$$

In the German scientific tradition, only molar distributions are described as "frequency distributions." This terminology can lead to confusion since weight and z distributions can be given in terms of frequency distributions.

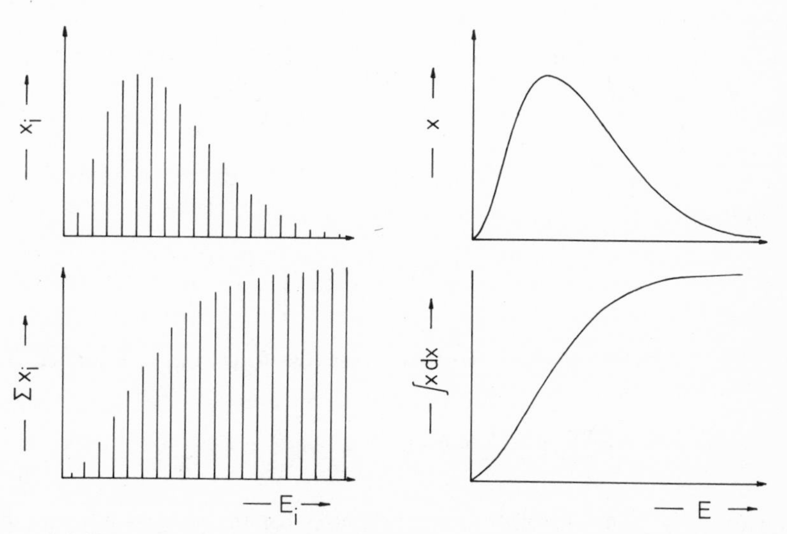

Figure 8-1. Frequency distribution (top left) and a cumulative distribution (bottom left), together with a differential distribution (top right) and an integral distribution (bottom right) of the mole fraction x of a property E (diagrammatic).

8.3.2. Types of Distribution Functions

Distribution functions are mostly named after the person who discovered them. Only the mathematical consequences of the distribution functions will be discussed in this chapter. The application to mechanisms or molecular weight averages will be discussed in Chapters 16–20.

8.3.2.1. Gaussian Distribution

The Gaussian distribution is the best known distribution. It represents the error law about the arithmetic mean. Because of its frequent appearance, the Gaussian distribution is also called the normal distribution in mathematics. In contrast, a certain form of the Schulz–Flory distribution is often called the normal distribution in macromolecular science.

The mole fraction differential distribution of the property E is given in the form of a Gaussian function as

$$x(E) = \frac{1}{\sigma_n(2\pi)^{1/2}} \exp\left[\frac{(E - \bar{E}_m)^2}{2\sigma_n^2}\right] \tag{8-15}$$

where $\bar{E}_m$ is the median value, i.e., that value for which

$$\int_{-\infty}^{+E_m} dx = 0.5$$

Since the Gaussian distribution is symmetric about the median (see Figure 8-2), the median is equivalent to the number-average $\bar{E}_n$ of the property.

The term σ_n is a parameter which is proportional to the width of the distribution, and thereby also describes the deviation from the mean value. It is known as the standard deviation. The deviation of an E_i value from the mean value is given as the mean "error" s_n of the individual value:

$$s_n = \left[\frac{n_i(E_i - E_n)^2}{\sum_i n_i}\right]^{0.5} \tag{8-16}$$

If equation (8-16) is solved and summed, the following is obtained with $\sum_i s_n^2 = \sigma_2^2$:

$$\sigma_n^2 \sum_i n_i = \sum_i n_i E_i^2 - 2E_n \sum_i n_i E_i + E_n^2 \sum_i n_i \tag{8-17}$$

Dividing by $\sum_i n_i$, inserting the expressions for the weight- and number-average degrees of properties [see equations (8-40) and (8-41)], and solving, we obtain

$$\sigma_n = (\bar{E}_w \bar{E}_n - \bar{E}_n^2)^{0.5} \tag{8-18}$$

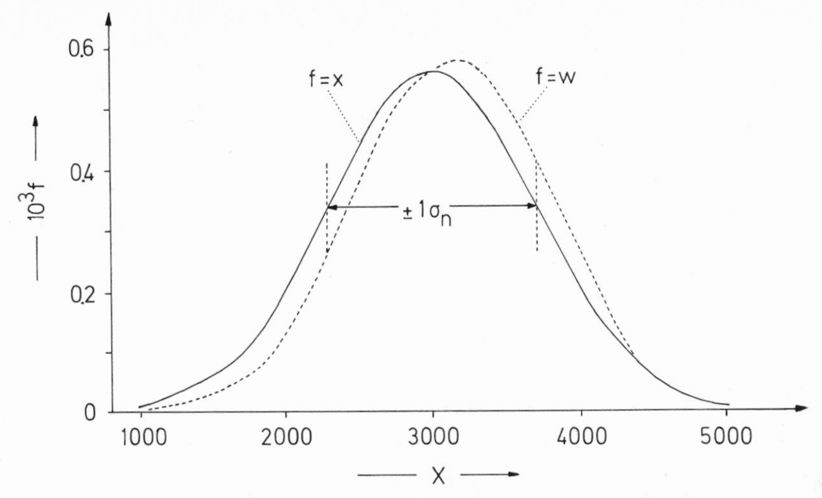

Figure 8-2. The mole fraction x (—) or mass fraction w (- -) as a function of the degree of polymerization X for a Gaussian molar differential distribution of the degree of polymerization. Calculations based on $\overline{X}_w = 3170$ and $\overline{X}_n = 3000$.

An analogous equation is used in literature for the number standard deviation of degrees of polymerization X, although degrees of polymerization (or molecular weights) can never possess a Gaussian distribution because they cannot exhibit negative values:

$$\sigma_n = (\overline{X}_w \overline{X}_n - \overline{X}_n^2)^{0.5} \tag{8-18a}$$

Thus, the standard deviation of the molar distribution of the degree of polymerization can be calculated from the number and weight averages. Further, the standard deviation is an absolute measure of the width of a Gaussian distribution (and only of a Gaussian distribution), since the mole fraction of 0.6826 always lies within the limits $\overline{X}_n \pm 1\sigma_n$, and the mole fraction of 0.9544 lies within $\overline{X}_n \pm 3\sigma_n$. Thus, for $\overline{X}_w = 3170$ and $\overline{X}_n = 3000$ in the case of the example shown in Figure 8-2, $\sigma_n = 714$, according to equation (8-18). Consequently, 68.26% of all molecules lie within the range of $\overline{X}_n = 3000 \pm 714$.

The differential molar Gaussian distribution loses its symmetry when the distribution is plotted in terms of mass fractions instead of mole fractions (Figure 8-2). The curve symmetry occurs again if a mass fraction Gaussian distribution is plotted in terms of the mass fraction. Of course, this differential mass fraction Gaussian distribution must be given in terms of the mass weight-average degree of polymerization and the standard deviation σ_w:

$$w(\overline{X}) = \frac{1}{\sigma_w (2\pi)^{1/2}} \exp\left[- \frac{(\overline{X} - \overline{X}_w)^2}{2\sigma_w^2} \right] \tag{8-19}$$

and the standard deviation is given as

$$\sigma_w = (\overline{X}_z \overline{X}_w - \overline{X}_w)^{0.5} \tag{8-20}$$

Consequently, when giving the type of a distribution, it is always necessary to state the statistical weight.

Gaussian distributions rarely occur in macromolecular science. The Gaussian distribution allows negative values of a given property, which is nonsense with respect to the molecular weight.

8.3.2.2. Logarithmic Normal Distribution

The differential logarithmic normal distribution has the same mathematical form as the Gaussian distribution with the small difference that the logarithm of the property occurs in place of the property itself:

$$x(X) = \frac{1}{(2\pi)^{0.5} X \sigma_n^*} \exp\left[- \frac{(\ln X - \ln \overline{X}_M)^2}{2(\sigma_n)^2} \right] \tag{8-21}$$

Here, the curve is symmetric about $\ln \overline{X}_M$. The median of the curve $\overline{X}_M$ is not identical with the number average $\overline{X}_n$ (see below). The function corresponds to the error distribution about the geometric mean. The *ratio* of the degrees of polymerization is therefore important with the logarithmic normal distribution, in contrast to the Gaussian distribution, where the *difference* is important.

Differential logarithmic normal distributions can be generalized, for example, for the mass distribution of the degree of polymerization:

$$w(X) = \frac{1}{(2\pi)^{0.5} \sigma_w^*} \frac{X^A}{B \overline{X}_M^{A+1}} \exp\left[- \frac{(\ln X - \ln \overline{X}_M)^2}{2(\sigma_w^*)^2} \right] \tag{8-22}$$

with $B = \exp[0.5(\sigma_w^*)^2 (A + 1)^2]$.

Two special cases are used in macromolecular science:

Lansing–Kraemer distribution:	$A = 0;$	$B = \exp[0.5(\sigma_w^*)^2]$
Wesslau distribution:	$A = -1;$	$B = 1$

Neither the Wesslau nor the Lansing–Kraemer distribution has as yet been related to a definite polymerization mechanism.

A molar logarithmic normal distribution as described by equation (8-21) is shown in Figure 8-3. The logarithmic normal distribution is thus a skewed distribution when the degree of polymerization is chosen as the abscissa. The curve maximum does not occur at the same point as the number-average degree of polymerization.

If the molar logarithmic normal distribution is plotted in terms of the mass fraction, and not the mole fraction, the shape of the curve remains

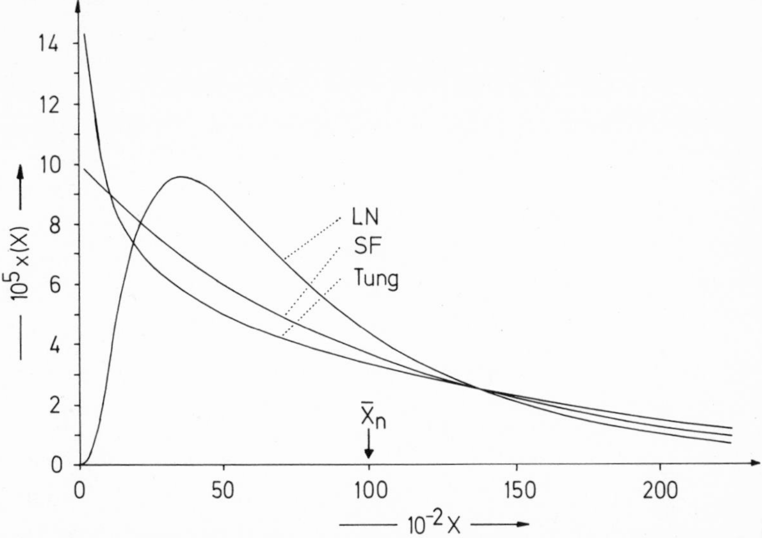

Figure 8-3. Differential molar distribution of the degree of polymerization for three different distribution functions: logarithmic normal (LN), Schulz–Flory (SF), and Tung (Tung). Calculations based on $\overline{X}_n = 10{,}000$ and $\overline{X}_w/\overline{X}_n = 2$.

essentially unchanged (Figure 8-4). Equally, the curve maximum is not identical with either the number- or weight-average degree of polymerization.

Logarithmic normal distributions give straight-line plots when the integral distribution is plotted on paper where the ordinate is graduated in summation probability units and the abscissa is logarithmic (Figure 8-5).

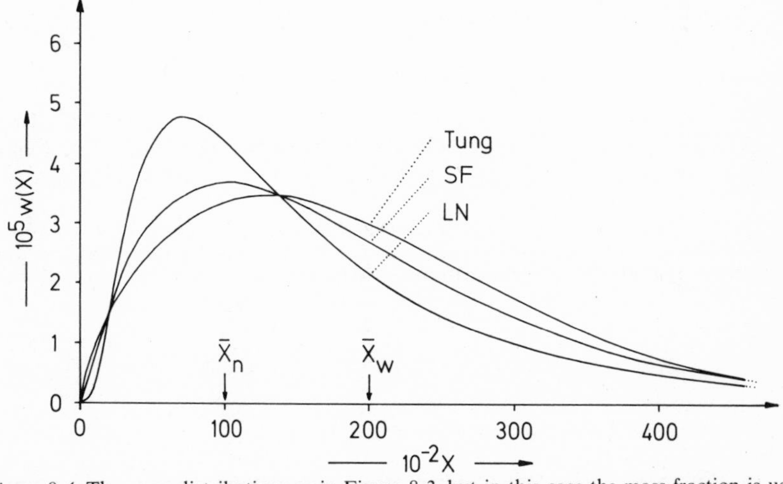

Figure 8-4. The same distributions as in Figure 8-3, but in this case the mass fraction is used instead of the mole fraction.

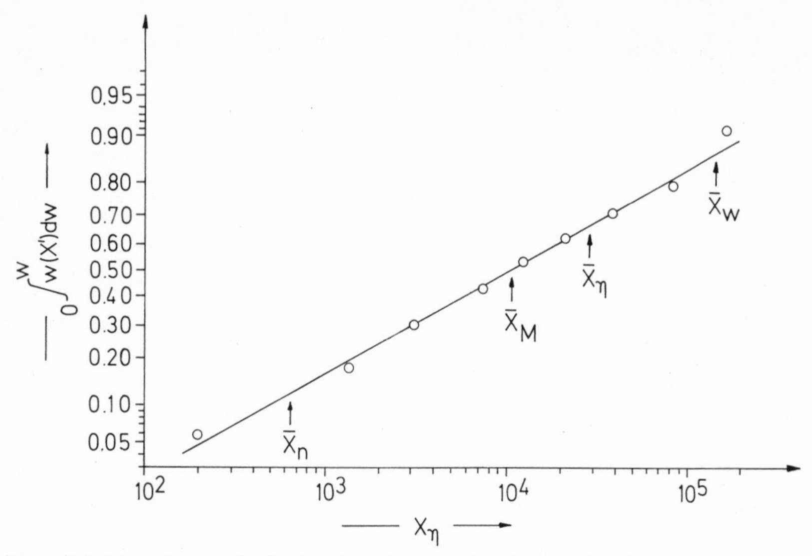

Figure 8-5. Integral mass distribution for a Wesslau distribution. The viscosity-average degree of polymerization $(\bar{X}_\eta)_i$ and the mass fraction were measured. The viscosity-average degree of polymerization $\bar{X}_\eta$ of the original material with the already known exponent a_η, the median value $\bar{X}_M$, and equation (8–23) and (8–24) were used to calculate the number- and weight-average molecular weights.

The relationships between the median value and the various averages can be derived from equation (8-22):

$$\bar{X}_n = \bar{X}_M \exp[(2A + 1)(\sigma_w{}^*)^2/2] \tag{8-23}$$

$$\bar{X}_w = \bar{X}_M \exp[(2A + 3)(\sigma_w{}^*)^2/2] \tag{8-24}$$

$$\bar{X}_z = \bar{X}_M \exp[(2A + 5)(\sigma_w{}^*)^2/2] \tag{8-25}$$

and analogously for the viscosity-average degree of polymerization (with the exponent a_η of the viscosity–molecular weight relationship, see Chapter 9):

$$\bar{X}_\eta = \bar{X}_M \exp\{[2(A + a_\eta) + 1](\sigma_w{}^*)^2/2\} \tag{8-26}$$

Equations (8-23)–(8-25) consequently lead to

$$\exp(\sigma_w{}^*)^2 = \frac{\bar{X}_w}{\bar{X}_n} = \frac{\bar{X}_z}{\bar{X}_w} \tag{8-27}$$

8.3.2.3. Poisson Distribution

A Poisson distribution occurs when a constant number of polymer chains begin to grow simultaneously and when the addition of monomeric units is random and occurs independently of the previous addition of other monomeric units. Consequently, Poisson distributions may occur with what are called "living polymers" (see Section 16.5).

For the differential molar degree of polymerization distribution, the Poisson distribution gives

$$x = \frac{\nu^{X-1}\exp(-\nu)}{\Gamma(X)} \tag{8-28}$$

where $\nu = \overline{X}_n - 1$ and $\Gamma(X)$ is the gamma function. Further, the number-average degree of polymerization is related to the weight-average degree by (see Chapter 18)

$$\frac{\overline{X}_w}{\overline{X}_n} = 1 + \left(\frac{1}{\overline{X}_n}\right) - \left(\frac{1}{\overline{X}_n}\right)^2 \tag{8-29}$$

Consequently, the ratio $\overline{X}_w/\overline{X}_n$ in the Poisson distribution depends on the number-average degree of polymerization and on no other parameter. The ratio $\overline{X}_w/\overline{X}_n$ tends to a value of unity with increasing degree of polymerization. The Poisson distribution is consequently a very narrow distribution.

8.3.2.4. Schulz–Flory Distribution

The process whereby growing chains whose number is constant in time randomly add monomer till the growth center of individual chains is destroyed by termination provides the basis for the Schulz–Flory distribution. Thus, the originally existing individual chain growth centers do not necessarily remain active. This is in contrast to the case leading to a Poisson distribution. Also, individual chain growth centers need not all commence growth at the same time in the Schulz–Flory distribution. It is only required that the growth center concentration remain constant. Polycondensation and most free radical polymerizations are processes of this type. Such processes were considered to be the normal processes in the early days of polymer science, and thus the distributions so produced were called "normal distributions." This terminology is different from mathematical terminology with respect to a "normal distribution" (see Section 8.3.2.1). The "most probable distribution" often referred to in the literature is concerned with a Schulz–Flory distribution where $\overline{X}_w/\overline{X}_n = 2$.

What is known as the degree of coupling k must be known before the Schulz–Flory distribution can be calculated. The degree of coupling is defined as the number of independently growing chains required to form one dead chain. The degree of coupling is two when two growing free radical chains form one dead chain by recombination:

$$P_i^{\cdot} + P_{x-i}^{\cdot} \rightarrow P_x \tag{8-30}$$

The differential molar distribution is given as (see Chapter 19)

$$x = \frac{\beta^{k+1} X^{k-1} \overline{X}_n \exp(-\beta X)}{\Gamma(k+1)} \tag{8-31}$$

and, from this, the differential mass fraction distribution is given by

$$w = \frac{\beta^{k+1} X^k \exp(-\beta X)}{\Gamma(k+1)} \tag{8-32}$$

where

$$\beta = \frac{k}{\overline{X}_n} \tag{8-33}$$

An "exponentially" decreasing curve is obtained for $k = 1$ when the mole fraction is plotted against the degree of polymerization (see Figure 8-3). For this reason, and not because an exponential function appears in equation (8-31), the Schulz–Flory distribution is called an exponential distribution.

The simple one-moment degree of polymerization averages are related to each other by

$$\frac{\overline{X}_n}{k} = \frac{\overline{X}_w}{k+1} = \frac{\overline{X}_z}{k+2} \tag{8-34}$$

Consequently, the distributions become increasingly narrower for increasing degrees of coupling. The Schulz–Flory distribution can be distinguished from the logarithmic normal distributions via equations (8-27) and (8-34). For this, however, three different averages must be known.

8.3.2.5. *Tung Distribution*

The Tung distribution was found empirically. It gives the differential mass distribution of the degrees of polymerization as

$$w = DB(X-1)^{B-1} \exp[-D(X-1)^B] \tag{8-35}$$

where D and B are empirical constants. The Tung distribution consequently represents a modified Schulz–Flory distribution.

8.4. *Moments*

The term "moment" comes from mechanics. The first moment of a force $v^{(1)}$ is defined as the product of the force (e.g., g) and the distance (e.g., E) along the axis from the point of application of the force. The second moment $\nu^{(2)}$ is the product of the force and the square of the distance. If several forces are applied at various distances, then the moments are determined by summation of the vector products. Also, the first and second moments, as well as any desired moment, may be defined with respect to any desired reference point E_0:

$$\nu_g^{(q)}(E) \equiv \frac{\sum_i g_i(E_i - E_0)^q}{\sum_i g_i} \tag{8-36}$$

$\nu_g^{(q)}(E)$ is consequently the qth moment of E values with respect to E_0. Moments can, of course, be used not only for the relationships between force and distance, but also for the relationships between any desired quantities.

The statistical weight g can, for example, be a number or a mass (see also Section 8.2).

The order q can be positive or negative, as desired. It may be an integer or a fraction and can take real or imaginary values. Consequently, a moment generally is expressed in physical units different from the property. The property may be the degree of polymerization, the molecular weight, the sedimentation coefficient, molecular dimensions, or any other desired property.

In principle, E_0 may take on any chosen value. However, since, for example, negative degrees of polymerization do not exist, it is frequently convenient to use moments with respect to a reference value of zero and to give such moments a special symbol:

$$\mu_g^{(q)}(E) = \frac{\sum_i g_i E_i^q}{\sum_i g_i} \tag{8-37}$$

The introduction of moments of degrees of polymerization or molecular weights considerably simplifies equations describing complicated averages of these quantities.

8.5. Averages

8.5.1. General Relationships

In contrast to moments, averages always possess the same physical units as the properties on which they are based. Averages are consequently first-order moments or such combinations of moments of different order that the resulting physical units are the same as those of the property.

Most of the averages so far considered are composed of one or two moments. They can be described by the generalized formula

$$\overline{X}_{g(p,q)} = \left[\frac{\mu_g^{p+q-1}(X)}{\mu_g^{q-1}(X)}\right]^{1/p} = \left(\frac{\sum_i G_i X_i^{p+q-1}}{\sum_i G_i X_i^{q-1}}\right)^{1/p} \tag{8-38}$$

Equation (8-38) contains four important special cases:

1. When $p = q = 1$, equation (8-38) reduces to that of a simple one-moment average.
2. When $q = 1$ and $p \neq q$, a one-moment exponent average is obtained.
3. When $q \neq 1$ and $p \neq q$, a two-moment exponent average results.
4. When $p = 1$ and $p \neq q$, an average of two-moment order results.

8.5.2. Simple One-Moment Averages

Simple one-moment averages are defined by

$$\overline{X}_g = \frac{\sum_i g_i X_i}{\sum_i g_i} = \sum_i G_i X_i \tag{8-39}$$

According to the nature of the argument, they are called the number-average ($g = n$), the "weight"-average ($g = m$), the z-average ($g = z$), etc. For historic reasons, and to avoid confusion with the index m for "mol," the weight-average degree of polymerization is usually given the symbol $\overline{X}_w$ and not the symbol $\overline{X}_m$. Thus, the number-average molecular weight is given by

$$\overline{M}_n = \frac{\sum_i n_i (\overline{M}_i)_n}{\sum_i n_i} = \sum_i x_i (\overline{M}_i)_n = \frac{\sum_i c_i}{\sum[c_i/(\overline{M}_i)_n]} = \frac{c}{\sum[c_i/(\overline{M}_i)_n]} \tag{8-40}$$

and the weight-average molecular weight is given by

$$\overline{M}_w = \frac{\sum_i m_i (\overline{M}_i)_w}{\sum_i m_i} = \frac{\sum_i n_i (\overline{M}_i)_n (\overline{M}_i)_w}{\sum_i n_i (\overline{M}_i)_n} = \frac{\sum_i c_i (\overline{M}_i)_w}{\sum_i c_i} = \sum_i w_i (\overline{M}_i)_w$$
$$= \sum_i x_i (\overline{M}_i)_n (\overline{M}_i) \tag{8-41}$$

The z-average molecular weight is given by

$$\overline{M}_z = \frac{\sum_i z_i (\overline{M}_i)_z}{\sum_i z_i} = \frac{\sum_i m_i (\overline{M}_i)_w (\overline{M}_i)_z}{\sum_i m_i (\overline{M}_i)_w} = \frac{\sum_i n_i (\overline{M}_i)_n (\overline{M}_i)_w (\overline{M}_i)_z}{\sum_i n_i (\overline{M}_i)_n (\overline{M}_i)_w}$$
$$= \sum_i Z_i (\overline{M}_i)_z = \sum_i w_i (\overline{M}_i)_w (\overline{M}_i)_z = \sum_i x_i (\overline{M}_i)_n (\overline{M}_i)_w (\overline{M}_i)_z \tag{8-42}$$

According to these equations, the following must always hold:

$$\overline{M}_z \geq \overline{M}_w \geq \overline{M}_n \tag{8-43}$$

In the early days of polymer chemistry, the number-average degree of polymerization was known as the "average" and the weight-average degree was known as the "mean" in German-speaking countries. The term *DP* is consequently still used to mean the number-average degree of polymerization, especially in industry. In English-speaking countries, *DP* is the abbreviation for the degree of polymerization, and can be DP_n, DP_w, or DP_z, etc.

8.5.3. One-Moment Exponent Averages

The general formula for a one-moment average is

$$\overline{X}_g = \left(\sum_i G_i X_i^q\right)^{1/q} \tag{8-44}$$

The best known of these one-moment exponent averages is what is known as the viscosity-average molecular weight:

$$\overline{M}_\eta = \left(\sum_i w_i M_i^{a_\eta}\right)^{1/a_\eta} \tag{8-45}$$

where a_η is the exponent in the Staudinger index–molecular weight relationship $([\eta] = K_\eta M^{a_\eta})$. Strictly speaking, the viscosity-average is a mass-viscosity-average, since the argument on which it is based is a mass. Analogous averages with various arguments exist for sedimentation, diffusion, etc.

8.5.4. Multimoment Averages

According to equation (8-38), averages from two moments are also possible. The order of the moments [that is, $(p + q - 1)$ in the numerator and $(q - 1)$ in the denominator] must be combined with the exponent $1/p$ in such a manner that the total expression has the same physical units as the property. Since the physical units on both sides of the equation must be the same, the so-called exponent rule can be directly obtained from equation (8-38),

$$1 = (p + q - 1)\frac{1}{p} - (q - 1)\frac{1}{p} \tag{8-46}$$

or, in general form: The product sum of the exponents must always equal unity. The rule is based on dimensional analysis and is consequently independent of whatever shape the macromolecules may adopt.

The exponent rule is especially significant in combination with the rule that states that the relationships between two variables can always, at least over a limited range of values, be written as an exponential relationship. It has been found empirically that over wide ranges of the molecular weight the following relationships between the molecular weight and the sedimentation coefficient s, the diffusion coefficient D, or the Staudinger index $[\eta]$ are valid:

$$s = K_s M^{a_s} \tag{8-47}$$

$$D = K_D M^{a_D} \tag{8-48}$$

$$[\eta] = K_\eta M^{a_\eta} \tag{8-49}$$

The molecular weight can be obtained from any pair of the three quantities (for derivation, see Chapter 9)

$$\overline{M}_{sD} = A_{sD} K_{sD} s D^{-1} \tag{8-50}$$

$$\overline{M}_{s\eta} = A_{s\eta} K_{s\eta} s^{3/2} [\eta]^{1/2} \tag{8-51}$$

$$\overline{M}_{D\eta} = A_{D\eta} K_{D\eta} D^{-3} [\eta]^{-1} \tag{8-52}$$

The quantities K_{sD}, K_s, and K_D are accessible through independent measurements and are independent of the molecular weight. They are consequently called physical constants. On the other hand, A_{sD}, $A_{s\eta}$, and $A_{D\eta}$ are model constants, since they are based on certain assumptions. If, for example, the frictional coefficients from sedimentation and diffusion are of equal magnitude (see Chapter 9), then $A_{sD} = 1$. The model constants can, of course, influence the numerical value of the molecular weight, but they have no effect on the composition of the average from the various individual molecular species contributions. Consequently, model constants can always be assumed to have a value of unity until evidence to the contrary is obtained.

If it is assumed that the properties s, D, and $[\eta]$ can each be obtained as simple weight averages, then equations (8-47)–(8-52) give

$$\overline{M}_{s_w D_w} = A_{sD} \left(\sum_i w_i M_i^{a_s}\right) \left(\sum_i w_i M_i^{a_D}\right)^{-1}; \qquad a_s - a_D = 1 \tag{8-53}$$

$$\overline{M}_{s_w \eta_w} = A_{s\eta} \left(\sum_i w_i M_i^{a_s}\right)^{3/2} \left(\sum_i w_i M_i^{a_\eta}\right)^{1/2}; \qquad \tfrac{3}{2} a_s + \tfrac{1}{2} a_\eta = 1 \tag{8-54}$$

$$\overline{M}_{D_w \eta_w} = A_{D\eta} \left(\sum_i w_i M_i^{a_D}\right)^{-3} \left(\sum_i w_i M_i^{a_\eta}\right)^{-1}; \quad -3a_D - a_\eta = 1 \tag{8-55}$$

According to dimensional analysis, the product of the physical constants must be unity.

From equations (8-53)–(8-55), we obtain the following for the exponent rule:

$$a_\eta = 2 - 3a_s = -(1 + 3a_D) \tag{8-56}$$

The averages appearing in equations (8-53)–(8-55) can be transformed into other averages or moments with the help of these relationships. It can be seen from Table 8-1 that the averages of such two-moment averages are determined by the composition of the property averages as well as by the nature of the molecular-weight dependence of the property. For example, if $a_\eta = 2$, combining the weight average of the sedimentation coefficient with the weight average of the diffusion coefficient gives the number-average molecular weight. In some cases, the average of a two-moment average consists of a combination of two simple averages; in other cases it is simpler to describe a two-moment average via moments.

Consequently, for the same width of the distribution, the numerical values of these two-moment molecular-weight constants depend on the value of the constant a_η (Figure 8-6). For a given homologous series, a_η is in turn a function of the particle shape and its interaction with the solvent. For example, rigid rods have a value of $a_\eta = 2$; for spheres, $a_\eta = 0$. Values of between 0.5 and 0.9 are normally obtained for random coils (see Section 9.8).

Thus these two moment averages possess the apparent paradox that for polydisperse materials, an absolute method of determining the molecular weight gives different values according to the nature of the solvent (that is, according to a_η). A method of determining the molecular weight is considered to be absolute when all parameters can be directly measured and no assumptions need be made about the chemical and physical structure. This applies, for example, to equation (8-53), where the quantities s, D, and $K_{sD} = RT/(1 - v_2\rho_1)$ can be directly measured and A_{sD} can be set equal to one.

Table 8-1. Moments and Averages of Molecular Weights for Some Combinations of s, D, and [η]

Combination	Exponent a_η	Moment or average
s_n and D_n	As desired	$\mu_n^{(1)} \equiv \overline{M}_n$
s_w and D_w	2	$\mu_n^{(1)} \equiv \overline{M}_n$
s_w and D_w	0.5	$\mu_n^{(1)}\mu_w^{(0.5)}/\mu_n^{(0.5)} = \sum_i x_i M_i^{1.5} / \sum_i x_i M_i^{0.5}$
s_w and D_z	As desired	$\mu_w^{(1)} \equiv \overline{M}_w$
s_w and $[\eta]_w$	2	$(\mu_w^{(2)})^{0.5} = (\overline{M}_w \overline{M}_z)^{0.5}$
s_w and $[\eta]_w$	0.5	$(\mu_w^{(0.5)})^2 = (\sum_i w_i M_i^{0.5})^2 \equiv (\overline{M}_\eta)_\Theta$

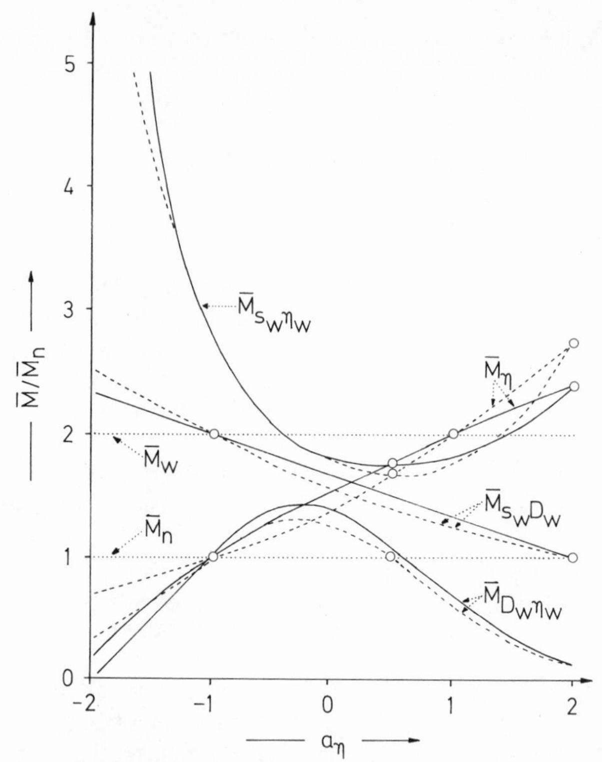

Figure 8-6. Calculated $\overline{M}/\overline{M}_n$ ratios as a function of the exponents a_η for a Schulz–Flory (—) or a generalized logarithmic normal (- -) molecular-weight distribution for, in each case, $\overline{M}_w/\overline{M}_n = 2$. $\overline{M}$ may be $\overline{M}_{s_w\eta_w}$, $\overline{M}_\eta$, $\overline{M}_{s_wD_w}$, or $\overline{M}_{D_w\eta_w}$. (After H.-G. Elias, R. Bareiss and J. G. Watterson.)

Two-moment averages can vary considerably with the exponent a_η even though the molecular-weight distribution remains the same (Figure 8-6). In some cases, different averages have identical values irrespective of the width of the distribution. For example, with $a_\eta = 1$, $\overline{M}_\eta = \overline{M}_w$ and with $a_\eta = -1$, $\overline{M}_{s_wD_w}$ is always equal to $\overline{M}_w$. When $a_\eta = 0.5$, $\overline{M}_{s_w\eta_w}$ always equals $\overline{M}_\eta$.

8.5.5. *Polydispersity Parameters*

The width of a molecular-weight distribution can, because of equation (8-43), always be described by the ratio of two molecular weight averages. The polydispersity index

$$Q = \overline{M}_w/\overline{M}_n \tag{8-57}$$

is very often used. Also used is the molecular inhomogeneity,

$$U = \overline{M}_w/\overline{M}_n - 1 \tag{8-58}$$

Combinations of other molecular weight averages are also possible to give analogous Q and U values. For a monodisperse material $Q = 1$ and $U = 0$.

The width of the molecular-weight distribution increases with increasing Q and U values. However, Q and U are not very sensitive to the distribution width for narrow molecular-weight distribution. Of course, according to equation (8-18), U is also given by

$$U = \frac{\overline{M}_w}{\overline{M}_n} - 1 = \left(\frac{\sigma_n}{\overline{M}_n}\right)^2 \tag{8-59}$$

Thus, the molecular inhomogeneity U and, consequently, the value of Q depend also on the number-average molecular weight. The standard deviation remains as a more sensitive measure of the distribution width than either Q or U, but with one exception, it is not an absolute measure of the width of the distribution (see also Section 8.3.2.1). To be an absolute measure of the width of the molecular-weight distribution, the standard deviation must encompass a fraction of the original material that is independent of the width of the distribution, and this only holds for a Gaussian distribution.

Literature

L. H. Peebles, *Molecular Weight Distributions in Polymers*, Wiley–Interscience, New York, 1971.

H.-G. Elias, R. Bareiss, and J. G. Watterson, Mittelwerte des Molekulargewichtes und anderer Eigenschaften, *Adv. Polym. Sci.—Fortschr. Hochpolym. Forsch.* **11**, 111 (1973).

Chapter 9

Determination of Molecular Weight and Molecular-Weight Distribution

9.1. Introduction and Survey

Molecular-weight determination methods can be classed as absolute, equivalent, or relative. Absolute methods allow the molecular weight to be calculated directly from the measured quantities; a knowledge of the physical and chemical structure of the molecules is not required. Absolute methods include colligative methods (membrane osmometry, ebulliometry, cryoscopy, and vapor-phase osmometry), light scattering, and equilibrium sedimentation. Some methods yield simple, others mixed, averages of the molecular weight. Colligative methods indicate the number of molecules, thus leading to a number-average molecular weight.

Equivalent methods always require some knowledge of the chemical structure of the molecules before the molecular weight can be calculated from the measured data. With end-group analysis, for example, it is necessary to know both the nature and the number per molecule of the end groups (see Section 2.3.4).

Relative methods, on the other hand, measure properties that depend on the chemical and physical structure of the macromolecule. The measurement of the viscosity of dilute polymer solutions is the most important of these methods. The viscosity is influenced by the constitution and configuration of the macromolecule and also depends on the macromolecular shape in solution and the interaction between polymer molecules and solvent. Relative methods, therefore, must always be calibrated against an absolute method.

The choice of the method depends primarily on the information required. For example, in polymer preparative and kinetic studies, a knowledge of the number of molecules produced is important, so a method giving the number-average molecular weight is chosen. On the other hand, mechanical properties often depend on the weight-average molecular weight.

Second, the operative region of the molecular-weight-determination method also has to be considered (Table 9-1). The operative region depends essentially on the magnitude of the effect that can be obtained at a given polymer concentration. With a molecular weight of 10^5 g/mol, membrane osmometry gives an osmotic pressure of $\Pi \approx 3.2$ cm column of water for a polymer at 10^{-2} g/cm^3 in an ideal solvent at 100°C. For the same system, vapor-pressure osmometry and ebulliometry yield temperature differences of $\Delta T \approx 4 \times 10^{-5}$ K and a relative lowering of the vapor pressure by $\Delta p/p \approx 3 \times 10^{-6}$, respectively. The choice of method is also dependent on the amount of polymer available. Finally, the experimental requirements of a given method, i.e., the standard of purity required of the test sample, as well as the time needed to carry out the experiment also play a role in the eventual choice of a method.

The equations used to calculate molecular weights are developed

Table 9-1. Approximate Working Ranges of the More Important Methods of Determining Molecular Weight

Molecular weight average	Method	Type[a]	Molecular weight range, g/mol molecule
$\overline{M}_n$	Ebulliometry, cryoscopy, vapor-phase osmometry, isothermal distillation	A	$<10^4$
$\overline{M}_n$	End-group analysis	E	10^2–3×10^4
$\overline{M}_n$	Membrane osmometry	A	5×10^3–10^6
$\overline{M}_n$	Electron microscopy	A	$>5 \times 10^5$
$\overline{M}_w$	Equilibrium sedimentation	A	10^2–10^6
$\overline{M}_w$	Light scattering	A	$>10^2$
$\overline{M}_w$	Equilibrium sedimentation in a density gradient	A	$>5 \times 10^4$
$\overline{M}_w$	Small-angle X-ray scattering	A	$>10^2$
$\overline{M}_{sD}$	Sedimentation combined with diffusion	A	$>10^3$
$\overline{M}_\eta$	Dilute solution viscometry	R	$>10^2$

[a]A = absolute method; E = equivalent method; R = relative method.

from thermodynamic, hydrodynamic, or other principles. With the exclusion of theta solvents, all the methods normally show a marked concentration dependence of the apparent molecular weight M_{app} over that calculated from an equation for ideal conditions ($c \rightarrow 0$).

Coil-shaped macromolecules at a concentration of $c = 10^{-2}$ g solute/cm^3 solution (often incorrectly termed "1 %" solution), with a number-average molecular weight of $\overline{M}_n = 10^2$ g/mol may, e.g., show in good solvent an $(\overline{M}_n)_{app}$ of 99.2 g/mol molecule, whereas for a molecular weight of $\overline{M}_n = 10^6$, an $(\overline{M}_n)_{app}$ of only 5.55×10^5 will be obtained. With 10% solutions, the corresponding figures are, for 10^2, 92.6, and for 10^6, 1.1×10^5. Therefore, the apparent molecular weights obtained at finite concentrations must be extrapolated to zero concentration to obtain the true molecular weight (see also Chapter 6).

9.2. *Membrane Osmometry*

9.2.1. *Semipermeable Membranes*

Equations applicable to membrane osmometry, as also in the case of ebulliometry, cryoscopy, and vapor-phase osmometry, can be rigorously derived from the second law of thermodynamics in the form

$$dG = V\,dp - S\,dT \tag{9-1}$$

In membrane osmometry, the pressure difference between a solution and the pure solvent is measured for the case where the solvent is separated from the solution by a semipermeable membrane, i.e., a membrane permeable only to solvent molecules. Since the experiment is carried out isothermally, equation (9-1), becomes, with $dT = 0$,

$$\Delta G = V\,\Delta p = V\Pi \tag{9-2}$$

when, for small pressure differences, differentials are replaced by differences. The manometrically determined pressure difference Δp is called the osmotic pressure Π. Differentiation of equation (8-2) with respect to amount of solvent (in moles) n_1 yields

$$\frac{\partial \Delta G}{\partial n_1} = \Pi \frac{\partial V}{\partial n_1} \tag{9-3}$$

From the well-known laws governing differences in chemical potential of the solvent $\Delta\mu_1$ and the partial molar volume $\tilde{V}_1^m$, one obtains

$$-\Delta\mu_1 = \mu_{1(p)} - \mu_1 = \Pi \tilde{V}_1^m \tag{9-4}$$

The chemical potential difference can be replaced by the solvent activity a_1 and, in very dilute solutions, the mole fraction x_1 of the solvent or that of the solute x_2:

$$\Pi \tilde{V}_1^m = -RT \ln a_1 \cong -RT \ln x_1 = -RT \ln (1 - x_2) \approx RTx_2 \quad (9\text{-}5)$$

In dilute solution, where $n_2 \ll n_1$ and $V_2 \ll V_1$, and consequently $x_2 = V_1^m c_2 / M_2$, since $x_2 = n_2/(n_2 + n_1)$ and $n_2 = m_2/M_2$, then $c_2 = m_2/V_2 + V_1$ and $V_1^m = V_1/n_1$, and we can derive from equation (9-5) the van't Hoff equation as a limiting law for infinite dilution where $V_1^m \approx \tilde{V}_1^m$:

$$\lim_{c_2 \to 0} \frac{\Pi}{c_2} = \frac{RT}{M_2} \quad (9\text{-}6)$$

For solutions of nonassociating nonelectrolytes at finite concentrations, Π/c_2 is given as an ascending series of positive powers of solute concentration [see also equation (6-54)]:

$$\frac{\Pi}{RTc_2} \equiv (M_2)_{\text{app}}^{-1} = A_1 + A_2 c_2 + A_3 c_2^2 + \cdots \quad (9\text{-}7)$$

where A_1, A_2, etc., are the first, second, etc., virial coefficients. Comparison of the coefficients of equations (9-6) and (9-7) gives A_1 as $(M_2)^{-1}$. The coefficient A_1 is obtained by measuring the osmotic pressure at different concentrations. The factor RTA_1 is given as the ordinate intercept at $c_2 \to 0$ in a plot of reduced osmotic pressure Π/c_2 against c_2 (Figure 9-1). With associating solutes, complicated expressions occur on the right-hand side of equation (9-7) (see Section 6.5). Since the measured osmotic pressure Π is inversely proportional to the molecular weight, the method becomes more and more inaccurate with increasing molecular weight. The upper limit for reasonable accuracy is at molecular weights of 1–2 million.

In the case of polyelectrolytes, the membrane is not permeable to the polyions because of their size, or to the gegenions because of the need to preserve electroneutrality. As there are many gegenions per polyion, the following approximation of equation (9-6) holds for low concentrations, where $[M_E] = c_2/M_2$:

$$\Pi = RTN_z[M_E] \quad (9\text{-}8)$$

N_z is the effective degree of ionization, that is, it is the fraction of gegenions that contribute to the osmotic pressure. Consequently, N_z is smaller than the total number of gegenions. It is practically constant at high degrees of ionization.

With a polydisperse solute, the molecular weight M_2 that occurs in equation (9-6) is the number-average molecular weight of the solute. In a multicomponent system, the observed osmotic pressure Π is given as the

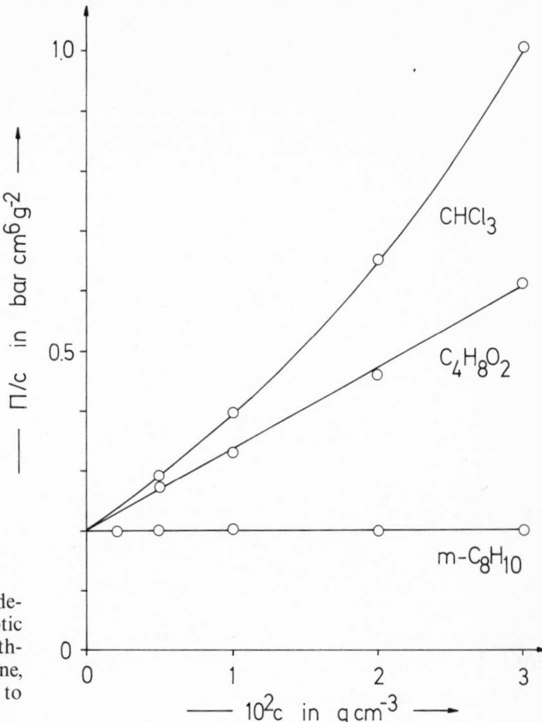

Figure 9-1. The concentration dependence of the reduced osmotic pressure Π/c of a poly(methyl methacrylate) in chloroform, dioxane, and *m*-xylene at 20°C (according to G. V. Schulz and H. Doll).

sum of all the osmotic pressures Π_i,

$$\Pi = \sum_i \Pi_i = RT \sum_i \frac{c_i}{M_i} \tag{9-9}$$

The sum $\sum_i (c_i/M_i)$ in equation (9-9) is also contained in the definition of the number-average molecular weight $\overline{M}_n = \sum_i c_i / \sum_i [c_i/(\overline{M}_i)_n]$ given in equation (8-40). If this expression is inserted into equation (9-9), it is seen that osmotic pressure measurements give the number-average molecular weight:

$$\Pi = \frac{RT \sum_i c_i}{\overline{M}_n} = \frac{RTc}{\overline{M}_n} \tag{9-10}$$

9.2.2. Experimental Methods

In the simplest case, the osmotic pressure Π is measured in a single-cell osmometer with a horizontally arranged membrane (Figure 9-2). Π is then identified as the manometrically measured difference in pressure Δp_e at equilibrium.

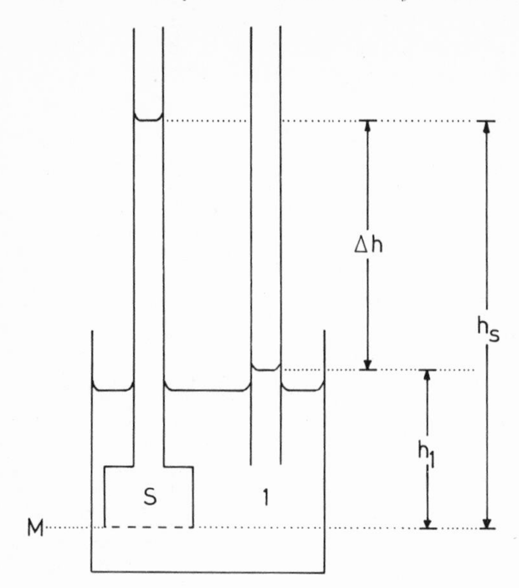

Figure 9-2. Calculation of the osmotic pressure Π from the heights of the solution h_s and of the solvent h_1 above a horizontally arranged membrane.

The osmotic pressure is obtained from h_s and h_1, the heights reached, and the densities ρ_s and ρ_1 of the solution S and solvent 1, with the notation $\Delta h = h_s - h_1$ and $\Delta\rho = \rho_s - \rho_1$, together with the relationship $\Pi = \Delta p_{\mathrm{eq}}$, which applies to semipermeable membranes, as

$$\Pi = \Delta p_{\mathrm{eq}} = h_s\rho_s - h_1\rho_1 = \Delta h\rho_1 - h_s\Delta\rho = \Delta h\rho_s + h_1\Delta\rho \quad (9\text{-}11)$$

To calculate the osmotic pressure Π it is necessary to know the absolute height h_2 (or h_1) and the density difference $\Delta\rho$, as well as the difference in heights reached Δh and the solvent density. When using osmometers with vertically arranged membranes, the center of the membrane, to a good approximation, can be taken as the reference point in measuring the absolute height h_2.

At the beginning of an osmotic experiment, the difference in heights Δh observed after filling both chambers of the osmometer does not correspond to the osmotic pressure at equilibrium. The equilibrium pressure is only observed after solvent molecules permeate the membrane. If Δh is greater than the equilibrium osmotic pressure, the solvent molecules permeate from the solution chamber into the solvent chamber, and in the reverse direction if Δh is smaller than the equilibrium osmotic pressure. The time taken to reach equilibrium increases with the amount of solvent that must be displaced, i.e., increases with the diameter of the capillaries. Since, experimentally, problems such as dirt in the capillaries, etc., limit the size of capillary that one can go down to, and since the membranes must be tight (semipermeability), the establishment of osmotic equilibrium can take days or weeks. Other problems such as poor solvent drainage in the capillaries, adsorption of solute on the membrane, partial permeation of solute through the membrane, etc., can interfere with the attainment of a true osmotic equilibrium. The absence or presence and allowance for these complications must be individually established.

The time required to reach equilibrium is much reduced through the use of novel technology in commercially available automatic membrane osmometers. If, for example, the capillary height in the solution chamber increases because solvent permeates from the solvent chamber, this is immediately compensated by the application, via a servomechanism, of a pressure on the solution chamber, such that the capillary heights above solvent and solution remain the same. Since this method involves the transport of only very small amounts of liquid, equilibrium is reached after only 10–30 min.

The osmotic pressure can, alternatively, be calculated from the rate of attaining equilibrium. The rate of approach to equilibrium is proportional to the displacement from equilibrium:

$$\frac{d(p - \Pi)}{dt} = -k(p - \Pi) \tag{9-12}$$

or, integrated,

$$\ln \frac{p_1 - \Pi}{p_2 - \Pi} = \frac{t_2 - t_1}{t_{0.5}} \ln 2 = \alpha \ln 2 \tag{9-13}$$

The terms p_1 and p_2 are the osmotic pressures at the times t_1 and t_2. Here $t_{0.5}$ is the half-time for solvent passage through the membrane and is determined in a preliminary experiment. The antilogarithmic form of equation (9-13) is solved for Π:

$$\frac{p_1 - \pi}{p_2 - \Pi} = 2^\alpha \tag{9-14}$$

$$\Pi = \frac{2^\alpha p_2 - p_1}{2^\alpha - 1} \tag{9-15}$$

Films of regenerated cellulose, for example, Cellophane 600, Gel cellophane, Ultracellafilter, grades fine and ultrafine, are used as membranes for organic solvent systems. For aqueous solutions, membranes of cellulose acetate (for example ultrafine filter) or nitrocellulose (collodium) are suitable. Corrosive solvents, such as formic acid, etc., require the use of glass membranes.

9.2.3. Nonsemipermeable or Leaky Membranes

According to the necessary theoretical assumptions, the membrane used should be strictly semipermeable, i.e., it should be permeable to solvent molecules and completely impermeable to solute. In the case of native proteins, for example, this requirement is easily fulfilled. Native proteins are predominantly monodisperse and have a compact structure. As long as the pore diameter of the membrane is less than the protein molecule diameter, the membrane is strictly semipermeable. Since protein molecules mostly have a molecular diameter of more than 5 nm, it is not too difficult to find a suitable membrane, e.g., in this case a cellulose acetate-based membrane for aqueous solutions. Coil-forming macromolecules, on the other hand, do have a large coil diameter, but only a very small chain diameter. Thus they can very easily pass through membranes with the relatively small pore size of 1–2 nm. This permeation occurs all the more readily the lower the molecular weight. In polydisperse substances, therefore, some of the substance can permeate. At osmotic equilibrium (with the so-called static techniques) all permeating species will distribute themselves according to their activities in a Donnan equilibrium on both sides of the membrane. The observed osmotic pressure does not correspond to the theoretical osmotic pressure of the original substance at any concentration. For the limiting value of the reduced osmotic pressure Π/c_2 for $c_2 \to 0$, one obtains the molecular weight of the nonpermeable part.

Partial or complete permeation of the solute can frequently be recognized when measurements are made "from below" (capillary height difference Δp_0 at $t = 0$ smaller than Π) since the observable pressure Δp goes through a maximum before reaching the equilibrium value (Figure 9-3). The effect is the result of the simultaneous permeation of the solvent molecules into the solution chamber and of the permeable solute material into the solvent chamber. Since at short observation times virtually no solute can permeate, it is often assumed that, even for permeating solutes, the true osmotic pressure is observed with automatic osmometers because of the short measuring times. This assumption is wrong.

With nonsemipermeable membranes, both the solute and the solvent can pass through the membrane. Let J_v be the volume flow and J_D be the flow due to diffusion (permeation). Both these flows can result from either a hydrostatic pressure difference Δp or an osmotic pressure Π:

$$J_v = L_p \Delta p + L_{pD} \Pi \tag{9-16}$$

$$J_D = L_{Dp} \Delta p + L_D \Pi \tag{9-17}$$

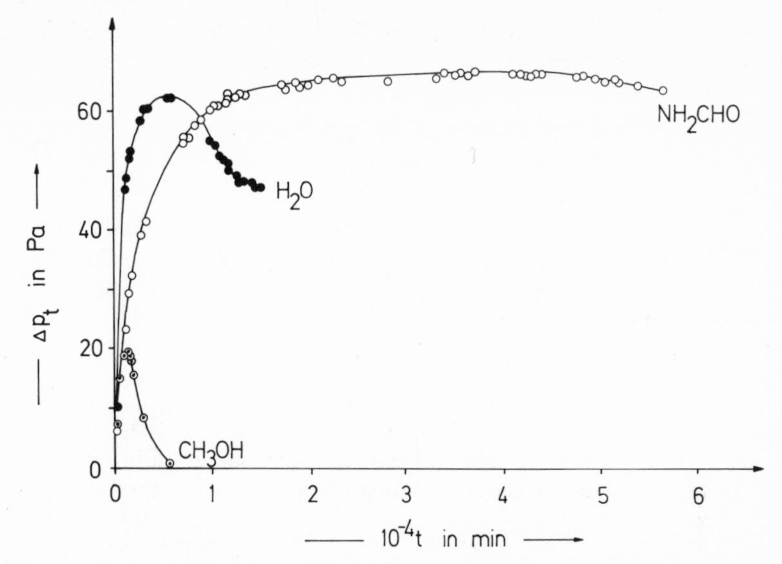

Figure 9-3. The time dependence of the hydrostatic heads Δp_t of solutions of poly(ethylene glycol) ($c = 2 \times 10^{-4}$ g/cm^3; $\overline{M}_n = 4000$, $\overline{M}_w = 4300$ g/mol molecule) in formamide, methanol, or water on cellophane 600 membranes (cellulose hydrate) at 25°C. The theoretically expected osmotic pressure in an ideal solution at this concentration is $\Pi_{id} = 127$ Pa. (According to H.-G. Elias.)

L_p, L_{pD}, L_{Dp}, and L_D are called the phenomenological coefficients. They have the following meaning:

1. With solutions of equal concentration (for example, even pure solvent) on both sides of the membrane, the osmotic pressure is zero. If a hydrostatic pressure Δp is now applied, a volume flow occurs. L_p is the permeability coefficient for equal concentration solutions. Thus, it follows from equation (9-16) for $\Pi = 0$ that

$$L_p = \left(\frac{J_v}{\Delta p}\right)_{\Pi=0} \tag{9-18}$$

2. The volume flow J_v can, however, also be produced by an osmotic pressure Π, i.e., in the case of two solutions at different concentrations with $\Delta p = 0$. We can therefore write

$$(J_v)_{\Delta p=0} = L_{pD}\Pi \tag{9-19}$$

3. If the hydrostatic pressure is $\Delta p = 0$, and if there are two solutions of different concentration, then permeation will take place as a result of

the osmotic pressure between two solutions of different concentrations separated by a nonsemipermeable membrane. It follows from equation (9-17) that

$$L_D = \left(\frac{J_D}{\Pi}\right)_{\Delta p = 0} \tag{9-20}$$

4. When the solutions on either side of the membrane are at the same concentration ($\Pi = 0$), however, then a diffusion flow can occur when a hydrostatic pressure is applied. However, with this "ultrafiltration," solute and solvent diffuse at different rates. We can write

$$(J_D)_{\Pi = 0} = L_{Dp}\Delta p \tag{9-21}$$

According to Onsager's reciprocity principle, for the steady-state condition of very small volume elements and close to equilibrium, we can write for the inverse-indexed phenomenological coefficients

$$L_{pD} = L_{Dp} \tag{9-22}$$

In dynamic osmometry, the pressure difference is determined for the volume flow $J_v = 0$. Equation (9-16) then becomes

$$(\Delta p)_{J_v = 0} = -\frac{L_{pD}}{L_p}\Pi = s\Pi \tag{9-23}$$

$s \equiv -L_{pD}/L_p$ is called the selectivity, or Staverman coefficient. Thus, the true osmotic pressure Π will never be obtained in dynamic osmometry for a volume flow of zero when membranes that are not ideally semipermeable are used. (The same is true even when measurements are made very soon after filling the osmometer.) In normal membrane osmometry, $-L_{pD} \leq L_p$; for semipermeable membranes, however, $-L_{pD} = L_p$. Therefore, the selectivity coefficient can only assume values between 1 and 0. For $s = 1$ the membrane is semipermeable. To date, it has not been possible to calculate s theoretically. According to measurements with the commonly used dried and reswollen Cellophane 600 membranes on samples with narrow molecular-weight distributions, $s = 1$ for $\overline{M}_n > 6000$. For $\overline{M}_n \lessdot 6000$, s decreases to zero for decreasing molecular weight.

In order to determine the molecular weight of a partially permeating solute, the sample is first dialyzed with the same membrane as is to be used for osmometry. The nondialyzing part is then studied by membrane osmometry, and the dialyzing part is studied, for example, by vapor-phase osmometry. The molecular weight of the original sample is then calculated from molecular weights and mass fractions of the two parts according to equation (8-40).

9.3. Ebulliometry and Cryoscopy

The boiling points of a solution and of the pure solvent are different because of the difference in activities. At equilibrium, equation (9-1) changes from the form $d\Delta G = \Delta V\,dp - \Delta S\,dt$ into (since $d\Delta G = 0$)

$$\Delta V\,dp = \Delta S\,dT \tag{9-24}$$

For a reversible isothermal isobaric process, on the other hand, the second law of thermodynamics applies in the form $\Delta S = (\Delta H)_{T,p}/T$. If this equation is inserted into equation (9-24), one obtains, after rearrangement

$$\Delta H = T\,\Delta V\,\frac{dp}{dT} \tag{9-25}$$

At the boiling point the volume of vapor is much greater than the volume of liquid: $\Delta V = V_{vap} - V_{liq} \approx V_{vap}$. On introducing this expression and the ideal gas law $pV_{vap} = RT_{bp}$, where T_{bp} is the boiling point, into equation (9-25), one obtains, with subscript bp for thermodynamic parameters at the boiling point,

$$\Delta H_{bp} = T_{bp}\,\frac{dp}{dt}\,\frac{RT_{bp}}{p} \tag{9-26}$$

or with Raoult's law $x_2 = \Delta p/p_1$ and with $x_2 = n_2/(n_1 + n_2) \approx n_2/n_1 = m_2M_1/m_1M_2 = m_2M_1/M_2\rho_1V_1 \cong c_2M_1/M_2\rho_1$, after rearranging,

$$\frac{\Delta T_{bp}}{c_2} = \left(\frac{RT_{bp}^2M_1}{\rho_1\Delta H_{bp}}\right)\frac{1}{M_2} = E\,\frac{1}{M_2} \qquad (\text{for } c_2 \to 0) \tag{9-27}$$

or, written analogously to equation (9-6),

$$\frac{\Delta T_{bp}}{c_2}\left(\frac{\rho_1\Delta H_{bp}}{T_{bp}M_1}\right) = \frac{RT_{bp}}{M_2} \qquad (\text{for } c_2 \to 0) \tag{9-28}$$

In order to obtain the largest possible boiling-point elevation ΔT_{bp} for a solute of given molecular weight M_2, the ebullioscopic constant E of the solvent must be large. The solvent should have a high boiling point T_{bp}, a large molecular weight M_1, and a low heat of vaporization ΔH_{bp}.

According to the derivation, equation (9-28) only applies to solutions at infinite dilution. For finite concentrations, one can, in analogy to the procedure adopted for membrane osmometry measurements, develop a series with virial coefficients. In polymeric solutes, the number-average molecular weight is measured in ebulliometry. (The proof is analogous to that given for osmotic-pressure measurements.)

An analogous expression can be derived for the lowering of the freezing point ΔT_M in cryoscopic measurements of infinitely dilute solutions:

$$\frac{\Delta T_M}{c_2} = \frac{R T_g^2 M_1}{\rho_1 \Delta H_M} \frac{1}{M_2} \qquad (\text{for } c_2 \to 0) \tag{9-29}$$

where T_g is the freezing point of the solvent and ΔH^m is the molar heat of fusion.

The elevation of the boiling point and the depression of the freezing point are relatively small effects; therefore these methods are too insensitive for measuring molecular weights of above 10,000–20,000. Errors in the determination of the elevation of the boiling point can be caused by superheating and by foaming, and in determinations of the depression of the freezing point by supercooling and by the formation of mixed crystals. Both methods measure the activities of *all* molecules present, i.e., even those of adsorbed water, for example. Since it becomes increasingly difficult to remove solvent inclusions from polymers of increasing molecular weight, and solvent inclusions contribute increasingly, numerically, to the results, the molecular weights determined for polymers of increasing molecular weight are increasingly lower than the true values.

9.4. Vapor-Phase Osmometry

Vapor-pressure osmometric (thermoelectric, vaporometric) measurements depend on the following principle: A drop of a solution with a nonvolatile solute resides on a temperature sensor, i.e., a thermistor. The surrounding region is saturated with solvent vapor. Initially, the drop and vapor are at the same temperature. Since the vapor pressure of the solution is lower than that of the pure solvent, solvent vapor condenses on the solution drop. Because heat of condensation is released, the temperature of the drop rises until the difference in temperature ΔT_{th} between the drop and the solvent vapor again eliminates the difference in vapor pressure, so that the chemical potential of the solvent in both phases is equal. An analogous equation to that which applies in ebulliometry is applicable in this case to the relationship between the temperature difference ΔT_{th} and the number-average molecular weight $\overline{M}_n = M_2$ of the solute

$$\frac{\Delta T_{\text{th}}}{c_2} = \frac{R T^2}{L_1 \rho_s} \frac{1}{\overline{M}_n} \qquad (\text{for } c_2 \to 0) \tag{9-30}$$

where L_1 is the latent heat of vaporization of the solvent per mass and ρ_s is the density of the solvent.

Thus, like ebulliometry or cryoscopy, the method would have a strong thermodynamic basis if heat transfer other than that due to vapor condensation could be prevented. Vapor and drop are, however, in contact with one another, and the temperature thus tends to equilibrate in time by convection, radiation, and conduction. This again causes renewed condensation of solvent vapor, which proceeds until a final steady state with a temperature difference ΔT is reached. Equation (9-30) becomes, with $\Delta T = k_E \Delta T_{\text{th}}$.

$$\frac{\Delta T}{c_2} = k_E \frac{RT^2}{L_1 \rho_s} \frac{1}{\overline{M}_n} = K_E \frac{1}{\overline{M}_n} \qquad (\text{for } c_2 \to 0) \tag{9-31}$$

Since k_E cannot be derived theoretically, K_E is usually determined by calibration with substances of known molecular weight. As with all the other molecular-weight-determination methods, only apparent molecular weights M_{app} are obtained for finite concentrations (see Sections 6.4 and 6.5) because of the effects of the virial coefficients and/or association, M_{app} must therefore be extrapolated to the concentration $c_2 \to 0$. In vapor-phase osmometry, low amounts of nonvolatile impurities interfere with the result, but volatile impurities do not, since they pass into the vapor phase.

9.5. Light Scattering

9.5.1. Basic Principles

Light scattering provides one of the most important methods of determining the molecular weights and dimensions of polymers. This method measures the light scattered at an angle to the incident light. The light scattered from large particles can be seen by the naked eye; it is called the Tyndall effect. Incident light of intensity L_0 on passing through a scattering medium, will have its intensity diminished by the amount I_s of scattered light according to Beer's law:

$$I_0 - I_s = I = I_0 \exp(-\tau r) \tag{9-32}$$

where r is the path length through the medium and τ is the extinction coefficient of the scattered light (radiation). The total intensity ($I_0 = I + I_s$) remains constant. This is a case of conservation of extinction and not of intensity loss as is the case with absorption by colored solutions. In pure liquids and dilute macromolecular solutions, the scattered light intensity I_s is only 1/10,000–1/30,000 of the incident intensity I_0. The intensity of light scattered I_s thus cannot be measured with sufficient accuracy simply by measuring the difference. Therefore I_s is directly measured by a photomultiplier–photocell system, and accuracy is achieved.

In light scattering, it is necessary to differentiate between "large" and "small" particles or molecules. The dimensions of small particles or molecules are much smaller than the wavelength λ of the incident light, i.e., smaller than about 0.05–0.07 λ.

9.5.2. Small Particles

Visible light has an electric vector $\mathbf{E}$ perpendicular to the propagation direction, and this vector varies sinusoidally with time. For the field strengths of the vertical ($\mathbf{E}_v$) and horizontal ($\mathbf{E}_h$) components of plane polarized light at one particular point in space, one can write (see texts on theoretical physics)

$$\mathbf{E} = \mathbf{E}_0 \cos \omega t$$

$$\mathbf{E}_v = \mathbf{E}_{0v} \cos \omega t, \qquad \mathbf{E}_{0v} = \mathbf{E}_0 \cos \phi \tag{9-33}$$

$$\mathbf{E}_h = \mathbf{E}_{0h} \cos \omega t, \qquad \mathbf{E}_{0h} = \mathbf{E}_0 \sin \phi$$

where $\mathbf{E}_0$ is the amplitude of the electric vector, ω is the angular frequency, t is the time, and ϕ is the angle between the vector $\mathbf{E}$ and the vertical.

The electric field interacts with every particle in the path of the light beam, producing a dipole moment $\mathbf{p}$ at the interaction site, since the electrons in the electron shells of atoms composing these particles are displaced in the opposite direction to the electric field. The electric field strength and the induced dipole moment are proportional to each other; the proportionality constant is called the polarizability α:

$$\mathbf{p} = \alpha \mathbf{E} \tag{9-34}$$

Assuming that the particles are small (no intramolecular interference, see Section 9.5.5), that they are independent of each other (ideal gas or infinitely dilute solution), and that there is no loss of light intensity due to absorption, equations (9-33) and (9-34) may be combined:

$$\mathbf{p} = \alpha \mathbf{E}_0 \cos \omega t \tag{9-35}$$

Equation (9-35) states that the induced dipole follows the oscillating electric field with the same frequency. An oscillating dipole also emits electromagnetic radiation, i.e., scattered light. According to equation (9-35), the scattered light has the same wavelength as the incident light. The energy of light is measured by its intensity, that is, the energy absorbed per second on a surface area. According to Poynting's theorem, this energy is proportional to the time-averaged mean square of the amplitude, i.e., it is proportional to $\overline{E^2} \equiv \langle E^2 \rangle$. From equation (9-33), accordingly, for the vertical and horizontal components of plane polarized light, we have

$$I_{0,v} = \text{constant} \times \langle E_v^2 \rangle = \text{constant} \times E_{0v}^2 \langle \cos^2 \omega t \rangle$$
$$I_{0,h} = \text{constant} \qquad (9\text{-}36)$$

The intensity $i_{s,v}$ of the vertically polarized light scattered by a molecule is analogously given by the amplitude $\mathbf{E}_{s,v}$ of the scattered light:

$$i_{s,v} = \text{constant} \times E_{s,v}^2 \qquad (9\text{-}37)$$

and also analogously for the horizontal component of the polarized scattered light. The field $\mathbf{E}_s$ of the vertical or horizontal components of the polarized scattered light can be obtained from the following considerations. The first derivative of the dipole moment with respect to time, $d\mathbf{p}/dt$, corresponds to an electric current that would produce a magnetic field of constant strength. The second derivative $d^2\mathbf{p}/dt^2$ corresponds to an oscillating field, such as that produced by an oscillating dipole. We have

$$\mathbf{E}_s = \text{constant}' \times \frac{d^2\mathbf{p}}{dt^2} \qquad (9\text{-}38)$$

An expression for $d^2\mathbf{p}/dt^2$ is obtained by double differentiation of equation (9-35):

$$\frac{d^2\mathbf{p}}{dt^2} = \alpha \mathbf{E}_0 \omega^2 \cos \omega t \qquad (9\text{-}39)$$

The proportionality constant, constant′, in (9-38) is composed of two factors, $1/r$ and $\sin \vartheta_x$. Here x indicates the horizontally (h) or the vertically (v) polarized component of scattered light.

The factor $1/r$ follows from the law of conservation of energy. The scattered light is distributed in all directions about the oscillating dipole. The total energy flux, i.e., the amount of energy scattered per time, must be constant. The energy flux per area is equal to the intensity; consequently the intensity varies with $1/r^2$. Since the intensity is proportional to the square of the field strength, or amplitude, the amplitude itself must be proportional to $1/r$.

The factor $\sin \vartheta_x$ is obtained by the following reasoning: Although the scattering envelope of the particle is spherical, the amplitude is direction dependent (Figure 9-4). Consider vertically polarized light of intensity $(I_v)_0$ falling in the x direction on the particle. A dipole is induced which oscillates in the z direction. The greatest amplitude of the scattered light is observed perpendicular to the dipole axis, whereas the amplitude is zero along the dipole axis. The amplitude is proportional to ($\sin \vartheta_v$), where ϑ_v is the angle between the dipole axis and the direction of observation. Consequently, in the xy plane, the intensity $i_{s,v}$ of vertically polarized light is independent of the observation angle.

Inserting these expressions and equation (9-39) into equation (9-38),

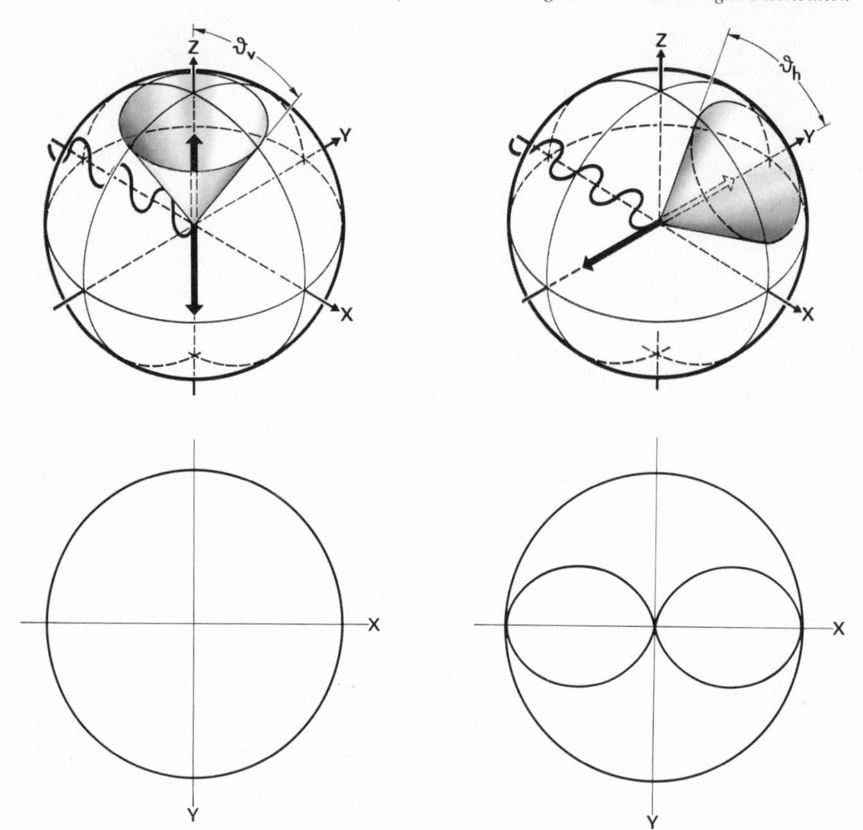

Figure 9-4. The scattering profile of small particles with vertically (v) and horizontally (h) polarized incident light. Upper row: the oscillating dipole position and definition of the angles ϑ_v and ϑ_h. Lower row: polar plot of the light-scattering intensity (indicated by arrows) in the xy plane.

we obtain the following for vertically polarized incident light:

$$\mathbf{E}_{s,v} = (1/r)(\sin \vartheta_v)(\hat{c})^{-2} \alpha \mathbf{E}_{0v} \omega^2 \cos \omega t \tag{9-40}$$

The right-hand side of this equation was also divided by the square of the speed of light $\hat{c}$ to maintain dimensionality. Combining equations (9-36), (9-37), and (9-40), considering that the frequency $\omega/2\tau = \hat{c}\lambda$, and using an average oscillation period, we obtain

$$i_{s,v}/I_{0,v} = 16\pi^4\alpha^2(\sin^2\vartheta_v)\, r^{-2}\lambda^{-4} \tag{9-41}$$

With horizontally polarized incident light, the dipoles oscillate in the y direction. The largest amplitude is again seen perpendicular to the dipole axis. In the y direction, the intensity is zero (Figure 9-4). The amplitude is proportional to $\sin \vartheta_h$, where ϑ_h is the angle between the dipole axis and the direction of observation. Analogously to equation (9-41), the scattered-light intensity of horizontally polarized light is

$$i_{s,h}/I_{0,h} = 16\pi^4\alpha^2(\sin^2\vartheta_h)\, r^{-2}\lambda^{-4} \tag{9-42}$$

or in terms of the total intensity, with $I_{0,v} = I_{0,h} = 0.5I_0$,

$$\frac{i_s}{I_0} = \frac{i_{s,v} + i_{s,h}}{I_0} = \frac{16\pi^4\alpha^2(\sin^2\vartheta_v + \sin^2\vartheta_h)}{2r^2\lambda^4} \tag{9-43}$$

and with $\sin^2\vartheta_v + \sin^2\vartheta_h = 1 + \cos^2\vartheta$, where ϑ is the angle between the observer and the incident light,

$$i_s r^2/I_0 = 16\pi^4\alpha^2\lambda^{-4}[(1 + \cos^2\vartheta)/2] \tag{9-44}$$

The factor $(1 + \cos^2\vartheta)/2$ describes the angular function for the reduced light scattering $i_s r^2/I_0$ for unpolarized light. Part of its contribution comes from the vertical component (1/2) and part from the horizontal component $(\cos^2\vartheta)/2$. The reduced intensity of scattered light for horizontally polarized light will be zero at an observation angle of 90°; consequently only vertically polarized or unpolarized light is used for light-scattering measurements.

The derivations given above assumed small, isotropic particles. With isotropic particles, vertically polarized incident light leads only to vertically polarized scattered light, and horizontally polarized incident light gives only horizontally polarized scattered light. With anisotropic particles (e.g., benzene molecules), however, depolarization of the scattered light occurs. In this case, therefore, vertically polarized incident light gives both vertically and horizontally polarized scattered light. One corrects for this effect by including a correction factor, the so-called Cabannes factor. With macromolecular solutions, the Cabannes factor is usually very close to one.

All the parameters in equation (9-44) except the polarizability α can be measured directly. α is the excess polarizability, that is, the difference between the polarizability of the solute and that of the displaced solvent in dilute solutions. The polarizability of gases is related to the dielectric constant ε via $\varepsilon - 1 = 4\pi\alpha(N/V)$, where N is the number of molecules in the volume V. Correspondingly, the difference in relative permittivity (dielectric constant) for solution and solvent must be considered in dilute solutions.

$$\varepsilon - \varepsilon_1 = 4\pi\alpha(N/V) = \Delta\varepsilon \tag{9-45}$$

From this and the Maxwell relation $\varepsilon = n^2$, with the definition $N/V \equiv cN_L/M_2$, we obtain

$$\alpha = \frac{M_2(n^2 - n_1^2)}{4\pi c N_L} \tag{9-46}$$

The refractive index n of dilute solutions can be expanded as a series in concentration c: $n = n_1 + (dn/dc)\,c + \cdots$, where n_1 is the refractive index of solvent. With $(dn/dc)^2 c^2 \ll 2n_1(dn/dc)\,c$, the square of the refractive index is given as

$$n^2 = n_1^2 + 2n_1 \frac{dn}{dc} c \tag{9-47}$$

On combining equations (9-44)–(9-47) and with $c = (N/V)\,M_2/N_L$, we obtain

$$R_\vartheta = \frac{i_s r^2 (N/V)}{I_0} = \frac{4n_1^2\pi^2(dn/dc)^2\,[(1+\cos^2\vartheta)/2]\,cM_2}{N_L\lambda^4} \tag{9-48}$$

The left-hand side of equation (9-48) corresponds to the reduced scattered light intensity by all N molecules in the volume V, and is called the Rayleigh ratio R_ϑ. Defining the optical constant κ as

$$\kappa \equiv 4\pi^2 n_1^2 \left(\frac{dn}{dc}\right)^2 N_L^{-1}\lambda^{-4}\frac{(1+\cos^2\vartheta)}{2} \tag{9-49}$$

allows equation (9-48) to be written as

$$R_\vartheta = \kappa c M_2 \qquad (\text{for } c \to 0) \tag{9-50}$$

According to the derivation, equation (9-50) is valid for infinitely dilute solutions. It is the basis of molecular-weight determinations by the light-scattering method. The molecular weight M_2 obtained here is the weight-average molecular weight $\overline{M}_w$, as shown in the following derivation:

The Rayleigh ratio for a mixture of i polymer homologous macromolecules of different molecular weights is, with equation (9-50),

$$\overline{R}_\vartheta = \sum_i (R_\vartheta)_i = \sum_i \kappa c_i M_i = \kappa \sum_i c_i M_i \tag{9-51}$$

since the refractive index increment dn/dc is independent of molecular weight for molecular weights above ~20,000, and, according to equation (9-49), κ is independent of both M and c. A comparison of the sum $\sum_i c_i M_i$ with the definitions of average molecular weights [equations (8-40) and (8-41)] shows that the following can be written at high molecular weights $(M_E \ll M_w)$ with $\sum c_i = c$ and $c_i/c = w_i$:

$$\overline{R}_\vartheta = \kappa c \overline{M}_w \tag{9-52}$$

9.5.3. Copolymers

Copolymers show in general both a molecular-weight and a sequence distribution. Since the individual molecules i do not have the same composition, they will also have different refractive index increments $Y_i = (dn/dc)_i$. The summation of equation (9-48) proceeds in this case not as in equation (9-51), but as

$$\bar{R}_\vartheta = \kappa' \sum_i Y_i^2 c_i M_i \tag{9-53}$$

κ' is defined analogously to κ [see equation (9-49)]:

$$\kappa' = 4\pi^2 n_1^2 N_L^{-1} \lambda^{-4} [(1 + \cos^2 \vartheta)/2] \tag{9-54}$$

The summation must proceed both over molecules of the same average composition and different molecular weights, and as molecules of the same molecular weight but different average composition. On conventional analysis of data according to equation (9-50) an apparent molecular weight $(M_w)_{\text{app}}$ would, on account of equation (9-53), be obtained instead of the true molecular weight $\bar{M}_w$, even when $c \to 0$, i.e.,

$$\bar{R}_\vartheta = \kappa c (M_w)_{\text{app}} = \kappa' Y_{\text{cp}}^2 c (M_w)_{\text{app}} \tag{9-55}$$

Here, Y_{cp} is the refractive index increment for the whole polymer. On combining equations (9-53) and (9-55), we obtain the following after introducing the mass contribution $w_i = c_i/c$ of the molecular species i:

$$(M_w)_{\text{app}} = Y_{\text{cp}}^{-2} \sum_i Y_i^2 w_i M_i \tag{9-56}$$

The refractive index increment Y_i of the molecular species i must now be related to the refractive index increments Y_A and Y_B of the unipolymers A and B. The refractive index n_{cp} of a copolymer of the monomeric units A and B depends on the refractive indices n_A and n_B of the unipolymers, as well as on the mass contributions w_A and w_B:

$$n_{\text{cp}} = n_A w_A + n_B w_B, \qquad w_A + w_B \equiv 1 \tag{9-57}$$

For a copolymer molecule of composition i, analogously,

$$n_i = n_A w_{A,i} + n_B w_{B,i} \tag{9-58}$$

The difference between the refractive indices and the refractive index of the solvent n_1 can be used instead of the refractive indices n_i:

$$n_{\text{cp}} - n_1 = (n_A - n_1) w_A + (n_B - n_1) w_B \tag{9-59}$$

or, after dividing both sides by the copolymer concentration c,

$$\frac{n_{\text{cp}} - n_1}{c} = \frac{n_A - n_1}{c} w_A + \frac{n_B - n_1}{c} w_B \tag{9-60}$$

The fractions represent the refractive index increments $Y = dn/dc$, assuming that the refractive indices of the solutions vary linearly with concentration:

$$Y_{\text{cp}} = Y_A w_A + Y_B w_B \tag{9-61}$$

Analogously, for the ith molecular species

$$Y_i = Y_A w_{A,i} + Y_B w_{B,i} \tag{9-62}$$

With $w_B = 1 - w_A$ and $w_{B,i} = 1 - w_{A,i}$, combining equations (9-61) and (9-62), we obtain

$$Y_i - Y_{cp} = (Y_A - Y_B)(w_{A,i} - w_A) = \Delta Y \, \Delta w_{A,i} \tag{9-63}$$

If equation (9-63) is inserted into equation (9-56), one obtains

$$(M_w)_{app} = \sum_i w_i M_i + 2\frac{\Delta Y}{Y_{cp}} \sum_i w_i M_i (\Delta w_{A,i}) + \left(\frac{\Delta Y}{Y_{cp}}\right)^2 \sum_i w_i M_i (\Delta w_{A,i})^2 \tag{9-64}$$

In this equation, the first sum corresponds to the weight-average molecular weight $\overline{M}_w = \sum_i w_i M_i$. The second and third sums contain the first and second moments of the z distribution $v_z^{(1)}$ and $v_z^{(2)}$ of $\Delta w_{A,i}$ (see Section 8.4), since, with $w_i = m_i / \sum m_i$, the following can be written after multiplying numerator and denominator by $\sum_i Z_i$, remembering that $\Delta w_{A,i} = E_i - \overline{E}$ (see Section 2.3.2.2 and 8.4):

$$\begin{aligned} \sum_i w_i M_i (\Delta w_{A,i}) &= \frac{\sum_i m_i M_i (\sum w_{A,i})}{\sum_i Z_i} \frac{\sum_i Z_i}{\sum_i m_i} = v_z^{(1)} \overline{M}_w \\ \sum_i w_i M_i (\Delta w_{A,i})^2 &= v_z^{(2)} \overline{M}_w \end{aligned} \tag{9-65}$$

With this relationship, equation (9-64) becomes

$$(M_w)_{app} = \overline{M}_w \left[1 + 2 v_z^{(1)} \frac{Y_A - Y_B}{Y_{cp}} + v_z^{(2)} \left(\frac{Y_A - Y_B}{Y_{cp}} \right)^2 \right] \tag{9-66}$$

This equation states that light scattering measurements on chemically nonuniform copolymers or polymer mixtures do not give a weight-average, but an apparent weight-average molecular weight $(M_w)_{app}$. The apparent weight-average molecular weight also depends on the refractive index increments Y_A, Y_B, and Y_{cp} of the unipolymers A and B and the copolymer. Since, in a series of solvents, these refractive index increments differ from one another, one can carry out a series of light scattering measurements in solvents with widely different refractive indices. Then, when $(M_w)_{app} = f((Y_A - Y_B)/Y_{cp})$ is plotted, the true weight-average molecular weight $\overline{M}_w$ is obtained as intercept for $(Y_A - Y_B)/Y_{cp} = 0$ (see Figure 9-5). The moments $v_z^{(1)}$ and $v_z^{(2)}$ can be calculated from the shape of the curve. In the case of constitutionally uniform copolymers, such as are obtained in an azeotropic copolymerization (see Chapter 22), the first and second moments $v_z^{(1)}$ and

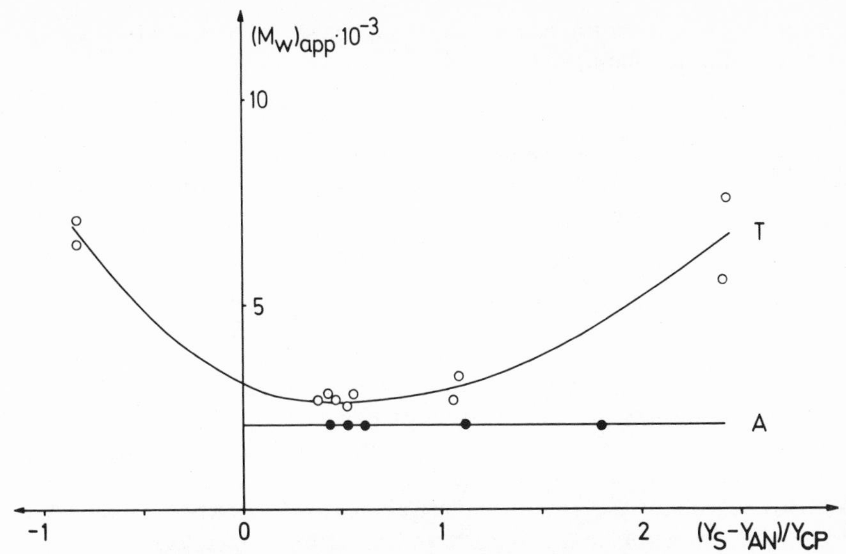

Figure 9-5. The dependence on solvent of the apparent weight-average molecular weight (at $c \to 0$) of one industrially (T) and one azeotropically (A) produced copolymer of styrene and acrylonitrile with refractive index increments of the polystyrene (Y_S), the polyacrylonitrile (Y_{AN}), and the copolymer (Y_{CP}) (according to H. Benoit). The $(\overline{M}_w)_{app}$ used here is extrapolated to $c \to 0$, and so is not the $(M_w)_{app}$ of Section 6.5 or the M_{app} of Section 9.5.4.

$v_z^{(2)}$ will be zero since $\Delta w_{A,i} = w_{A,i} - w_A = 0$. Copolymers of constitutional uniformity thus give molecular weights that do not depend on the refractive index of the solvent when light scattering measurements are carried out in various solvents (see Figure 9-5).

9.5.4. Concentration Dependence

The derivations given in Section 9.5.2 refer to small, isotropic, randomly distributed molecules that move independently of one another, e.g., in a vacuum. The total intensity of scattered light is here given as the sum of the intensities scattered by the individual molecules. In liquids, the Brownian motions of the molecules are not independent of each other. Because of intermolecular interference, the measured total intensity of scattered light is less than the sum of the individual intensities.

In pure liquids, Brownian (i.e., thermal) motion of the molecules leads to fluctuations in time and place of the density of the liquid. In solutions, there is also a fluctuation in solute concentration. It can be assumed that fluctuations in solvent density and solute concentration are independent of each other. In this case, the intensity of scattered light i_s by the solute is given simply by subtracting the intensity of scattered light by the pure solvent i_{solv} from that by the solution i_{soln},

$$i_s = i_{\text{soln}} - i_{\text{solv}} \tag{9-67}$$

Equation (9-44) contains the light scattering intensity emitted by one molecule. The Rayleigh ratio R_ϑ for a system consisting of N scattering molecules in a volume V is defined as $R_\vartheta = i_s r^2 (N/V)/I_0$ [see also equation (9-48)]. Correspondingly, equation (9-44) can be written as

$$R_\vartheta = \frac{i_s r^2 (N/V)}{I_0} = \frac{16\pi^4 \alpha^2 (N/V)}{\lambda^4} \left(\frac{1 + \cos^2\vartheta}{2} \right) \tag{9-68}$$

The total volume is then divided into q volume elements. Each volume element should possess dimensions smaller than the wavelength of light, but still be large enough to contain several scattering molecules. Every volume element should possess a polarizability of $\alpha^\S$, which varies by a certain amount $\Delta\alpha$ about the average polarizability $\bar{\alpha}$ of the whole system. For the square of the polarizability of a volume element, therefore,

$$(\alpha^\S)^2 = (\bar{\alpha} + \Delta\alpha)^2 = (\bar{\alpha})^2 + 2\Delta\alpha(\bar{\alpha}) + (\Delta\alpha)^2 \tag{9-69}$$

The average polarizability $\bar{\alpha}$ is the same for all volume elements and does not, therefore, contribute to the polarizability arising from random fluctuations. The average fluctuation $\Delta\alpha$ is zero; consequently, the contribution to the light scattering of the total system is only the mean square of the fluctuation, i.e., $(\Delta\alpha)^2$. The q volume elements make a contribution q times as large, so that equation (9-68) becomes

$$R_\vartheta = 16\pi^4 \overline{(\Delta\alpha)^2}\, q\lambda^{-4} [(1 + \cos^2\vartheta)/2] \tag{9-70}$$

According to equation (9-45), the polarizability is related to the optical relative permittivity. So, using the considered number of volume elements q instead of the concentration N/V, we find $(\Delta\varepsilon)^2 = (4\pi q)^2 (\Delta\alpha)^2$, and equation (9-70) becomes

$$R_\vartheta = \pi^2 \overline{(\Delta\varepsilon)^2}\, q^{-1} \lambda^{-4} [(1 + \cos^2\vartheta)/2] \tag{9-71}$$

The mean square of the fluctuation in relative permittivities can be expressed in terms of the corresponding concentration fluctuations:

$$\overline{(\Delta\varepsilon)^2} = \left(\frac{\partial\varepsilon}{\partial c} \right)^2 \overline{(\Delta c)^2} \tag{9-72}$$

The mean square of the fluctuation in concentrations is given by the probability p that the individual squares occur:

$$\overline{(\Delta c)^2} \equiv \frac{\int_0^\infty p (\Delta c)^2 \, d(\Delta c)}{\int_0^\infty p \, d(\Delta c)} \tag{9-73}$$

The probabilities p are obtained from the concentration dependence of the fluctuation in the Gibbs energy ΔG. For fluctuations that are not too large, ΔG can be developed in a Taylor series that is terminated after the second term:

$$\Delta G = \left(\frac{\partial G}{\partial c}\right)_{p,T}(\Delta c) + \frac{1}{2!}\left(\frac{\partial^2 G}{\partial c^2}\right)_{p,T}(\Delta c)^2 + \cdots \tag{9-74}$$

The fluctuations occur at constant temperature and pressure about the equilibrium concentration. It follows, therefore, that $\partial G/\partial c = 0$. The probability p of finding a given value of Δc is thus given from equation (9-74) as

$$p = \exp\left(\frac{-\Delta G}{kT}\right) = \exp\left(\frac{-(\partial^2 G/\partial c^2)(\Delta c)^2}{2kT}\right) \tag{9-75}$$

One obtains the following after inserting equation (9-75) in (9-73) and replacing sums by integrals:

$$\begin{aligned}\overline{(\Delta c)^2} &= \frac{\int_0^\infty \{\exp[-(\partial^2 G/\partial c^2)(\Delta c)^2/2kT]\}(\Delta c)^2\, d(\Delta c)}{\int_0^\infty \exp[-(\partial^2 G/\partial c^2)(\Delta c)^2/2kT]\, d(\Delta c)} \\ &= \frac{\int_0^\infty x^2 \exp(-ax^2)\, dx}{\int_0^\infty \exp(-ax^2)\, dx} = \frac{A}{B}\end{aligned} \tag{9-76}$$

with $x = \Delta c$ and $a = (\partial^2 G/\partial c^2)/2kT$. The solution to both integrals is known: $A = (1/4a)(\pi/a)^{0.5}$ and $B = (1/2)(\pi/a)^{0.5}$. So, equation (9-76) becomes

$$\overline{(\Delta c)^2} = \frac{kT}{(\partial^2 G/\partial c^2)_{p,T}} \tag{9-77}$$

Further, the following holds:

$$\left(\frac{\partial^2 G}{\partial c^2}\right)_{p,T} = \frac{-\partial \mu_1/\partial c}{V_1^m c q} \tag{9-78}$$

Combination of equations (9-71), (9-72), (9-77), and (9-78) leads to

$$R_\vartheta = \frac{\pi^2 kT V_1^m c\,(\partial \varepsilon/\partial c)^2}{\lambda^4(-\partial \mu_1/\partial c)}\left(\frac{1 + \cos^2\vartheta}{2}\right) \tag{9-79}$$

With Maxwell's relationship $\varepsilon = n^2$, it is possible to replace $\partial\varepsilon/\partial c$ by its equivalent $\partial n^2/\partial c$. From equation (9-47) this then gives $\partial\varepsilon/\partial c = 2n_1(dn/dc)$.

The change in chemical potential with concentration is obtained from equations (6-50) and (6-51) with $c = c_2$:

$$\frac{-\partial \mu_1}{\partial c} = RT\tilde{V}_1^m(A_1 + 2A_2 c + 3A_3 c^2 + \cdots) \tag{9-80}$$

The following is obtained from equations (9-79), (9-80), and (6-50) with $\tilde{V}_1^m \approx V_1^m$ for the angle $\vartheta = 0$

$$\frac{4\pi^2 n_1^2 (\partial n/\partial c)^2}{N_L \lambda^4} \frac{c}{R_0} = \frac{\kappa c}{R_0} = \frac{1}{M_2} + 2A_2 c + 3A_3 c^2 + \cdots \qquad (9\text{-}81)$$

With solutions of nonassociating solutes, the apparent molecular weight $M_{app} \equiv R_0/\kappa c$, according to equation (9-81), decreases regularly with increase in concentration c. In the case of polydisperse solutes, M_2 is a weight average [see equation (9-52)]. The virial coefficients A_2 and A_3 in equation (9-81) are complex average values, and are only identical with the virial coefficients determined using number-average methods when the solute is monodisperse. In solutions of associating solutes, the right-hand side of equation (9-81) becomes a complicated expression (see Section 6.5).

9.5.5. Large Particles

All the derivations presented so far related to molecules with small dimensions in comparison to the wavelength λ of incident light. If the dimensions are greater than $\sim(0.1\text{–}0.05)\,\lambda$, then the molecule behaves as if it has many scattering centers. The ratio of phases emitted by these centers is fixed because the light is coherent. Interference can therefore occur between the light waves emitted by the different scattering centers. A schematic representation of this effect is shown in Figure 9-6. The light waves scattered by the centers A and B with the same angle ϑ at any given instance lead to a path length difference Δ which depends on the cosine of the scattering angle, ϑ:

$$\Delta = \overrightarrow{DB} = \overrightarrow{AB} - \overrightarrow{AD} = \overrightarrow{AB}(1 - \cos\vartheta) \qquad (9\text{-}82)$$

The path length difference is thus zero for $\vartheta = 0$ and increases with increasing ϑ (Figure 9-7). The ratio z of the scattering intensity at two dif-

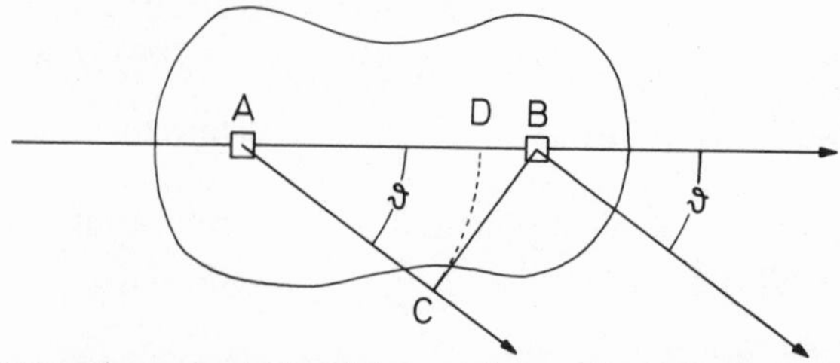

Figure 9-6. Diagrammatic representation of the out-of-phase displacement of the light scattered from two scattering centers A and B of a large particle.

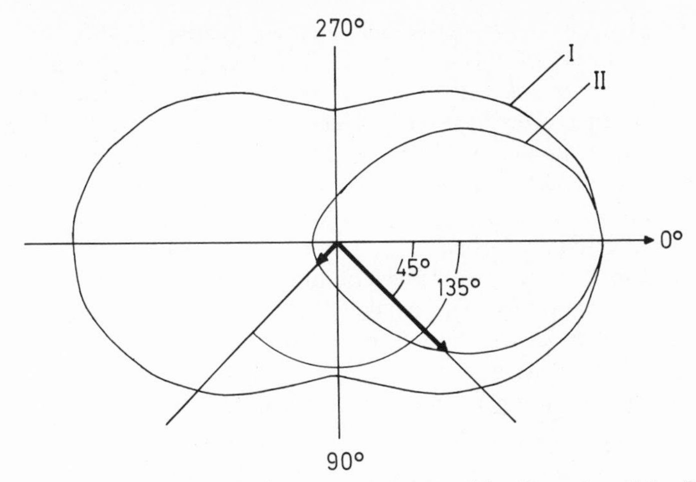

Figure 9-7. A plot of the scattering of unpolarized incident light; (I) small particles, (II) dilute solution of monodisperse spheres with diameter $\lambda/2$.

ferent observation angles ϑ and $180 - \vartheta$ is thus a measure of the interference that occurs. z is called the dissymmetry, and, experimentally, it is usually measured at the angles 45° and 135°. In this case, $z = R_{45}/R_{135}$. The dissymmetry is a measure of the size of the particles, but it also depends on their shape and molecular-weight distribution, and so a quantitative evaluation of the effect of interference requires additional information or assumptions (Figure 9-8). According to Figure 9-8, however, the influence of molecular-weight distribution can usually be ignored for coil-shaped macromolecules of not too broad molecular-weight distribution.

The scattering function $P(\vartheta)$ is more useful. $P(\vartheta)$ is defined as the angular dependence of the scattering intensity of large particles relative to small particles. One can also write $P(\vartheta) = R_\vartheta/R_0$. With equation (9-82), according to the definition, $P(\vartheta) = 1$ for $\vartheta = 0$. Thus, the equations derived in Sections 9.5.2–9.5.4 are also applicable to large particles when $\vartheta = 0$. The weight-average molecular weight $\overline{M}_w$ for large particles can also be obtained from equation (9-52) when the light-scattering intensity at $\vartheta = 0$ is extrapolated to zero concentration.

Experimentally, the intensities of scattered light or the Rayleigh ratios R_ϑ are measured at different angles ϑ and then extrapolated to zero angle. The derivation of the correct mathematical expression for the variation of light-scattering intensity with observation angle ϑ for any desired particle is complicated and not given here. The result for unpolarized incident light is

$$P(\vartheta) = 1 - \frac{1}{3}\left(\frac{4\pi}{\lambda'}\right)^2 \langle R_G^2 \rangle \sin^2\frac{\vartheta}{2} + \cdots \qquad (9\text{-}83)$$

Here $\lambda' = \lambda/n$ is the wavelength of light in the scattering medium. According to equation (9-83), the mean square radius of gyration $\langle R_G^2 \rangle$ is obtained from $P(\vartheta)$ measurements at small observation angles. Increasingly lower observation angles must be used for increasingly larger particles. Of course, a value of $\langle R_G^2 \rangle$ alone contributes nothing to our knowledge of the shape of the particle. However, since the dissymmetry is affected by both the size and the shape of the particle (Figure 9-8), a comparison of $P(\vartheta)$ [or $\langle R_G^2 \rangle$, which can be calculated from $P(\vartheta)$] with z leads to elucidation of the shape of the particle. If the molecular weight and the specific volume are known, the radius of gyration for rigid particles can be calculated (see Section 4.5), and deductions about the shape can be made from a comparison of calculated and observed results.

With the scattering function $P(\vartheta)$, the concentration dependence of

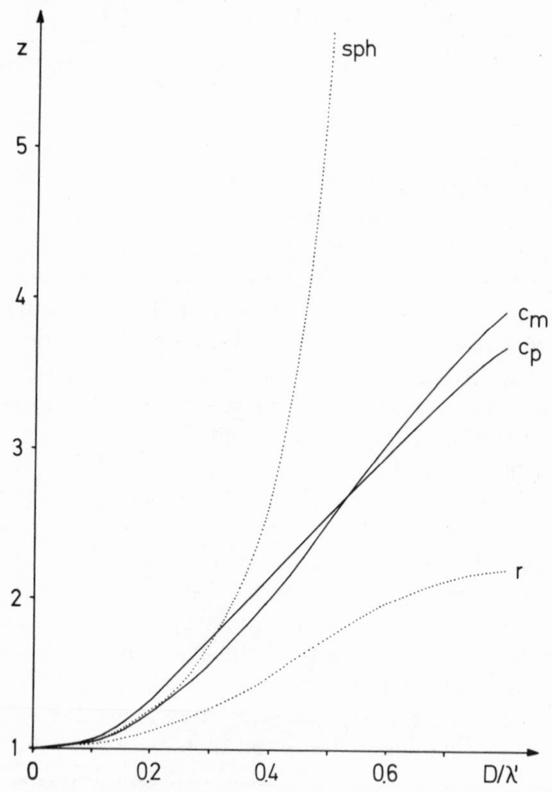

Figure 9-8. Dependence of the dissymmetry coefficient z of scattered light at angles of 45° and 135° on the ratio D/λ' for spheres (sph), monodisperse random coils (c_m), polydisperse ($\overline{M}_w/\overline{M}_n = 2$) random coils ($c_p$), and rods (r). Here λ' is the wavelength of light in the medium of refractive index n, and D corresponds to the diameter of the spheres, the length of the rods, and the chain end-to-end distance $\langle L^2 \rangle^{0.5}$ of coiled macromolecules.

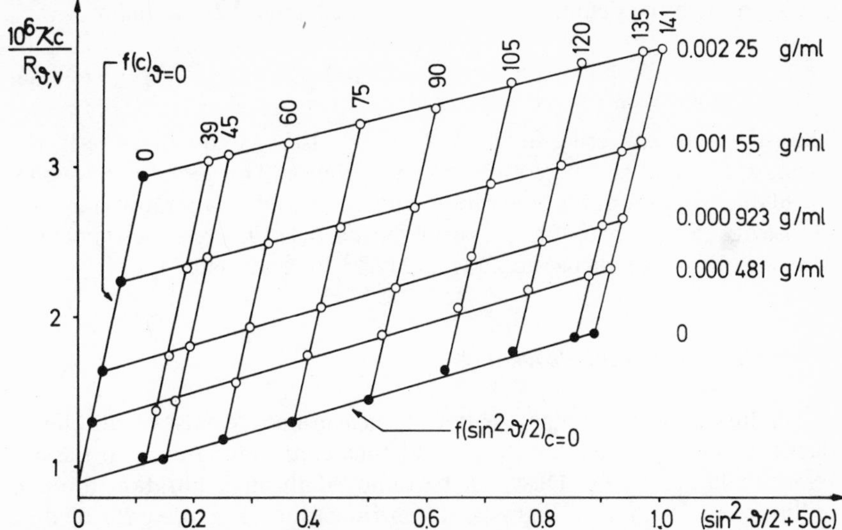

Figure 9-9. The Zimm plot of a poly(vinyl acetate) in butanone at 25°C (1 ml = 1 cm³).

the reduced intensities of scattered light is given as

$$\frac{\kappa c}{R_\vartheta} = \frac{1}{\overline{M}_w P(\vartheta)} + \frac{2A_2}{Q(\vartheta)} c + \cdots \tag{9-84}$$

Here, $Q(\vartheta)$ is another scattering function applicable to finite concentrations c. The second virial coefficient A_2 can thus be found from the dependence on concentration c of the $\kappa c/R_\vartheta$ values at zero angle, and the light scattering function $P(\vartheta)$ (and from it the radius of gyration) can be obtained from the angular dependence of $\kappa c/R_\vartheta$ at zero concentration. Both extrapolations, at $c \to 0$, or $\vartheta \to 0$, yield the weight-average molecular weight.

According to Zimm, both extrapolations can be carried out in the same plot. In a Zimm plot, $\kappa c/R_\vartheta$ is plotted against $\sin^2(\vartheta/2) + kc$, where k is an arbitrarily chosen constant whose sole purpose is to give a good spread to the grid-shaped plot (see Figure 9-9). Quite often, Zimm plots do not possess the simple grid shape shown in Figure 9-9. In particular, linearity of $\kappa c/R_\vartheta = f(\sin^2 \vartheta/2)$ for $c = 0$ is only to be expected for random coils having a most probable Schulz–Flory distribution ($\overline{M}_w/\overline{M}_n = 2$).

With a polydisperse sample, $P(\vartheta)$ and, therefore, $\overline{R_G^2}$ are mean values. According to equation (9-84), the mean value of the light scattering function $\overline{P}(\vartheta)$ is $\overline{P}(\vartheta) = R_\vartheta[(\kappa c\overline{M}_w)^{-1}]$. If the corresponding expression for the i species is inserted and summed, then we obtain

$$\overline{P}(\vartheta) = \frac{R_\vartheta}{\kappa c \overline{M}_w} = \frac{\sum_i \kappa c_i M_i P_i(\vartheta)}{\sum_i \kappa c_i M_i} = \frac{\sum_i c_i M_i P_i(\vartheta)}{\sum_i c_i M_i} = \frac{\sum_i m_i M_i P_i(\vartheta)}{\sum_i m_i M_i} \tag{9-85}$$

and then, with the definition $z_i \equiv m_i M_i$ (see Chapter 8.2), we find

$$\bar{P}(\vartheta) = \frac{\sum_i z_i P_i(\vartheta)}{\sum_i z_i} \equiv \bar{P}_z(\vartheta) \tag{9-86}$$

Thus, the light scattering function $\bar{P}(\vartheta)$ and also, via equation (9-83), the mean square radius of gyration are z averages. The molecular weights calculable from the radii of gyration with the aid of a calibration function represent, according to the shape of the particle, different averages: i.e., $\overline{M}_z$ for coils in the theta state, $(\overline{M}_{z+1}\overline{M}_z)^{0.5}$ for rods, etc.

9.5.6. *Experimental Procedure*

Solutions used for light scattering measurements must be absolutely dust-free. Dust particles are large, and thus contribute greatly to the observed light scattering. Dust can be removed through filtration through millipore filters and/or high-speed centrifugation. The presence of dust is usually recognized from the marked deviations observed in the Zimm plot at angles below ~ 45–$60°$.

Commercially available light scattering photometers use mercury lamp light of a definite wavelength, which is selected by the use of color filters. Light rays are made parallel by passing them through lens systems (Figure 9-10). Recently, laser beams have been used. The beam is then incident on the solution-containing cell, which has parallel exit and entry "windows." The intensity is measured with a photomultiplier–photocell arrangement. As can be seen from Figure 9-10, the incident rays "sees" scattering volumes of different size at varying observation angles ϑ. Correction of the observed intensity for this effect is accomplished by normalization of the "seen" volume by multiplying by the sine of the observation angle. For precise measurements, other corrections, which depend on

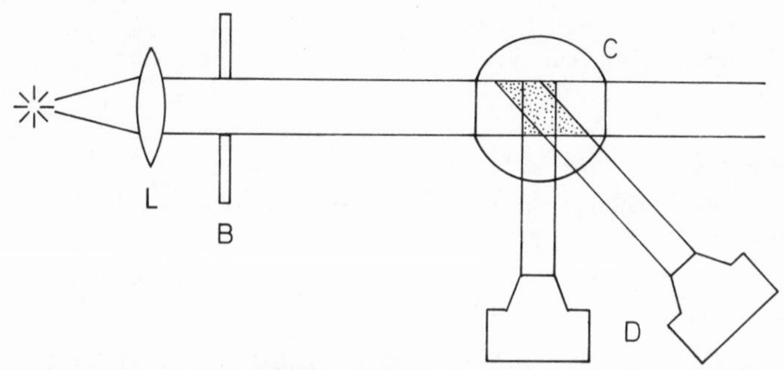

Figure 9-10. Diagram of a light-scattering photometer with light source; lens, L; collimator, B; measuring cell, C; and detector, D (photocell with secondary photomultiplier).

instrument design, must be made, e.g., corrections for shape of cell and collimator, for stray light (scattered from the walls), for multiple scattering, etc. Corrections may also be necessary for depolarization and fluorescence. In order to obtain the scattering intensity of the solute, the scattering intensity of the solvent is subtracted from the corrected light scattering intensity of the solution. Here, it is assumed that density and concentration fluctuations occur independently.

Scattering intensities of sufficient magnitude are only observed when the absolute values of the refractive index increments $Y = dn/dc$ are more than $\sim 0.05 \text{ cm}^3/\text{g}$. To a rough approximation, refractive index increments increase linearly with the refractive index n_1 of the solvent, the slope being given by the partial specific volume $\tilde{v}_2$ of the solute according to the Gladstone–Dale rule

$$\frac{dn}{dc} = \tilde{v}_2 n_2 - \tilde{v}_2 n_1 \tag{9-87}$$

In general, refractive index increments of polymer solutions rarely exceed $0.2 \text{ cm}^3/\text{g}$. For solutions where $c = 0.01 \text{ g/cm}^3$, therefore, even in the most favorable circumstances, the difference in the refractive index for solution and solvent is only 0.002 unit. To determine molecular weights to $\pm 2\%$, it is necessary to know the refractive index increments to within $\pm 1\%$, since they appear squared in equation (9-81). The difference in refractive indices must therefore be known to better than $\pm 2 \times 10^{-5}$. Because of temperature fluctuations during individual measurements, the refractive indices of solution and solvent are not measured separately; instead the difference is measured directly in special differential refractometers.

9.6. *Small-Angle X-Ray and Neutron Scattering*

The theory of light scattering applies to all wavelengths. Consequently, it is also valid for small-angle X-ray scattering (SAXS) and neutron scattering (SANS). The form of equation (9-81) remains the same in each case. Only the expression for the optical constant κ requires alteration. Whereas light scattering is concerned with the different polarizabilities of the molecules, X-ray scattering and neutron diffraction deal with differing electron densities and differing collision cross sections of atoms, respectively. The term κ is given for the various methods as

$$\kappa_{\text{LS}} = \frac{4\pi^2 n_1^2 (dn/dc)^2}{N_{\text{L}} \lambda_0^4} \tag{9-88}$$

$$\kappa_{\text{SAXS}} = \frac{e^4 (\Delta N_e)^2}{m_e^2 \hat{c}^4 N_{\text{L}}} \tag{9-89}$$

$$\kappa_{\text{SANS}} = \frac{N_L N_p^2 (b_H - b_D)^2}{M_u^2} \tag{9-90}$$

Here, e and m_e are the charge and mass of the electron; ΔN_e is the difference between the number of electrons in 1 g of polymer and the number of electrons in the same volume of solvent; $\hat{c}$ is the velocity of light; N_p is the number of exchanged protons per monomeric unit of formula molecular weight M_u, and b_H and b_D are the coherent collision or scattering amplitudes of the hydrogen and the deuterium atoms, respectively. It is assumed in equation (9-90) that the experiments are carried out with hydrogen-containing polymers in their deuterium-containing analogs.

With light scattering, the wavelength of the incident light is greater than the molecular dimensions. For X-ray scattering, the wavelength is smaller. According to equation (9-83), the light-scattering intensity of a given particle at the angle ϑ is reduced by the factor $[\sin^2(\vartheta/2)]/(\lambda')^2$. Equation (9-83) is also valid for X-ray scattering. Consequently, the light scattering intensity reduction for an incident wavelength of $\lambda = 436$ nm and a solvent with $n_1 = 1.45$ (i.e., $\lambda' = 436/1.45 = 300$ nm) at an angle of 90° would be the same as that observed with X-ray scattering ($\lambda = 0.154$ nm) at $\vartheta = 0.03°$.

According to Guinier, the scattering function for small-angle X-ray scattering measurements can be approximated by

$$P(s) = \exp\left(-\frac{4\pi^2}{3\lambda^2} \langle R_e^2 \rangle \vartheta^2 \right) \tag{9-91}$$

$\langle R_e^2 \rangle$ is the mean square radius of gyration of the distribution of electrons and not of the weight. Like the scattering function $P(\vartheta)$, $P(s)$ is also normalized to a value of unity at the angle 0°. The scattering function is Gaussian in form. Experimentally, it is often only applicable at $\vartheta \to 0$, with large deviations occurring at greater angles. In contrast to light scattering, large associates do not influence the accuracy of small-angle X-ray scattering, since they only cause extremely small-angle scattering. Small-angle X-ray scattering can be used to determine the radius of gyration of molecules down to a molecular weight of about 300.

9.7. Ultracentrifugation

9.7.1. Phenomena and Methods

Dissolved particles of density ρ_2 travel through a solvent of density ρ_1 under the influence of a centrifugal field. They sediment in the direction of the centrifugal field when $\rho_2 > \rho_1$, and move to the center of rotation when $\rho_2 < \rho_1$. Under otherwise constant conditions, the rate of sedimentation (or flotation) depends on the mass and shape of the particles as well as on the solution viscosity. Therefore all of these quantities can, in theory, be determined from the rate of sedimentation.

Sedimentation works against diffusion caused by Brownian motion. With a sufficiently weak centrifugal field (relative to particle and density differences), a stage will be reached where the rate of sedimentation equals the rate of diffusion, and a state of equilibrium sedimentation occurs. For given experimental conditions, the sedimentation equilibrium depends on the mass of the dissolved molecule and is therefore a method for determining molecular weights.

Sedimentation rate and sedimentation equilibrium experiments are carried out on an instrument first developed by Th. Svedberg (see Figure 9-11). Such ultracentrifuges reach speeds of ~70,000 revolutions per minute, which corresponds to a gravitational field of ~350,000g (g is the gravitational field due to the earth). Rotational velocities v_r (in revolutions per minute) can be converted into angular velocities ω (in radians per second) with the relation $\omega = 2\pi v_r/60$. Solutions under study are placed in special cells having quartz or sapphire windows, and these fit into a rotor

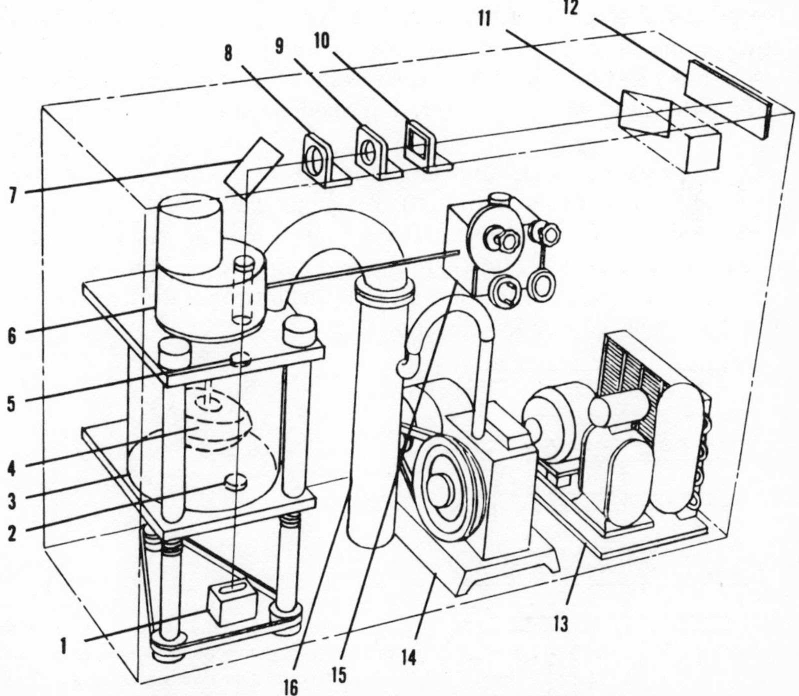

Figure 9-11. Diagram of an analytical ultracentrifuge. The numbers are identified in the text. (Spinco ultracentrifuge made by Beckman Instruments.)

(Figure 9-11; 4) made of duraluminum or titanium. The rotor is driven by an electric motor via a gear system (6). The rotor speed is continuously compared with a reference speed of a synchronized motor with differential gear drive (15). In this way, the rotor speed is kept constant. For protection against accidental damage, the rotor is suspended in a steel chamber (3) by a thin steel axis. In order to keep frictional heating to a minimum, the steel chamber is evacuated to $\sim 10^{-6}$ bar by means of a rotational (14) and an oil-diffusion (16) vacuum pump. With the aid of a cooling system (13) and a heating system (not shown), the temperature of the rotor is regulated at $\sim \pm 0.1$°C. Changes occurring in the cells can be monitored by an optical system whereby light from a source (1) passes through lenses (2, 5, 9, 10) and is then registered directly on photographic plates (12), observed directly after being reflected by a mirror (11) in the required direction. Three optical monitoring systems are in use: interference optics, Schlieren optics, and optical absorption. In interference optics the number or displacement of interference lines is measured. The number of interference lines is proportional to the difference in refractive indices and, thereby, the difference in concentrations. In Schlieren optics, the change in concentration c with distance r is optically differentiated with the aid of a special optical system, so that the concentration gradient dc/dr is observed as a function of distance r. The optical absorption of visible or UV light is measured and registered in absorption optics. In the newer absorption optics systems, the optical absorption is measured directly at each point by a photoelectric cell, thus obviating the need for the tedious intermediate use of a photographic plate.

Experiments are carried out in sector cells in order to avoid convection currents during sedimentation. If, for example, the walls of the cell are not arranged radially to the rotational center, moving particles collide with the wall (Figure 9-12; I), where they are reflected and form a layer of higher concentration close to the wall (II). Such a concentration distribution leads to radial convection currents (III).

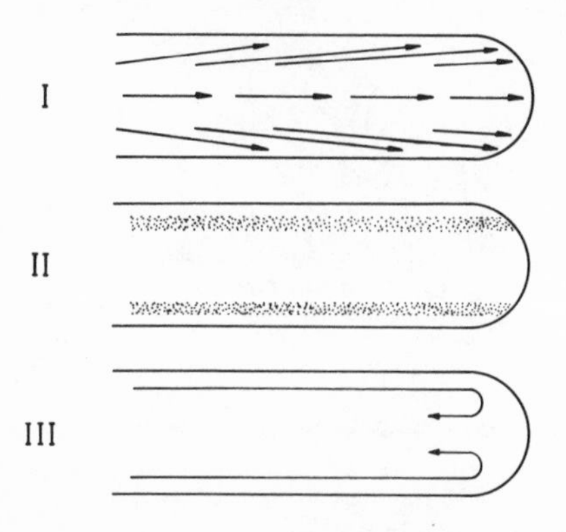

Figure 9-12. Schematic representation of the formation of convection currents in parallel-sided ultracentrifuge cells at the times I, II, and III.

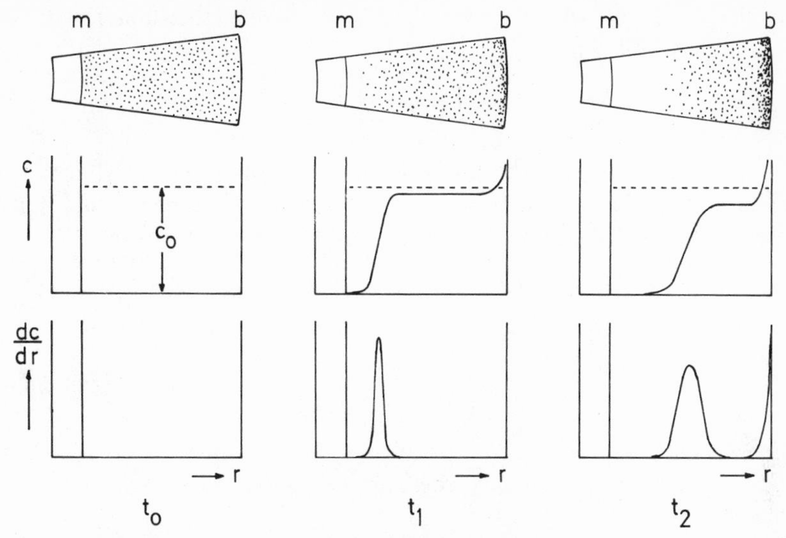

Figure 9-13. Schematic representation of the course of sedimentation in sectored cells at the times t_0, t_1, and t_2.

At time t_0, the cell is filled with a solution that is homogeneous with respect to concentration (Figure 9-13). All the molecules begin to move under the influence of gravity. After the time t, there is a layer of pure solvent at the meniscus (m) and molecules settle at the bottom (b). The boundary layer between solvent and sedimenting solution is not sharp, because of back-diffusion. Thus, a curve $c = f(r)$ is obtained instead of a concentration jump. The sectoring of the cell causes a dilution effect. With increasing duration of the experiment, the concentration in the constant-concentration zone of the cell becomes increasingly smaller. A gradient curve is obtained on differentiation (third row in Figure 9-13). The rate of displacement of the gradient curve is a measure of the sedimentation velocity.

9.7.2. Basic Equations

At every point of the cell a flow of $J = cv$ (i.e., the product of concentration, c and molecular mobility v) occurs during sedimentation. The amount of solute that flows from a volume element A at distance r_A from the rotational center into a volume element B at distance r_B must be the same as the change with time of the concentration of the rest of the solute:

$$(rJ)_A - (rJ)_B = \frac{\partial}{\partial t} \int_{r_A}^{r_B} rc\, dr \tag{9-92}$$

On dividing both sides by $\Delta r = r_B - r_A$ and using the limit for $\Delta r \to 0$, we obtain the following:

$$\left(\frac{\partial c}{\partial t}\right) = -\frac{1}{r}\left[\frac{\partial(rJ)}{\partial r}\right]_t \tag{9-93}$$

Sedimentation causes a flow J_s in the direction of the centrifugal field. The flow J_d due to diffusion attempts to maintain concentrational homogeneity, and works against J_s. For the resulting flow, we have

$$J = J_s + J_d = cv_s + cv_d \tag{9-94}$$

The flow due to diffusion J_d is given by equation (7-13):

$$J_d = -D\frac{\partial c}{\partial r} \tag{9-95}$$

The molecular sedimentation velocity v_s is proportional to the centrifugal field, $\omega^2 r$; the proportionality coefficient is called the sedimentation coefficient s:

$$v_s = s\omega^2 r \tag{9-96}$$

The term s is defined here as the sedimentation velocity in a unit field:

$$s \equiv \frac{1}{\omega^2 r}\frac{dr}{dt} \tag{9-97}$$

A sedimentation coefficient of the value 1×10^{-13} s is called a Svedberg unit (1 S).

Combination of equations (9-93)–(9-96) gives what is called the Lamm differential ultracentrifuge equation:

$$\left(\frac{\partial c}{\partial t}\right)_r = \frac{-\partial[s\omega^2 r^2 c - rD(\partial c/\partial r)]}{r\partial r} \tag{9-98}$$

9.7.3. Sedimentation Velocity

For sedimentation velocity experiments, angular velocities ω are chosen to be so high that the diffusion term $rD(\partial c/\partial r)$ in equation (9-98) is much smaller than the sedimentation term $s\omega^2 r^2 c$. A forced migration of 1 mol of molecules with the velocity dr/dt produces a resistance F_s:

$$F_s = f_s N_L \frac{dr}{dt} \tag{9-99}$$

The proportionality constant f_s is called the frictional coefficient. Additionally, an effective centrifugal force F_r acts on the molecule of hydrodynamic volume V_h. The force F_r is the resultant of the centrifugal field force $m_h\omega^2 r$ and the buoyancy $V_h\rho_1\omega^2 r$ due to the solvent

$$F_r = m_h\omega^2 r - V_h\rho_1\omega^2 r \tag{9-100}$$

If $F_s = F_r$ and equations (7-1) and (7-6) are inserted for m_h and V_h, respectively, then with $\rho_1 = 1/v_1$ and equation (9-99), we have

$$M_2 = \frac{f_s s N_L}{1 - \tilde{v}_2\rho_1} \tag{9-101}$$

Thus, s values alone are not a measure of the molecular weight, since the molecular weight also depends on the frictional coefficient f_s and the buoyancy term $(1 - \tilde{v}_2\rho_1)$ (see Table 9-2). Frictional coefficients are determined by the shape and degree of solvation of the particles.

The frictional coefficient can be eliminated by the following procedure. If frictional coefficients are equal for both sedimentation and diffusion, as is observed experimentally, then the Svedberg equation is obtained from the combination of equation (9-101) with equation (7-21):

$$M_2 = \frac{sRT}{D(1 - \tilde{v}_2\rho_1)} \tag{9-102}$$

Another means of eliminating the frictional coefficient f_s comes from viscosity measurements. According to equation (7-25), the frictional coefficient f_D is related to an asymmetry factor f_A and the Stokes frictional coefficient for spheres. On extending equation (7-25) with $M_2^{0.5}$, we obtain the form

$$f_s = f_D = f_A 6\pi\eta_1 \left(\frac{\langle R_G^2\rangle}{M_2}\right)^{0.5} M_2^{0.5} \tag{9-103}$$

Table 9-2. Sedimentation Coefficient s and Frictional Coefficient f_s (as the ratio f_s/f_{sphere}) of Macromolecules with Molecular Weight M_2

Substance	$10^{-4} M_2$, g/mol	Solvent	Temperature, °C	$10^{13}s$, s	f_s/f_{sphere}
Poly(styrene)	9	Butanone	20	12	1.38
Poly(styrene)	96	Butanone	20	22	3.75
Poly(styrene)	500	Butanone	20	45	5.24
Poly(vinyl alcohol)	6.5	Water	25	1.54	3.5
Cellulose	590	Cuoxam	20	17.5	13.1
Ribonuclease	1.27	Dilute salt solution	20	1.85	1.04
Myoglobin	1.67	Dilute salt solution	20	2.04	1.11
Tobacco mosaic virus	5900	Dilute salt solution	20	17.4	2.9

which is similar to the expression (discussed in Section 9.9.6) for the intrinsic viscosity $[\eta]$:

$$[\eta] = \Phi\left(\frac{\langle R_G^2 \rangle}{M_2}\right)^{3/2} M_2^{0.5} \tag{9-104}$$

The Mandelkern–Flory–Scheraga equation is obtained by combining equations (9-101), (9-103), and (9-104):

$$M_2 = \left[\frac{N_{\mathrm{L}}\eta_1}{\Phi^{1/3}(6\pi f_A)^{-1}(1 - \tilde{v}_2\rho_1)}\right]^{3/2} [\eta]^{1/2}\, s^{3/2} \tag{9-105}$$

The following symbols are very often used in the literature:

$$P = 6\pi f_A \tag{9-106}$$

$$\beta = \Phi^{1/3} P^{-1} \tag{9-107}$$

The factor f_A describes, first, the relationship between the radius of gyration and the radius most suitable in describing the molecule, and, second, all deviations from the Stokes frictional coefficient of an unsolvated sphere. The relationship $\langle R_G^2 \rangle = (3/5)\, r^2$ is valid for spheres. Thus, $f_A = (5/3)^{0.5}$ for unsolvated spheres. Consequently, P takes on a value of 24.34. The Einstein equation allows Φ to be calculated for spheres:

$$[\eta] = \frac{2.5}{\rho_2} = \frac{2.5V_2}{m_2} = \frac{2.5(4\pi r^3/3)}{M_2/N_{\mathrm{L}}}$$

$$= \frac{2.5 \cdot 4\pi N_{\mathrm{L}}(5/3)^{3/2}}{3}\left(\frac{R_G^3}{M_2}\right) = \Phi\frac{R_G^3}{M_2} \tag{9-108}$$

Thus, Φ is given as 13.57×10^{24} (mol macromolecule)$^{-1}$ and is based on the radius of gyration. $[\eta]$ is measured in cm^3/g. If Φ is based on the radius of gyration, then $\Phi = 6.30 \times 10^{22}$ is obtained with $[\eta]$ measured in $100\ \mathrm{cm}^3/\mathrm{g}$. Other numerical values for P and β are obtained with other molecular shapes (see Table 9-3).

Both equations (9-102) and (9-105) yield mixed averages of the molecular weight (see Section 8.5.4).

In these derivations, it was implicitly assumed that s and D are independent of the concentration. Equation (9-102), therefore, only applies at infinite dilution. However, sedimentation and diffusion coefficients are measured at finite concentrations, and so must be extrapolated to $c \to 0$. The extrapolation formula for sedimentation coefficients is obtained from the reasoning that, according to equation (9-101), the s values, and so also the s_c values, are inversely proportional to the frictional coefficients f_s. Since f_s is proportional to the viscosity η, and this in turn is proportional to the concentration, $1/s_c$ must be directly proportional to the concentra-

Table 9-3. Calculated Constants Φ, P, and β [a]

Molecular shape	r_a/r_b	$10^{22}\Phi$, 100/mol	P	$10^6\beta$
Spheres, unsolvated	1	13.57	24.34	2.11
Ellipsoids, oblate	1	—	—	2.12
	2	—	—	2.13
	3–300	—	—	$1.81\,(r_a/r_b)^{0.126}$
Ellipsoids, prolate	1–0.067	—	—	2.12–2.14
	0.05–0.0033	—	—	2.15
Coils, theta state	—	4.21	5.20	2.73
Semirigid chains	—	—	—	2.81

[a] Φ is given relative to the radius of gyration, whereas β is relative to the chain end-to-end distance. r_a is the axial rotation axis radius and r_b is the equatorial radius of the ellipsoid. $[\eta]$ is in 100 cm³/g.

tion. This dependence is usually formulated as

$$\frac{1}{s_c} = \frac{1}{s}(1 + k_s c) \tag{9-109}$$

Since the sedimentation velocity depends on the molecular weight, for paucimolecular substances with two, three, etc., kinds of molecules but different molecular weights, two, three, etc., gradient curves are observed in the sedimentation plot. Sedimentation measurements are therefore used very frequently in protein chemistry to test the homogeneity of materials. In polydisperse substances, on the other hand, the distribution of sedimentation coefficients can be determined from the time-broadening of the gradient curves. For this, gradient curves are obtained at various times for a given initial concentration, and the sedimentation coefficients corresponding to weights of 5, 10, 20, ..., 80, 90, 95% materials are determined. Since the gradient curves can also be broadened by diffusion, however, the sedimentation curves thus obtained are extrapolated to infinite time. At infinite time, only the different sedimentation velocities play a role, and back-diffusion is no longer important. A function $w_i = f(s_i)$ results which is converted into the molecular weight distribution $w_i = f(M_i)$ by means of the empirical relation $s = K_s M^{a_s}$. Such measurements are best made in Θ-solvents; otherwise too many corrections are required for thermodynamic nonideality, etc.

9.7.4. Equilibrium Sedimentation

In equilibrium sedimentation, the concentration at every point in the ultracentrifuge cell no longer varies with time, i.e., $(\partial c/\partial t)_r = 0$. Equation

(9-98) therefore becomes

$$\frac{s}{D} = \frac{\partial c/\partial r}{\omega^2 rc} \tag{9-110}$$

Combining equations (9-110) and (9-102) gives the following for the molecular weight at equilibrium sedimentation:

$$M_2 = \frac{RT}{\omega^2(1 - \tilde{v}_2\rho_1)} \frac{dc/dr}{rc} \tag{9-111}$$

The molecular weight M_2 can be determined from equation (9-111) if the concentrations c and the concentration gradients dc/dr at a distance r from the rotational center are known. With polydisperse substances, this molecular weight M_2 is a weight-average $\overline{M}_w$. That is, after transforming equation (9-111), the mean concentration gradient is given by

$$\overline{\left(\frac{dc}{dr}\right)} = \frac{\omega^2 r(1 - \tilde{v}_2\rho_1)}{RT} \sum_i c_i M_i \tag{9-112}$$

and with the definition of the weight-average molecular weight $\overline{M}_w \equiv \sum_i c_i M_i / \sum_i c_i$ and $\sum_i c_i = c$, we have

$$\overline{M}_w = \frac{RT}{\omega^2(1 - \tilde{v}_2\rho_1)} \frac{\overline{dc/dr}}{rc} \tag{9-113}$$

Equation (9-113) is, strictly speaking, only applicable at infinite dilution. At finite concentrations, it yields an apparent molecular weight $(M_w)_{app}$, which is extrapolated in the usual manner by plotting $1/(M_w)_{app} = f(c)$ to $c \to 0$.

9.7.5. Sedimentation Equilibrium in a Density Gradient

Up to now, it has been assumed that the solvent in the sedimentation experiment consists of a single component. If, however, the solvent system consists of a mixture of two substances of widely different densities (e.g., CsCl in water or mixtures of benzene and CBr_4), the solvent components will sediment to different extents. At equilibrium, the solvent system possesses a density gradient. One density ρ_b, applies at the bottom of the cell, and another, ρ_m, at the meniscus. The density of the solute, ρ_2, should lie between these two densities ($\rho_m < \rho_2 < \rho_b$). The macromolecules will then sediment from the meniscus toward the base of the cell and float from the base toward the meniscus (Figure 9-14). At equilibrium, the macromolecules will take up a position (designated by §) at which the density ρ_g exactly

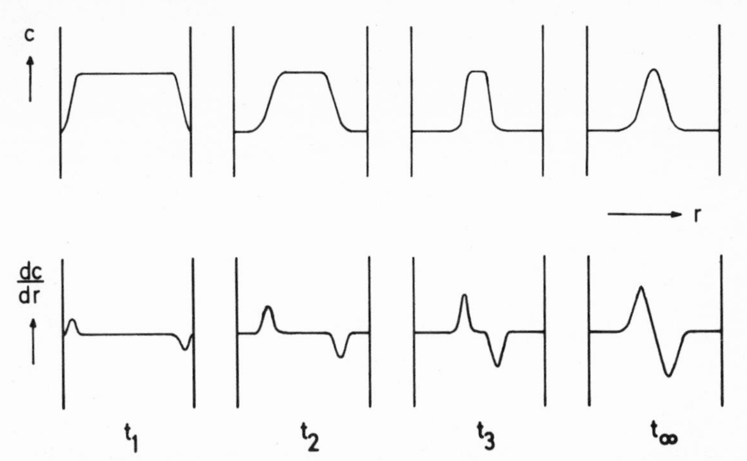

Figure 9-14. Diagrammatic representation of the establishment of a sedimentation equilibrium in a density gradient (c is the concentration of the macromolecular solute).

corresponds to the density of the macromolecule in solution ($\rho_g = \rho_2 \approx 1/v_2^{\S}$). This position is at a distance $r^{\S}$ from the center of rotation.

Quantitative evaluation begins with the equilibrium sedimentation equation (9-111) in the form for $r \approx r^{\S}$:

$$\frac{d \ln c}{dr} = M_2^{\S} \omega^2 r^{\S} \frac{1 - \tilde{v}_2 \rho}{RT} \tag{9-114}$$

where ρ is now the density of the variable-density system at any given point r. Reference is made, not to the distance from the rotational center, but to the distance from point r, so that dr is replaced by $d(r - r^{\S})$. Equation (9-114) then becomes

$$\frac{d \ln c}{d(r - r^{\S})} = M_2^{\S} \omega^2 r^{\S} \frac{1 - (\rho/\rho^{\S})}{RT} \tag{9-115}$$

To a first approximation, the density in the variable-density system varies in the neighborhood of r as

$$\rho = \rho^{\S} + \left(\frac{d\rho}{dr}\right)^{\S} (r - r^{\S}) \tag{9-116}$$

Since the concentrations of the components forming the density gradient are much higher than the macromolecular concentration, the density gradient is practically unaffected by the presence of the macromolecules. Combination of (9-115) and (9-116) leads to

$$\frac{d \ln c}{d(r - r^{\S})} = -\frac{M_2^{\S} \omega^2 r^{\S} (d\rho/dr)^{\S} (r - r^{\S})}{RT\rho^{\S}} \tag{9-117}$$

or, on integration,

$$\ln \frac{c}{c^{\S}} = -\frac{M_2^{\S}\omega^2 r^{\S}(d\rho/dr)^{\S}(r - r^{\S})^2}{2RT\rho^{\S}} \tag{9-118}$$

or, on transforming,

$$c = c^{\S}\exp\left[-\frac{M_2^{\S}\omega^2 r^{\S}(d\rho/dr)^{\S}(r - r^{\S})^2}{2RT\rho^{\S}}\right] = c^{\S}\exp\left[-\frac{(r - r^{\S})^2}{2\sigma^2}\right] \tag{9-119}$$

where we have

$$\sigma^2 = \frac{RT\rho^{\S}}{M_2^{\S}\omega^2 r^{\S}(d\rho/dr)^{\S}} \tag{9-120}$$

Equation (9-119) corresponds to a Gaussian distribution function (see also Chapter 8.3.2.1). The molecular weight can be calculated from the position of the inflection point of the function $c = f(r - r^{\S})$. For proteins in $CsCl/H_2O$, the lower molecular weight limit giving a meaningful measurement is 10,000–50,000 g/mol molecule. The limit is essentially governed by the length of the ultracentrifuge cell ($\sim$1.2 cm) and the optimal values of $r - r^{\S}$ for this length.

The molecular weight given in equation (9-120), however, is not the molecular weight of the unsolvated molecule, which appears in equations (9-111) and (9-102). The measurements are, of course, made in mixed solvents, where one component solvates the molecule more than the other. If only component 1 solvates, then the mass of the solvated macromolecule is composed of the mass of the "dry" macromolecules m_2 and the mass of the solvating solvent $m_1^{\square}$. With the definition $\Gamma_1 = m_1^{\square}/m_2$, it follows from $m_2^{\S} = m_2 + m_1^{\square}$ that $m_2 = m_2^{\S}(1 + \Gamma_1)$, and on recalculating on the basis of molecular weight, with $M_2 = m_2 N_L$, we have

$$M_2^{\S} = M_2(1 + \Gamma_1) \tag{9-121}$$

The parameter Γ_1 can be evaluated from the partial specific volumes of the solute $\tilde{v}_2$ or solvent $\tilde{v}_1$ and the density $\rho^{\S}$ via

$$\Gamma_1 = \frac{\tilde{v}_2\rho^{\S} - 1}{1 - \tilde{v}_1\rho^{\S}} \tag{9-122}$$

To a good approximation, $\tilde{v}_2$ can be determined from density measurements for the macromolecule in one-component solvents by using $\rho = \rho_1 + (1 - \tilde{v}_2\rho_1)c$.

Equilibrium sedimentation studies are mostly used to investigate density differences in different macromolecules. The method has been used, for example, in research on the replication of ^{15}N-labeled deoxyribonucleic acids. Theoretically, it is also suitable for distinguishing between polymer blends and true copolymers. In such studies, problems usually arise from the considerable back-diffusion and the wide molecular-weight distribution. The gradient curves are so strongly influenced by these two effects that there is considerable overlap in the curves for substances of differing densities.

9.7.6. Preparative Ultracentrifugation

Substances of differing molecular weights can be separated in sedimentation experiments. The method is the preferred method in protein and nucleic acid chemistry, since it is especially suitable for the separation of compact molecules. Three procedures can be distinguished: In the normal procedure, the particles sediment in pure solvent or in relatively dilute salt solutions. The density of the solvent or the salt solution is practically constant over the whole cell (Figure 9-15). The more rapidly moving particles tend to collect at the bottom. However, they are always contaminated with the more slowly sedimenting particles. In contrast, slower moving components cannot be isolated quantitatively.

Band ultracentrifugation is also a sedimentation velocity method, but here the particles sediment in a mixed solvent system (e.g., a salt solution). First a density gradient is produced by ultracentrifugation of the solvent

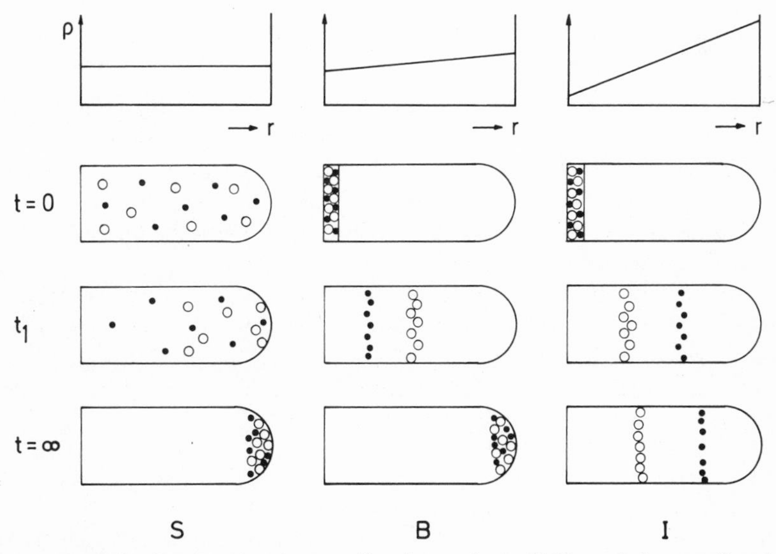

Figure 9-15. Kinds of preparative ultracentrifugation methods. (S) Normal ultracentrifugation, (B) band ultracentrifugation in a stabilized gradient, (I) isopycnic zone centrifugation; ρ is the density of the gradient-forming substance; (○) high molecular weight, low density; (●) low molecular weight, high density.

system alone. Then the solute solution is added at the cell meniscus. The individual components sediment in bands which are no longer contaminated by the other components. The relatively weak density gradient is solely intended to stabilize the moving bands. As soon as the bands have formed, the sedimentation experiment is terminated. The isolated components are of high purity.

Isopycnic zone ultracentrifugation, on the other hand, deals with the preparative analog to equilibrium sedimentation in a density gradient. Here, large density gradients are used, and one waits until equilibrium is attained. Both band and isopycnic zone ultracentrifugation can be carried out with very small solute concentrations. These methods have become, therefore, particularly important in the separation of biological macromolecules (biopolymers). Both methods are also described in the literature under a variety of other names.

9.8. Chromatography

Macromolecules can be fractionated according to their constitution, configuration, or molecular weight by chromatographic methods. Adsorption chromatography is used rarely. Elution chromatography and gel-permeation chromatography are used much more often.

9.8.1. Elution Chromatography

The material to be separated is placed as a thin layer on an inert carrier and then eluted. Metal foil or quartz sand, for example, are suitable inert carriers. The metal foil is dipped into the macromolecular solution and then dried. The surface film is then eluted at constant temperature with solvent–precipitant mixtures of increasing solvent power. Thus, the lower molecular weight fractions are removed first.

An elegant variant of this procedure is known as the Baker–Williams method. In this method, the chromatographic column is also surrounded by a thermostated heating jacket which ensures that an adequate thermal gradient is maintained. The separation efficiency is enhanced by the simultaneous concentration and temperature gradients.

9.8.2. Gel-Permeation Chromatography

In gel-permeation chromatography, the separating column consists of a solvent-swollen gel with pores of various diameters. An $\sim 0.5\%$ solution is added to the column, which is then eluted with a steady stream of solvent. In a homologous series of molecules of similar shape, molecules with the highest molecular weight appear first, i.e., they have the smallest elution volume. The effect is interpreted as follows: Large molecules cannot easily, if at all, penetrate the pores of the gel material, so they have the shortest retention time (Figure 9-16). The method is known in the literature under numerous other names (gel filtration, gel chromatography, exclusion chromatography, molecular-sieve chromatography, etc.) It is an improved form of liquid–liquid chromatography.

Elution takes place under pressures of up to ~10 bar. The gel used must therefore not be compressed under these conditions. Cross-linked poly(styrenes) or porous glass is usually used for organic solvent systems, and cross-linked dextrans, poly(acrylamides), or celluloses for aqueous solutions. The concentration of the eluting solutions is mostly registered automatically as a function of the volume, for example, by refractive index or by spectroscopy (Figure 9-17). The maximum in the $n_{sol} = f(V_e)$ plot is known as the elution volume. An empirical relationship exists between the elution volume V_e and the molecular weight for substances of similar molecular shape and similar interaction with the solvent (Figure 9-18). For both low and high molecular weights, however, elution volumes are independent of the molecular weight. The exclusion limits depend on the gel material as well on the dissolved macromolecules.

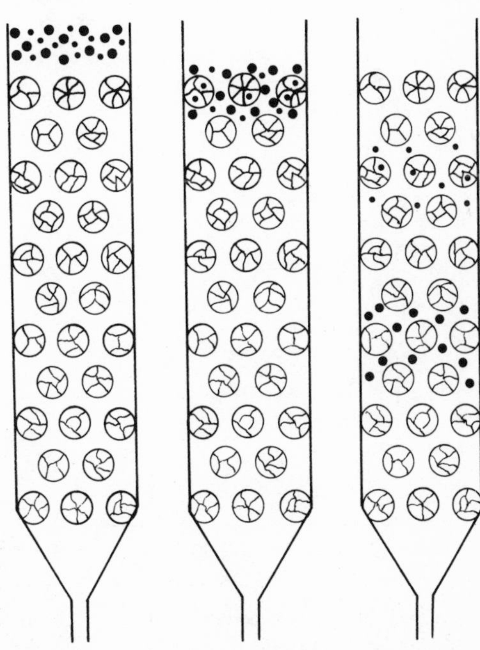

Figure 9-16. Schematic representation of the separation of molecules of different sizes on macroporous gels by gel-permeation chromatography.

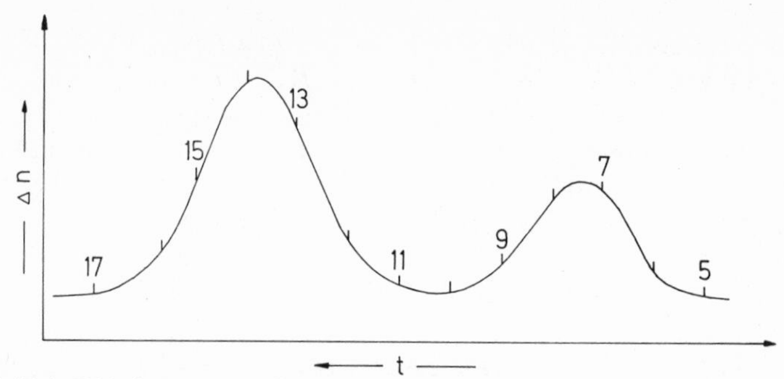

Figure 9-17. A GPC diagram. The numbers give the fraction numbers, which are proportional to the eluted volume. The refractive index difference Δn is generally measured as a function of time.

The elution volumes of a homologous polymer series depend on the molecular weight because the hydrodynamic volumes of macromolecules vary regularly with molecular weight. The hydrodynamic volume is proportional to the product $[\eta]M$. A "universal" calibration curve of $\log([\eta]M) = f(V_e)$ for various linear and branched polymers actually does exist for each gel material (Figure 9-19).

The molecular weight associated with the elution volume is probably

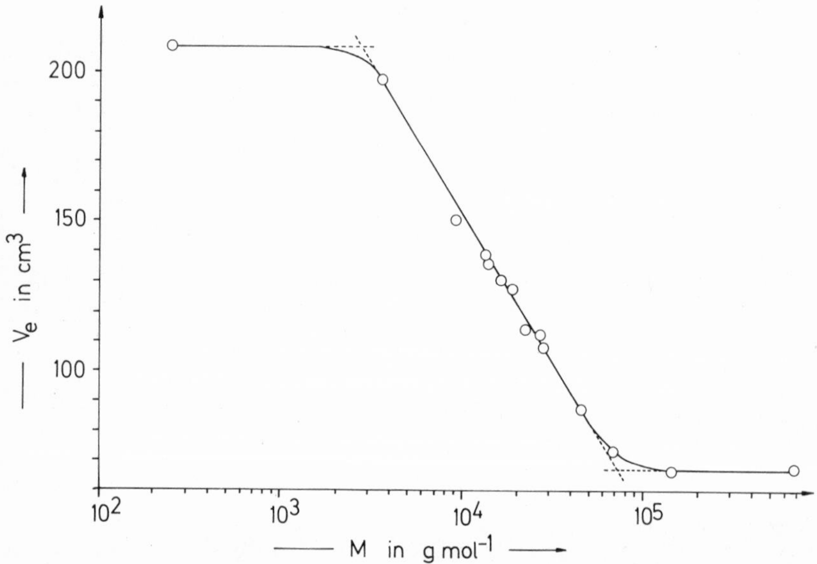

Figure 9-18. Elution volume V_e as a function of the molecular weight M of dilute salt solutions of saccharose and various proteins on cross-linked dextrane (Sephadex G-75) (according to P. Andrews).

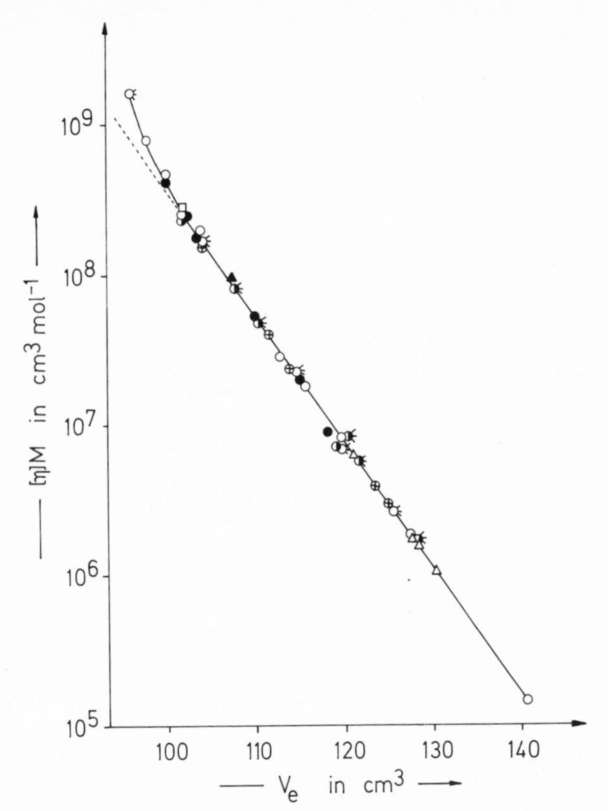

Figure 9-19. A "universal" gel-permeation chromatography calibration curve obtained from measurements on linear poly(styrene) (○), comb-branched poly(styrene) (○⋲), star-branched poly(styrene) (⊕), poly(methyl methacrylate) (●), poly(vinyl chloride) (△), *cis*-1,4-poly-(butadiene) (▲), poly(styrene)–poly(methyl methacrylate) block copolymer (◑⋲), random copolymer from styrene and methyl methacrylate (◑), and ladder polymers of poly(phenyl siloxanes) (□) (according to Z. Grubisic, P. Rempp, and H. Benoit).

represented by the following average:

$$\overline{M}_{\text{GPC}} = \frac{\sum_i m_i M_i^{1+a_\eta}}{\sum_i m_i M_i^{a_\eta}} \tag{9-123}$$

This average gives, for a Schulz–Flory distribution of degree of coupling k,

$$\overline{M}_{\text{GPC}} = \frac{\overline{M}_w(k + a_\eta + 1)}{k + 1} = \frac{\overline{M}_n(k + a_\eta + 1)}{k} \tag{9-124}$$

Consequently, $\overline{M}_{\text{GPC}} = \overline{M}_w$ for spheres, where $a_\eta = 0$, and $\overline{M}_{\text{GPC}} = \overline{M}_{z+1}$ for rigid rods. An average close to the weight average is obtained for coil-shaped molecules with the usual distributions when the experiments are run in theta solutions.

The width of the elution curve increases with the width of the distribution. But monodisperse substances do not give a sharp signal; an elution curve is also obtained. This is produced by what is known as the axial dispersion. According to the model described above, it signifies a distribution of residence times in the pores. The effect must also be taken into account when calculating molecular-weight distributions. For this, it is assumed that the total standard deviation σ_{tot} is composed of the standard deviation associated with molecular inhomogeneity σ_{mol} and the standard deviation resulting from axial dispersion σ_{ad}:

$$\sigma_{tot}^2 = \sigma_{mol}^2 + \sigma_{ad}^2 \tag{9-125}$$

σ_{ad} is determined by reversing the flow of the liquid after development of the elution curve. The σ_{mol} can be calculated, and the true elution curve can be subsequently constructed.

9.8.3. Adsorption Chromatography

Adsorption chromatography is based on the variation in strong interactions between adsorbent and adsorbate. It is suitable for effecting a separation according to differences in constitution and configuration. In real systems, however, the adsorption–desorption equilibrium is submerged beneath a series of other effects.

Four regions can be distinguished according to the proportion of precipitant in the developer in the thin-layer chromatography of polymers (Figure 9-20). Adsorption predominates at low precipitant contents. The

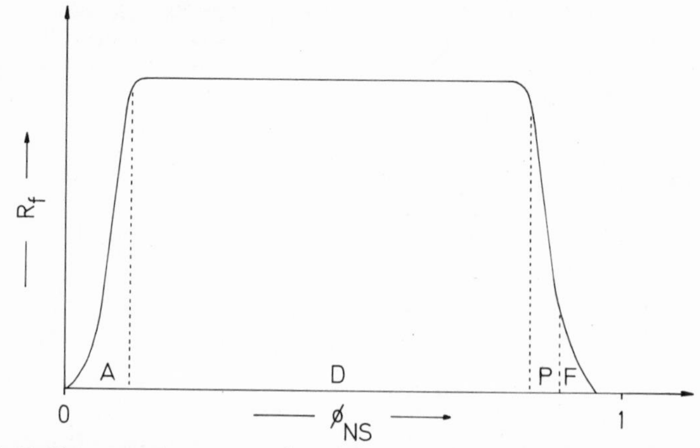

Figure 9-20. Diagrammatic representation of R_f values as a function of the volume fraction ϕ_{NS} of nonsolvent in thin-layer chromatrography. A, adsorption; D, desorption; P, phase separation; F, precipitation.

R_f values increase with increasing precipitant content until the desorption region is reached, when they become independent of the precipitant content. At still higher precipitant contents, a phase separation first starts, followed by precipitation of the polymer. Separation according to constitution and configuration occurs in the adsorption–desorption region, and separation according to molecular weight occurs in the precipitation region. A molecular-sieve effect from the moderately coarse-pored materials is superimposed on these effects.

9.9. Viscometry

9.9.1. Basic Principles

A relationship is shown to exist in viscometry experiments between particle size or molecular size and the viscosity of dispersions of inorganic colloids or the viscosity of macromolecular solutions. It is therefore possible to determine the molecular weight from the viscosity of dilute macromolecular solutions. Since this experiment can be rapidly performed with simple equipment, it is, in practice, the most important molecular-weight-determination method. However, the method is not an absolute one, since the viscosity depends on other molecular properties (for example, on the shape of the molecule), as well as on the molecular weight.

As early as 1906, in fact, Einstein derived a relationship between the viscosity η of solutions of unsolvated spheres, the volume fraction ϕ_2 of the spheres, and the viscosity of the pure solvent η_1,

$$\eta_{sp} \equiv \frac{\eta}{\eta_1} - 1 = 2.5\phi_2 \qquad (\phi_2 \to 0) \tag{9-126}$$

η_{sp} is called the specific viscosity and $\eta/\eta_1 = \eta_{rel}$ is called the relative viscosity. The constant 2.5, results from hydrodynamic calculations. Equation (9-126) only applies in the absence of interactions between the components of the solution, i.e., in infinitely dilute solutions. The influence of finite concentrations can be allowed for by a series expansion in powers of the volume fraction, as was found experimentally with measurements on dispersions of glass spheres or gutta percha:

$$\eta_{sp} = 2.5\phi_2 + \alpha\phi_2^2 + \beta\phi_2^3 + \cdots, \qquad \alpha, \beta = \text{constants} \tag{9-127}$$

Equation (9-127) can be generalized to other particles besides unsolvated spheres, e.g., to coils or rods. The volume fraction of the solute is defined by $\phi_2 \equiv V_2/V$. The volume V_2 of all the solute molecules in a solution of volume $V = V_{\text{soln}}$ (ml) is related to the hydrodynamically effective volume V_h of the individual molecule by the number N_2 of solute molecules: $V_2 = N_2 V_h$. The molar concentration $[M_2]$ of solute molecules (in mol/dm^3) is related to N_2 by $[M_2] = 10^3 N_2/N_L V_{\text{soln}}$ and to the solute concentration c_2 (in g/cm^3) by $[M_2] = 10^3 c_2/M_2$, where M_2 is the molecular weight of the solute. If these relationships are inserted into equation (9-126), this gives on rearrangement

$$\frac{\eta_{\text{sp}}}{c_2} = 2.5 N_L \frac{V_h}{M_2} \qquad (c_2 \rightarrow 0) \tag{9-128}$$

where the limiting value

$$\lim_{c_2 \rightarrow 0} \frac{\eta_{\text{sp}}}{c_2} \equiv [\eta] \tag{9-129}$$

is called the Staudinger index, or intrinsic viscosity $[\eta]$. According to equation (9-128), $[\eta]$ depends on both the molecular weight M_2 and the hydrodynamic volume V_h of the solute. V_h itself is a function of the mass, shape, and density of the solute molecules. The Staudinger index of material consisting of flexible macromolecules can differ by up to a factor of five, depending on the degree of interaction between the material and its solvent (Table 9-4).

The term $[\eta]$ has the physical units of reciprocal concentration, and so those of specific volume. In the current literature the dimensions are usually quoted as cm^3/g. Older literature uses the dimensions of dl/g or liter/g (the latter with the special symbol Z_η), so the numerical values are 100 or 1000 times smaller than when the units are in cm^3/g. In solutions of low-molecular-weight material, $[\eta]$ can in some cases have negative

Table 9-4. Influence of Solvent on Staudinger Index $[\eta]$ of Poly(isobutylene) (PIB), Poly(styrene) (PS), and Poly(methyl methacrylate) (PMMA) at 34°C

	$[\eta]$, cm^3/g		
Solvent	PIB	PS	PMMA
Cyclohexane	478	44	Nonsolvent
CCl_4	462	100	305
n-Hexane	327	Nonsolvent	Nonsolvent
Chlorobenzene	250	107	—
Toluene	247	—	403
Benzene	119	114	640
Butyl acetate	Nonsolvent	—	195

Table 9-5. Staudinger Indices $[\eta]$ *of Poly(styrene) and Poly(methyl methacrylate) Mixtures in* CCl_4 *at 25°C*

Mass fractions[a]		$[\eta]$, cm^3/g	
w_{PS}	w_{PMMA}	Calculated[b]	Observed
1.0	0.0	—	120.0
0.7	0.3	90.8	102.0
0.5	0.5	76.4	84.0
0.3	0.7	58.9	65.0
0.1	0.9	41.4	44.0
0.0	1.0	—	32.7

[a] PS = poly(styrene); PMMA = poly(methyl methacrylate).
[b] Calculations were done according to equation (9-130).

values, namely, if the solution viscosity (and thus that of the solute) is lower than the solvent viscosity.

In polymer mixtures where there is no special interaction between the polymers, e.g., in a polymer homologous series, the Staudinger index is obtained as a weight-average (Philippoff equation):

$$\overline{[\eta]}_w = \frac{\sum_i W_i[\eta]_i}{\sum_i W_i} = \frac{\sum_i c_i[\eta]_i}{\sum_i c_i} = \frac{\sum_i c_i[\eta]_i}{c} = \sum_i w_i[\eta]_i = [\eta] \qquad (9\text{-}130)$$

Because of the attraction and repulsion forces, the conventionally determined Staudinger index of mixtures of different polymers may appear higher or lower than that calculated from equation (9-130) from the mass contributions w_i and Staudinger indices $[\eta]_i$ of the components. Examples of this are the values for mixtures of poly(styrene) and poly(methyl methacrylate) (Table 9-5).

9.9.2. Experimental Methods

To calculate the Staudinger index $[\eta]$, the viscosities of the solvent and of solutions of different concentrations must be measured. The polymer solution concentrations must not be too high, since this makes the extrapolation of the viscosity data to infinite dilution difficult. Experience shows that solute concentrations are best chosen such that η/η_1 lies between about 1.2 and 2.0.

The upper limit of $\eta_{rel} \approx 2.0$ results from the fact that the relationship between η_{sp}/c and c becomes increasingly nonlinear with increasing concentration. The lower limit of $\eta_{rel} \approx 1.2$ arises from the fact that at low concentrations anomalies in the function $\eta_{sp}/c = f(c)$ begin to appear. These anomalies are usually considered to be apparatus-dependent and to result from adsorption of macromolecules on the capillary walls.

To determine $\eta_{sp} = \eta_{rel} - 1$ for $\eta_{rel} = 1.2$ to an accuracy of $\sim \pm 1\%$, the viscosity ratio must be determined to an accuracy better than $\pm 0.2\%$, and the viscosities themselves to an accuracy better than $\pm 0.1\%$ (compounding of errors). Capillary viscometers are particularly suitable for such determinations. The usual rotation viscometer of the Couette type at best gives an accuracy of $\pm 1\%$. Falling-ball viscometers are even less exact.

The frequently used viscometer types for macromolecular chemistry are shown in Figure 9-21. In all types, the time taken for a given quantity of liquid to pass between two points is taken as an indication of the viscosity. (The pressure head under which these experiments are carried out is the pressure head of the liquid under study in the capillary. This pressure head varies during the experiment, but one can assume that an average pressure head is effective during the whole of the experiment.) As long as liquid flow is not infinitely slow, potential energy is partially lost in overcoming frictional forces during flow. Some of the potential energy is also lost through conversion to kinetic energy, which in turn is dissipated through the formation of vortices on exit from the capillary (Hagenbach). Additionally, a given amount of initial work is done on forming the parabolic shear-rate gradient (Couette). The apparent increase in viscosity caused by these two nonfrictional effects is, according to Hagenbach and Couette, taken into account by the incorporation of a correction term in the Hagen–Poiseuille equation [for derivation, see equation (7-47)]:

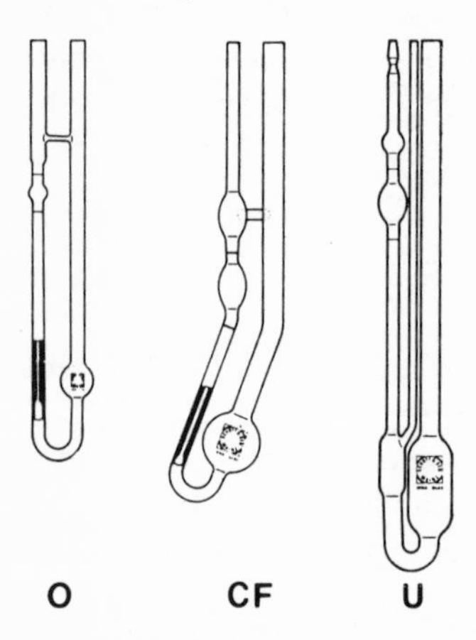

Figure 9-21. Ostwald (O), Cannon–Fenske (CF), and Ubbelohde (U) capillary viscometers.

$$\eta = \frac{\pi r^4 pt}{8LV} - \frac{k\rho V}{8\pi Lt} = \text{constant} \times pt - \text{constant}' \times \rho t^{-1} \quad (9\text{-}131)$$

where r is the radius, p is the pressure, L is the capillary length, and V is the volume of liquid with density ρ. The constant k depends on the geometrical shape of the capillary ends. This cannot be obtained theoretically, and must be determined by calibration measurements on liquids with various viscosities.

According to equation (9-131), the Hagenbach–Couette correction can be ignored if the viscometer capillary is sufficiently long. With commercial viscometers, the Hagenbach–Couette correction values are supplied by the manufacturer in the form of correction times. Measurement times should, preferably, never be less than 100 s; otherwise one incurs too high a percentage error. Also, the viscometer must hang vertically for every reading; otherwise the effective length of the capillary varies from one reading to another. The temperature should be constant to $\sim \pm 0.01°C$, since a temperature difference of 0.01°C usually means a viscosity change of $\sim 0.02\%$.

Solution and solvent almost always have different densities. Therefore, the average pressure head p in capillary viscometers filled to the same height $h = h_0$ will vary in a series of measurements involving different solute concentrations. Thus, according to equation (9-131), the following can be given for the relative viscosity in the case of a vanishingly small Hagenbach–Couette correction:

$$\eta_{\text{rel}} = \frac{\eta}{\eta_1} = \frac{\text{const} \times pt}{\text{const} \times p_1 t_1} = \frac{(h\rho)\,t}{(h_1 \rho_1)\,t_1} = \frac{\rho t}{\rho_1 t_1} \quad (9\text{-}132)$$

With relatively low-molecular-weight substances, relatively high concentrations have to be used in order to obtain η_{rel} values of 1.2–2, so that in these cases particular attention should be paid to the differences in density between solution and solvent during viscosity measurements.

The capillary viscometers shown in Figure 9-21 differ in their areas of application. Because of their low price and the fact that they only require low amounts, $\sim 3\ \text{cm}^3$, of liquid, Ostwald viscometers are by far the most frequently used. The amount of liquid used must be measured in very exactly and maintained constant; otherwise the pressure head will vary with the different solutions.

Suspended-level Ubbelohde viscometers are so constructed that, during a measurement, the pressure head of the suspended liquid at the capillary outlet is independent of the amount of liquid originally introduced into the viscometer. This also means that solution originally introduced into the viscometer can be diluted with solvent to provide a series of concentrations without having, as in the case of the Ostwald viscometer, to empty and clean the viscometer after every measurement. On the other hand, the amount of liquid required for a measurement is higher in the Ubbelohde than in the Ostwald viscometer. Another advantage of the Ubbelohde suspended-level viscometer is that the stress on the suspended liquid exactly compensates the surface-tension effects on the upper meniscus, and this is particularly important in the case of surface-active substances.

The Cannon–Fenske viscometer has two bulbs, and since the pressure head is different for each bulb, this allows qualitative assessment of the effect of shear stress σ_{ij} on the viscosity of the solution. The shear stress is given by

$$\sigma_{ij} = pr/2L \tag{9-133}$$

where p is the pressure head, and r and L are the capillary radius and length, respectively (for derivation, see Section 7.6.2). With Newtonian liquids, the viscosity, and thus also the Staudinger index, is independent of σ_{ij}. In some cases, solutions of macromolecules of high molecular weight show non-Newtonian behavior even in dilute solutions, i.e., we have $\eta = f(\sigma_{ij})$. According to theoretical studies and experimental evidence, the Staudinger index measured at a certain shear stress decreases with increasing σ_{ij} according to

$$[\eta]_{\sigma_{ij}} = [\eta](1 - A\beta^2 \cdots) \tag{9-134}$$

β is a generalized shear stress function,

$$\beta = [\eta]\,\eta_1 \frac{\overline{M}_n}{RT} \sigma_{ij} \tag{9-135}$$

that takes into account the solvent viscosity η_1, the number-average molecular weight $\overline{M}_n$, and the temperature T. In (9-134), A is a constant that remains dependent on the goodness of the solvent. Poly(styrene) solutions exhibit non-Newtonian behavior when $\beta > 0.1$ (Figure 9-22), which corresponds, roughly, to a molecular weight of $\overline{M}_n > 500{,}000$.

Since the effects of shear stress are particularly strong in the case of rodlike macromolecules, rotation viscometers are frequently used for measurements on substances such as deoxyribonucleic acid (Figure 9-23). With a sufficiently low rotational speed and a narrow gap between rotor and stator, a linear shear-rate gradient can be produced between the rotor and stator of a rotation viscometer. With such narrow gaps between rotor and stator, the centering of rotor in the stator (or, in some viscometers, vice versa) is particularly important. In rotation viscometers of the Couette type, centering is achieved by use of a mechanical axis. A much better centering system is used in the Zimm–Crothers viscometer. This utilizes

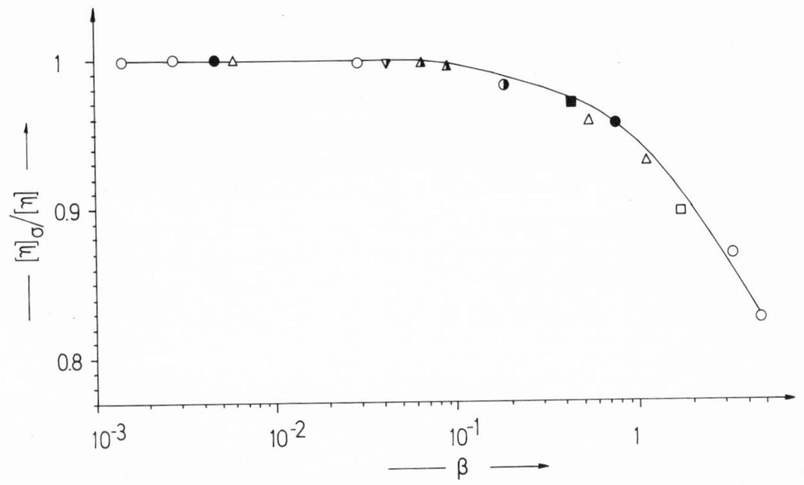

Figure 9-22. Dependence of the reduced Staudinger index $[\eta]_\sigma/[\eta]$ on the reduced shear stress $\beta = ([\eta]\,\eta_1 \overline{M}_\eta/RT)\,\sigma_{ij}$ of polystyrene with molecular weight $\overline{M}_w = 7.1 \times 10^6$ (○, □, △), 3.2×10^6 (○), and 1.4×10^6 (◑, ◨, ▲, ▼) in good (○, ◑, ●, □, ◨) and theta (△, ▲, ▼) solvents.

the principle that, when the viscometer contains the liquid to be measured, a suitably buoyant rotor will automatically center perfectly in the stator because of the surface tension of the liquid to be measured. The rotor contains an iron rod. The thermostated stator resides between the poles of a magnet. The magnet is connected to a motor that can be driven at a regulatable constant speed of rotation. The interaction between the external magnetic field and the magnetic field induced in the iron rod produces a weak torque. Shear stresses of down to 40 nN/cm^2 can thus be obtained.

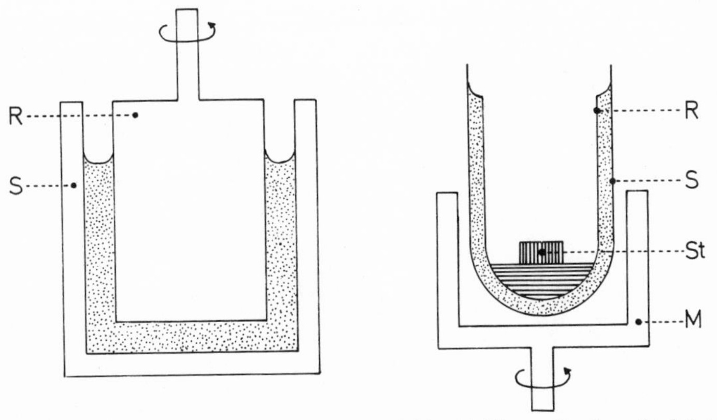

Figure 9-23. Rotation viscometers of the Couette (left) and Zimm–Crothers (right) type. R, rotor; S, stator; St, steel plate; M, magnet.

9.9.3. Concentration Dependence for Nonelectrolytes

The Staudinger index $[\eta]$ is defined as the limiting value of the reduced viscosity η_{sp}/c $(=\eta_{sp}/c_2)$ at infinite dilution. Since measurements are made at finite concentrations, a suitable equation is required in order that values of η_{sp}/c, or related quantities, may be extrapolated to $c \to 0$. The extrapolation should be as linear as possible over the range $\eta_{rel} = 1.2$–2.

All the extrapolation formulas introduced to date are empirical. The much used expressions of Schulz and Blaschke, Huggins, and Kraemer start from the relationship

$$\frac{\eta_{sp}}{c} = \frac{[\eta]}{1 - k[\eta]c} \tag{9-136}$$

The extrapolation formula that is obtained by transforming equation (9-136)

$$\frac{c}{\eta_{sp}} = \frac{1}{[\eta]} - kc \tag{9-137}$$

is, to date, practically unused. If equation (9-136) is solved for c one obtains $c = \eta_{sp}/([\eta] + \eta_{sp}k[\eta])$. If this expression is then inserted into the right-hand side of equation (9-136) in place of c, then the formula known as the Schulz–Blaschke equation is obtained:

$$\frac{\eta_{sp}}{c} = [\eta] + k[\eta]\eta_{sp} \tag{9-138}$$

For low values of $k[\eta]c$, the expression in the denominator in equation (9-136) can be developed in a series: $(1 - k[\eta]c)^{-1} = 1 + k[\eta]c + \cdots$. On inserting this expression in equation (9-136), we obtain the Huggins equation

$$\frac{\eta_{sp}}{c} = [\eta] + k[\eta]^2 c \tag{9-139}$$

If one develops $\ln \eta_{rel} = \ln(1 + \eta_{sp})$ in a Taylor series

$$\ln \eta_{rel} = \eta_{sp} - (1/2)\eta_{sp}^2 + (1/3)\eta_{sp}^3 - \cdots \tag{9-140}$$

and inserts this in equation (9-139), the Kraemer equation is obtained:

$$\frac{\ln \eta_{rel}}{c} = [\eta] + [\eta]^2\{k - 0.5 + [[\eta]c(0.333 - k)]\}c + \cdots \tag{9-141}$$

which is mostly written without the component $(0.333 - k)[\eta]^3 c^2$ in the shortened form

$$\frac{\ln \eta_{rel}}{c} = [\eta] + (k - 0.5)[\eta]^2 c \tag{9-142}$$

The omitted term comes from the mathematical development of equation (9-140) in a series. Because of its magnitude, it cannot be simply ignored. Figure 9-24 shows examples of concentration dependence.

The numerical evaluation of experimental data shows that equations (9-137)–(9-139) and (9-142) yield different values of $[\eta]$ and k. Since equations (9-139) and (9-142) are approximations of equation (9-136), they must, a priori, extend over a narrower concentration range. This argument naturally assumes that equation (9-136) really describes the concentration dependence of η_{sp}/c adequatly. For wider concentration ranges, the Martin (sometimes referred to as the Bungenberg–de Jong) equation is used in the form

$$\log \frac{\eta_{sp}}{c} = \log [\eta] + kc \tag{9-143}$$

One often makes do with a viscosity measurement at a single concentration (usually 0.5 %) to give what is called the inherent viscosity $\{\eta\}_c = (\ln \eta_{rel}/c)_c$ for this concentration. Fikentscher constants K are also used, particularly in the German literature, to characterize the classic polymers such as polystyrene and polyvinyl chloride. [The K here is not to be

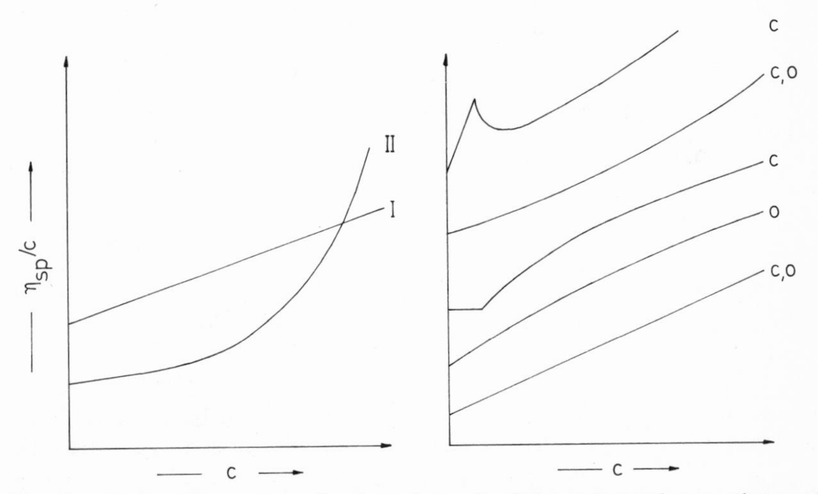

Figure 9-24. A plot of the concentration dependence of η_{sp}/c for various polymer–solvent and polymer–polymer interactions. Left diagram: nonassociating polymers; right diagram: associating polymers. (I) Good solvent, (II) poor solvent. (o) Open association, (c) closed association. The multiplicity of curves derives from the number of possible models as well as well as from the relative influences of the equilibrium association constant, the association number, the molecular weight, the size and shape of the molecules and associates, and the solvent interaction.

confused with the K, defined by equation (9-156), of the modified Staudinger equation.] K is evaluated from the relative viscosity at relatively high concentrations from tabular data and the equation

$$\log \eta_{\mathrm{rel}} = \left(\frac{75k_F^2}{1 + 1.5k_F c} + k_F \right) c \tag{9-144}$$

with the definition $K = 1000k_F$. At the time it was introduced, K was considered, on the basis of limited experimental data, to be a concentration-independent constant that was related to the molecular weight. However, K is concentration dependent, and at high molecular weights it becomes increasingly less sensitive to changes in the molecular weight.

The molecular significance of the coefficient k has not yet been elucidated. Theoretical studies indicate that for coils k is composed of a hydrodynamic factor k_h and a thermodynamic factor $(3A_2M/[\eta])\,f(\alpha)$, where $f(\alpha)$ is a function of the expansion factor α:

$$k = k_h - \frac{3A_2M}{[\eta]} f(\alpha) \tag{9-145}$$

The hydrodynamic factor k_h probably has values of between 0.5 and 0.7. In theta solvents, where $A_2 = 0$, a value of $k = 0.5$–0.7 is expected; in good solvents, since $A_2 > 0$, the value of k is <0.5–0.7. Experimentally, values of k between 0.25 and 0.35 are found for good solvents. As the molecular weight increases, the second term in equation (9-145) should, according to this equation, increase strongly initially, and subsequently increase less strongly, since, with random coils, $[\eta]$ and $f(\alpha)$ increase less than proportionally with the molecular weight, and A_2 decreases less than proportionally with M. The major proportion of the experimental data confirm these expectations.

9.9.4. Concentration Dependence for Polyelectrolytes

In the absence of foreign salts, η_{sp}/c increases sharply with decreasing polyelectrolyte concentration in solutions of polyelectrolytes (Figure 9-25). At increasing foreign salt concentration, this increase in η_{sp}/c becomes weaker. At low polymer concentrations, the function $\eta_{\mathrm{sp}}/c = f(c)$ passes through a maximum, but this phenomenon is not observed with all polyelectrolyte solutions. At high polyelectrolyte concentrations, the increase of η_{sp}/c with concentration is similar to that observed with solutions of nonelectrolytes.

The effect is explained in the following manner: With decrease in polyelectrolyte concentration, the degree of ionization increases. In the

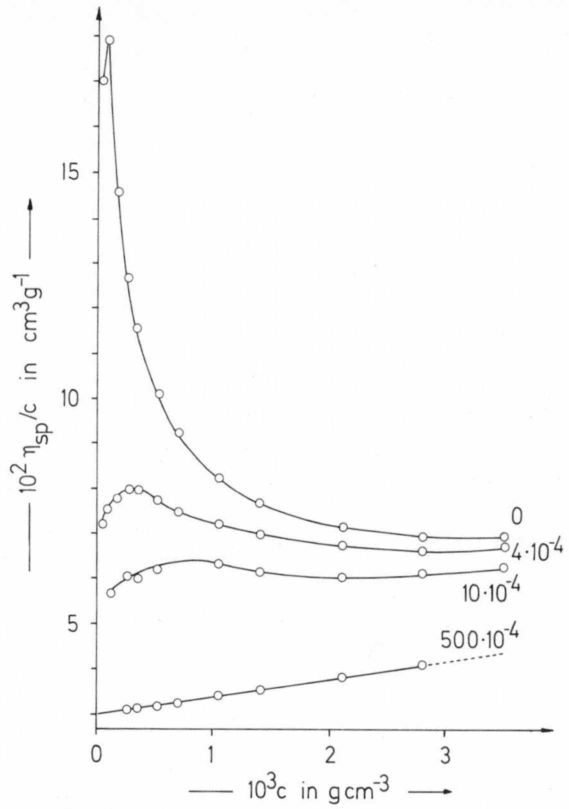

Figure 9-25. Concentration dependence of the reduced viscosity η_{sp}/c of sodium pectinate in solutions of different NaCl concentrations (in mol/liter) at 27°C (according to D. T. F. Pals and J. J. Hermans).

case of polysalts [e.g., sodium pectinate, the sodium salt of poly(acrylic acid), etc.] the gegenions form an ion atmosphere around the chains of the polyelectrolyte macromolecule. In very dilute, foreign-salt-free solutions, the diameter of the ion atmosphere is greater than the diameter of the coiled molecule. The carboxylate ions, $-COO^-$, repel each other, increasing the chain rigidity and expanding the polymer coil, with consequent increase in the viscosity. At medium polyelectrolyte concentrations, the gegenions reside partly within and partly outside the polymer coil. At very high polyelectrolyte concentrations, the gegenions reside more within than outside the polymer coil. The osmotic effect that this produces causes more water to penetrate the coil and expand it. Thus at low concentrations the electrostatic effect predominates, and at high concentrations the osmotic effect predominates. The introduction of foreign salts into the polyelectrolyte solution causes the ionic strength outside the polymer coil to be increased relative to that inside the polymer coil, and the diameter of the ion atmosphere is decreased. Both effects decrease the diameter of the polymer coil and, thus, also η_{sp}/c.

The Staudinger index can be empirically determined via the Fuoss equation

$$\frac{c}{\eta_{sp}} = \frac{1}{[\eta]} + Bc^{0.5} - \cdots \tag{9-146}$$

where the values of c/η_{sp} for $c > c_{max}$ are plotted against $c^{0.5}$ and extrapolated to $c^{0.5} \rightarrow 0$.

9.9.5. The Staudinger Index and the Molecular Weight of Rigid Molecules

Unsolvated Spheres. The Staudinger index $[\eta]$, according to equation (9-128), depends on both the molecular weight and the hydrodynamic volume, and the latter can also be a function of the molecular weight. Unsolvated spheres present the simplest case. These spheres can be defined by their type and a density that is independent of its environment and equal to the density of the dry material. The mass m_{mol} of an individual molecule is related to its hydrodynamic volume via $m_{mol} = V_h \rho_2$, or with the molecular weight M_2 via $M_2 = N_L m_{mol} = N_L V_h \rho_2$. Equation (9-128) therefore changes, for unsolvated spheres, to

$$[\eta] = 2.5/\rho_2 \tag{9-147}$$

Solvated Spheres. To a good approximation, dispersions [e.g., poly(styrene) latices] can be considered as unsolvated sphere systems, but the approximation no longer holds for isolated macromolecules. Certain proteins, however, are in the form of solvated spheres in aqueous solutions. With these proteins, some of the amino acid residues are in a helical conformation; the rest are in a random-coil conformation (cf. Figure 4-19). Under the influence of water, the helix and coil portions mutually align themselves via hydrophobic bonds, salt bonds, etc., to form solvated spherical particles. The mass of the hydrodynamically effective individual molecule is consequently composed of the mass of the protein component and the mass of the water of hydration, i.e., $m_h = m_2 + m_1^{\square}$. On average, the density of every segment of the sphere is the same. If the ratio of the masses $\Gamma = m_1^{\square}/m_2$ is sufficiently low, no water of hydration is exchanged with the surrounding water during the measurement. The hydrated sphere is then nondraining, and all of the water of hydration has to be considered as part of the mass and the volume of the hydrodynamically effective particle.

If the hydrodynamic volume V_h in equation (9-128) is replaced by the expression in equation (7-6), the following is obtained:

$$[\eta] = 2.5(\tilde{v}_2 + \Gamma v_1) \tag{9-148}$$

Thus, the Staudinger index of solvated spheres depends on the partial specific volume $\tilde{v}_2$ of the solute, the specific volume v_1 of water, and the mass ratio $\Gamma = m_1^{\square}/m_2$ (degree of solvation) of both components in the interior of the sphere. Therefore, it is not possible to calculate the molecular weight of a solvated sphere from the Staudinger index alone. The Staudinger indices of spherical protein molecules are low, and for equal degrees of hydration are independent of the molecular weight (Table 9-6). Admittedly, the proteins included in Table 9-6 are not perfectly spherical, since their coefficients of friction f are somewhat larger than those expected for a perfect sphere, f_0.

Unsolvated Rods. For unsolvated rods, the relationship between the Staudinger index and the molecular weight can be deduced as follows: $[\eta]$ depends on the hydrodynamic volume [cf. equation (9-128)] and on the radius of gyration. One can write, in place of equation (9-128), therefore,

$$[\eta] = \frac{\Phi^* \langle R_G^2 \rangle_{st}^{3/2}}{M_2} \tag{9-149}$$

where Φ^* is a general proportionality constant.

According to equation (4-53), the radius of gyration of rods is related to the chain end-to-end distance: $\langle L^2 \rangle = 12 \langle R_G^2 \rangle_{rod}$. For macromolecules in the all-*trans* conformation, the end-to-end distance is equal to the maximum chain length L_{max}. However, L_{max} is proportional to the degree of polymerization X [see also equation (4-11)]. With helix-forming, rodlike

Table 9-6. Staudinger Index $[\eta]$, and Frictional Coefficient of Approximately Spherical Proteins in Dilute Salt Solutions at 20°C

Protein	Molecular weight M_2	$[\eta]$, cm^3/g	f/f_0
Ribonuclease	13,683	3.30	1.14
Myoglobin	17,000	3.1	1.11
β-Lactoglobulin	35,000	3.4	1.25
Serum albumin	65,000	3.68	1.31
Hemoglobin	68,000	3.6	1.14
Catalase	250,000	3.9	1.25

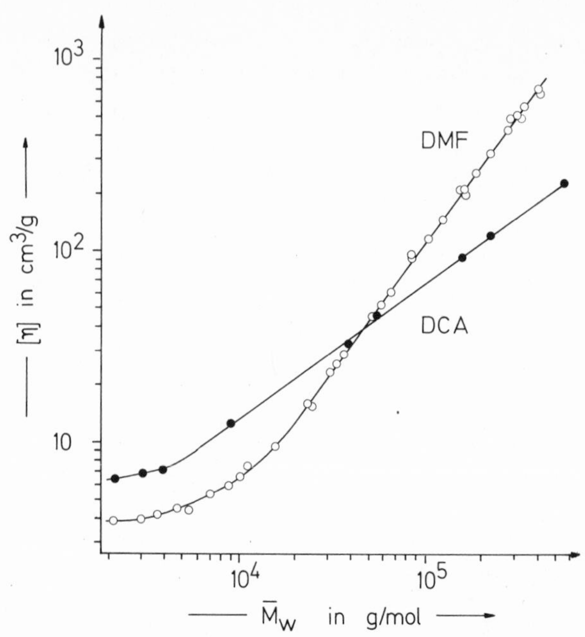

Figure 9-26. A plot according to equation (9-156) of the molecular-weight dependence of the Staudinger indices of poly(γ-benzyl-L-glutamate) in dichloroacetic acid (DCA) and dimethyl formamide (DMF) at 25°C. Coils occur in DCA and helices occur in DMF. (According to P. Rohrer and H.-G. Elias.)

macromolecules, the end-to-end distance is equal to the length of the helix and is also proportional to the degree of polymerization. As a general rule, then, L^2 = constant × X^2 for rodlike macromolecules. With $X = M_2/M_u$, it thus follows that

$$[\eta] = \Phi^* \left(\frac{\text{constant}}{12 M_u^2} \right)^{3/2} M_2^2 = K M_2^2 \qquad (9\text{-}150)$$

According to this relationship, the Staudinger index of rods is proportional to the square of the molecular weight. For a homologous series of rodlike molecules, i.e., a series with constant rod cross section, this functionality only applies over a limited molecular-weight range. At low molecular weights, the rod increasingly takes on the character of a sphere, and the exponent decreases below 2. At high molecular weights, a real rod is no longer inflexible, since a rod of infinite length behaves like a random coil (cf. the flexibility of thin steel wires of various lengths). Thus, at high molecular weights, the exponent also falls below 2 (Figure 9-26).

9.9.6. The Molecular Weight and Staudinger Index of Coil-like Molecules

Nondraining Coils. Nondraining coils are defined as those in which the solvent molecules within the coil move with the same velocity as the nearest segments of the coil itself during transport processes. The hydrodynamic volume V_h of such a coil is greater in thermodynamically good solvents than in theta solvents. This extension can be described by a coefficient of expansion α_η in a way analogous to that used in the case of the radius of gyration (see, for example, Chapter 4.5.2.1), i.e., as $V_h = (V_h)_\Theta \alpha_\eta^3$. The quantity α_η differs, however, from the expansion coefficient used for the radius of gyration, since the non-Gaussian segment distribution in the coil causes the hydrodynamic radius to vary somewhat differently with the molecular weight than is the case with the radius of gyration. As a general rule, however, one can write

$$V_h = (V_h)_\Theta \alpha_\eta^3 = (V_h)_\Theta \alpha_R^q, \qquad q \neq 3 \tag{9-151}$$

According to model calculations, the factor q possesses a value of 2.43 for a sphere-like coil and a value of 2.18 for an ellipsoid-like coil. Insertion of equation (9-15) in (9-128) gives

$$[\eta] = 2.5 N_L (V_h)_\Theta \frac{\alpha_R^q}{M_2} \tag{9-152}$$

In the theta state, the hydrodynamic volume $(V_h)_\Theta$ is proportional to the third power of the radius of gyration, i.e., it is possible to write $(V_h) = \Phi' \langle R_G^2 \rangle_0^{3/2}$. Equation (9-152) then becomes

$$[\eta] = 2.5 N_L \Phi' \frac{\langle R_G^2 \rangle_0^{3/2} \alpha_R^q}{M_2} = 2.5 N_L \Phi' \left(\frac{\langle R_G^2 \rangle_0}{M_2} \right)^{3/2} M_2^{0.5} \alpha_R^q$$
$$= \Phi \left(\frac{\langle R_G^2 \rangle}{M_2} \right)^{3/2} M_2^{0.5} = \Phi \frac{(\langle R_G^2 \rangle)^{3/2}}{M_2} \tag{9-153}$$

where the relationship $\langle R_G^2 \rangle_0 = \langle R_G^2 \rangle_0 \alpha_R^2$ is used and $\Phi = 2.5 N_L \Phi' (\alpha_R / \alpha_R^3)$ is assumed.

Equation (9-153) describes the relationship between the Staudinger index $[\eta]$ of nondraining coils as a function of the molecular weight and the radius of gyration. Two methods can be used to obtain $[\eta]$ as a function of the molecular weight alone:

1. According to equation (4-49), $\langle R_G^2 \rangle = \text{constant}' \times M_2^{1+\varepsilon}$. Equation

(9-153) can thus be written as

$$[\eta] = \Phi(\text{constant}')^{3/2} M_2^{0.5(1+3\varepsilon)} \tag{9-154}$$

or, with $K = \Phi(\text{constant}')^{3/2}$ and the definition

$$a_\eta \equiv 0.5(1 + 3\varepsilon) \tag{9-155}$$

also as

$$[\eta] = K M_2^{a_\eta} \tag{9-156}$$

Since ε rarely takes on values above 0.23 in the case of nondraining coils, a_η values up to a maximum of ~ 0.9 are obtained for nondraining coils. In the theta state, $\varepsilon = 0$ and equation (9-156) reduces to

$$[\eta]_\Theta = K_\Theta M_2^{0.5} \tag{9-157}$$

Equation (9-156) is known as the modified Staudinger equation (originally with $a_\eta = 1$) or as the Kuhn–Mark–Houwink–Sakurada equation. It was originally found empirically. K and a_η are empirical constants obtained by calibration (see also Chapters 9.9.7 and 9.9.8 and Figure 9-26).

2. According to equation (4-50), $(\langle R_G^2 \rangle / M_2)^{3/2}$ can be expressed as

$$\left(\frac{6 \langle R_G^2 \rangle}{M_2} \right)^{3/2} = A^3 + 0.632 B M_2^{0.5} \tag{9-158}$$

If equation (9-153) is inserted into (9-158), the Burchard–Stockmayer–Fixman equation (often just known as the Stockmayer–Fixman equation) is obtained:

$$\frac{[\eta]}{M_2^{0.5}} = K_\Theta + \frac{0.632}{6^{3/2}} \Phi B M_2^{0.5} \tag{9-159}$$

where it is assumed that

$$K_\Theta = \frac{\Phi}{6^{3/2}} A^3 \tag{9-160}$$

By plotting $[\eta]/M_2^{0.5} = f(M_2^{0.5})$, the quantity K_Θ can, according to equation (9-159), be obtained, and K_Θ contains the steric hindrance contribution in the factor A (see also Figure 9-27). Besides the steric hindrance parameter σ, A also contains, according to equation (4-51), the bond length L between the main-chain atoms, the valence angle ϑ, and the average formula molecular weight $\overline{M}_u$ of the chain units:

$$A = \left[\left(\frac{1 - \cos \vartheta}{1 + \cos \vartheta} \right) \frac{\sigma^2 l^2}{M_u} \right]^{0.5} \tag{9-161}$$

Equation (9-159) thus allows σ to be determined from viscometric data

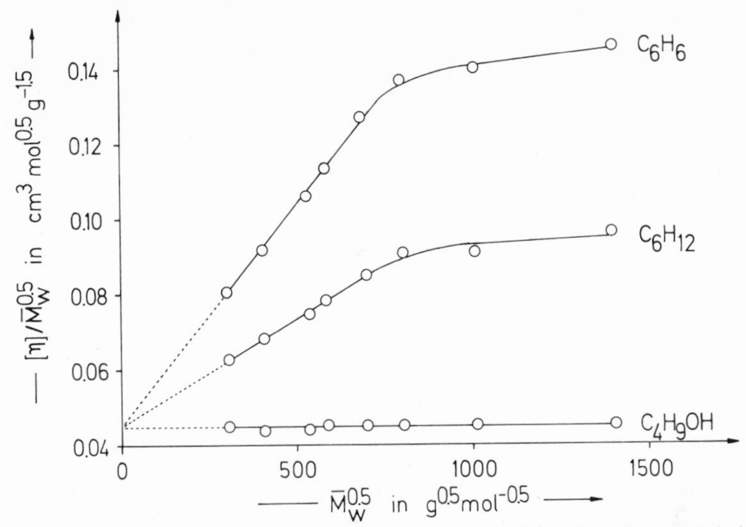

Figure 9-27. Burchard–Stockmayer–Fixman plot for poly(cyclohexyl methacrylate) in benzene and cyclohexane at 25°C and in butanol at 23°C (according to N. Hadjichristidis, M. Devaleriola, and V. Desreux).

alone. Like σ, K_Θ is only a substance-specific constant when measurements are carried out in single, apolar solvents (see also Section 4.4.2.5).

Both equations (9-157) and (9-159) give the Staudinger index as a function of the molecular weight. It has been shown experimentally that equation (9-156) is valid over a wider range of molecular weights, temperature, and solvents than equation (9-159). Equation (9-159) is fulfilled satisfactorily, but at higher molecular weights it yields values of $[\eta]/M^{0.5}$ that are relatively too small when $T > \Theta$ (see Figure 9-27). Many other functions have been proposed to achieve better linearity. The Berry equation has shown itself to be very useful (see Figure 9-28):

$$\left(\frac{[\eta]}{\overline{M}_w^{0.5}}\right)^{0.5} = K_\Theta^{0.5} + D\frac{\overline{M}_w}{[\eta]} \tag{9-162}$$

where D is an adjustable constant.

In equations (9-153), (9-154), and (9-159), Φ appears everywhere as a constant. Theoretical and experimental investigations have shown that Φ does not depend on either the constitution or the configuration of the polymer or on the chemical nature of the solvent used. As a proportionality factor between the radius of gyration and the hydrodynamic volume, Φ is only related to the expansion of the coil in the relevant solvent, i.e., to the values of α or ε. The theoretical calculation leads to

$$\Phi = \Phi_0(1 - 2.63\varepsilon + 2.86\varepsilon^2) \tag{9-163}$$

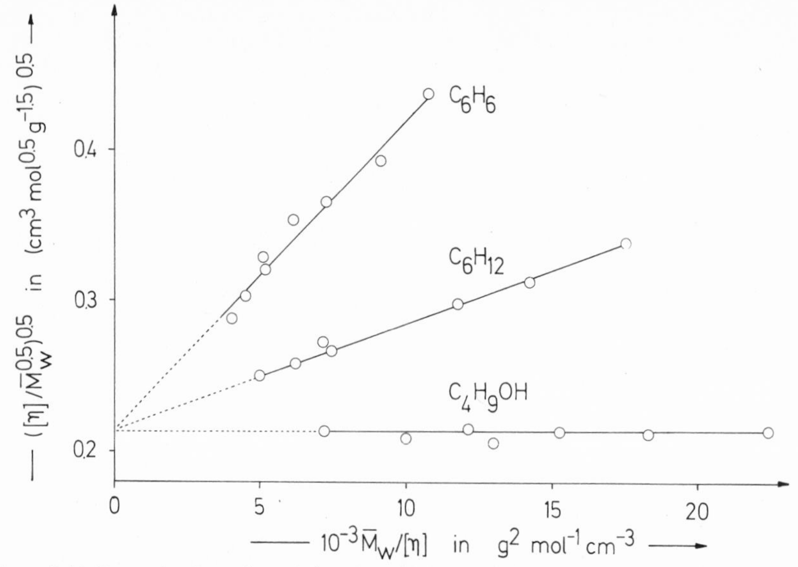

Figure 9-28. Berry plot for poly(cyclohexyl methacrylate) in benzene and cyclohexane at 25°C and in butanol at 23°C (same data as in Figure 9-27) (according to N. Hadjichristidis, M. Devaleriola, and V. Desreux).

where Φ_0 is the value in the theta state ($\varepsilon = 0$ or $a_\eta = 0.5$). If Φ_0 is related to the radius of gyration, then $\Phi_0 = 4.18 \times 10^{24}$ (mol macromolecule)$^{-1}$. If one uses the chain end-to-end distance, then, because $\langle R_G^2 \rangle_0 = \langle L^2 \rangle_0/6$, $\Phi_0 = 2.84 \times 10^{23}$ (mol macromolecule)$^{-1}$ (see also Table 9-3).

Free-Draining Coils. In the case of free-draining coils, the relative velocity of solvent movement with respect to the coil is the same both within and outside the coils. Free-draining coils may be expected in the case of relatively rigid chains in good solvents, whereas nondraining coils are to be expected in the limiting case of very flexible chains in poor solvents. All variations between these two extremes are possible. However, the concept of a partially draining coil can be described theoretically only with difficulty. In the scientific literature, two principal theories are discussed.

In the Kirkwood–Riseman theory, the perturbation of the rate of flow of the solvent by $N - 1$ chain elements is calculated for the Rth chain element and summed over all possible conformations. Suitable parameters that are used are the effective bond length b and the frictional coefficient ζ of the base unit.

The Debye–Bueche theory, on the other hand, considers the partially draining coil as a sphere that is more or less permeable, within which a number of smaller beads is homogeneously distributed. The beads correspond to the monomeric units. The drag which one bead produces on the others is calculated, and this resistance is then expressed in terms of a length L, which corresponds to the distance from the surface of the sphere to where the flow rate of the solvent is reduced to $1/e$ times what it is at the surface of this sphere. The shielding ratio, or "shielding factor," ζ is given by

$$\zeta = R_s/L \tag{9-164}$$

where R_s is the sphere radius and L is the depth of shielding. ζ can also be calculated from the shielding function $F(\zeta)$ derived by Debye and Bueche:

$$F(\zeta) = 2.5 \frac{1 + (3/\zeta^2) - (3/\zeta)\cot\zeta}{1 + (10/\zeta^2)[1 + (3/\zeta^2) - (3/\zeta)\cot\zeta]} \tag{9-165}$$

which is, in turn, related to the Staudinger index:

$$[\eta] = F(\zeta)\, N_L \left(\frac{4\pi R_s^3}{3}\right) M_2^{-1} \tag{9-166}$$

Real macromolecules tend, however, to occur in the form of ellipsoids (cf. Section 4.4.2.5). With ellipsoid-shaped coils, R_s corresponds to the major rotational axis of the ellipsoid. For such coils, a simple empirical relationship between the shielding factor ζ and the quantity ε has been found when the coils are flexible [e.g., in the case of poly(styrene)]. The relationship is $\zeta\varepsilon = 3$. The effect of the solvent on the coil expansion is described by ε [cf. equation (9-163)]. Thus, equation (9-164) becomes

$$L = R_s\varepsilon/3 \tag{9-167}$$

According to equation (9-155), ε can be calculated from the exponent a_η of the viscosity–molecular weight relation $[\eta] = KM^{a_\eta}$. For example, with $a_\eta = 0.8$, $\varepsilon = 0.2$, and the shielding depth L is $0.067R_s$ according to equation (9-167). For a coil that is expanded as in this particular case, the shielding depth is only 6.7% of the major rotational axis of the ellipsoid. The solvent, therefore, only partially penetrates the coil. For the more rigid cellulose chains, the shielding depth is greater.

Molecular Dimensions from Viscosity Measurements. According to equation (9-153), the Staudinger index is related to the radius of gyration. Furthermore, in very dilute solutions, $[\eta] \approx \eta_{sp}/c$, and, consequently,

$$\frac{\eta_{sp}}{c} \approx [\eta] = \frac{\Phi\langle R_G^2\rangle^{3/2}}{M} = \frac{\Phi E}{M} \tag{9-168}$$

where $E \equiv \langle R_G^2\rangle^{3/2}$. Correspondingly, for a polydisperse solute with i components, with $m_i = n_i M_i$, $w_i = c_i/c$, and $w_i = m_i/\sum_i m_i$, one has

$$\sum_i (\eta_{sp})_i = \Phi \sum_i E_i \frac{c_i}{M_i} = \Phi c \frac{\sum_i n_i E_i}{\sum_i m_i} \tag{9-169}$$

According to equation (9-130), the Staudinger index $[\eta]$ of a polydisperse solute is the weight average of the Staudinger indices of the individual components:

$$[\eta] = \frac{\sum_i c_i([\eta]_i}{\sum_i c_i} \approx \frac{\sum_i c_i(\eta_{sp})_i/c_i}{\sum_i c_i} = \frac{\sum_i (\eta_{sp})_i}{c} \tag{9-170}$$

Inserting equation (9-170) into (9-169), and recalling the definition of the number-average molecular weight $\overline{M}_n = \sum_i m_i / \sum_i n_i$, one obtains, after expanding with $\sum_i n_i / \sum_i n_i$,

$$[\eta] = \Phi \frac{\sum_i n_i E_i}{\sum_i m_i} = \Phi \frac{\sum_i n_i E_i}{\sum_i n_i} \frac{\sum_i n_i}{\sum_i m_i} = \Phi \frac{\overline{E}_n}{\overline{M}_n} = \Phi \frac{\langle R_G^2 \rangle_n^{3/2}}{\overline{M}_n} \tag{9-171}$$

Equation (9-171) shows that the number average of the 1.5 power of the mean square radius of gyration is obtained from viscosity measurements.

9.9.7. *Calibration of the Viscosity—Molecular Weight Relationship*

A comparison of the expressions giving the molecular-weight dependence of the Staudinger index shows that the relationships can be generalized for all macromolecular types in the form of the modified Staudinger equation $[\eta] = KM^{a_\eta}$ (see also Table 9-7). Both K and a_η are usually unknown. For each polymer homologous series, therefore, the modified Staudinger equation must be empirically determined. To do this, the molecular weights and Staudinger indices are determined for a number of samples (solvent, temperature = constant), and then $\log [\eta]$ is plotted against $\log M_2$ according to equation (9-156) (Figures 9-26 and 9-29). The slope is a_η and the intercept at $\log M_2 = 0$ is equal to K. A smaller slope is obtained at low molecular weights (end-group influence, deviation from

Table 9-7. Theoretical Exponents a_η of the Viscosity–Molecular Weight Relationship [Equation (9-169)]

Shape	Homology	a_η
Rods	Constant diameter; height proportional to M; no rotational diffusion	2
Rods	Same as above, but with rotational diffusion	1.7
Coils	Unbranched; free draining; no excluded volume	1
Coils	Unbranched; none draining; excluded volume	0.51–0.9
Disks	Diameter proportional to M, height constant	0.5
Spheres	Constant density, unsolvated or uniformly solvated	0
Disks	Diameter constant, height proportional to M	−1
Rods	Diameter proportional to $M^{0.5}$, height constant	−1
Rods	Diameter proportional to M, height constant	−2

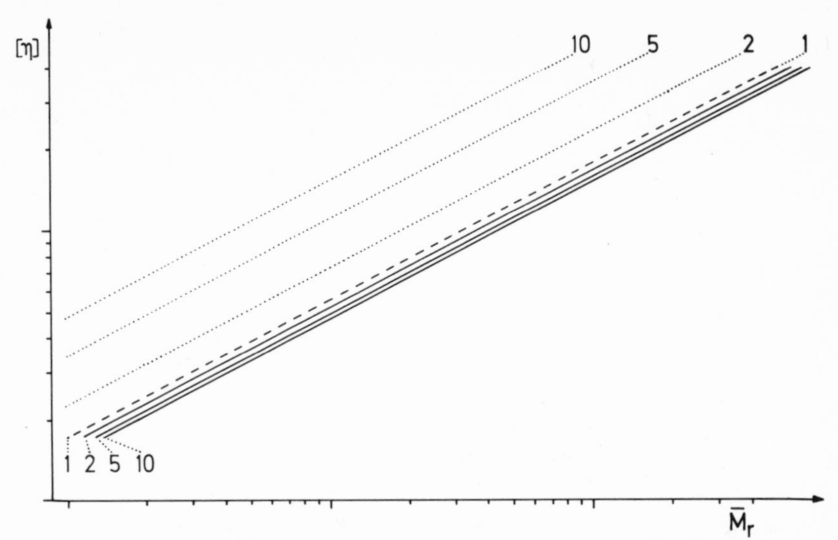

Figure 9-29. Influence of the molecular-weight distribution (expressed as $\overline{M}_w/\overline{M}_n$) on the molecular weight (M_r)–Staudinger index ([η]) relationship at $a_\eta = 0.5$. (···) $M_r = M_n$, (—) $M_r = M_w$.

random-flight statistics, etc.); this is overemphasized in Figure 9-26 because of the double logarithmic plot. With random-coil macromolecules, a_η increases with the thermodynamic goodness of the solvent. K decreases with increasing a_η (Table 9-8).

Since it is only valid to compare like averages with each other in a

Table 9-8. Constants K and a_η for Coil-like Macromolecules

Substance	Temperature, °C	Solvent	$10^3 K$, cm^3/g	a_η
at-Poly(styrene)	34	Benzene	9.8	0.74
at-Poly(styrene)	34	Butanone	28.9	0.60
at-Poly(styrene	35	Cyclohexane	78	0.50
Nylon 6,6	25	90% HCOOH	13.4	0.87
Nylon 6,6	25	*m*-Cresol	35.3	0.79
Nylon 6,6	25	2 *m* KCl in 90% HCOOH	142	0.56
Nylon 6,6	25	2.3 *m* KCl in 90% HCOOH	253	0.50
Cellulose tricaproate	41	Dimethyl formamide	245	0.50
Cellulose tricarbanilate	20	Acetone	4.7	0.84
Amylose tricarbanilate	20	Acetone	0.81	0.90
Poly(γ-benzyl-L-glutamate)	25	Dichloroacetic acid	2.8	0.87
Poly(γ-benzyl-L-glutamate)	25	Dimethyl formamide	0.00029	1.70

calibration of this kind, the question arises as to what kind of molecular-weight average is obtained from equation (9-156). According to equation (9-130), the Staudinger index of a polydisperse substance is a weight average. Equation (9-156) can thus be written as

$$KM_2^{a_\eta} = [\eta] = \frac{\sum_i W_i[\eta]_i}{\sum_i W_i} = \frac{\sum_i W_i K_i M_i^{a_\eta}}{\sum_i W_i} \tag{9-172}$$

At sufficiently high molecular weights, K_i is independent of M_2, so $K = K_i$. Solving equation (9-156) in the form of $[\eta] = K(\overline{M}_\eta)^{a_\eta}$ for $\overline{M}_\eta$ and inserting equation (9-172), we obtain

$$\overline{M}_\eta = \left(\frac{[\eta]}{K}\right)^{1/a_\eta} = \left(\frac{\sum_i W_i M_i^{a_\eta}}{\sum_i W_i}\right)^{1/a_\eta} \tag{9-173}$$

The molecular weight obtained by viscosity measurements is a viscosity average, which differs from the number average and the weight average (see also Figure 8-6). The viscosity-average molecular weight is only equal to the weight-average molecular weight when $a_\eta = 1$. For $a_\eta < 1$, $\overline{M}_\eta < \overline{M}_w$. Thus, for substances with identical molecular weight distributions, if $\log [\eta]$ is plotted against $\log \overline{M}_w$ instead of $\log \overline{M}_\eta$, the constant K thus obtained is too low for $a_\eta < 1$ (Figure 9-29). a_η is unaffected. If $\log [\eta]$ is plotted against $\log \overline{M}_n$, the K obtained is, conversely, too large. If the width of the molecular-weight distribution, or the distribution type, varies with the molecular weight, then incorrect values of K and a_η are obtained with plots of $\log [\eta]$ against either $\log \overline{M}_w$ or $\log \overline{M}_n$.

$\overline{M}_w$, $\overline{M}_n$, and a_η must be known before the viscosity-average $\overline{M}_\eta$ can be calculated for calibration. Then $\log [\eta]$ is plotted against $\log \overline{M}_w$, and K and a_η are determined. The viscosity average is then determined from the following approximation, which holds for distributions that are not too wide ($\overline{M}_w/\overline{M}_n < 2$):

$$\frac{\overline{M}_\eta}{\overline{M}_n} = \frac{1 - a_\eta}{2} + \frac{1 + a_\eta}{2}\frac{\overline{M}_w}{\overline{M}_n} \tag{9-174}$$

Finally, $\log [\eta]$ is plotted against $\log \overline{M}_\eta$, and K and a_η are again evaluated. This procedure is continued until K and a_η cease to change on further iteration.

9.9.8. Influence of the Chemical Structure on the Staudinger Index

The Staudinger index is a measure of the macromolecular dimensions. Thus, for flexible macromolecules, chain skeleton parameters (bond lengths, valence angles, degree of polymerization, mass of the monomeric unit), the steric hindrance parameter σ, as a measure of the hindrance to rotation, and the expansion factor α as a measure of the interaction with the solvent, determine the magnitude of the Staudinger index. In theta solvents, $\alpha = 1$. Thus, from equations (9-153) and (9-151)

$$K_\Theta = \Phi_0 \left(\frac{\langle R_G^2 \rangle_\Theta}{M_2} \right)^{3/2} = \Phi_0 \left(\frac{\langle R_G^2 \rangle_\Theta}{X_2} \right)^{3/2} (M_u)^{-3/2} \qquad (9\text{-}175)$$

According to equation (4-14), $\langle R_G^2 \rangle_\Theta$ is proportional to σ^2. Thus, if the hindrance parameter is independent of the type of substituents, log K plotted against log M_u would have a slope of $-3/2$. In compounds of the type $(\text{—CH}_2\text{—CR}^1\text{R}^2\text{—})_n$, the slope is more positive (Figure 9-30). Thus, the hindrance parameter increases with the formula molecular weight M_u of the monomeric unit, i.e., σ increases with the size of the substituents R^1 and R^2. Values that are far too high are obtained, for example, with poly(vinyl N-carbazole). Figure 9-30 can also serve to estimate K values for polymers for which the viscosity–molecular weight relationship is not known.

K_Θ values for copolymers cannot be obtained from a linear extra-

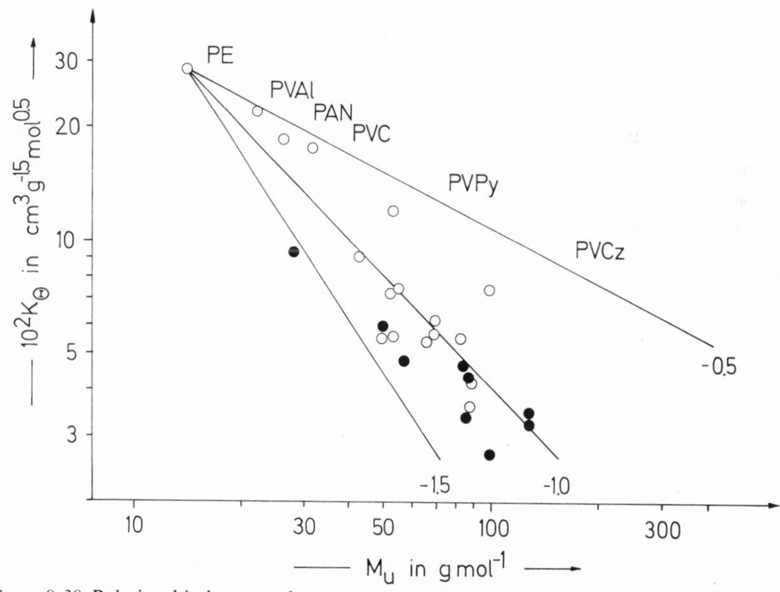

Figure 9-30. Relationship between the constant K_Θ of the molecular weight relationship $[\eta] = K_\Theta M^{0.5}$ and the formula molecular weight M_u of the monomeric units in polymers of the type $(\text{—CH}_2\text{—CHR—})_n$ (○) or $(\text{—CH}_2\text{—CRR'—})_n$ (●). PE, poly(ethylene); PVAl, poly(vinyl alcohol); PAN, poly(acrylonitrile); PVC, poly(vinyl chloride); PVPy, poly(2-vinyl pyridine), PVCz, poly(N-vinyl carbazole). The figures give the slope.

polation of the K_Θ values for the unipolymers. For example, with poly(*p*-chlorostyrene) and poly(methyl methacrylate), K_Θ values of 0.050 and 0.049 cm^3/g, respectively, are found. For a copolymer with $x_{mma} = 0.484$, however, a value of $K_\Theta = 0.064$ cm^3/g is found. The copolymer coil has therefore been expanded because of the mutual repulsion of the polar groups.

Branching reduces the hydrodynamic volume relative to the mass of the coil. The Staudinger index of branched macromolecules is thus lower than that of their unbranched (linear) counterparts. The effect is particularly marked in the case of long-chain branching. If the number of branch points in a polymer homologous series increases with the molecular weight, then the $[\eta]$ values also decrease relative to those of linear molecules. Thus, the slope of the $\log [\eta] = f(\log M_2)$ curve continuously decreases with increasing molecular weight, and such an observation can be taken as evidence of branching. The observation is not unambiguous, and one must ascertain that the observation does not come from the effect of shear stress on the Stuadinger index, which also increases with molecular weight (see Section 9.9.2).

Polymers of higher tacticity have more rigid chains than atactic polymers, since more base units are fixed in conformational sequences of the coil, which corresponds formally to an increase in the steric hindrance parameter σ. With a predominantly isotactic poly(methyl methacrylate), for example, a $K_\Theta = 0.087$ cm^3/g was obtained in heptanone-3 ($\Theta = 40.0°C$), whereas an at-PMMA in the same solvent gave only $K_\Theta = 0.063$ cm^3/g ($\Theta = 33.7°C$). At higher temperatures, potential barriers to rotation can be more readily surmounted. The K_Θ values then become practically identical: 0.057 for it-PMMA ($\Theta = 152.1°C$) and 0.058 for at-PMMA ($\Theta = 159.7°C$) in *p*-cymene.

9.9.9. Temperature Dependence of the Staudinger Indices

The Staudinger index is a function of the hydrodynamic volume, and this itself depends on the mass, shape, and density of the particle. The density of the particle is dependent on the flexibility of the molecular chain and the solvation. The flexibility and solvation of rigid particles such as spheres, ellipsoids, and rods vary only slightly with temperature. Thus, the Staudinger index of these particles is practically temperature independent.

With flexible macromolecules, the coil expands with increasing temperature, probably because solvation increases with temperature. The expansion occurs, however, only up to a limiting value that is given by the skeletal parameters and the optimum degree of solvation. At still higher temperatures, the flexibility of the chain increases because of a decrease in the hindrance to rotation, and the Staudinger index again decreases. Thus, with rising temperature, $[\eta]$ first increases sharply, and then possibly passes through a maximum.

The increase in Staudinger index with temperature is technologically made use of in the case of the so-called viscosity improvers (VI). The viscosity of lubricating oils decreases as the temperature rises. This increase in fluidity is not always desirable, so macromolecules are added to the lubricant such that they dissolve in the oil and are in the theta state, approximately, at the lowest application temperature. With increasing temperature, the viscosity η_1 of the lubricating oil indeed falls. The Staudinger index $[\eta]$ of the added polymer, on the other hand, increases. With a suitable choice of one or more polymers as additives, a compensation of both effects can be attained, as seen by writing equation (9-139), with $\eta_{sp} = (\eta - \eta_1)/\eta_1$, in the form

$$\eta = \eta_1 + \eta_1[\eta]c + \eta_1 k_H[\eta]^2 c^2 \quad (9\text{-}176)$$

and so a solution viscosity η which is predominantly temperature independent is obtained. Good lubricating oil improvers are, for example, copolymers of methyl methacrylate and lauryl methacrylate. The lauryl residues aid the solubilizing of the poly(methyl methacrylate), which is too polar to dissolve directly in the apolar lubricating oil.

Literature

Section 9.1. Molecular Weight and Molecular-Weight Distribution

R. U. Bonnar, M. Dimbat, and F. H. Stross, *Number Average Molecular Weights*, Interscience, New York, 1958.

P. W. Allen, ed., *Techniques of Polymer Characterization*, Butterworths, London, 1959.

Ch'ien Jên-Yuen, *Determination of Molecular Weights of High Polymer*, Oldbourne Press, London, 1963.

S. R. Rafikov, S. Pavlova, and I. I. Tverdokhlebova, *Determination of Molecular Weights and Polydispersity of High Polymers*, Acad. Sci., USSR, Moscow 1963; Israel Program for Scientific Translation, Jerusalem, 1964.

Characterization of Macromolecular Structure, Natl. Acad. Sci. U.S.A. Pub. 1973, Washington DC., 1968.

Section 9.2. Membrane Osmometry

H. Coll and F. H. Stross, Determination of molecular weights by equilibrium osmotic pressure measurements, *in: Characterization of Macromolecular Structure*, Natl. Acad. Sci. U.S. Publ. 1573, Washington, D.C., 1968.

H.-G. Elias, Dynamic osmometry, *in: Characterization of Macromolecular Structure*, Natl. Acad. Sci. U.S. Publ. 1573, Washington, D.C., 1968.

H. Coll, Nonequilibrium osmometry, *J. Polymer Sci. D (Macromol. Rev.)* **5**, 541 (1971).
M. P. Tombs and A. R. Peacock, *The Osmotic Pressure of Biological Macromolecules*, Clarendon Press, Oxford, 1975.

Section 9.3. Ebulliometry and Cryoscopy

R. S. Lehrle, Ebulliometry applied to polymer solutions, *Prog. High Polym.* **1**, 37 (1961).
M. Ezrin, Determination of molecular weight by ebulliometry, *in: Characterization of Macromolecular Structure*, Natl. Acad. Sci. U.S. Publ. 1573, Washington, D.C., 1968.

Section 9.4. Vapor-Phase Osmometry

W. Simon and C. Tomlinson, Thermoelektrische Mikrobestimmung von Molekulargewichten, *Chimia* **14**, 301 (1960).
K. Kamide and M. Sanada, Molecular weight determination by vapor pressure osmometry, *Kobunshi Kagaku (Chem. High Polym. Japan)* **24**, 751 (1967) (in Japanese).
J. van Dam, Vapor-phase osmometry, *in: Characterization of Macromolecular Structure*, Natl. Acad. Sci. U.S. Publ. 1573, Washington, D.C., 1968.

Section 9.5. Light Scattering

D. MacIntyre and F. Gornick, eds., *Light Scattering from Dilute Polymer Solutions*, Gordon and Breach, New York, 1964.
K. A. Stacey, *Light Scattering in Physical Chemistry*, Butterworths, London, 1956.
M. Kerker, *The Scattering of Light and Other Electromagnetic Radiation*, Academic Press, New York, 1969.
M. B. Huglin, ed., *Light Scattering from Polymer Solutions*, Academic Press, London, 1972.

Section 9.6. Small-Angle X-ray Scattering and Neutron Diffraction

H. Brumberger, ed., *Small Angle X-ray Scattering*, Gordon and Breach, New York, 1967.
T. L. Cottrell and I. C. Walker, Interaction of slow neutrons with Molecules, *Q. Rev. (London)* **20**, 153 (1966).
G. Allen and C. J. Wright, Neutron scattering studies of polymers, *in: Macromolecular Science* (Vol. 8 of Physical Chemistry Series 2) (C. E. H. Bawn, ed.), MTP International Review of Science, 1975, p. 223.

Section 9.7. Ultracentrifugation

T. Svedberg and K. O. Perdersen, *Die Ultrazentrifuge*, D. Steinkopff, Dresden, 1940; The Ultracentrifuge, Clarendon Press, Oxford, 1940.
H. K. Schachmann, *Ultracentrifugation in Biochemistry*, Academic Press, New York, 1959.
R. L. Baldwin and K. E. van Holde, Sedimentation of high polymers, *Fortschr. Hochpolym. Forschg.* **1**, 451 (1960)
H.-G. Elias, *Ultrazentrifugen-Methoden*, Beckman Instruments, München, 1961.
H. Fujita, *Mathematical Theory of Sedimentation Analysis*, Academic Press, New York, 1962.
J. Vinograd and J. E. Hearst, Equilibrium sedimentation of macromolecules and viruses in a density gradient, *Fortschr. Chem. Org. Naturstoffe* **20**, 372 (1962).
J. W. Williams, ed., *Ultracentrifugal Analysis in Theory and Experiment*, Academic Press, New York, 1963.
H. Fujita, *Foundations of Ultracentrifugal Analysis*, Wiley, New York, 1975.

Section 9.8. Chromatography

G. M. Guzmàn, Fractionation of high polymers, *Prog. High Polym.* **1**, 113 (1961).
R. M. Screaton, Column fractionation of polymers, *in: Newer Methods of Polymer Characterization,* (B. Ke, ed), Interscience, New York, 1964.
J. F. Johnson, R. S. Porter, and M. J. R. Cantow, Gel permeation chromatography with organic solvents, *Rev. Macromol. Chem.* **1**, 393 (1966).
M. J. R. Cantow, ed., *Polymer Fractionation,* Academic Press, New York, 1967.
H. Determann, *Gelchromatographie,* Springer, Berlin, 1967.
J. F. Johnson and R. S. Porter, Gel permeation chromatography, *Prog. Polym. Sci.* **2**, 201 (1970).
K. H. Altgelt and L. Segal, *Gel Permeation Chromatography,* Dekker, New York, 1971.
N. Friis and A. Hamielec, Gel permeation chromatography—review of axial dispersion phenomena, their detection and correction, *Adv. Chromatog.* **13**, 41 (1975).

Section 9.9. Viscometry

G. Meyerhoff, Die viskosimetrische Molekulargewichtsbestimmung von Polymeren, *Fortschr. Hochpolym. Forschg.—Adv. Polym. Sci.* **3**, 59 (1961/64).
M. Kurata and W. H. Stockmayer, Intrinsic viscosities and unperturbed dimensions of long chain molecules, *Fortschr. Hochpolym. Forschg.—Adv. Polym. Sci.* **3**, 196 (1961/64).
H. van Oene, Measurement of the Viscosity of Dilute Polymer Solutions, *in: Characterization of Macromolecular Structure,* Natl. Acad. Sci. U.S. Publ. 1573, Washington, D.C., 1968.
H. Yamakawa, *Modern Theory of Polymer Solutions,* Harper & Row, New York, 1971.

Other Methods

D. V. Quayle, Molecular weight determination of polymers by electron microscopy, *Br. Polym. J.* **1**, 15 (1969).

Part III
Solid-State Properties

Chapter 10
Thermal Transitions

10.1. Basic Principles

10.1.1. Phenomena

Low-molecular-weight materials change their physical state as the temperature increases; at the melting point they change visibly from a crystal to a liquid, and at the boiling point from a liquid to a vapor. Each true phase transition is defined thermodynamically by a marked change in the enthalpy. However, since changes in enthalpy can only be determined with expensive instruments, other methods are generally employed to determine the transition temperatures. For example, in organic chemistry melting points are measured via the formation of the liquid state in the melting-point tube. This method can be used for the determination of the melting point because the viscosity at the melting point changes by several orders of magnitude and the viscosity of the melt is very low. Thus, the melting-point really represents a primitive viscometer.

The method must fail when the viscosity of the melt is so high that flow can no longer be perceived within the observation time of the experiment. This is the case with highly fused aromatic rings such as coronene, and even more so in crystalline macromolecular substances. In the case of coronene, the melting point cannot be determined unambiguously with the melting-point tube. In macromolecules, the "melting point" determined in the melting-point tube is really a flow point, which, in certain cases, can lie well above the true melting point because of the high melt viscosity.

Noncrystalline materials can also show flowing in the melting-point tube. For example, radically polymerized styrene changes on heating from

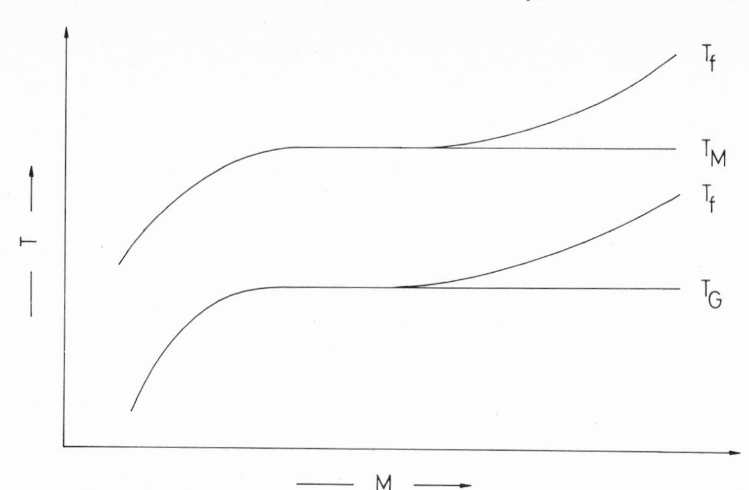

Figure 10-1. Schematic representation of the dependence of the melting, glass-transition, and flow temperatures T_M, T_G, *and* T_f on the molecular weight for a polymer homologous series.

a brittle, rather, glassy material into one which is softer and rubbery. As X-ray analysis shows, this poly(styrene) is amorphous, and there is no doubt that this is *not* a melting temperature. It is more likely an indication of a glass-transition temperature: a transition from a glassy state into a rubbery one, the "melt." For low-molecular-weight amorphous materials, the glass-transition and flow temperatures are practically identical (Figure 10-1). Naturally, glass-transition temperatures cannot be observed in 100% crystalline polymers. In partially crystalline polymers, on the other hand, both melt and glass temperatures occur.

As well as melt, glass, and flow temperatures, there is a series of other transition points, but, in contrast to the aforementioned three transition temperatures, it is not immediately possible to assign a molecular phenomenon to the other transitions. When there is uncertainty about the physical significance of the experimentally observed transition points, the transition that occurs at the highest temperature is termed the α transition, the next lowest the β transition, etc.

10.1.2. Thermodynamics

First-order thermodynamic transitions are defined as a sharp discontinuity in the *first* derivative of the Gibbs energy G, i.e., a sharp discontinuity in the enthalpy H, the entropy S, or the volume V (Figure 10-2):

$$H = G - T\left(\frac{\partial G}{\partial T}\right)_p \tag{10-1}$$

$$S = -\left(\frac{\partial G}{\partial T}\right)_p \tag{10-2}$$

$$V = \left(\frac{\partial G}{\partial p}\right)_T \tag{10-3}$$

If the first derivative shows a sharp discontinuity, the second derivative must also show a sharp discontinuity for the transition, i.e., a sharp discontinuity in the heat capacity C_p, the cubic expansion coefficient α, and the isothermal compressibility κ (Figure 10-2):

$$C_p = \left(\frac{\partial H}{\partial T}\right)_p = T\left(\frac{\partial S}{\partial T}\right)_p = -T\left(\frac{\partial^2 G}{\partial T^2}\right)_p \tag{10-4}$$

$$\alpha = V^{-1}\left(\frac{\partial V}{\partial T}\right)_p \tag{10-5}$$

$$\kappa = -V^{-1}\left(\frac{\partial V}{\partial p}\right)_T \tag{10-6}$$

The melting point is a typical first-order transition.

Conversely, second-order thermodynamic transitions are defined as those transitions that first show a sharp discontinuity for the *second* derivative of the Gibbs energy, i.e., for C_p, α, and κ. In second-order transitions,

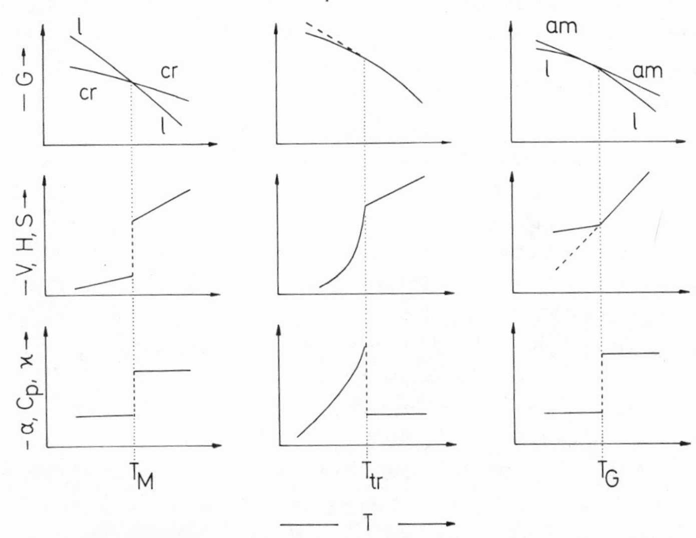

Figure 10-2. Schematic representation of various thermal transitions: the melting process as a first-order thermodynamic process, a rotational transition as a second-order thermodynamic process, and a glass transition; l, liquid; cr, crystal; am, amorphous state (after G. Rehage and W. Borchard).

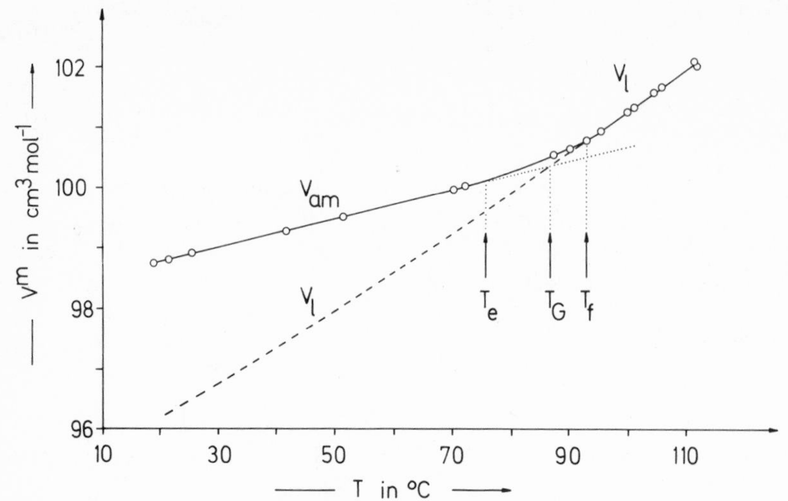

Figure 10-3. The molar volume of an atactic poly(styrene) with $\overline{M}_n = 20{,}000$ g/mol as a function of temperature. T_e, softening point; T_G, glass-transition temperature; T_f, freezing-in temperature (after G. Rehage).

the first derivative of the Gibbs energy and the Gibbs energy itself are continuous through the transition point (Figure 10-2). Genuine first- and second-order thermodynamic transitions are characterized by the fact that a thermodynamic equilibrium exists between each side of the transition point. All definitely known second-order thermodynamic transitions are single-phase transitions. Examples are rotational transitions in crystals and the disappearance of ferromagnetism at the Curie point.

Several aspects of a second-order thermodynamic transition can be observed with a glass transition, i.e., discontinuities in C_p, α, and κ (Figure 10-2). But the glass transition is not a genuine thermodynamic transition since there is no equilibrium between *both* sides of the glass-transition point. The position of the glass-transition point depends essentially on the rate of cooling, the transition temperature lying at lower temperatures for lower rates of cooling. At very low cooling rates, no sharp discontinuity at all is obtained for the volume as a function of the temperature (Figure 10-3), that is, the glass-transition point has disappeared.

The curve for very slow coolings cannot be obtained by direct measurement. To obtain such a curve, one can, for example, define a time $t_{1/e}$ for which the distance from the equilibrium curve is $1/e$ times the initial deviation. For the poly(styrene) data shown in Figure 10-3 ($T_G = 89$°C), this time is 1 s at 95°C, 5 min at 89°C, but already 1 year at 77°C. The dotted curve in Figure 10-3 was consequently obtained by extrapolation of the values for a poly(styrene) solution in malonic ester. This extrapolation is possible since the cubic expansion coefficient of the polymer in the solvent is a linear function of the concentration up to a poly(styrene) concentration of 90%.

The lowering of the glass-transition temperature by slower cooling indicates a kinetic cause. In genuine thermodynamic second-order transitions, the transition temperature does not depend on the rate of cooling.

Another point against the glass-transition temperature being a thermodynamic transition point is the fact that C_p, α, and κ are smaller below the glass-transition temperature than above it. This is exactly the reverse of what would be expected of a genuine second-order thermodynamic transition (Figure 10-2).

As already mentioned in Section 5.5.1, chain-segment mobility becomes frozen in at and below the glass-transition temperature. Perfect crystals do not have any mobile segments, since all segments are firmly set in the crystal lattice. Consequently, highly crystalline macromolecules do not show a glass-transition temperature. Conversely, amorphous polymers do not have a melting temperature since a melting point requires a crystal lattice.

On the other hand, a partially crystalline polymer possesses both a glass-transition temperature and a melting temperature (Figure 10-4). Since the chain segments still retain some mobility even below the glass-transition temperature (see also Section 10.5), crystallization can start even below the

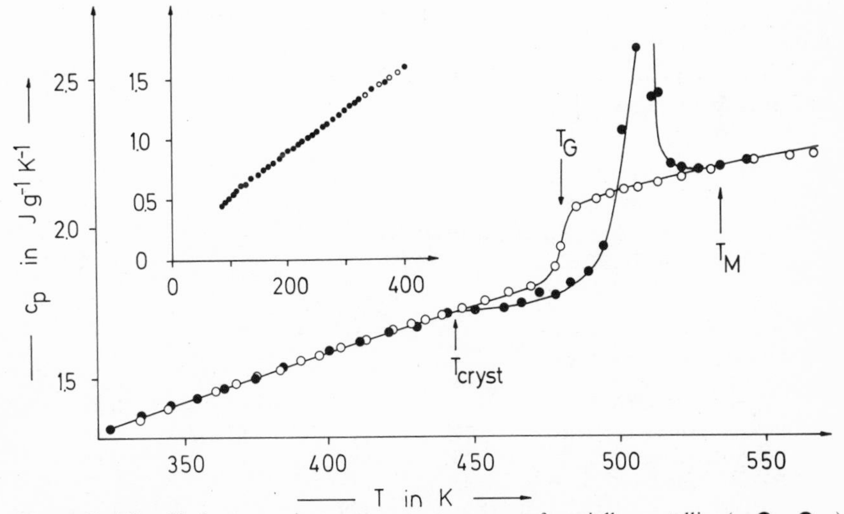

Figure 10-4. Specific heat capacity c_p at constant pressure of partially crystalline (—●—●—) and amorphous (—○—○—) poly[oxy-(2,6-dimethyl)-1,4-phenylene]. T_{cryst} denotes the beginning of recrystallization, T_G is the glass-transition temperature, and T_M is the melting temperature (after F. R. Karasz, H. E. Bair, and J. M. O.Reilly).

glass-transition temperature (Figure 10-4) because of the ever-present crystallization nucleators (see Section 10.3.2). On heating, there is a constant interchange of chain segments between the crystalline and noncrystalline regions so that they are constantly being redistributed. Consequently, a melting *region* is observed instead of a melting point. The upper end of the melting region is defined as the melting temperature T_M of the sample. This melting temperature T_M lies lower than the thermodynamic melting temperature of the perfect crystal.

10.2. Special Parameters and Methods

10.2.1. Expansion

Thermal expansion depends on variations in the interatomic forces with temperature. These forces are strong for covalent bonds and weak for dispersion forces. For example, for quartz, all atoms are three-dimensionally fixed in space: The thermal expansion is consequently very small. On the other hand, in liquids, intermolecular forces are dominant: The thermal expansion is large. The main-chain atoms of organic polymers are covalently bonded in one direction only; in the two other directions in space only intermolecular forces are operative. Thus, polymers lie between liquids and quartz (or metals) as far as thermal expansion is concerned (Table 10-1).

Because of the great difference in the thermal expansion of polymers and metals or glass, significant problems can arise when thermal stress is

Table 10-1. Density ρ, Specific Heat Capacity at Constant Pressure c_p, Linear Coefficient of Expansion β, and Thermal Conductivity λ of Polymers, Metals, and Glass at 25°C

Material	ρ, g/cm^3	c_p, J g^{-1} K^{-1}	$10^5\beta$, K^{-1}	λ, J m^{-1} s^{-1} K^{-1}
Poly(ethylene)	0.92	2.1	20	0.35
Poly(styrene)	1.05	1.3	7	0.16
Poly(vinyl chloride)	1.39	1.2	8	0.18
Poly(methyl methacrylate)	1.19	1.5	8.2	0.20
Poly(caprolactam)	1.13	1.9	8	0.29
Poly(oxymethylene)	1.42	1.5	9.5	0.23
Copper	8.9	0.39	2	350
Cast iron	7.25	0.54	1	58
Jena glass 16 III	2.6	0.78	1	0.96
Quartz (spatial average)	2.65	0.72	0.1	10.5

Table 10-2. Cubic or Volume Expansion Coefficients α_{am} and α_l

Polymer	T_G, K	$10^4\alpha_l$, K^{-1}	$10^4\alpha_{am}$, K^{-1}	f_{exp} eq. (10-10)	f_{exp} eq. (5-6)
Poly(ethylene)	203	8.9	4.7	0.94	—
Poly(isobutylene)	200	6.2	1.5	0.097	0.125
Poly(styrene)	373	5.5	2.5	0.118	0.127
Poly(vinyl acetate)	301	6.0	1.8	0.134	0.14
Poly(methyl methacrylate)	383	4.6	2.2	0.100	0.13
Poly(acrylonitrile)	377	3.4	1.8	0.065	—

applied to composites of these materials. The so-called dimensional stability of the polymer is also of technological importance. Dimensionally stable polymers must not only possess a small coefficient of thermal expansion: They should also not exhibit recrystallization phenomena. Recrystallizations lead to distortions because of the difference in densities between amorphous and crystalline regions.

Expansion coefficients above and below the glass-transition temperature are different (Figures 10-2, 10-3, and Table 10-2). A relationship between these two quantities and the glass-transition temperature exists.

To a first approximation, volumes change linearly with temperature. Consequently, with the definition of the cubic expansion coefficient α, the following is obtained for the liquid state and the amorphous state:

$$(V_l^0)_T = (V_l^0)_0 + (V_l^0)_T \alpha_l T \tag{10-7}$$

$$(V_{am}^0)_T = (V_{am}^0)_0 + (V_{am}^0)_T \alpha_{am} T \tag{10-8}$$

To a first approximation, the volume elements of liquid and amorphous material are equal at the glass-transition temperature. Consequently, $(V_l^0)_G = (V_{am}^0)_G$. Equating equations (10-7) and (10-8) and reintroducing equation (10-8), we obtain

$$\left[\frac{(V_{am}^0)_0 - (V_l^0)_0}{(V_{am}^0)_0}\right](1 - \alpha_{am} T_G) = (\alpha_l - \alpha_{am})\, T_G \tag{10-9}$$

The volumes of liquid and crystal must be equal at 0 K. The term in square brackets in equation (10-9) must give the free-volume fraction [see equation (5-10)]. We obtain

$$f_{exp} = (\alpha_l - \alpha_{am})\, T_G/(1 - \alpha_{am} T_G) \tag{10-10}$$

The free-volume fraction f_{exp} calculated from equation (10-10) (see Table 10-2) agrees very well with the values given in Table 5-6. It has been empirically found that the product $(\alpha_l - \alpha_{am}) T_G$ generally has a value of about 0.11 for a large number of amorphous polymers. With crystalline polymers, on the other hand, the values are lower.

10.2.2. Heat Capacity

The heat capacity ("specific heat") C_p of macromolecular substances at constant pressure is the only heat capacity readily accessible experimentally. For theoretical considerations, however, the heat capacity at constant volume C_V is important. According to the laws of thermodynamics, these two quantities are related to each other via the cubic expansion coefficient α and the isothermal compressibility κ:

$$C_p = C_V + \frac{TV\alpha^2}{\kappa} \tag{10-11}$$

The molar heat capacity of crystalline polymers at constant volume C_v^m can be theoretically calculated when the frequency spectrum is known. Atoms oscillate harmonically about their equilibrium positions in the crystalline state. In accordance with the Einstein function, each individual oscillation contributes

$$E\left(\frac{\Theta}{T}\right) = \Theta^2 \frac{\exp(\Theta/T)}{1 - \exp(\Theta/T)} \tag{10-12}$$

to the total heat capacity. Here $\Theta = h\nu/k$ is the Einstein temperature. The molar heat capacity is simply the sum of all these contributions

$$C_V^m = R \sum E\left(\frac{\Theta}{T}\right) \tag{10-13}$$

At very low temperatures, these lattice oscillations comprise almost all of the heat capacity. At higher temperatures, a correction for the inharmonicity of the lattice oscillations must be considered. In addition, contributions from group oscillations and rotations about main-chain bonds must also be added at higher temperatures. Finally, a contribution from defects may also be needed.

In fact, the heat capacities of amorphous and crystalline polymers are practically the same below the glass-transition temperature (Figure 10-4). Because of the onset of new oscillations at the glass-transition temperature, the heat capacity increases more or less sharply. Since such oscillations can start below the glass-transition temperature (see Section 10.2.4), crystallizable amorphous polymers can occasionally even start to recrystallize below the glass-transition temperature. The heat capacity then passes through a maximum when melting occurs. The melting temperature is then the upper end of the melting range.

10.2.3. Differential Thermal Analysis

In differential thermal analysis (DTA), a test sample and a comparison sample are heated at a constant rate. The temperature difference between the two samples is measured. The comparison sample should not show any chemical or physical changes in the range of temperatures studied. When, for example, the temperature reaches the melting point of the test sample, then a fixed amount of heat (heat of fusion) must be applied until the whole of the test sample has melted. The temperature of the test sample does not change at the melting point, whereas that of the comparison sample continues to rise. At the melting point, therefore, an endothermal process is observed, i.e., a negative ΔT (Figure 10-5). As heating continues, the test sample finally attains the same temperature as the comparison sample again, and ΔT becomes zero. Since the substances generally show different heat capacities above and below the melting point, however, the baseline at both sides of the signal for the melting point is usually not at the same level. In some cases it is not even parallel to the temperature axis.

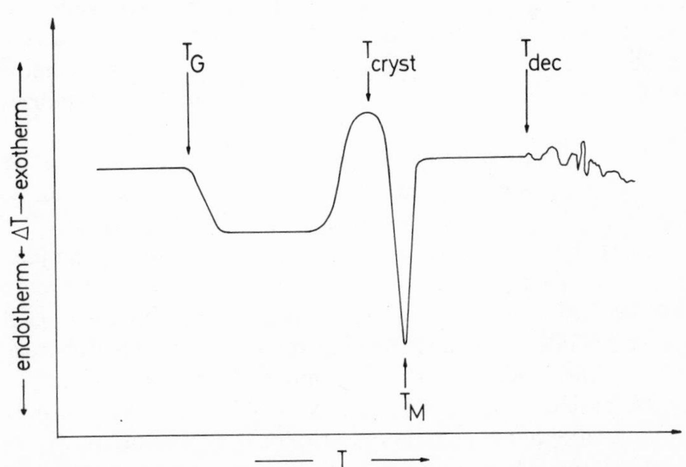

Figure 10-5. Illustration of the thermogram of a partially crystalline polymer with the glass-transition temperature T_G, the recrystallization temperature T_{cryst}, the melting point T_M, and the decomposition temperature T_{dec}.

Exotherms occur during many chemical reactions and during recrystallization below the melting point. The elimination of volatile components is usually apparent from a fluctuation of the baseline since the average heat capacity of the sample changes constantly because of gas loss. Glass-transition temperatures usually are revealed by a jump in the $\Delta T = f(T)$ curve. According to the rate of heating, quantity of sample used, and the heat capacity involved, however, similar signals are often observed for glass-transition and melting points.

DTA is particularly suitable for routine analysis, as it can be carried out simply and quickly. On the other hand, measurements on unknown samples can often be interpreted only with difficulty and sometimes only if data from other methods are available. Quantitative information is also not readily accessible by this technique. Uncertainties in quantitative interpretation result because signal size and shape depend on the experimental conditions. For example, a signal with a minimum in the $\Delta T = f(T)$ curve is usually ascribed to a melting point, although the same signal can indicate a glass-transition temperature, especially at high rates of heating. A distinction between a melting and a glass-transition temperature can be made by observing the thermal behavior of the samples under a polarizing microscope. Crystalline polymers are birefringent below the melting point (if the crystallite size is greater than the wavelength of light), the birefringence vanishing at the melting point. If the DTA signal corresponds to a melting process, the minimum of the DTA signal is taken as the melting point because the minimum in the DTA signal at zero rate of heating corresponds to the temperature at which the most abundant crystallite size melts. The melting point thus defined usually lies higher for greater rates of heating, since heat flow into the sample is faster than heat removal from the sample. The reverse effect is observed if the sample recrystallizes during heating.

The glass-transition temperature is taken as the onset of a deviation either from the baseline or from the inflection point. Here, too, for more accurate measurements it is always necessary to determine to what extent the results depend on the rate of heating and the amount of sample. Larger quantities of sample lead to a greater temperature gradient and to a smaller heat exchange, causing the signal to broaden and possibly even to be shifted to higher temperatures. Higher heating rates produce greater areas under the signals, as more heat of fusion is eliminated per unit of time.

In differential scanning calorimetry (DSC), the sample is not heated at a constant rate, but a definite quantity of heat is either added or taken away isothermally. This method is particularly suited to measure the heats of fusion at crystallization or to follow the course of crystallization at a given temperature.

10.2.4. Nuclear Magnetic Resonance

Atomic nuclei with uneven numbers of protons possess a magnetic moment and therefore precess about a magnetic field from an oscillating radio frequency. If the oscillating frequency of the electromagnetic field equals the precession frequency of the atomic nuclei, a resonance signal is observed whose frequency depends on the ratio of the nuclear magnetic field to the rotational torque on the nucleus and on the strength of the external, steady, magnetic field.

High-resolution nuclear magnetic resonance detects the shielding from neighboring protons in the same molecule. Such high-resolution spectroscopy methods can be used to elucidate the configuration and constitution of molecules. Much higher concentrations of, and consequently stronger interactions between, the magnetic dipoles of different nuclei exist in solids below the glass temperature or in melts. The magnetic dipoles of these neighboring nuclei have a distribution of orientations relative to a given nucleus. A broad signal thus results.

Chain segments become more mobile with increasing temperature. The distribution of orientations around a given nucleus thus becomes more and more random. The increased anisotropy of dipole interactions leads to a sharpening of the signals. A line width is thus a measure of the mobility of the molecules, and thence the glass-transition temperature. Since the measurements are carried out at frequencies in the MHz range, the glass-transition temperatures from NMR measurements are higher than those from static measurements of the temperature dependence of heat capacities or by DTA (see Section 10.5.2). Broad-line NMR also measures side-group motion, but not that from short chain segments below the glass-transition temperature. The method is not very suitable for determining melting points. In fact, the resonance signals continue to sharpen even well below the melting point of crystalline polymers, whereas the X-ray crystallinity remains constant. Therefore, there must be a certain segmental mobility within the crystal lattice even below the melting point.

A magnetic polarization results from the sudden imposition of a magnetic field on a sample. The magnetization generally follows an exponential function. The time constant T_1 is known as the spin–lattice relaxation time. Consequently, the nuclear magnetic resonance experiment corresponds macroscopically to a dielectric relaxation experiment. Differences exist, however, on the molecular level. The nuclear magnetization is, of course, equal to the sum of all the individual nuclear magnetic moments. The orientation of these nuclear magnets, however, is only loosely coupled with the molecular positions. Consequently, T_1 is generally much larger than the molecular relaxation time from dielectric relaxation measurements (see also Section 10.5.2).

10.2.5. Dynamic Methods

The mechanical and dielectric loss methods are based on the different mobilities of the chain segment or dipoles bound to them in the solid state and in the melt. These differences lead to an anomalous dispersion of the modulus of elasticity (see Section 11.4.4) or the relative permittivity (see Section 14.1.2) and to corresponding losses in the mechanical or the electrical alternating fields.

Part of the work performed on a sample will be converted irreversibly into random thermal motion by movement of the molecules or molecule segments. This loss passes through a maximum at the appropriate transition temperature or relaxation frequency in the associated alternating mechanical field (torsion pendulum test). A similar effect is obtained by the delayed response of the dipoles with dielectric measurements. Therefore, dielectric measurements can be made only on polar polymers. According to the frequency used, the glass-transition temperatures measured with dynamic methods lie higher than those obtained by quasistatic methods (see Section 10.5.2).

Of course, such dynamic mechanical test methods are only suitable for such probes as can support their own weight. In examining paints, lacquers, and non-self-supporting films, a glass fiber cord is impregnated with a solution of the test material and the solvent is evaporated. The impregnated cord is then subjected to periodic oscillations (torsional braid analysis).

10.2.6. Industrial Testing Methods

A whole series of empirical test methods are used in industry to determine the physical transitions. These methods are usually simultaneously influenced by various physical quantities and must therefore be standardized. The standards vary among countries.

The temperature at which a sample breaks when struck is measured in brittleness determinations. At this temperature, larger chain segments can no longer be displaced. In contrast, the mobility of much smaller chain segments is effective at the glass-transition temperature, so the brittle temperature always lies higher than the glass-transition temperature. The brittleness temperature does not depend solely on the mobility of larger chain segments; it also depends on the elasticity of the test sample, since the break behavior is influenced by the deformability of the sample. Thin samples are, however, more elastic than thick ones. Brittleness temperatures decrease with increasing molecular weight up to a limiting molecular-weight value, since increased molecular length leads to increased strength (Figure 10-6).

The dimensional stability of plastics on heating is characterized (in Germany) by Martens numbers (or temperatures) or Vicat temperatures. The Martens number is defined as the temperature at which a standard test specimen at a standard rate of heating bends to a fixed extent under a fixed load. Thus, elasticity also influences the result in this method. The ASTM heat-distortion temperature is measured in a similar way. The temperatures are not comparable, however, because of differences in these two methods. The Vicat softening point measures the penetration of a needle to a fixed depth under otherwise constant conditions (load, specimen dimensions, etc.). Thus, this method also depends on the surface hardness. The Vicat softening point for amorphous polymers generally differs by about 5–10°C from the "static" glass-transition temperature. Martens numbers of amorphous polymers are about 20–25°C lower than the glass-transition temperature.

All the methods named are suitable only for studies on samples with a single transition temperature. For this reason, they cannot be used effectively with partly crystalline substances or with polymer blends (e.g., high-impact plastics).

In laboratories where preparative work is done, "softening points" are often determined with the Kofler bar, which consists of a metal plate with a temperature gradient along it. The sample is moved with a brush

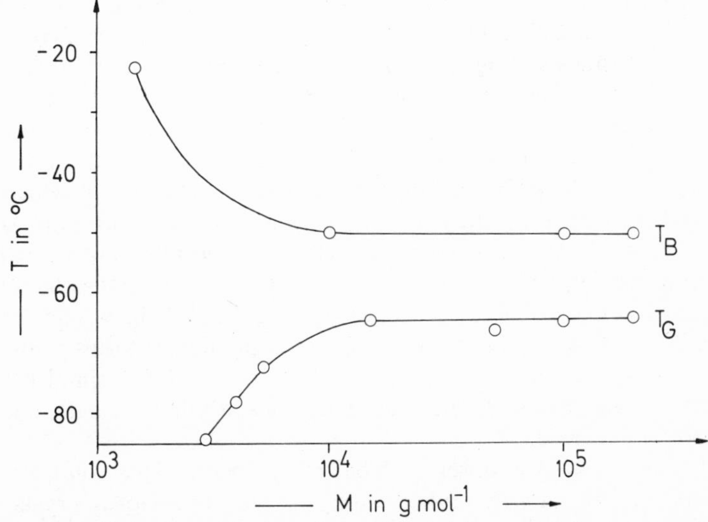

Figure 10-6. Molecular weight dependence of the brittleness and the glass-transition temperatures T_B and T_G of poly(isobutylene) (after A. X. Schmidt and C. A. Marlies).

from the colder to the warmer points on the metal plate. At a given point, the sample will remain stuck to the plate; the temperature associated with this point is taken as the "softening point." Since this temperature depends on both the viscosity of the sample and its adhesion to the metal surface, the softening point thus determined is very approximate. It often bears no simple relationship to the glass-transition or melt temperature.

10.3. Crystallization

10.3.1. Morphology

The states of order which occur when polymer solutions or melts are cooled depend on the configuration and constitution of the polymer. They also depend very much on the external conditions such as the concentration, type of medium, temperature, method by which induction of crystallization occurs, etc. The more well known types of crystal growth can be classed as single crystal, dendritic, and fibrillar.

Single crystals were first discovered when 0.01% solutions of poly(ethylene) were cooled, whereupon lozenge-shaped structures were produced (Figure 5-14). Lamellar single-crystal growth occurs only in very dilute solutions. The lamellar-type structures resulting from single-crystal-type growth show very few surfaces. From these very dilute solutions the polymer molecules tend to grow individually and in one dimension along the side surfaces of the lamellae, but they can also grow two-dimensionally along fold surfaces, thus forming steps or spirals. The lamellar height (also called fold length) for a given temperature depends, for very low-molecular-weight material, on the degree of polymerization, and then, above a certain degree of polymerization, remains constant (see Section 5.3.1). When the polymer molecule has different macroconformations, the stabilities of which vary from solvent to solvent, then the chain conformation varies with solvent. For example, amylose tricarbanilates crystallize from a dioxane–ethanol mixture in the form of folded chains, whereas from pyridine–ethanol they grow in the form of folded helices. For very high-molecular-weight material in equally dilute solutions, more complex morphological forms result since in these solutions the viscosity is much higher and the simpler folded structures aggregate. At higher concentrations, fibrillar, or even network, structures are formed.

Dendritic growth is observed from more concentrated solutions. It is also produced by sharply undercooling the melt of a slowly crystallizing polymer.

Fibrillar growth leads to spherulites. Although dendritic growth from moderately concentrated polymer solutions or from the melt in polymers of low stereoregularity can lead to primitive spherulites, well-developed spherulites are the normal mode of crystallization of linear polymers from the melt. Spherulites consist of a spherically symmetric array of fibrils which radiate out from the center (see Figure 5-24).

Transcrystallization is the name given to a crystallization in which the formation of a large number of adjacent nuclei occurs on the surfaces of the melt. It is considerably influenced by nucleation density.

Other types of crystalline physical structures can be observed if crystallization occurs under stress (for example, under shear stress during processing). In this case, a continuous series of crystallization nucleation centers is formed by the induction of order along the lines of flow. A "row structure" of oriented lamellae is consequently produced. These row structures are related to the "shish-kebab" structures shown in Figure 5-27.

Another special crystallization case is the crystallization immediately after polymerization. For example, poly(oxymethylene) molecules in an all-*trans* conformation are produced when formaldehyde crystals are polymerized. The same polymer crystallizes out of dilute solution in the form of a 9_5 helical macroconformation with tgtg sequences. Elementary filaments of native cellulose produced by biosynthesis contain extended chain cellulose molecules. Regenerated cellulose has a different unit cell and consequently probably occurs in a different macroconformation.

Crystallization takes place in two stages: nucleation and the growth of nuclei. Nucleation is important not only because it induces crystallization, but because it may also determine the various crystal structures. If, for example, one adds 10^{-5}–10^{-4}% by weight *p-t*-butylbenzoic acid to it-poly(propylene), the polymer crystallizes monoclinically, whereas with permanent red E3B (a quinacridone dye) the polymer crystallizes pseudohexagonally.

10.3.2. Formation of Nuclei

Crystallization nuclei can be formed either homogeneously or heterogeneously. Homogeneous (or thermal) nucleation results spontaneously from the molecules of the crystallizing material, whereas heterogeneous nucleation occurs on exposed foreign surfaces (container wall, dust particles and, intentionally, with added nucleating agents). Between homogeneous and heterogeneous nucleation is the possibility of an athermal process in which crystallization is induced by still-remaining crystalline remnants of the same molecule which survived melting, i.e., ordered regions which are not fully destroyed during melting. This type of delayed or heterogeneous nucleation is responsible for the "memory" effect of polymer

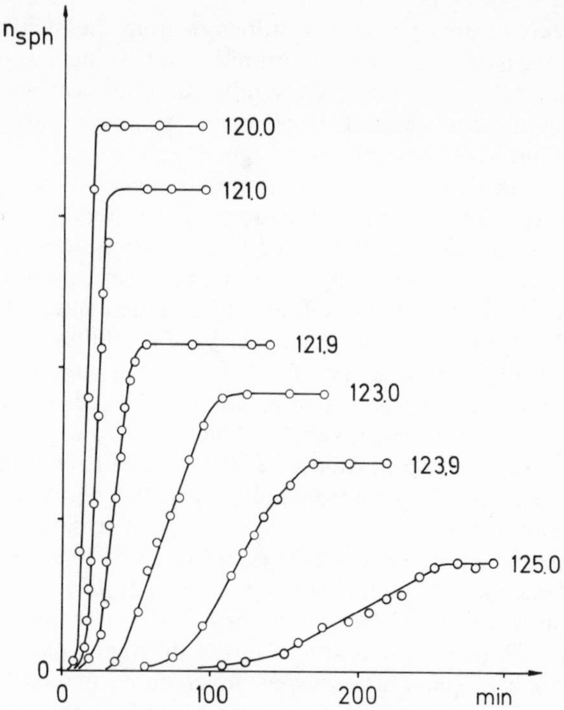

Figure 10-7. Time dependence of the number n_{sph} of spherulites formed in melts of poly(decamethylene terephthalate) at different temperatures (after A. Sharples).

melts, i.e., for the phenomenon that, after melting and recooling, the nuclei often reappear at exactly the same position as before melting. The nuclei reappear at the same place because the very high viscosity of the melt is not favorable to their diffusion.

In what is considered "heterogeneous" nucleation, the concentration of nucleating centers varies, according to the crystallizing material, by several orders of magnitude. In poly(ethylene), for example, nucleation center concentrations of $<10^{12}$ nuclei/cm^3 were found, whereas concentrations of >1 nuclei/cm^3 were observed in poly(ethylene oxide).

Homogeneous formation of nuclei has not yet been established with certainty in macromolecular solutions. If it does in fact exist, then it is more probable that folded-chain nuclei will be formed in this way than micellar nuclei. In contrast to folded chain nuclei, in micellar nuclei several molecules must come together in order to form a stable nucleus. Micellar nuclei thus lead to a greater entropy loss. A homogeneous nucleation is more probable in melts than in solutions, since the concentration of centers is potentially greater in the former. Even in melts, however, it is uncertain whether homogeneous nucleation makes any significant contribution to nucleus formation, since heterogeneous nucleation is a much more rapid process.

Heterogeneous nucleation can only occur when the growing crystal can wet the surface of the foreign material. Thus, a specific interaction between nucleus and melt is necessary. If this interaction is very strong, then all the crystallization nuclei become effective immediately. The number of nuclei remains constant. If, on the other hand, the interaction is weak, crystallization is induced by a number of nuclei which initially increases with time and then remains constant.

Nucleus formation can be followed through the development and propagation of spherulites. When poly(decamethylene terephthalate) is crystallized, an induction or nucleation period first occurs (Figure 10-7). During this period, either homogeneous or weak-interaction heterogeneous stable nuclei are formed.

In the case of heterogeneous nucleation with strong interactions, on the other hand, the number of spherulites remains constant with time. In addition, all the spherulites are of equal size. In purely homogeneous nucleation, the number of nuclei is constantly increasing, and a spherulite size distribution is observed.

In secondary nucleation, chain molecules deposit onto the surface of an already existing nucleus by a chain-folding mechanism (Figure 10-8). The Gibbs energy of nucleation is given by the surface energies of the fold surface σ_f and of the side surfaces σ_s, as well as the crystallization energy

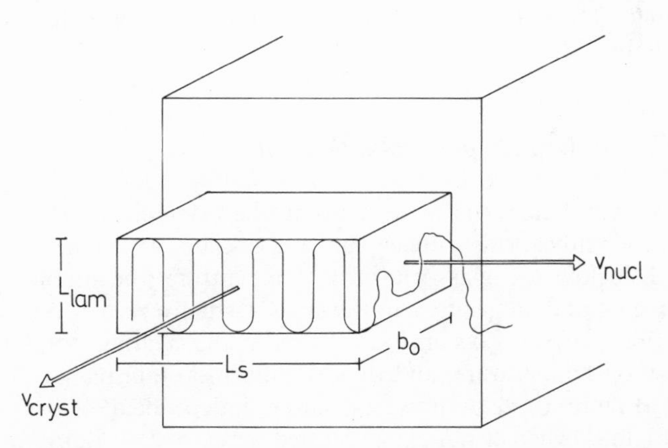

Figure 10-8. Illustration of the formation of secondary nuclei through growth on an already-existing nucleus. L_{lam}, lamellar height; L_s, lamellar length; b_0, lamellar thickness (about molecular diameter thickness). The nucleus forms in the L_s direction at a rate of v_{nucl}. The crystal grows through deposition of other molecules in the direction b_0 and at the rate v_{cryst}.

per unit volume ΔG_{cryst}:

$$\Delta G = 2L_s b_0 \sigma_f + 2L_{\text{lam}} b_0 \sigma_s - b_0 L_s L_{\text{lam}} \Delta G_{\text{cryst}} \tag{10-14}$$

Only two of the four possible side surfaces need to be considered, since, in the b_0 direction, no new surface is created. Differentiating equation (10-14) with respect to L_s and equating to zero, we obtain

$$(L_{\text{lam}})_{\text{theor}} = 2\sigma_f / \Delta G_{\text{cryst}} \tag{10-15}$$

for the theoretical critical lamellar height. The critical lamellar thickness or side-surface length is correspondingly given by differentiation with respect to L_{lam}:

$$(L_s)_{\text{theor}} = 2\sigma_s / \Delta G_{\text{cryst}} \tag{10-16}$$

The Gibbs energy of crystallization, however, depends on the crystallization temperature:

$$\Delta G_{\text{cryst}} = (\Delta H_M)_u - T_{\text{cryst}} (\Delta S_M)_u \tag{10-17}$$

The melting point, on the other hand, is given by

$$T_M^0 = \frac{(\Delta H_M)_u}{(\Delta S_M)_u} = \frac{\Delta H_M^0}{\Delta S_M^0} \tag{10-18}$$

Inserting equations (10-17) and (10-18) into (10-15) leads to

$$(L_{\text{lam}})_{\text{theor}} = \frac{2\sigma_f T_M^0}{(\Delta H_M)_u (T_M^0 - T_{\text{cryst}})} \tag{10-19}$$

Thus, the critical theoretical lamellar height decreases with increasing supercooling $(T_M^0 - T_{\text{cryst}})$. This has been observed for actual lamellar heights (Figure 5-21).

10.3.3. Crystal Growth or Crystallization

Immediately below the melting point, the crystallization rate is very small, since already formed nuclei can disassociate. At a temperature T_{ch} about 50 K below the glass-transition temperature, the mobility of the chain segments and molecules, however, is virtually zero. Consequently, crystallization generally occurs only between the melting point and the glass-transition temperature, and the crystallization rate passes through a maximum (Figure 10-9). A universal curve, independent of the polymer type, is obtained when $\ln(v/v_{\max})$ is plotted against $(T - T_{\text{ch}})/(T_M^0 - T_{\text{ch}})$. The maximum occurs at about a value of 0.63. If T_{ch} is expressed in terms of $T_G - 50$, and considering that the values of T_M^0/T_G lie between 2 and 1.5, then it can be shown that the maximum rate of crystallization occurs at about 0.8–$0.87 T_M^0$.

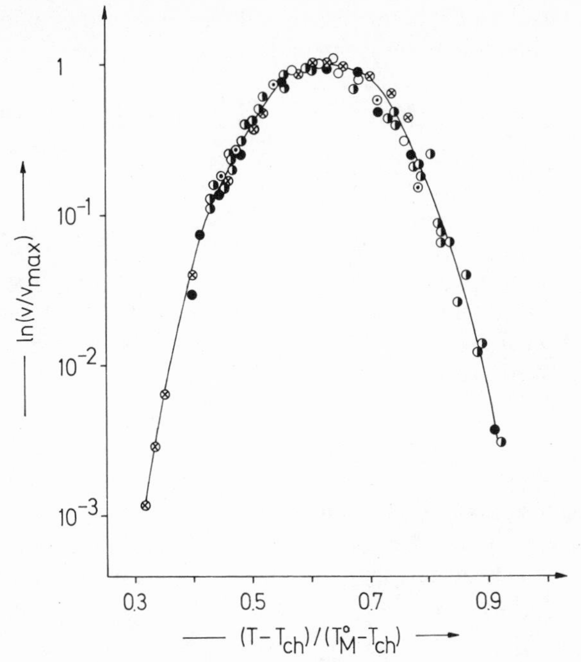

Figure 10-9. Plot of the natural logarithm of the reduced growth rate of spherulites of different polymers as a function of a reduced temperature. T_{ch} is the characteristic temperature, about 50 K below the glass-transition temperature, at which all molecular motion ceases. T_M^0 is the thermodynamic melting point (after A. Gandica and J. H. Magill).

The time dependence of the overall (primary) crystallization is described by the Avrami equation, which was originally derived for the crystallization of metals. The crystallinity is expressed as the volume fraction ϕ of the crystalline material in the total sample. For the derivation, it is assumed that each nucleus leads to an entity (e.g., a rod, disk, or sphere). After an infinitely long time the whole sample is filled with these shapes. The crystallinity of the sample is then ϕ_∞. This, however, is the crystallinity of a single entity which is assumed not to change during crystallization. At the time t, the fraction of the sample volume filled by these entities is ϕ/ϕ_∞. With randomly distributed nuclei the probability p that a point does not lie in any one entity is proportional to this fraction, i.e.,

$$p = 1 - \frac{\phi}{\phi_\infty} \tag{10-20}$$

The probability p_i that a point does not lie in a given entity with the volume V_i is

$$p_i = 1 - \frac{V_i}{V} \tag{10-21}$$

The probability that a point lies outside all the entities is equal to the product of all the individual probabilities:

$$p = p_1 p_2 \cdots p_n = \prod_{i=1}^{n} \left(1 - \frac{V_i}{V}\right) \tag{10-22}$$

$$\ln p = \sum_{i=1}^{n} \ln\left(1 - \frac{V_i}{V}\right) \tag{10-23}$$

Should the volume of every entity be very much smaller than the total volume ($V_i \ll V$), then the logarithmic expression can be developed into a series $\ln(1 - x) = -x + x^2/2 - \cdots$, whereby terms in x^2 and higher can be neglected. We have

$$\ln p = -\sum_{i=1}^{n} \frac{V_i}{V} = -V^{-1} \sum_{i=1}^{n} V_i \tag{10-24}$$

The mean volume $\overline{V_i}$ of a single entity is given by $\overline{V_i} = (\sum_{i=1}^{n} V_i)/N$, where N is the number of entities. The concentration ν of entities per unit volume is $\nu = N/V$. Accordingly, when the antilogarithm has been evaluated, equation (10-24) becomes $p = \exp(-\nu\overline{V_i})$, and equation (10-20) becomes

$$\phi = \phi_\infty[1 - \exp(-\nu\overline{V_i})] \tag{10-25}$$

If all the entities are formed simultaneously, then the density of nuclei is constant ($\nu = k_0$). Furthermore the growing entities all have the same volume $\overline{V_i}$, which naturally increases with time. For rodlike entities with constant cross section A, this increase results entirely from the increase with time of the length L:

$$\overline{V_i} = AL = Ak_1 t \qquad \text{(rods)} \tag{10-26}$$

With disk-shaped entities, the thickness d remains constant, and the radius grows with time; thus, for average volume,

$$\overline{V_i} = \pi d r^2 = \pi d(k_2 t)^2 \qquad \text{(disks)} \tag{10-27}$$

With spheres, the radius also increases proportionately with time, giving

$$\overline{V_i} = (4/3)\,\pi(k_3 t)^3 \qquad \text{(spheres)} \tag{10-28}$$

The nuclei are not formed spontaneously, but appear randomly with time in sporadic nucleation. Therefore ν is a time-dependent entity, increasing with time. When it is assumed that the nuclei appear randomly in both time and space, the mathematics becomes simpler. The final result is not altered by this double assumption because when the treatment concerns the whole volume, including that already filled with growing forms, the fictitious nuclei within the already crystallized regions do not affect the volume fraction of space available for growth. With sporadic nucleation, it is possible that the concentration ν of nuclei increases with time:

$$\nu = kt \tag{10-29}$$

The mean volume $\overline{V}_i$ of a nucleus is then given by reasoning that every nucleus has the same probability of being formed in the same interval of time. For the time interval from $t - \tau$ to $t - \tau + d\tau$ this interval is $d\tau/t$. Equations (10-26)–(10-28) then become

$$\overline{V}_i = k_1 A \int_0^t (t - \tau)\frac{d\tau}{t} = 0.5k_1 At \qquad \text{(rods)} \tag{10-30}$$

$$\overline{V}_i = \pi d k_2^2 \int_0^t (t - \tau)^2 \frac{d\tau}{t} = \frac{1}{3}\pi d k_2^2 t^2 \qquad \text{(disks)} \tag{10-31}$$

$$\overline{V}_i = \frac{4}{3}\pi k_3^3 \int_0^t (t - \tau)^3 \frac{d\tau}{t} = \frac{1}{3}\pi k_3^3 t^3 \qquad \text{(spheres)} \tag{10-32}$$

If these expressions for ν and $\overline{V}_i$ are inserted into equation (10-25), one obtains equations of the general type

$$\phi = \phi_\infty[1 - \exp(-zt^n)] \tag{10-33}$$

Equation (10-33) is known as the Avrami equation. The physical significances of the constants z and n is given in Table 10-3.

A double logarithmic form of equation (10-33) is used to evaluate crystallization kinetic data:

$$\ln[-\ln(1 - \phi\phi_\infty^{-1})] = \ln Z + n \ln t \tag{10-34}$$

The constant n can be found from a plot of the left-hand side of equation (10-34) against $\log t$.

It is sometimes difficult to attribute a real physical significance to the actual values of n obtained. For example, when poly(chlorotrifluoroethylene) is crystallized, values for n of 1 or 2 are found according to crystallization

Table 10-3. Constants z and n of the Avrami Equation

	z		n	
Shape	Instantaneous	Sporadic	Instantaneous	Sporadic
Rod	$k_0 k_1 A$	$0.5kk_1 A$	1	2
Disk	$k_0 k_2^2 \pi d$	$\frac{1}{3}kk_2^2 \pi d$	2	3
Sphere	$\frac{4}{3}k_0 k_3^3 \pi$	$\frac{4}{3}kk_3^3 \pi$	3	4

conditions. A value of $n = 3$ is obtained for poly(hexamethylene adipamide). Poly(ethylene terephthalate) gives values of between 2 and 4 according to the crystallization temperature. Extra difficulties in the interpretation of values of n can be seen from the fact that nonintegral numbers and values of $n = 6$ are also known. Different methods can yield different values of n when the methods measure different aspects of the crystallization. For example, dilatometry generally measures the growth of spherulites, whereas calorimetry also measures the growth of lamellae in spherulites. The Avrami equation, however, is, of course, suitable in all cases up to the point where growth forms impinge on each other.

The rates of crystallization vary markedly from polymer to polymer (Table 10-4). Polymers such as poly(ethylene terephthalate) that crystallize slowly can in fact be obtained in almost completely amorphous form by rapid supercooling from the melt. An amorphous form has never been achieved with the very rapidly crystallizing poly(ethylene), even when it is supercooled from the melt with liquid nitrogen. The crystallization rate depends on polymer constitution and configuration. Symmetric polymers crystallize rapidly. Bulky substituents and chain units lower the crystallization rate.

Spherulitic growth is a special case in crystallization. Spherulites form only within a specific temperature range; for example, with it-poly(propylene) with a melting point of 170°C, they are first formed below 115°C. With spherulites, the rate of advance of the spherulite boundary is followed. This boundary encloses the crystalline portion of the spherulite. Since spherulites also contain noncrystalline material, however, the spherulite growth rate thus corresponds to the linear crystal growth rate. As the molecular weight increases, the rate of crystallization falls, since the rate of diffusion of segments and molecules decreases.

Table 10-4. Linear Crystallization Rates of Various Polymers from the Melt When Supercooled to Approximately 30°C Below the Melting Point

Polymer	Crystallization rate, μm/min
Poly(ethylene)	5000
Poly(hexamethylene adipamide)	1200
Poly(oxymethylene)	400
Poly(caprolactam)	150
Poly(trifluorochloroethylene)	30
it-Poly(propylene)	20
Poly(ethylene terephthalate)	10
it-Poly(styrene)	0.25
Poly(vinyl chloride)	0.01

The fibrillar structure within a spherulite (Figure 5-23) results from the chemical and molecular nonuniformity of the crystallizing macromolecule. A polymer of low molecular weight and excessive branching needs a larger degree of supercooling because of its lower melting point; it therefore crystallizes less readily. The growth of spherulites thus leads to crystallization with fractionation. The poorly crystallizing parts are excluded from the propagation zones and are transferred into interlamellar spaces. Here, they may suppress crystallization, which leads to preferential growth in the propagation zone, and hence to a fibrillar structure.

10.3.4. The Influence of Additives

When pure polymers are crystallized, crystallization nuclei are produced more or less at random according to thermal history, impurity content, etc., of the polymer. Since there are usually only relatively few nuclei produced, large spherulites are formed. Such large spherulites have an unfavorable effect on mechanical properties. Crystallization is therefore controlled in some cases by the addition of nucleating agents, e.g., alkali metal salts of long-chain fatty acids in the case of poly(olefins). The effectiveness of the nucleating agents is greatly dependent on their solubilities in the polymer melt. Soluble additives act simply as diluting agents, retarding the rate of crystallization. Insoluble additives, on the other hand, have no effect on, or increase the rate of, crystallization according to whether the additive is completely inert to, or can be wetted by, the polymer melt.

Effects of this kind obviously play a part in the coloring of polymers with pigments. Inorganic pigments are insoluble in, but can probably be wetted by, the polymer melt, in which case heterogeneous nucleation is induced and smaller spherulites are obtained. Organic pigments are mostly somewhat soluble, which leads to a dilution effect with retardation of nucleation and crystallization. This means that in practice a postcrystallization takes place during the cooling of injection-molded specimens, leading to shrinking and, because of the anisotropy of cooling in different sections of the specimen, to distortion.

10.3.5. Recrystallization

According to Section 5.3.3, crystalline polymers can occur in different polymorphic forms. When a polymer melt is cooled to a given crystallization temperature, the crystal modification that is thermodynamically stable at this temperature is not always formed (because of kinetic reasons). In many cases, the modifications only transform into one another very slowly, which

Table 10-5. Thermodynamic Parameters for the Three Modifications of it-Poly(butene-1)

Modification	T_M, °C	ΔH_M, J/mol	ΔS_M, J K^{-1} mol^{-1}
1	138	6700	16.3
2	130	4200	10.4
3	106.5 (?)	6300	16.5

is one of the reasons for the poor durability or inadequate low-temperature stability of some polymers. With it-poly(butene-1), for example, the modification 3 is obtained with crystallization from solution, whereas modification 2 crystallizes from the melt. Both modifications are metastable. Modification 2 reverts to the stable modification 1 via a solid–solid transition. Modification 3, however, is stable at room temperature, but converts at higher temperatures into modification 1 or 2. The observed differences in the melt enthalpies and entropies are relatively small (Table 10-5).

10.4. Melting

10.4.1. Melt Processes

Long-chain polymers generally crystallize in lamellae with chain folding (see also Section 5.4.2). The top and bottom layers of these lamellae are amorphous; the interior is crystalline (see Figure 5-15). It has been suggested that if such lamellae are heated, the thickness of both the amorphous layers and the crystalline layer remains constant over a wide temperature range (Figure 10-10). Above a certain temperature, however, the mean thickness of the crystalline layers decreases sharply and that of the amorphous layers increases strongly. Consequently, melting must begin at the surface and then proceed to the interior.

However, the melting point is defined as that temperature at which the crystalline layer is in thermodynamic equilibrium with the melt. It must of necessity depend on the lamellar thickness *before* the onset of the melting process. Each monomeric unit contributes an enthalpy of fusion $(\Delta H_M)_u$ to the observed enthalpy of fusion ΔH_M. The enthalpy of fusion is also lowered by an amount equal to the surface enthalpy ΔH_f of both sides of the lamellae. Thus, for a lamella of N_u monomeric units

$$\Delta H_M = N_u(\Delta H_M)_u - 2\,\Delta H_f \tag{10-35}$$

The melting point observed for such a lamella is

$$T_M = \frac{\Delta H_M}{\Delta S_M} \tag{10-36}$$

whereas the melting point of an infinitely thick lamella has the value

$$T_M^0 = \frac{N_u(\Delta H_M)_u}{N_u(\Delta S_M)_u} = \frac{(\Delta H_M)_u}{(\Delta S_M)_u} \tag{10-37}$$

Here, the entropy of fusion of a lamella of N_u monomeric units is

$$\Delta S_M = N_u(\Delta S_M)_u \tag{10-38}$$

Inserting equations (10-35)–(10-37) into each other leads to

$$T_M = T_M^0 - \frac{2\Delta H_f}{(\Delta S_M)_u} \frac{1}{N_u} \tag{10-39}$$

or, with equation (10-37),

$$T_M = T_M^0 \left[1 - \frac{2\Delta H_f}{(\Delta H_M)_u} \frac{1}{N_u} \right] \tag{10-40}$$

Consequently, the thermodynamic melting point T_M^0 can be obtained by extrapolation from a plot of the melting temperature T_M against the number of monomeric units per lamella, that is, the lamellar height (Figure 10-11).

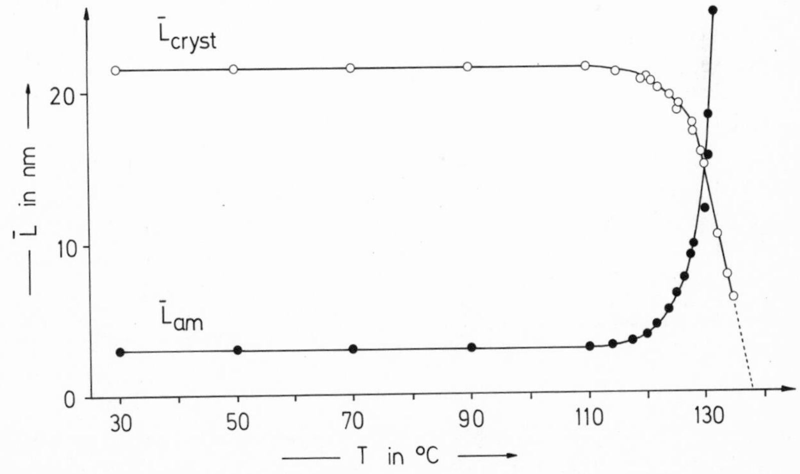

Figure 10-10. Mean lengths of the crystalline and amorphous portions of the lamellar height of poly(ethylene) as inferred from small-angle X-ray measurements carried out at different temperatures. All experiments were carried out on the same material (after K. V. Fulcher, D. S. Brown, and R. E. Wetton).

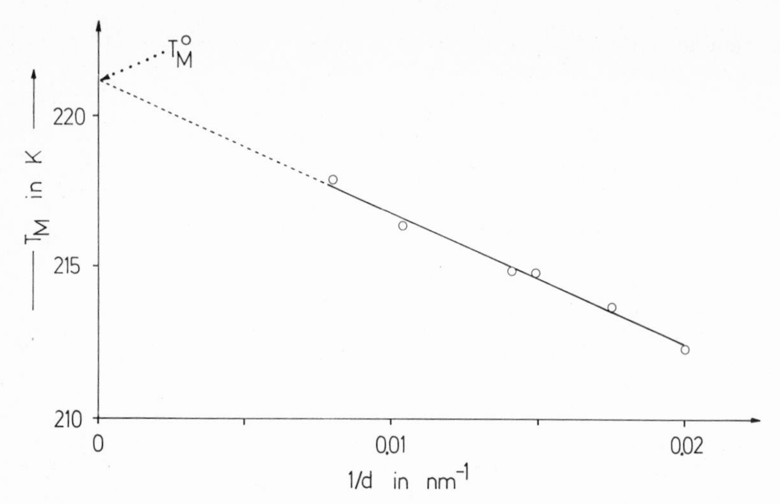

Figure 10-11. The melting-temperature dependence of the reciprocal lamellar thickness $1/d$ of lamellae for poly(trifluorochloroethylene). The lamellar thickness was measured as the interlamellar distance by small-angle X-ray analysis, and thus contains both the crystalline component and the amorphous surface layer (after J. D. Hoffman from data of P. H. Geil and J. J. Weeks).

The number of monomeric units per chain section in the lamella is related to the observed lamellar thickness L_{obs} via the crystallographic monomeric unit length L_u:

$$L_{obs} = N_u L_u \tag{10-41}$$

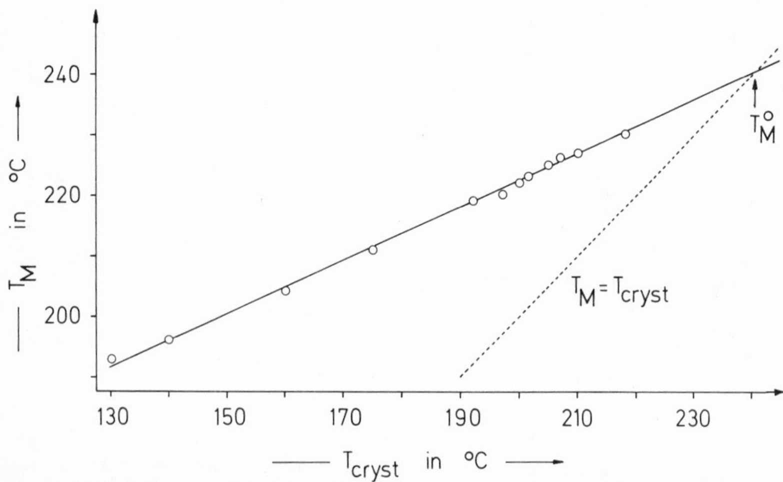

Figure 10-12. Influence of the crystallization temperature T_{cryst} on the melting point of it-poly(styrene) (after N. Overbergh, H. Bergmans, and G. Smets).

The observed lamellar thickness is, however, always greater than the lamellar thickness calculated from crystal growth theory $(L_{\text{lam}})_{\text{theor}}$ by a factor of γ:

$$L_{\text{obs}} = \gamma (L_{\text{lam}})_{\text{theor}} \tag{10-42}$$

Insertion of equations (10-18), (10-19), (10-41), and (10-42) into equation (10-39) leads to (with $\sigma_f = L_u \Delta H_f$)

$$T_M = T_M^0 (1 - \gamma^{-1}) + \gamma^{-1} T_{\text{cryst}} \tag{10-43}$$

Consequently, a plot of the melting temperature T_M of the crystals against their crystallization temperature should give a straight line (Figure 10-12). The intersection of this line with the line $T_M = T_{\text{cryst}}$ gives the thermodynamic melting point T_M^0. The γ value for many polymers is found to be about 2.

10.4.2. *Melting Temperature and Molecular Weight*

The thermodynamic melting point increases in a polymer homologous series with increasing molecular weight up to a point where the influence of the end groups can be ignored (Figure 1-1). Consequently, extrapolation to infinitely high molecular weight leads to the thermodynamic melting point of a perfect crystal of infinitely long chains (Figure 10-13).

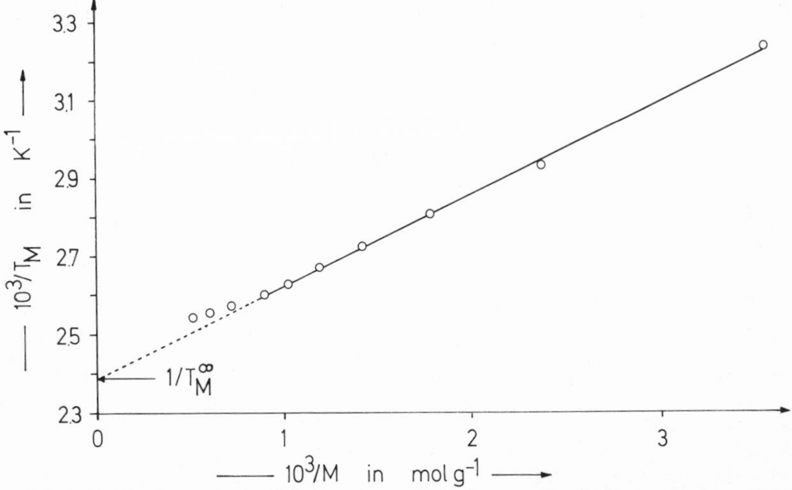

Figure 10-13. The dependence of the melting points of zone-refined alkanes on the molecular weight. The melting point of a 100% crystalline poly(ethylene) of infinite molecular weight is given as 146.3°C. (From data of Figure 1-1 and from W. Heitz, R. Peters, G. Strobl, and E. W. Fischer.)

Most of the melting points reported in the literature do not apply to perfect crystals of infinitely long chains (T_M^{∞}) or even to perfect crystals of finite chains (T_M^0). They are more often the melting points T_M of polymers of finite molecular weight, indefinite lamellar height, and containing defects, that is, not perfect crystals.

10.4.3. Melting Point and Constitution

The heats and entropies of fusion per mole of monomeric unit can vary over wide limits (Table 10-6). However, the entropy of fusion per mole of chain unit is moderately constant for many polymers. Since the melting point is essentially determined by the enthalpy of fusion, it was presumed that the cohesive energy is the definitive quantity.

But the cohesive energy is a measure of the intermolecularly operative forces in a liquid–gas phase transition, whereas one is considering a solid–liquid transition during melting. Thus the cohesive energies for both types of transition are not necessarily comparable. Infrared measurements on polyamide melts, for example, have shown that the majority of hydrogen bonding units remain bonded above the melting point. Thus, the cohesive energy, as defined for a liquid–gas transition, must be relatively unimportant.

If the melting points were primarily determined by a cohesive energy related to intermolecular interactions, then melting points should increase with increasing numbers of groups of high cohesive energy per monomeric unit. The conventionally defined cohesive energy of a methylene group is

Table 10-6. Entropy of Fusion per Monomeric Unit ΔS_u and per Chain Unit ΔS_{cu} and Enthalpy of Fusion per Monomeric Unit ΔH_u and per Chain Unit ΔH_{cu} for Various Polymers[a]

Polymer	T_M, °C	N	ΔH_u, kJ/mol	ΔH_{cu}, kJ/mol	ΔS_u, J/K mol	ΔS_{cu}, J/K mol
Poly(methylene)	144	1	3.29	3.29	7.87	7.87
Poly(oxymethylene)	180	2	7.45	3.73	16.50	8.25
Poly(oxyethylene)	67	3	8.29	2.76	24.37	8.12
Poly(decamethylene adipate)	80	16	42.71	2.67	121.42	7.59
Poly(decamethylene sebacate)	80	20	50.24	2.51	142.36	7.12
Poly(decamethylene terephthalate)	138	13	46.06	3.54	112.21	8.63
Poly(ethylene terephthalate)	267	5	23.02	4.60	42.71	8.54
Poly(caprolactam)	225	6	21.35	3.56	42.71	7.12
Poly(hexamethylene adipamide)	267	12	43.13	3.60	79.97	6.66
Cellulose tributyrate	207	2	12.56	6.28	26.17	13.08

[a] The groups COO, NHCO, and C_6H_5 are each taken to be a chain unit. T_M is the melting point and N the number of chain units per monomeric unit.

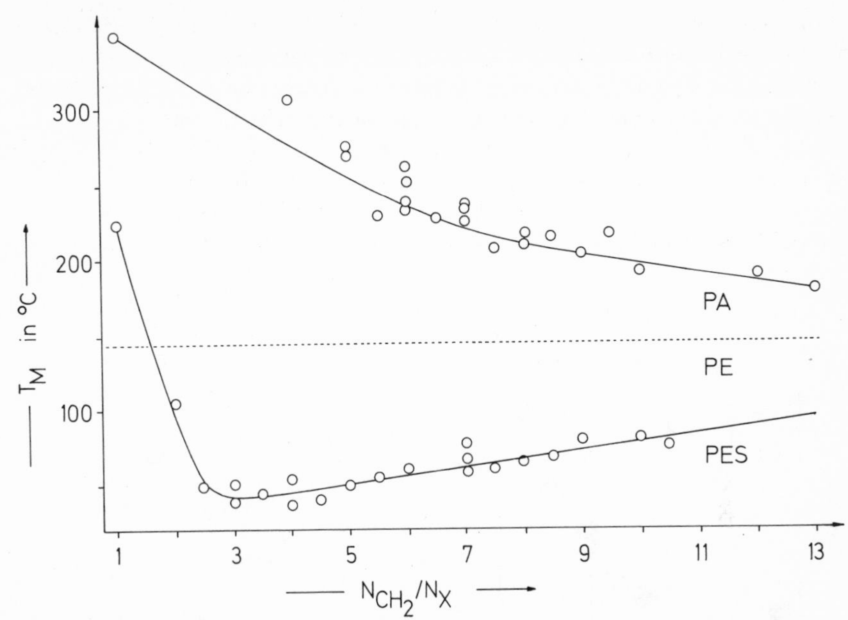

Figure 10-14. Dependence of the melting point T_M of aliphatic polyamides (PA) and aliphatic polyesters (PES) with X = amide or ester groups as a function of the group content. Result for polyethylene is given by the lashed line.

2.85 kJ/mol; of an ester group, 12.1 kJ/mol; and of an amide group, 35.6 kJ/mol. Consequently, the melting point of polyamides and polyesters should increase with decreasing number of methylene groups per monomeric unit. The reverse behavior occurs (Figure 10-14 and Table 10-7) for polyesters. On the other hand, ester groups possess a lower potential energy barrier for rotation than do methylene groups and amide groups. The flexibility of the individual molecule, and not the intermolecular interaction between chains, is the primary factor in determining the melting point.

The flexibility of a molecule depends on the constitution and configuration of the chains and the conformation produced by these. Given the same conformation, the larger the distance between the chain atoms and the greater their valence angles, the greater will be the flexibility. It is higher when steric hindrance to rotation is lower. Poly(ethylene) ($T_M = 144°C$), with its relatively high rotational potential energy barrier for the methylene–methylene bond, thus has a higher melting point than the ether–oxygen-containing poly(tetrahydrofuran), $(\text{—}CH_2\text{—}CH_2\text{—}CH_2\text{—}CH_2\text{—}O\text{—})_n$, with $T_M \approx 35°C$ (see Section 4.2). Rigid groups (phenylene residues, etc.) raise the melting temperature.

According to conformational sequence, helices are more tightly or more loosely constructed. The helices of poly(oxymethylene), with the conformational sequence gg, for example, have a much smaller diameter than poly(ethylene oxide) helices, which possess the ttg sequence conformation and have roughly the same number of main-chain atoms per unit length of chain along the chain direction of the helix (Figure 10-15). The poly(oxymethylene) sequences are therefore more rigid. Consequently, the melting point of poly(oxymethylene) is higher than that of poly(oxyethylene).

The tendency of the helix to become more rigid because of closer packing can also be affected by substituents. Increased substitution in

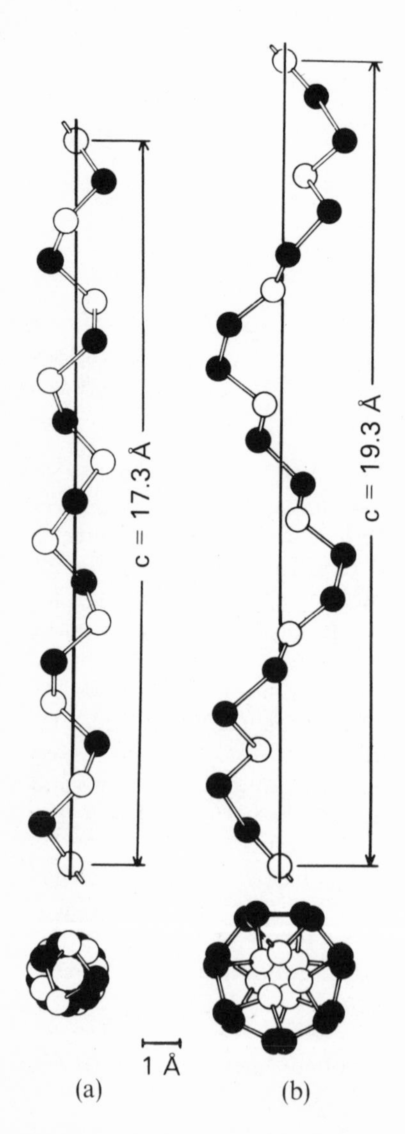

Figure 10-15. Crystal structure of (a) poly(oxymethylene) and (b) poly(oxyethylene) (after H. Tadokoro, Y. Chatani, M. Kobayashi, T. Yoshihara, K. Murahashi, and K. Imada). 1 Å = 0.1 nm.

direct proximity to the main chain of helix-forming macromolecules widens the helix and lowers the melting point. The melting point of it-poly(butene-1) is therefore lower than that of it-poly(propylene). Because of tighter intermolecular packing, poly(3-methylbutene-1) possesses a higher melting point than poly(butene-1).

$\text{+CH}_2\text{—CH(CH}_3\text{)+}_n$	$\text{+CH}_2\text{—CH(CH}_2\text{—CH}_3\text{)+}_n$	$\text{+CH}_2\text{—CH(CH(CH}_3\text{)—CH}_3\text{)+}_n$	$\text{+CH}_2\text{—CH(C(CH}_3\text{)}_2\text{—CH}_3\text{)+}_n$
it-poly(propylene)	it-poly(butene-1)	it-poly(3-methylbutene-1)	it-poly(3.3′-dimethyl-butene-1)
3_1 helix	3_1 helix	4_1 helix	?
T_M = 186°C	T_M = 136°C	T_M = 304°C	T_M > 320°C

In the series of poly(α-olefins),

$$[\text{—CH}_2\text{—CH(CH}_3\text{)—}]_n, \quad [\text{—CH}_2\text{—CH(CH}_2\text{CH}_3\text{)—}]_n,$$
$$[\text{—CH}_2\text{—CH(CH}_2\text{CH}_2\text{CH}_3\text{)—}]_n, \text{ etc.}$$

the substituent bonded directly to the main chain backbone is always a $\text{—CH}_2\text{—}$ group. Thus the chain conformation is retained. The longer side chains decrease the efficiency with which the chains can pack together, however, and so the melting point falls (Figure 10-16). Only when the side chains are very long does their ordering give additional order, so that the melting point again rises as the number of carbon atoms increases.

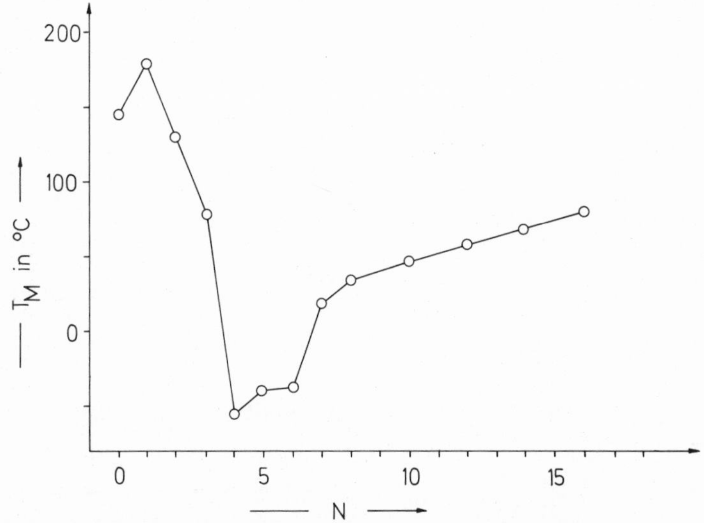

Figure 10-16. Melting point T_M of isotactic poly(α-olefins), $\text{+CH}_2\text{—CHR+}_n$ as a function of the number of methylene groups N in the unit $\text{R} = \text{+CH}_2\text{)}_N\text{H}$.

Table 10-7. Comparison of Cohesive Energies and Melting Points of Polymers

Chain units		Polymeric examples		
Group	Cohesive energy, kJ/mol group	Monomeric unit	Mean cohesive energy, kJ/mol group	T_M, °C
$—CH_2—$	2.85	$—CH_2—$	2.85	144
$—CF_2—$	3.18	$—CF_2—$	3.18	327
—O—	4.19	$—CH_2—O—$	3.52	188
		$—CH_2—CH_2—O—$	3.31	67
$—C(CH_3)_2—$	8.00	$—CH_2—C(CH_3)_2—$	5.40	44
$—CCl_2—$	13.0	$—CH_2—CCl_2—$	7.91	198
$—CH(C_6H_5)—$	18.0	$—CH_2—CH(C_6H_5)—$	10.4	250[a]
—CHOH—	21.4	$—CH_2—CHOH—$	12.1	265[b]
—COO—	12.1	$—(CH_2)_5—COO—$	4.4	55
—CONH—	35.6	$—(CH_2)_5—CONH—$	8.3	228

[a] Isotactic.
[b] Probably syndiotactic.

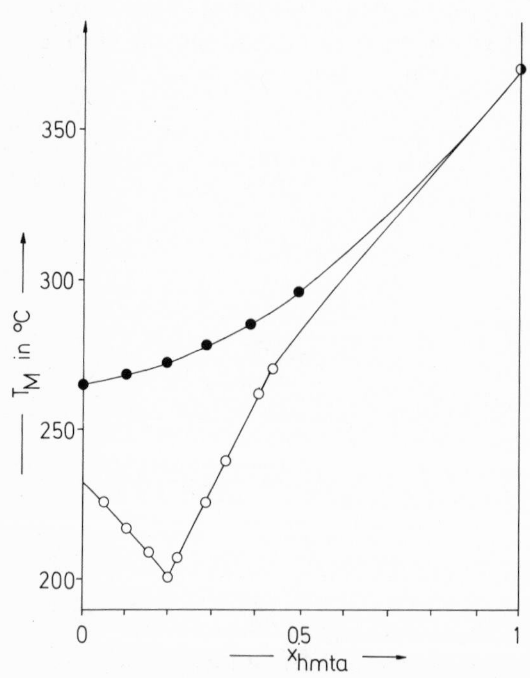

Figure 10-17. Melting point T_M of copolymers of hexamethylene terephthalamide (HMTA) and hexamethylene adipamide (●) or hexamethylene sebacamide (○) (after O. B. Edgar and R. Hill).

10.4.4. Melting Point of Copolymers

The different base units in a copolymer can be isomorphous. If, in addition, they occur randomly along the chain, then the melting points of such copolymers increase regularly with the mole fraction of the comonomer that melts at the highest temperature. An example of this is the copolymer of hexamethylene terephthalamide and hexamethylene adipamide (Figure 10-17). With nonisomorphous base units, on the other hand, the lengths of the crystallite regions in the solid polymer are decreased as the proportion of the second comonomer is increased. The melting points fall, reaching a minimum at a specific copolymer composition, as shown in Figure 10-17 for the copolymers of hexamethylene terephthalamide and hexamethylene sebacamide.

10.5. Glass Transitions

10.5.1. Phenomena

In the middle of the 1920s, it was observed that the viscosity of some low-molecular-weight substances (glycerol, brucine, etc.) and silicate melts suddenly increased with decreasing temperature by several orders of magnitude at a quite specific temperature. At this "freezing" temperature, the viscosity was about 10^{12} Pa s and independent of the substance. An "isoviscous" process was considered to characterize this freezing-in process. Today, the freezing-in temperature is considered to be that temperature at which all substances have the same free volume (see also Section 10.2.1). Which free volume is involved is still a matter of dispute. But it is agreed that the process is concerned with the freezing of segmental motion, for it was observed that cross-linking polymerizations lead to the same freezing-in temperature as long as there were about 30–50 chain units between cross-link points.

The terms "freezing-in temperature," "glass-transition temperature," and "softening-point temperature" are mostly phenomenologically defined today. The freezing-in temperature is consequently the temperature at which measured parameters first deviate from the behavior "normal" to higher temperatures (see also Figure 10-3). The softening-point temperature is similarly defined for experiments carried out with increasing temperature. The glass-transition temperature is defined as the intersection of the "linear" portion of the curve above the freezing-in temperature with the "linear" portion of the curve below the softening-point temperature. When the glass-transition temperature is so defined, it lies between the other two temperatures. In many cases, the difference between these three quantities is numerically and conceptually insignificant.

10.5.2. Static and Dynamic Glass-Transition Temperatures

The numerical value of the glass-transition temperature depends on the rate of measurement (see Section 10.1.2). The techniques are therefore subdivided into "static" and "dynamic" measurements. The static methods include determinations of heat capacities (including differential thermal analysis), volume change, and, as a consequence of the Lorentz–Lorenz volume–refractive index relationship, the change in refractive index as a function of temperature. Dynamic methods are represented by techniques such as broad-line nuclear magnetic resonance, mechanical loss, and dielectric-loss measurements. Static and dynamic glass transition temperatures can be interconverted. The probability p of segmental mobility increases as the free volume fraction f_{WLF} increases (see also Section 5.5.1). For $f_{\text{WLF}} = 0$, of necessity, $p = 0$. For $f_{\text{WLF}} \rightarrow \infty$, it follows that $p = 1$. The functionality is consequently

$$p = \exp\left(\frac{-B}{f_{\text{WLF}}}\right), \qquad B = \text{constant} \tag{10-44}$$

The extent of the deformation depends on the time t. To a good approximation, it can be assumed that pt = constant, and equation (10-44) becomes

$$\log pt = -\frac{B}{f_{\text{WLF}}}\log e + \log t = \log(\text{constant}) \tag{10-45}$$

For the differences in the logarithms of the times t_2 and t_1, therefore, we have, with the corresponding free volumes,

$$\log t_2 - \log t_1 = \Delta(\log t) = B(\log e)\left[\frac{1}{(f_{\text{WLF}})_2} - \frac{1}{(f_{\text{WLF}})_1}\right] \tag{10-46}$$

A change in the time scale consequently corresponds to a change in the free volume (a smaller time corresponds to a larger volume). On the other hand, the free volume fraction must increase with increasing temperature. This increase is linear in the vicinity of the glass-transition temperature

$$(f_{\text{WLF}})_2 = (f_{\text{WLF}})_1 + (\alpha_l - \alpha_{\text{am}})(T_2 - T_1) \tag{10-47}$$

where α_l and α_{am} are the expansion coefficients of the liquid and amorphous state of the polymer. The free volume fractions of these two states, however, are of the equal magnitude at the glass-transition temperature. With $T_1 = T_G$ and any desired temperature $T_2 = T$, equations (10-46) and (10-47) give

$$\Delta(\log t) = \frac{[B(\log e)/(f_{\mathrm{WLF}})_G](T - T_G)}{[(f_{\mathrm{WLF}})_G/(\alpha_l - \alpha_{\mathrm{am}})] + (T - T_G)} \tag{10-48}$$

or, solved for T, and with $\Delta(\log t) = -\log a_t$,

$$T = T_G + \frac{[(f_{\mathrm{WLF}})_G/(\alpha_l - \alpha_{\mathrm{am}})] \log a_t}{(B \log e)/(f_{\mathrm{WLF}})_G - \log a_t} \tag{10-49}$$

It has been found empirically that $(f_{\mathrm{WLF}})_G \approx 0.025$ (see Section 5.5.1). B can, to a good approximation, be made equal to unity. Since $\alpha_l - \alpha_{\mathrm{am}}$ is about $4.8 \times 10^{-4}\ \mathrm{K}^{-1}$ for many materials (see also Table 10-2), equation (10-49) can be given as

$$T = T_G + \frac{51.6 \log a_t}{17.4 - \log a_t} \tag{10-50}$$

Equation (10-50) or (10-49) is known as the William–Landels–Ferry (WLF) equation. It applies to all relaxation processes, and therefore also for the temperature dependence of the viscosity (see Section 7.6.4). Its validity is limited to a temperature range from T_G to about $T_G + 100$ K. Outside this temperature range the expansion coefficient α_l varies, not linearly, but with the square root of temperature.

The WLF equation enables static glass-transition temperatures T_G and various dynamic glass-transition temperatures T to be interconverted. To do this, the deformation times for the various individual methods must be known (Table 10-8). The shift factor a_t for the calculation is obtained as the difference between the logarithms of the deformation times.

Table 10-8. Deformation Times or Periods (Reciprocal Effective Frequencies) of Various Methods and the Glass-Transition Temperature Observed with Poly(methyl methacrylate)

Method	Deformation period, s	Glass-transition temperature, °C
Thermal expansion	10^4	110
Penetrometry	10^2	120
Mechanical loss[a]	1–10^{-4}	—
Rebound elasticity	10^{-5}	160
Dielectric loss[a]	10^2–10^{-8}	—
Broad-line NMR	10^{-4}–10^{-5}	—
NMR spin-lattice relaxation	10^{-7}–10^{-8}	—

[a] The frequency can be altered with this method.

According to the method used, then, one and the same material can exhibit a quite different mechanical behavior. Poly(methyl methacrylate), according to Table 10-8, is a glass at 140°C with respect to measurement of the rebound elasticity of spheres, but a rubbery elastic body with respect to penetrometric measurements. Static and dynamic glass-transition temperatures thus also have direct practical significance. In the case of the static glass-transition temperature, the body changes from the brittle to the tough state under slow stresses such as drawing, bending, etc., whereas the dynamic glass-transition temperature is important over shorter periods of stress (shocks, jolts).

10.5.3. Glass-Transition Temperature and Constitution

Chain segment mobility is drastically reduced by chemical or physical cross-linking when the distance between cross-link points is smaller than the segment length. In fact, the glass-transition temperature of many partially crystalline polymers also increases with increasing crystallinity (see Figure 10-18). Examples of such behavior are seen with it-poly(styrene), st-1,2-poly(butadiene), poly(vinyl chloride), poly(ethylene oxide), and poly(ethylene terephthalate). On the other hand, no influence of crystallinity on the glass-transition temperature is seen with it-poly(propylene) and poly(chlorotrifluorethylene). Consequently, the influence of the constitu-

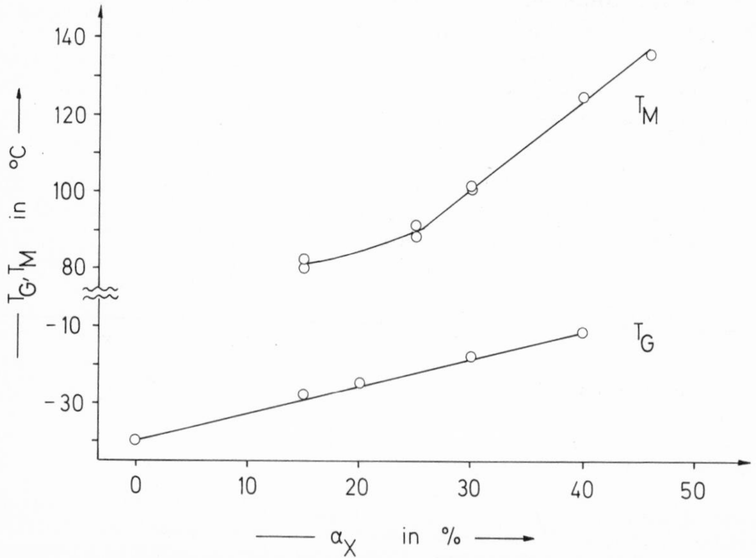

Figure 10-18. Dependence of the melting point T_M and glass-transition temperature T_G of a 90% syndiotactic 1,2-poly(butadiene) on the crystallinity α_X.

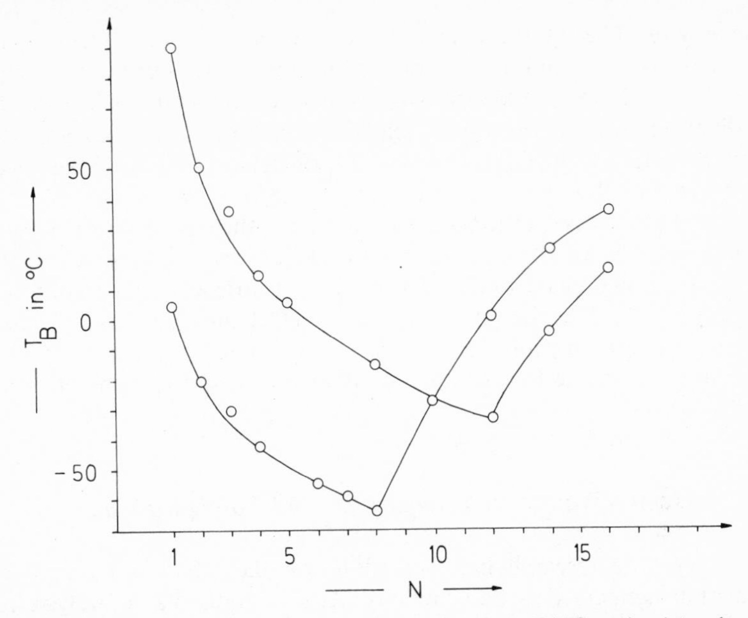

Figure 10-19. Brittleness temperature T_B of poly(acrylic esters) (PAES) and poly(methacrylic esters) (PMES) as a function of the number N of carbon atoms in the aliphatic side chains.

Table 10-9. Glass-Transition and Melting Temperatures of Polymers

Polymer	Monomeric unit	T_M, °C	T_G, °C	T_G/T_M, K/K
Poly(ethylene)	—CH_2—CH_2—	144[a]	−70	0.48
Poly(tetrafluoroethylene)	—CF_2—CF_2—	327	?	?
Poly(vinylidene chloride)	—CH_2—CCl_2—	198	−17	0.54
Poly(isobutylene)	—CH_2—$C(CH_3)_2$—	44	−73	0.63
Poly(oxymethylene)	—O—CH_2—	188	−85	0.42
Poly(oxyethylene)	—O—CH_2—CH_2—	67	−67	0.61
Poly(thioethylene)	—S—CH_2—CH_2—	205	−50	0.47
it-Poly(propylene)	—CH_2—$CH(CH_3)$—	208[a]	−15	0.53
Poly(ethyl vinyl ether)	—CH_2—$CH(OC_2H_5)$—	86	−25	0.69
st-Poly(acrylonitrile)	—CH_2—CH(CN)—	317	104	0.64
it-Poly(4-methyl pentene-1)	—CH_2—$CH[CH_2CH(CH_3)_2]$—	250	17	0.55
Poly(caprolactam)	—NH—$(CH_2)_5$—CO—	228	49	0.64
Selenium	—Se—	220	36	0.63

[a] T_M^0.

tion can only be discussed in terms of completely amorphous polymers.

The glass-transition temperature and the related brittleness temperature T_B are affected in the same way as melting temperatures by side chains of differing lengths. In poly(acrylic) and poly(methacrylic esters), longer alkyl residues (up to C_8H_{17}) lessen the efficiency of the packing together of the chains, thus lowering the glass transition and the brittleness temperatures. If the alkyl residues are even longer, the side chains of a polymer molecule will be rigidly fixed relative to one another, causing the flexibility of the individual chain to decrease, and the brittleness temperature to rise (Figure 10-19). Fluorine atoms in the main chain considerably raise the glass-transition temperature, whereas when situated in the side chains they have little effect (see for example numbers 6, 7, and 8 or numbers 1 and 2 in Table 10-9).

10.5.4. Glass-Transition Temperature and Configuration

Very little research has been done on the relations between glass-transition temperatures and tacticity. Atactic and isotactic poly(styrenes) almost always have the same glass-transition temperatures, and this is also the case for at- and it-poly(methacrylate). The glass-transition temperature of it-poly(methyl methacrylate) (42°C), on the other hand, is distinctly lower than that of the atactic product (103°C).

10.5.5. Glass-Transition Temperature of Copolymers

The glass-transition temperatures of copolymers can vary linearly or nonlinearly with increasing proportion of one component in the polymer (Figure 10-20). Since the glass-transition temperature depends on the chain flexibility, and consequently depends on the conformational energy about a bond, the glass-transition temperature depends on the conformational energies and the proportions of all the different types of bonds in the chain. The following can be used empirically:

$$\frac{1}{T_G} = \frac{w_A p_{AA}}{(T_G)_{AA}} = \frac{w_A p_{AB} + w_B p_{BA}}{(T_G)_{AB}} + \frac{w_B p_{BB}}{(T_G)_{BB}} \tag{10-51}$$

where w_A and w_B are the mass fractions of A and B monomeric units; p_{AA}, p_{AB}, p_{BA}, and p_{BB} are the probabilities of occurrence of AA, AB, BA, and BB diads; and T_G, $(T_G)_{AA}$, $(T_G)_{AB} = (T_G)_{BA}$, and $(T_G)_{BB}$ are the glass-transition temperatures of the copolymer and the corresponding A unipolymer, AB alternating copolymer, and B unipolymer, respectively. For

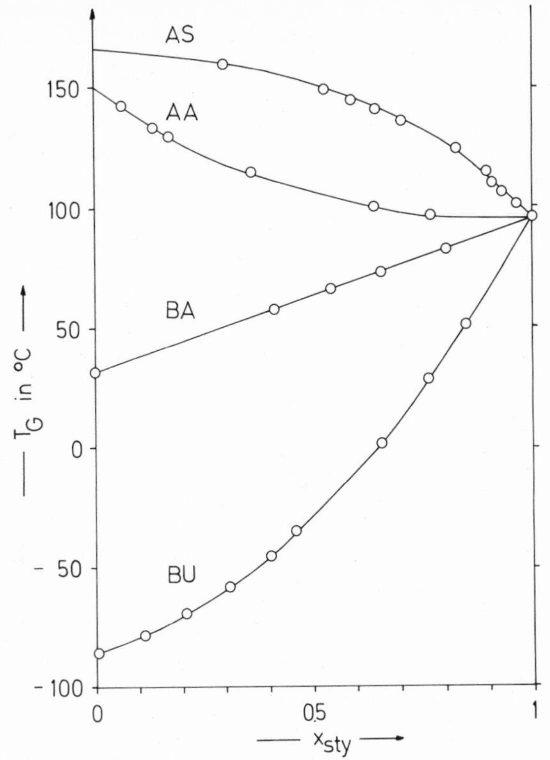

Figure 10-20. Glass-transition temperatures T_G of free-radically-polymerized copolymers of styrene and acrylic acid (AS), acrylamide (AA), *t*-butyl acrylate (BA), and butadiene (BU) as a function of the mole fraction x_{sty} of styrene monomeric units (after K. H. Illers).

vanishingly small **AB** bond fractions (i.e., relatively long blocks), equation (10-51) reduces to

$$\frac{1}{T_G} = \frac{w_A}{(T_G)_A} + \frac{w_B}{(T_G)_B} \quad (10\text{-}52)$$

If the glass-transition temperature is lowered by polymerizing in a second component, this is referred to as "internal plasticization." Copolymers of styrene with butadiene are examples of this (Figure 10-20).

There is quite a good relationship between the glass-transition temperature and the steric hindrance parameter σ (see Section 4.4.2.3) (see Figure 10-21). Since the hindrance parameter is a measure of the flexibility of an individual chain, the glass-transition temperature consequently depends primarily on the flexibility of the individual chain and only secondarily on forces operating between chains. Since the melting point also depends primarily on the flexibility of the individual chain, a relationship between the melting point and the glass transition temperature can be presumed.

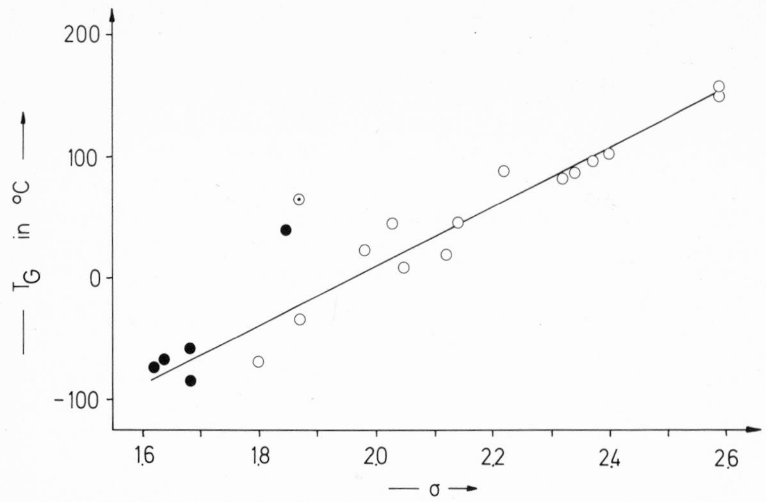

Figure 10-21. The relationship between the glass-transition temperature T_G and the steric hindrance parameter σ for carbon–carbon chains (○), carbon–oxygen chains (●), and carbon–nitrogen chains (⊙).

In actual fact, the ratio T_G/T_M values only vary between about 0.42 and 0.7 (Table 10-9).

10.5.6. Plasticizers

The glass-transition temperature can also be lowered by the addition of low-molecular-weight compounds. In analogy to "internal plasticizing" by copolymerization, the effect is known as "external plasticizing," or, more commonly, as plasticizing. The effectiveness of the plasticizer is measured by the extent of the lowering of the glass-transition temperature, which increases with plasticizer content. The specific plasticizing effectiveness is greater, therefore, the lower the plasticizer content (Figure 10-22). Plasticizers are added to industrial polymers in amounts up to 10–20%. Greater amounts of plasticizer are rarely used because (1) such large amounts can have a deleterious effect on mechanical properties and (2) the plasticizer is usually more expensive than the polymer.

The effectiveness of a plasticizer is determined by its chemical constitution and the interaction forces which this entails. In order to lower the glass-transition temperature, the mobility of the chain segments must be increased. For the plasticizer to create such an effect, it must, of course, be compatible (or at least metastable) with the polymer, i.e., it must form a thermodynamically stable mixture with the polymer. Good solubility, for example, results from strong interactions between the base units of the polymer and the plasticizer. Solvation, however, stiffens the chain. Plasticizers must therefore be the poorest solvents possible, as long as they are not nonsolvents for the polymer. Furthermore, the larger the plasticizer molecules, the greater will be the stiffening effect. The plasticizer molecules should therefore be as small as possible to give good plasticizing effectivity. Very small molecules have, however, high vapor pressures, and thus greater volatility, which is commercially undesirable. Also, interactions between the plasticizer molecules themselves should be low, since strong interactions lessen the possible interactions with the polymer. Such strongly self-interacting plasticizer molecules tend to form a kind of "network" against which polymer segment movement must occur. More energy is therefore required, and the result is that the glass-transition temperature is again raised. Little interaction between plasticizer molecules means a lower viscosity. The viscosity of the plasticizer should be as low as possible (Leilich's rule).

These observations show that the demands for the highest possible plasticizing effect per quantity of plasticizer through poor solubility, small molecules, and low viscosity are not in accord with the actual industrial

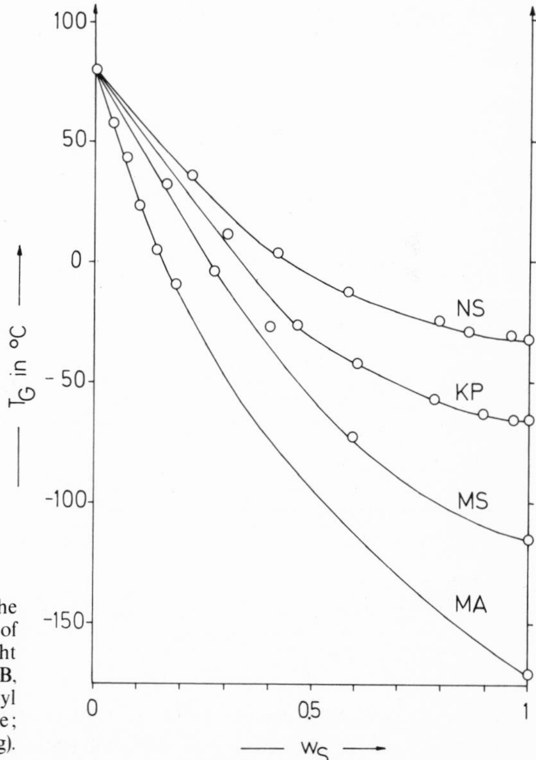

Figure 10-22. Lowering of the glass-transition temperature T_G of poly(styrene) by different weight fractions w_s of plasticizers. NB, naphthyl salicylate; KP, tricresyl phosphate; MS, methyl salicylate; MA, methyl acetate (after G. Kanig).

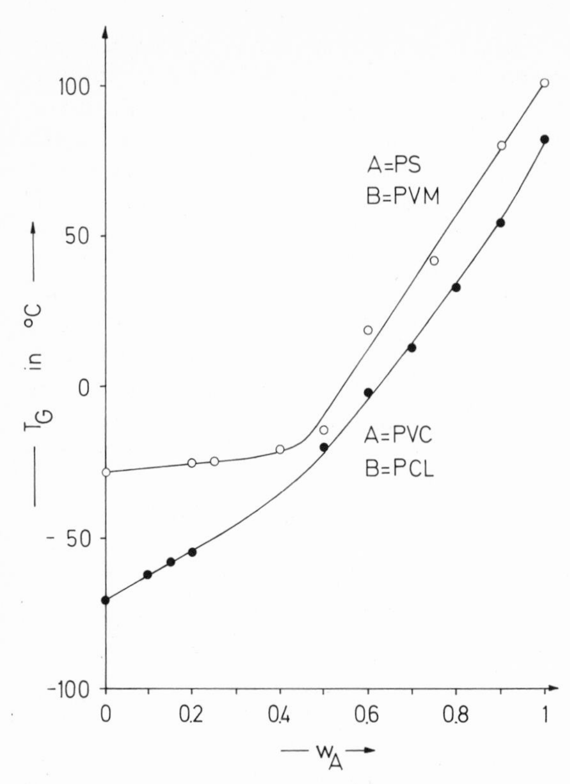

Figure 10-23. The effect of the polymeric plasticizers poly(vinyl methyl ether) (PVM) or poly(ε-caprolactone) (PCL) on the glass-transition temperature of poly(styrene) (PS) or poly(vinyl chloride) (PVC) (from data of M. Bank, J. Leffingwell, and C. Thies with J. V. Koleske and R. L. Lundberg).

usefulness of plasticizers. Therefore commercially available plasticizers always represent a compromise between desirable and just tolerable properties. Demand for low volatility, for example, can be satisfied by adding what are called polymer plasticizers. These are aliphatic polyesters and polyethers (chains with high flexibility). Their molecular weights are about 2000–4000. Because of molecular size, however, they have to be added in greater quantities in order to obtain roughly the same effect as given by low-molecular-weight plasticizers (see, e.g., Figure 10-23). As the molecular weight rises, however, the thermodynamic incompatibility also increases (Section 6.6.6). On the other hand, plasticizer migration, for kinetic reasons, will be comparatively insignificant, since the diffusion coefficients of the polymer plasticizer in the polymer are rather low because of the high molecular weights.

10.6. Other Transitions

Polymers may possess other thermal transitions besides the melt and glass-transition temperatures. Not all of these transitions can be observed with a single experimental method. Several techniques are consequently combined; for example, creep experiments (see Section 11.4.3) over long periods are combined with short-duration oscillation experiments and all data evaluated according to

$$t = \frac{1}{\omega} = \frac{1}{2\pi\nu} \tag{10-53}$$

on a common time or frequency scale. An example of this is the determination of the mechanical loss factor tan δ (for definition, see Section 11.4.5). The reciprocal values of the various frequencies for which a temperature maximum occurs (Figure 10-24) are plotted as a logarithmic function of the measuring frequency (Figure 10-25). The temperatures at which frequency maxima occur in Figure 10-24 are the dynamic transition temperatures for these frequencies.

The central frequencies from 1 to 10,000 Hz in Figure 10-24 can be covered experimentally, but another experimental method, as mentioned above, must be used for the very low and the very high frequencies. It is also possible to carry out the same experiments at various temperatures. All the data are then used to produce a master curve with the aid of the WLF equation.

The loss maxima reproduced in Figure 10-24 must result from cyclohexyl group motion. Values for poly(cyclohexyl methacrylate) and poly-

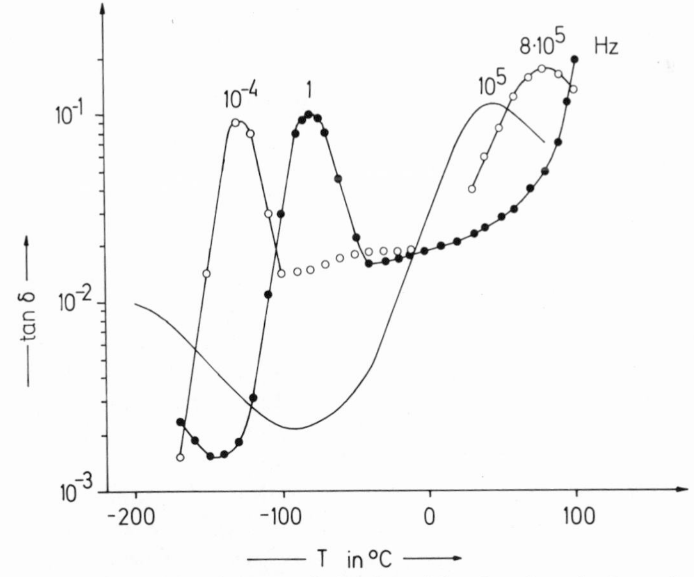

Figure 10-24. Mechanical loss factor tan δ of poly(cyclohexyl methacrylate) as a function of temperature for various frequencies (after J. Heijboer).

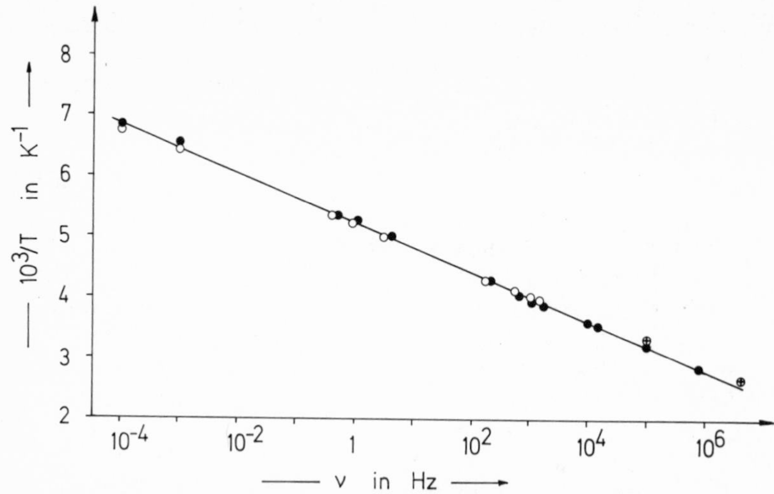

Figure 10-25. Temperature dependence of the loss maxima of poly(cyclohexyl methacrylate) (●), poly(cyclohexyl acrylate) (○), and cyclohexanol (⊕) (after J. Heijboer).

(cyclohexyl acrylate) as well as for cyclohexanol can be superimposed on the same curve (Figure 10-25), but values of, for example, poly(phenyl acrylate) do not fit on this curve. The observed loss maxima must consequently result from the chair–boat cyclohexyl ring transition.

On the other hand, for the most part, the nature of other transitions below the glass transition are not known. The motion of very short segments (for example, the coupled movement of four methylene groups) has been discussed. It is also known, however, that observed transitions are often due to impurities (e.g., water) and can also depend on the thermal history of the sample.

10.7. *Thermal Conductivity*

The usual polymers do not conduct electricity. Consequently, heat cannot be transferred by electrons in these polymers. Heat must be mostly transported by elastic waves (phonons in the corpuscular picture). The distance at which the intensity has decreased to $1/e$ is known as the free path length. This free path length is comparatively independent of temperature for glasses, amorphous polymers, and liquids and is about 0.7 nm. From this, it can be concluded that the weak decrease in thermal conductivity observed for amorphous polymers below the glass-transition temperature is essentially due to a decrease in the heat capacity with temperature (see Figures 10-26 and 10-4).

Since, above about 150 K, heat is essentially transported by inter-

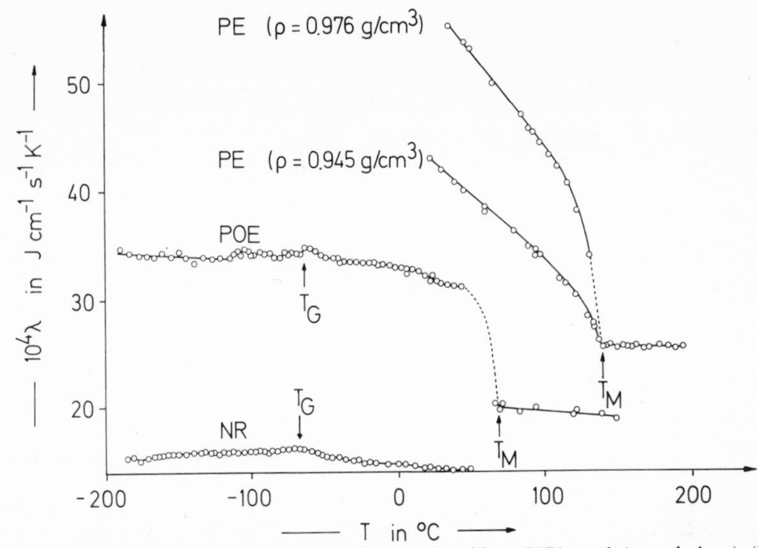

Figure 10-26. Thermal conductivity λ of natural rubber (NR), poly(oxyethylene) (POE), and poly(ethylene) (PE) of various densities as a function of temperature. T_G is the glass-transition temperature, T_M is the melting point. (From data of various authors in the compilation of W. Knappe.)

molecular collisions, a decrease in the thermal conductivity above the glass-transition temperature can be expected because of the increasingly loose arrangement of the molecules (Figure 10-26). The thermal conductivities above and below the glass-transition temperature do not differ very much, because the molecular packing above and below this temperature is also not very different. Consequently, the thermal conductivity exhibits only a weak maximum at the glass-transition temperature.

On the other hand, the packing density of crystalline polymers changes drastically at the melting point. In fact, a very sharp decrease in the thermal conductivity is also observed at the melting point (Figure 10-26). The decrease becomes stronger with increasing crystallinity of the polymer involved. The decrease begins well below the temperatures at which the onset of melting is first observed by other methods (e.g., specific volume, heat capacity).

Literature

General

P. E. Slade and L. T. Jenkins, eds., *Techniques and Methods of Polymer Evaluation*, M. Dekker, New York, Vol. 1 (Thermal Analysis) 1966; Vol. 2 (Thermal Characterisation Techniques) 1970.

W. Wrasidlo, Thermal Analysis of Polymers, *Adv. Polym. Sci.* **13**, 1 (1974).

G. M. Bartenev and Yu. V. Zelenev, eds., *Relaxation Phenomena in Polymers,* Halsted, New York, 1974.

A. M. North, Relaxation in polymers, *in: Macromolecular Science* (Vol. 8 of Physical Chemistry Series 2) (C. E. H. Bawn, ed.), MTP International Review of Science, 1975, p. 1.

Journals

Thermochimica Acta (Vol. 1 from 1970)
Journal of Thermal Analysis (Vol. 1 from 1969)
Thermal Analysis Abstracts (Vol. 1 from 1972)

Section 10.2. Special Parameters and Methods

M. Dole, Calorimetric studies of state and transitions in solid high polymers, *Fortschr. Hochpolym. Forschg.* **2,** 221 (1960).

B. Ke, Differential thermal analysis, *in: Newer Methods of Polymer Characterisation,* (B. Ke, ed.), Interscience, New York, 1964.

D. Schulze, *Differentialthermoanalyse,* Verlag Chemie, Weinheim, 1969.

B. Wunderlich, *Crystal Structure, Morphology, Defects, Macromolecular Physics,* Vol. 1, Academic Press, New York and London, 1973.

B. Wunderlich and H. Bauer, Heat capacities of linear high polymers, *Adv. Polym. Sci.* **7,** 151 (1970).

J. K. Gillham, Torsional braid analysis, a semimicro thermomechanical approach to polymer characterisation, *Crit. Rev. Macromol. Sci.* **1,** 83 (1972).

A. M. Hassan, Application of wide-line NMR to polymers, *Crit. Rev. Macromol. Sci.* **1,** 399 (1972).

J. Chiu, Dynamic thermal analysis of polymers, an overview, *J. Macromol. Sci. A (Chem.)* **8**, 1 (1974).

W. J. Smothers and Y. Chiang, *Handbook of Differential Thermal Analysis,* Chemical Publishing Company, New York, 1966.

R. C. MacKenzie, ed., *Differential Thermal Analysis,* Academic Press, New York, Vol. 1, 1970; Vol. 2, 1972.

Section 10.3. Crystallization

L. Mandelkern, *Crystallization of Polymers,* McGraw-Hill, New York, 1964.

A. Sharples, *Introduction to Polymer Crystallization,* E. Arnold, London, 1966.

B. Wunderlich, Crystallization during polymerization, *Adv. Polym. Sci.* **5,** 568 (1968).

J. D. Hoffman, J. I. Lauritzen, E. Passaglia, G. S. Ross, L. J. Frolein, and J. J. Weeks, Kinetics of polymer crystallisation from solution and from the melt, *Kolloid-Z. Z.-Polym* **231** (1–2), 564 (1969).

J. N. Hay, Application of the modified Avrami equation to polymer crystallisation kinetics, *Br. Polym. J.* **3**, 74 (1971).

M. Iguchi, H. Kanetsuna, and T. Kawai, Formation of polymer crystals during polymerisation, *Kobunshi (High Polym. Japan)* **19,** 577 (1970); *Br. Polym. J.* **3,** 177 (1971).

R. H. Marchessault, B. Fisa, and H. D. Chanzy, Nascent morphology of polyolefins, *Crit. Rev. Macromol. Sci.* **1,** 315 (1972).

I. C. Sanchez, Modern theories of polymer crystallization, *J. Macromol. Sci. C (Rev. Macromol. Chem.)* **10**, 113 (1974).

Section 10.4. Melting

H. G. Zachmann, Das Kristallisations- und Schmelzverhalten hochpolymerer Stoffe, *Fortschr. Hochpolym. Forschg.—Adv. Polym. Sci.* **3,** 581 (1961/64).

Section 10.5. Glass Transitions

R. F. Boyer, The relation of transition temperatures to chemical structure in high polymers, *Rubber Rev.* **36,** 1303 (1963).

A. J. Kovacs, Transition vitreuse dans les polmères. Etude phénoménologique, *Fortschr. Hochpolym. Forschg.—Adv. Polym. Sci.* **3,** 394 (1961/64).

M. C. Shen and A. Eisenberg, Glass transitions in polymers, *Rubber Chem. Technol.* **43,** 95 (1970).

W. A. Lee and G. J. Knight, Ratio of the glass transition temperature to the melting point in polymers, *Br. Polym. J.* **2**, 73 (1970).

Section 10.6. Other Transitions

A. Hiltner and E. Baer, Relaxation processes at cryogenic temperatures, *Crit. Rev. Macromol. Sci.* **1,** 215 (1972).

Section 10.7. Thermal Conductivity

D. R. Anderson, Thermal conductivity of polymers, *Chem. Rev.* **66,** 677 (1966).

W. Knappe, Wärmeleitung in Polymeren, *Adv. Polym. Sci.* **7,** 47 (1971).

D. Hands, K. Lane, and R. P. Sheldon, Thermal conductivities of amorphous polymers, *J. Polym. Sci. C (Polym. Symp.)* **42**, 717 (1973).

Chapter 11
Mechanical Properties

11.1. Phenomena

Macromolecular materials react quite differently to mechanical stresses. Beakers of conventional polystyrene are very brittle, and a short, quick blow will break them. In contrast, beakers of nylon 6 are very tough. Weakly cross-linked natural rubber expands on stretching by several hundred percent; after being released, it returns to what is practically its original form. When plasticine is deformed, on the other hand, it completely retains its new shape.

The reaction of a material to stress often seems related intuitively to the aggregate state. Accordingly, low-molecular-weight substances can be solid, liquid, or gaseous. With low-molecular-weight substances, a classification according to the aggregate state is usually also a classification according to the states of order. A classification according to the three classic aggregate states proves to be too restrictive, however, in the case of macromolecular substances.

Low-molecular-weight materials are termed "solid" when they show a high order and a high resistance toward deformation. Iron and common salt are solid in this sense. The order is brought about by their high crystallinity. A small stress displaces the atoms from their rest position and the atomic distances increase. The atoms again take up their original rest positions when the stress is released. In order for the deformation to be reversible, extension should not exceed amounts of $\sim 1-2\%$. This type of body is called *ideal-elastic* or *energy-elastic.*

Wood and glass, however, are also solids in the usual sense of the word. Both materials exhibit a high resistance to deformation at room temperature, but are not X-ray crystalline. On the other hand, there are "liquid crystals" which do show an order that can be detected optically, but exhibit little resistance to deformation.

True liquids, in contrast, do not show an extensive order. Even with very slight stresses applied for a short time, they deform so completely that they very quickly adopt the form of the surrounding container. Low-molecular-weight liquids thus behave in a purely viscous way under normal conditions. When stress is applied, the molecules are displaced irreversibly in relation to one another. In high-molecular-weight substances above the glass-transition temperatures, flow can be produced relatively easily. Deformations are much more difficult below the glass-transition temperature of amorphous polymers. For this reason, and because of their lack of order, amorphous substances below their glass-transition temperatures are often termed "supercooled liquids."

According to rheology (the science of flow), viscous flow and energy elasticity are only two extreme forms of the possible types of behavior of matter. It is appropriate to consider the entropy-elastic (or rubber elastic), viscoelastic, and plastic bodies as other special cases.

Plastic bodies only show irreversible deformation above a given shear stress (see Section 7.6.1).

Entropy-elastic or highly elastic bodies, contrary to the energy-elastic materials, can be extended reversibly by very large amounts (see Section 11.3.1). With the exception of large elongations of several hundred percent, entropy-elastic bodies therefore behave like solid, low-molecular-weight substances under stress. On the other hand, like liquids, elastic bodies show a limited dimensional stability, but a high volume stability. The coefficients of expansion and contraction are smaller, however, than those of liquids. Weakly cross-linked rubbers, for example, are almost purely entropy-elastic bodies. Entropy-elasticity is therefore often termed rubber or elastomeric elasticity. The cause of the behavior is a displacement of the molecular segments out of their rest position such that they take up new conformations. After relaxation, the most probable distribution of conformations reappears. The elasticity thus depends on a change in entropy. Since, in this kind of material, the molecular chains are fixed in relation to one another by cross-linking, extensive slipping of the molecular chains past one another, i.e., viscous flow, cannot occur.

Linear macromolecular substances also show entropy-elastic behavior to a certain extent. One can imagine, therefore, that the molecular chains are partly entangled. The entanglements cannot free themselves during short periods of stress. Thus, the entanglements behave here as cross-links, and the body shows an entropy-elastic behavior. During long periods of stress, the chains disentangle; the substance flows. Materials showing comparable entropy-elastic and viscous behavior are called *viscoelastic*.

The scientific classification of materials according to their flow behavior corresponds, in a limited sense, to the classification according to their commercial application. A distinction is made here between thermoplasts, fibers, elastomers, and thermosets. This classification naturally only applies at the application or processing temperature under consideration.

Thermoplasts are non-cross-linked materials whose application temperatures lie below and whose processing temperatures lie above their glass-transition temperatures (if amorphous) or melting temperatures (if partially crystalline). Above these temperatures, their viscosities are lower by orders of magnitude; the materials can consequently be heat-formed. At the processing temperatures, however, they still show elastic characteristics, and are, therefore, viscoelastic substances. As a rule, they show no plasticity, so that the name thermoplast is incorrect. In order to be used as a thermoplast, a material must consist of non-cross-linked or, at the most, weakly cross-linked molecules. Typical thermoplasts are, according to their behavior, poly(ethylene) and other poly(olefins), poly(styrene), poly(vinyl chloride), poly(methyl methacrylate), and polyamides.

Fibers, on the other hand, are, so to speak, one-dimensionally oriented thermoplasts, provided that the macromolecular substance is produced first, and the fibers then made from this. In order to be able to process a substance into fibers of sufficient strength, they must be long enough, unbranched or slightly branched molecules. Long fibers are then drawn from the melt or the solution of this substance, and these are then stretched at temperatures below the melting or glass-transition temperatures. On being stretched, the molecular chains become oriented, causing an increase in tensile strength. In certain cases, the fibers are subsequently cross-linked chemically [poly(vinyl alcohol) fibers, graphite, and carbon fibers, etc.]. Thermoplasts and fibers should exhibit as high a glass-transition temperature (if amorphous) or melt temperature (if crystalline) as possible. Typical fiber-forming materials are cellulose, certain proteins, polyamides, aromatic polyesters, and poly(propylene).

Elastomers are weakly cross-linked macromolecular substances with glass-transition and melting temperatures well below the application temperature. The chains are thus very flexible, and flexible chains readily deform under stress. To prevent viscous flow, the chains are fixed in relation to one another by cross-linking. The extent of the cross-linking must be weak so that the segments between the cross-linking points are still mobile. The substances are frequently termed elastomers before cross-linking, even though a typical elastomeric character only appears after cross-linking. At one time, elastomers (cross-linked) were distinguished from rubbers (not cross-linked). Typical elastomers are *cis*-1,4-poly(isoprene), poly(butadiene-co-styrene), butyl rubber (copolymer of isobutylene with a little isoprene), and polysulfide rubber.

Thermosets are highly cross-linked substances. Cross-linking occurs simultaneously with or after the formation of the finished shape. After this cross-linking or curing, the material cannot be reshaped by heating alone. Because they are highly cross-linked, thermosets are dimensionally stable up to elevated temperatures. Typical thermosets are, e.g., phenolic, urea, melamine, and epoxy resins.

Between the limiting cases of thermoplasts, fibers, elastomers, and thermosets, of course, there are many intermediate substances. Classical fibers, for example, only exhibit a limited elongation at break and a high glass-transition temperature, whereas the classical elastomers have a high elongation at break and a low glass-transition temperature. Elastic fibers, on the other hand, combine a high elongation at break with a high glass-transition temperature. This combination of properties can be achieved by combining "rigid segments" (high glass-transition temperature) and "flexible segments" (low glass-transition temperature) in the same molecule. The so-called Spandex fibers with rigid segments of urethane residues and flexible segments of aliphatic polyesters or polyethers [poly(propylene oxide) or poly(tetrahydrofuran)] are typical of these.

Thermoplastic elastomers, as they are called, are constructed according to a similar principle. They are block copolymers in which a flexible block ($T_G <$ application temperature) occurs between two rigid blocks ($T_G >$ application temperature). The different blocks are mutually incompatible. The rigid blocks form physical cross-links. At the application temperature, the material behaves as a cross-linked elastomer. At higher temperatures, the rigid blocks are also above the glass-transition temperature, the physical cross-links are lost, and the material can be deformed like a thermoplast.

11.2. Energy-Elasticity

11.2.1. Basic Parameters

An ideal-elastic or energy-elastic body deforms under the influence of a force by a definite amount which does not depend on the duration of the force. For comparison purposes, reference is made, not to the force, but to the force per unit surface area, i.e., the stress. The deformation may be a stretching, shearing, turning, compression, or bending (see Table 11-1).

Hooke's law describes energy-elastic bodies under stress/strain. This law relates the tensile stress (stress) σ_{ii} to the deformation (strain) $\varepsilon = \Delta l/l_0$ by

$$\sigma_{ii} = E\varepsilon \tag{11-1}$$

The proportionality constant E in stress/strain measurements is known as

Table 11-1. Classification of Moduli According to the Deformation

Force	Deformation	Modulus
Tensile stress	Tensile strain	Modulus of elasticity
Shear stress (tangential)	Shear strain	Shear modulus, torsional modulus
Torque (rotational shear stress)	Torsional strain	Torsional modulus
Compressive stress	Compressive strain	Modulus of elasticity (if pressure on all sides, modulus of compression)
Flexural stress	Flexural strain	Modulus of elasticity (as average of the tension and the compression)

the modulus of elasticity, the E modulus, or Young's modulus. The E modulus is always related to the cross section of the sample *before* stretching. The E modulus assumes a particular significance for fibers; it is the force per unit area required to elongate the fiber by its own length.

A similar relationship exists between the shear stress σ_{ij} and the deformation γ:

$$\sigma_{ij} = G\gamma \tag{11-2}$$

Here, G is the shear modulus. If a compression occurs on all sides, then the proportionality constant between the pressure p and the compression $-\Delta V/V_0$ is the modulus of compression K:

$$p = K\frac{-\Delta V}{V_0} \tag{11-3}$$

The Poisson ratio μ is a measure of the relative lateral contraction induced by a tensile stress in an isotropic material:

$$E = 2G(1 + \mu) \tag{11-4}$$

$$E = 3K(1 + \mu) \tag{11-5}$$

For incompressible materials, $E = 3G$. This is also true for many elastomers. The reciprocal of a modulus is known as a compliance.

11.2.2. Structural Influences

On the molecular level, energy-elasticity is concerned with changes in bond lengths and valence angles. Consequently, with the aid of X-ray measurements, the deformation of the crystal lattice under strain can be determined. In the calculations, it must be assumed that the strain distribution is homogeneous. This assumption seems valid since the so-called lattice moduli E_{lat} are independent of the density crystallinity (Table 11-2). The moduli of elasticity of the whole sample depend on the crystallinity. Traditionally, the moduli of elasticity are given in units of mass/area (e.g., kg/cm^2 or $lb/in.^2$) instead of force/area (e.g., N/m^2).

The lattice moduli of elasticity of poly(ethylene) are quite high since a deformation can only result from a valence angle deformation because of the all-*trans* conformation in the chain direction. On the other hand, it-poly(propylene), poly(oxymethylene), and poly(oxyethylene) occur in helix conformations. Here, deformation occurs as a result of rotation about the valence angles. Consequently, the lattice moduli of elasticity of helix-forming polymers are smaller than those of polymers in the all-*trans* conformation, i.e., their compliance (extensibility) is higher.

In general, the moduli of elasticity are much lower than the lattice moduli of the same compounds. There are many reasons for this. For one thing, most chains of a sample do not normally lie in the strain direction. Consequently, deformation can occur through increase in the interchain distance, that is, through other conformational positions (entropy-elasticity). Deformations can also occur through irreversible slippage of chains past each other (viscoelasticity).

Thus, in contrast to lattice moduli, moduli of elasticity obtained from stress/strain measurements are not measures of the energy elasticity, because of the effects of entropy-elasticity and viscoelasticity. Moduli of elasticity have more the character of being solely proportionality constants in a Hooke-type law. The "proportionality limits" are 0.05% extension for steel and 0.1–0.2% for polymers. Above these so-called proportionality

Table 11-2. Crystal Lattice (in the Chain Direction) Modulus of Elasticity E_{lat} and Modulus of Elasticity E of the Whole Test Sample for Various Polymers[a]

Polymer	α_D, (%)	E_{lat}, (N/cm^2)	E (N/cm^2)
Poly(ethylene)	84	24,000	1500
	78	24,000	700
	64	24,000	240
	52	24,000	65
Poly(tetrafluoroethylene)	—	15,600	—
it-Poly(propylene)	—	4,200	—
Poly(oxymethylene)	—	5,400	—
Poly(oxyethylene)	—	1,000	—

[a] α_D is the density crystallinity. According to data from Sakurada *et al.*

Table 11-3. The Modulus of Elasticity of Various Materials at Room Temperature

Material	E, N/cm^2
Vulcanized rubber	10^2–10^3
Crystallized rubber	10^4
Unoriented partially crystalline polymers	10^4–10^5
Organic glasses	10^5–10^6
Fibers, glass-fiber-reinforced plastics	10^7
Inorganic glasses	10^7–10^8
Crystals	10^8–10^9

limits, the relationship between the stress and the strain can be completely different (Section 11.5.1). The moduli of elasticity of polymers are consequently normally measured for extensions of 0.2 % over times of about 100 s.

Moduli of elasticity also depend strongly on the environmental conditions. In many cases, water acts as a plasticizer and lowers the modulus of elasticity by increasing chain mobility. Since the diffusion of water into the material is time dependent, the modulus of elasticity also varies with time. For example, a polyamide sample in a dry state has a modulus of elasticity of $E = 2700$ N/mm^2. This decreases to 1700 N/mm^2 in humid air and to 860 N/mm^2 after 4 months in air.

A significant decrease in the modulus of elasticity is observed at the glass-transition temperature and at the flow temperature because of the increasing influence of entropy-elasticity with rising temperature. Within each physical state, however, only a relatively small variation of the modulus of elasticity with polymer structure is observed (Table 11-3).

11.3. Entropy-Elasticity

11.3.1. Phenomena

The behavior of weakly cross-linked rubber was described in Section 11.1 as entropy-elastic. If this material is deformed, the chains are displaced from their equilibrium positions and brought into a state which is entropically less favorable. Because of the weak cross-linking, the chains are unable to slip past one another. On relaxation, the chains return from the ordered position to a disordered one; the entropy increases. The phenomenon can be described in various ways. Seen thermodynamically, the rubber elasticity is related to a lowering of entropy on deformation. From the molecular point of view, the molecular particles are forced to adopt an

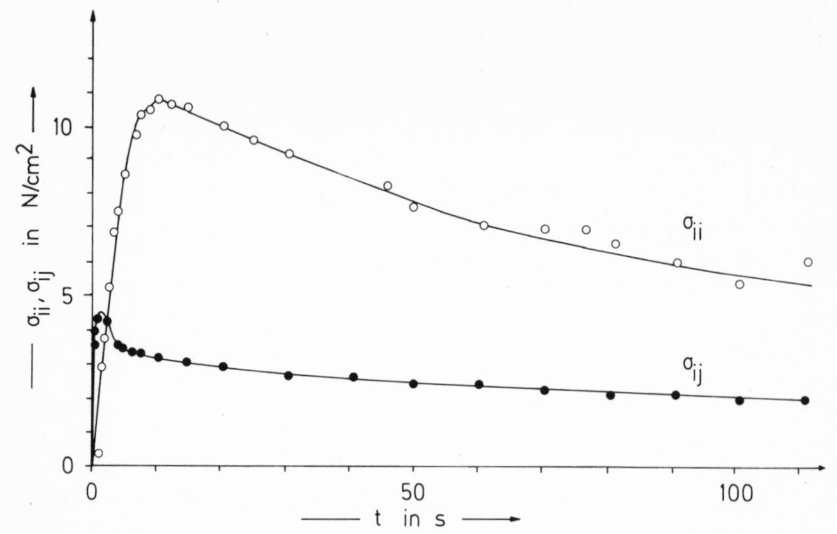

Figure 11-1. Shear stress σ_{ij} and normal stress σ_{ii} of a poly(ethylene) at 150°C and a shear gradient of 8.8 s^{-1} as a function of time in seconds. Measurements with a cone-and-plate viscometer. (After BASF.)

imposed orientation, i.e., there is a change of conformation, on being deformed. In the language of mechanics, it is considered as the occurrence of normal stress. This stress contribution is so called because it is effective at right angles (normal) to the direction of deformation.

Entropy- and energy-elastic bodies differ very characteristically in some phenomena:

1. Energy-elastic bodies show small, reversible deformations of a maximum of 0.1–1% at large moduli of elasticity (see, e.g., Table 1-5). The energy-elastic-body steel, for example, possesses an E modulus of $\sim 2.1 \times 10^7$ N/cm². The entropy-elastic-body soft rubber, on the other hand, shows a large reversible deformation of several hundred percent at low moduli of elasticity of 20–80 N/cm². The E moduli of entropy-elastic bodies thus lie at low values similar to those of gases (~ 10 N/cm²).

2. The energy-elastic-body steel cools on elongation, whereas the entropy-elastic-body rubber becomes warmer.

3. Steel under a load expands on heating, as does only very slightly ($\sim 10\%$) elongated rubber. More elongated rubbers, however, contract on heating. Their shear modulus thus increases on heating.

Non-cross-linked chain molecules, as well as chemically weakly cross-linked materials, show entropy-elastic behavior under given stresses. It is also conceivable that the chains of long, non-cross-linked, flexible macromolecules can become entangled or interpenetrate each other to a certain extent. When deformation by drawing is rapid, the entanglements act as cross-links. The parts of the chain adopt more random positions in an attempt to return to the original random positions. A normal stress is created. Both the normal stress and the shear stress can be measured in a cone and plate viscometer. During rotation there is shearing. It is possible to calculate the shear stress from the rotational moment (torque) applied to the rotor. The cone and plate are pushed away from one another, however, by the normal stress developed by the sample when sheared. In order to prevent this, it is necessary to apply a force, which is then proportional to the normal stress. The normal stress can be much greater than the shear stress (Figure 11-1).

Shear and normal stresses cannot be measured individually in a capillary viscometer. If material with entropy-elastic components is pressed through a nozzle, the macromolecules are deformed. If this time of loading is short and the load not too great, however, the segments cannot slip away from one another, because of entanglements. At the nozzle opening, the material again has more space at its disposal. The material will therefore expand when it leaves the nozzle. The effect is known as the Barus or memory effect in melts and the Weissenberg effect in solutions. In extrusion it is called extrusion swelling and in the blowing of hollow samples it is called swelling behavior. Of course, the effect is also time dependent, since, as the time increases, the chains can slip away from one another more readily.

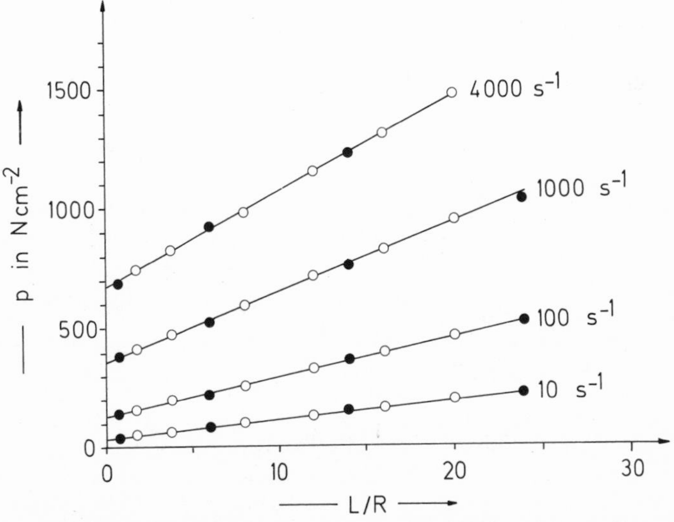

Figure 11-2. Bagley plot of a high-impact poly(styrene) at 189°C at shear gradients of 10, 100, 1000, and 4000 s^{-1} from measurements with capillaries of R = 1 mm in diameter (●) or R = 0.6 mm (○) and different lengths L (after BASF).

The Barus effect can be studied with the aid of the Bagley plot. Solving equation (7-43) for the pressure p and replacing the shear stress σ_{ij} for non-Newtonian liquids by $\sigma_{ij} = \eta_{\text{app}} D$ (see Section 7.6.2), we obtain

$$p = 2\sigma_{ij}\frac{L}{R} = 2\eta_{\text{app}} D \frac{L}{R} \tag{11-6}$$

In the Bagley plot, the pressure p is plotted against the nozzle geometry L/R at a constant shear rate D. With Newtonian liquids, according to equation (7-43), for $L/R = 0$, p will also be 0; the slope of the straight lines is given by $2\sigma_{ij}$. Behavior of this type is also found for concentrated polymer solutions at high values of L/R (Figure 11-3). Obviously, one is then working in the second Newtonian viscosity region, η_∞. At low L/R values, however, the function $p = f(L/R)$ for D = constant deviates from this straight line, and tends toward a new linear relationship. This straight line occurring at low L/R values does not intersect the p axis at $p = 0$, but at a finite value p_0 (Figures 11-2 and 11-3). Equation (11-6) thus becomes

$$p = p_0 + \text{constant}' \times \frac{L}{R} \qquad D = \text{constant} \tag{11-7}$$

Usually p_0 is identified with the loss of pressure caused by the elastically accumulated energy of the flowing liquid and by the formation of a

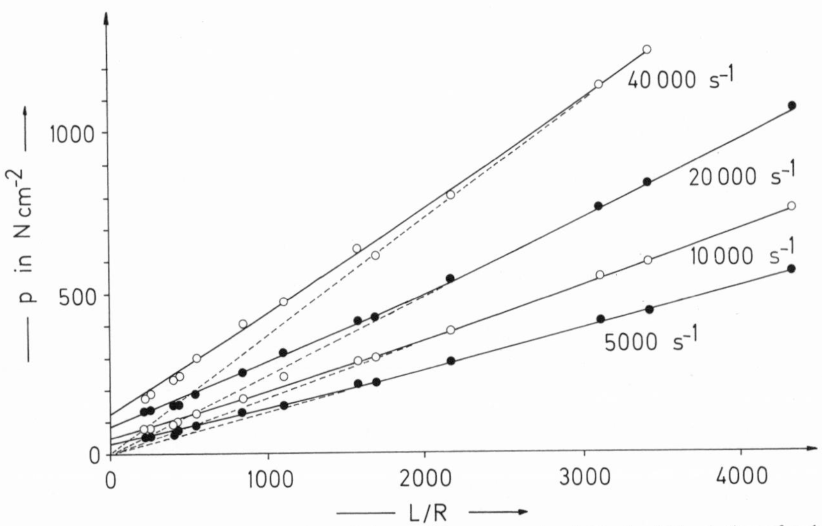

Figure 11-3. Pressure p as a function of the capillary geometry L/R of a 6% solution of poly(isobutylene) ($\overline{M}_w = 6 \times 10^6$, $\overline{M}_n = 0.55 \times 10^6$ g/mol) in toluene for various shear rates at room temperature. L is the length and R is the radius of the capillary. (After J. Klein and H. Fusser.)

stationary flow profile at both ends of the capillary. Since the loss in pressure disappears at large L/R values, it may be assumed that the elastic deformation can equilibrate in time in very long capillaries (disentangling). This is also confirmed by the fact that the p_0 values are greater, the larger the shear rates, i.e., the shorter the retention times. In addition, according to measurements on poly(ethylene) melts, the extrusion swelling falls off at very high L/R values. According to the general theory, however, the extrusion swelling is a measure of the elastic energy accumulated in the system.

Apart from these theoretical considerations, however, p_0 has a direct practical meaning in the extrusion and spinning of plastics. That is, the greater the nozzle ratio L/R, the higher the pressure that has to be applied. In practice, therefore, as small a nozzle length as possible is used.

11.3.2. Phenomenological Thermodynamics

The changes in the states of entropy-elastic bodies described in the previous section can be expressed quantitively by phenomenological thermodynamics, starting with one of the fundamental equations in thermodynamics. The relationship of interest here relates the pressure p with the internal energy U, the volume V, and the thermodynamic temperature T (see textbooks of chemical thermodynamics):

$$p = -\left(\frac{\partial U}{\partial V}\right)_T + \left(\frac{\partial p}{\partial T}\right)_V T \tag{11-8}$$

Instead of the change in volume dV, the change in length dl on application of a stretching force F is considered. F and p have opposite signs. Equation (11-8) thus becomes

$$F = \left(\frac{\partial U}{\partial l}\right)_T + \left(\frac{\partial F}{\partial T}\right)_l T \tag{11-9}$$

If the second law of thermodynamics $A = U - TS$ is differentiated with respect to the length l, this gives

$$\left(\frac{\partial A}{\partial l}\right)_T = \left(\frac{\partial U}{\partial l}\right)_T - T\left(\frac{\partial S}{\partial l}\right)_T \tag{11-10}$$

By setting the two expressions for $(\partial U/\partial l)_T$ in equations (11-9) and (11-10) equal, one finds

$$\left(\frac{\partial A}{\partial l}\right)_T + T\left(\frac{\partial S}{\partial l}\right)_T = F - T\left(\frac{\partial F}{\partial T}\right)_l \tag{11-11}$$

In thermodynamics, furthermore, the following is quite generally valid:

$$\left(\frac{\partial S}{\partial V}\right)_T = \left(\frac{\partial p}{\partial T}\right)_V \tag{11-12}$$

Since the force F is proportional to the pressure p, and the length l to the volume V, it is possible to write, in analogy to equation (11-12),

$$\left(\frac{\partial S}{\partial l}\right)_T = -\left(\frac{\partial F}{\partial T}\right)_l \tag{11-13}$$

If equation (11-13) is inserted into (11-11), one obtains

$$\left(\frac{\partial A}{\partial l}\right)_T - T\left(\frac{\partial F}{\partial T}\right)_l = F - T\left(\frac{\partial F}{\partial T}\right)_l \tag{11-14}$$

The second term on both sides of equation (11-14) is identical. What is called the equation of the thermal state for entropy-elastic bodies is therefore

$$\left(\frac{\partial A}{\partial T}\right)_l = F \tag{11-15}$$

It has been found experimentally that the force F is directly proportional to the temperature T in the case of weakly cross-linked natural rubber extended by less than 300%. From this it follows that F = constant × T or $\partial F/\partial T$ = constant, or

$$\frac{F}{T} = \frac{\partial F}{\partial T} \tag{11-16}$$

If equation (11-16) is inserted into equation (11-9), the result is

$$\left(\frac{\partial U}{\partial l}\right)_T = 0 \tag{11-17}$$

Thus, the internal energy U of an entropy-elastic body does not alter during elongation. An entropy-elastic body differs fundamentally from an energy-elastic one in this respect. The change of force F on heating can be obtained by combining equations (11-16) and (11-13):

$$dS = -\frac{F}{T}\,dl \tag{11-18}$$

$T\,dS$ and F are positive. Therefore, the change in length dl must be negative; entropy-elastic bodies have a negative thermal coefficient of expansion.

The total differential of the change in length of an entropy-elastic body is

$$dl = \left(\frac{\partial l}{\partial F}\right)_T dF + \left(\frac{\partial l}{\partial T}\right)_F dT \tag{11-19}$$

On heating at constant length, $dl = 0$. Equation (11-18) then becomes

$$\left(\frac{\partial F}{\partial T}\right)_l = -\frac{(\partial l/\partial T)_F}{(\partial l/\partial F)_T} \tag{11-20}$$

According to equation (11-18), the thermal coefficient of expansion $\gamma = (1/l)(\partial l/\partial T)$ of an entropy-elastic body is negative. The denominator $(\partial l/\partial F)_T$ in equation (11-20) corresponds to the increase in length when the stress is increased and is positive, so that $(\partial F/\partial T)_l$ must be positive in equation (11-20). When an entropy-elastic body is heated, the stress increases.

11.3.3. Statistical Thermodynamics

Phenomenological thermodynamics describes changes in energies, temperatures, volumes, etc. Unless additional assumptions are made, however, it cannot give any information about the molecular phenomena that lie behind these processes. Statistical thermodynamics attempts to obtain such information through the use of probability functions.

When a weakly cross-linked material is stretched, the molecular segments between two cross-linking points adopt a more unfavorable (less random) position. The ends of the segments move away from one another (Figure 11-4). With one end of the chain at the origin of the coordinate

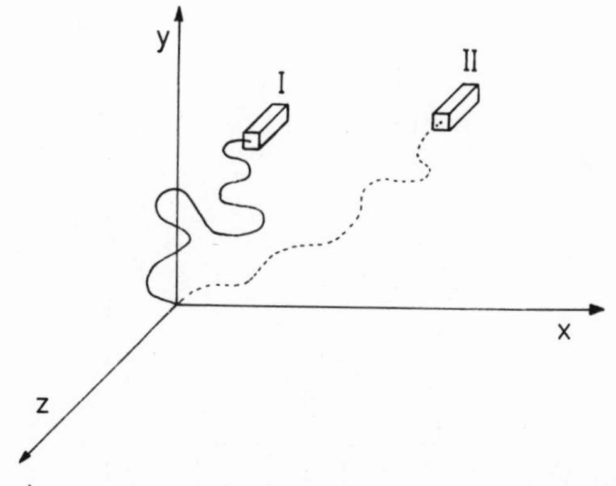

Figure 11-4. Schematic representation of the extension of a segment between two cross-linking points. One of the cross-linking points is fixed at the origin; the other moves without alteration of chain length from position I before cross-linking to position II during elongation.

system, the probability $\Omega_i(x, y, z)\,dx\,dy\,dz$ of finding the other end point in a volume element $dx\,dy\,dz$ is then [see also equation (A4-37) for the corresponding expression for *one* direction]

$$\Omega_i(x, y, z)\,dx\,dy\,dz = \left(\frac{b}{\pi^{0.5}}\right)^3 \exp\left[-b^2(x_i^2 + y_i^2 + z_i^2)\right] dx\,dy\,dz \qquad (11\text{-}21)$$

with

$$b^2 = \frac{3/2}{Nl^2} \qquad (11\text{-}22)$$

N is the number of segments in the chain and l is the segment length. Inserting equation (11-21) into the Boltzmann equation $s_i = k \ln \Omega_i$, we obtain the following for the entropy of a chain:

$$s_i = k \ln\left[\text{constant} \times \left(\frac{b}{\pi^{0.5}}\right)^3 dx\,dy\,dz\right] - kb^2(x_i^2 + y_i^2 + z_i^2) \qquad (11\text{-}23)$$

The constant serves to maintain dimensionality, and is, therefore, purely a factor of convenience (in the subsequent mathematical treatment it is canceled out). It is then assumed, in fact, that the coordinates of each individual segment change in the same proportion as the external coordinates of the test body. Thus if the external coordinates are extended by an elongation ratio α, the coordinate in the x direction of the i segment should be α times as large after elongation as the initial unextended value $(x_{i,0})$, etc.:

$$x_i = \alpha_x x_{i,0}; \qquad y_i = \alpha_y y_{i,0}; \qquad z_i = \alpha_z z_{i,0} \qquad (11\text{-}24)$$

Equation (11-24), together with equation (11-23) for the change in entropy of a segment becomes, for extension,

$$\Delta s_i = s_i - s_{i,0} = -kb^2\left[(\alpha_x^2 x_{i,0}^2 + \alpha_y^2 y_{i,0}^2 + \alpha_z^2 z_{i,0}^2) - (x_{i,0}^2 + y_{i,0}^2 + z_{i,0}^2)\right] \qquad (11\text{-}25)$$

The total change in entropy must be additive. With chains of equal length it then follows that

$$\Delta S = \sum_i \Delta s_i = -kb^2\left[(\alpha_x^2 - 1)\sum_i x_{i,0}^2 + (\alpha_y^2 - 1)\sum_i y_{i,0}^2 + (\alpha_z^2 - 1)\sum_i z_{i,0}^2\right] \qquad (11\text{-}26)$$

For the end-to-end chain distance [see equation (4-9)]

$$\langle L^2\rangle_{00} = Nl^2 = \langle x_0^2\rangle + \langle y_0^2\rangle + \langle z_0^2\rangle \qquad (11\text{-}27)$$

With the number N_i of chains, we have

$$\sum_i x_{i,0}^2 = N_i\langle x_0^2\rangle; \qquad \sum_i y_{i,0}^2 = N_i\langle y_0^2\rangle; \qquad \sum_i z_{i,0}^2 = N_i\langle z_0^2\rangle \qquad (11\text{-}28)$$

and for an isotropic material

$$\langle x_0^2 \rangle = \langle y_0^2 \rangle = \langle z_0^2 \rangle \tag{11-29}$$

It therefore follows from equations (11-26)–(11-28) and (11-22) that

$$\Delta S = -0.5kN_i(\alpha_x^2 + \alpha_y^2 + \alpha_z^2 - 3) \tag{11-30}$$

According to equation (11-30), the entropy change of a rubber on stretching is not due to a particular chemical structure, but simply to the number of chains and hence the number of cross-linking points linking the chains.

Equation (11-30) describes a special case, namely that of elongation without volume change ($\alpha_x\alpha_y\alpha_z = 1$). Various equations have been proposed for the general case, and they can all be formulated as

$$\Delta S = -A[(\alpha_x^2 + \alpha_y^2 + \alpha_z^2 - 3) - B] \tag{11-31}$$

where A and B have different significances according to the assumptions made in different theoretical treatments (for example, B is often taken as $B = \ln \alpha^3$).

According to equation (11-28), the elongation force F can be expressed by

$$F = -T\left(\frac{\partial S}{\partial l}\right)_{T,V} = -T\left(\frac{\partial \Delta S}{\partial l}\right)_{T,V} \tag{11-32}$$

ΔS can be found from equation (11-30). If, on stretching, the chain is elongated in one direction ($\alpha_x = \alpha$) and simultaneously contracted in the two other directions ($\alpha_y = \alpha_z = 1/\alpha_x^{0.5}$), then equation (11-30) becomes

$$\Delta S = -0.5kN_i\left(\alpha^2 + \frac{2}{\alpha} - 3\right) \tag{11-33}$$

and, with $\alpha = l/l_0$, we find, after differentiation,

$$\left(\frac{\partial \Delta S}{\partial l}\right)_{T,V} = -\frac{kN_i(\alpha - \alpha^{-2})}{l_0} \tag{11-34}$$

or, inserted into equation (11-32),

$$F = \frac{kTN_i(\alpha - \alpha^{-2})}{l_0} \tag{11-35}$$

If both sides are divided by the original cross-sectional area $A_0 = V_0/l_0$, then, with the definition of the tensile stress $\sigma_{ii} \equiv F/A_0$, the gas constant $R = kN_L$, and the molar concentration $[M_i] = N_i/V_0N_L$ of the network chains, equation (11-35) becomes

$$\sigma_{ii} = kT\frac{N_i}{V_0}(\alpha - \alpha^{-2}) = RT[M_i](\alpha - \alpha^{-2}) \tag{11-36}$$

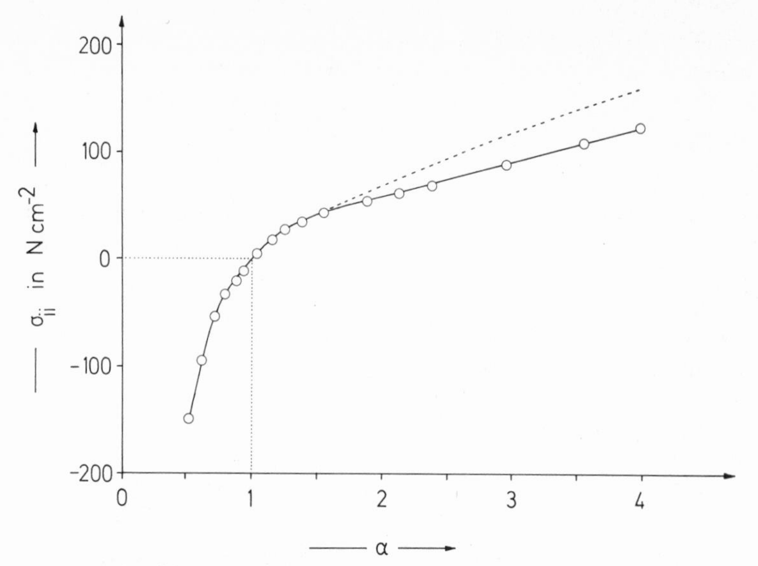

Figure 11-5. Relationship between the stress σ_{ii} and the draw ratio $\alpha = L/L_0$. (—○—○—) Experimental, (- - -) calculated according to equation (11–36). Measurements by elongation (at $\alpha > 1$) or compression ($\alpha < 1$). (After L. R. G. Treloar).

Experiment and theory agree well with regard to the relationship between σ_{ii} and α in the case of compression and small elongations (Figure 11-5). Deviations occur at large elongations, which could be due to beginning rubber crystallization, to a non-Gaussian distribution of cross-linking sites, or to time effects.
Equation (11-36) can be written as

$$\frac{\sigma_{ii}}{\alpha - \alpha^{-2}} = C_1 = RT[M_i] \qquad (11\text{-}37)$$

which resembles the empirical Mooney–Rivlin equation:

$$\frac{\sigma_{ii}}{\alpha - \alpha^{-2}} = C_1 + C_2\alpha^{-1} \qquad (11\text{-}38)$$

The physical significance of C_1 and C_2 is not established. The C_1 of equation (11-38) is often assumed to be identical with the C_1 of equation (11-37). Some authors suggest that the appearance of C_2 is due to time effects, i.e., that the measurements were not carried out under equilibrium conditions. This is partially confirmed by the fact that the value of C_2 falls as the rubber swells, i.e., as the segments become more mobile. Depending on the test conditions, C_2 either falls as the temperature rises (more rapid establishment of equilibrium), remains constant, or even increases (development of stresses).

In other ways, too, a real elastomer does not behave as ideally as was supposed in the preceding derivations. Some of the cross-linking points, for example, do not connect different chains to each other, but merely lead to more free chain ends. Only a fraction of chemical cross-linking sites is therefore physically effective. Conversely, additional physical "cross-linkings" occur due to chain entanglements. In addition, chain sections between the cross-linking sites may be of unequal length. Also, cross-linking is usually carried out at higher temperatures than the subsequent measurement of the deformation. At the cross-linking temperature, however, the rubber has other dimensions than at the deformation temperature, so that a perturbed initial state precedes deformation.

Shearing can be treated in the same way as drawing. When sheared, the sample is elongated in the x direction and correspondingly contracted in the y direction. The coordinates in the z direction remain constant. Thus, $\alpha = \alpha_x$, $\alpha = 1/\alpha_y$, and $\alpha_z = 1$. Equation (11-30) therefore becomes

$$\Delta S = -0.5kN_i(\alpha^2 + \alpha^{-2} - 2) \qquad (11\text{-}39)$$

and correspondingly, for the change in entropy ΔS_V which is related to the unit volume V_0, with $[\mathrm{M}_i] = N_i/V_0N_\mathrm{L}$, $R = kN_\mathrm{L}$, and $\Delta S_V = \Delta S/V_0$,

$$\Delta S_V = -0.5R[\mathrm{M}_i](\alpha^2 + \alpha^{-2} - 2) \qquad (11\text{-}40)$$

The shear strain γ produced by a shearing load is given by $\gamma = \alpha - \alpha^{-1}$. From equation (11-39), therefore, since $\gamma^2 = (\alpha - \alpha^{-1})^2 = \alpha^2 + \alpha^{-2} - 2$,

$$\Delta S_V = -0.5R[\mathrm{M}_i]\gamma^2 \qquad (11\text{-}41)$$

The relationship between shear stress σ_{ij} and shear strain γ is given, in analogy to equation (11-32), by

$$\sigma_{ij} = -T\frac{\partial \Delta S_V}{\partial \gamma} \qquad (11\text{-}42)$$

After differentiating equation (11-41), we obtain from equation (11-42)

$$\sigma_{ij} = RT\gamma[\mathrm{M}_i] = G\gamma \qquad (11\text{-}43)$$

According to equation (11-43), the shear stress σ_{ij} is directly proportional to the shear strain γ. The rubber, therefore, deforms on shearing according to Hooke's law with the shear modulus G [see equation (11-2)], but is non-Hookean during elongation [see equation (11-36)].

11.3.4. Elastoosmometry

In a swollen network (a gel) at equilibrium, the chemical potential of the solvent is exactly the same within and outside the gel. If the solvent outside the gel is replaced by a solution, then the chemical potential of the solvent within the gel will change. This will cause a change in the amount of solvent within the gel. This change in solvent content of the gel will further cause a change in volume, a change in length (under constant load), or a change in deformation load (under constant extension) in the gel.

To a first approximation, the change in deformation load or force F is given by

$$\Delta F = F - F_0 = \frac{\partial F}{\partial \mu_1} \Delta \mu_1^{\text{int}} = \frac{\partial F}{\partial \mu_1} \Delta \mu_1^{\text{ext}} \tag{11-44}$$

The chemical potential of the solvent outside the gel can be replaced by the expression for the osmotic pressure [see equations (9-4) and (9-7)]:

$$-\Delta \mu_1^{\text{ext}} = \Pi \tilde{V}_1^{\text{m}} = \tilde{V}_1^{\text{m}} \left(\frac{RTc_2}{\overline{M}_n} + RTA_2c_2^2 \right) \tag{11-45}$$

Combination of equations (11-44) and (11-45) leads to

$$\frac{\Delta F}{c_2} = -\frac{\partial F}{\partial \mu_1} \tilde{V}_1^{\text{m}} \left(\frac{RT}{\overline{M}_n} + RTA_2c_2 \right) \tag{11-46}$$

The number-average molecular weight of the solute and the second virial coefficient can be obtained from the concentration dependence of the change in the deformation load. The change in deformation load with chemical potential, which is also necessary for the calculation, can be derived from the general thermodynamic expression for the change in the Gibbs energy:

$$dG = -S\,dT + V\,dp + F\,dl + \mu\,dn \tag{11-47}$$

For an isothermal–isobaric process, $dT = 0$, $dp = 0$, and $dG = 0$,

$$F = -\mu \frac{dn}{dl} \tag{11-48}$$

Differentiation of this equation gives

$$\left(\frac{\partial F}{\partial \mu} \right)_{p,T,l} = -\frac{\partial(\mu\,dn/dl)}{\partial \mu} = -\left(\frac{dn}{dl} \right)_{p,T,\mu} \tag{11-49}$$

Consequently, the change in deformation load with chemical potential can be obtained from the change in length with amount (in moles).

11.4. Viscoelasticity

11.4.1. Basic Principles

In the preceding discussion about the energy- and entropy-elastic behavior of matter, it was tacitly assumed that the body returns immediately and completely to the original state when the load is removed. In actual fact, this process always takes a certain time in macromolecular substances. In addition, not all bodies return completely to the original position; in some cases they are partially irreversibly deformed.

In these bodies, then, there must be simultaneous cooperation between time-independent elastic and time-dependent viscous properties. Many phenomena can be described by different combinations of the basic equations for elastic and viscous behavior.

The two extreme cases of mechanical behavior can be reproduced very well by mechanical models. A compressed Hookean spring can serve as a model for the energy elastic body under load (Figure 11-6). On releasing the load, the compressed spring immediately returns to its original position. The relationship between the shear stress $(\sigma_{ij})_e = \sigma_e$, the shear modulus G_e, and the elastic deformation γ_e is given by Hooke's law [equation (11-2)]:

$$(\sigma_{ij})_e = \sigma_e = G_e \gamma_e \tag{11-50}$$

Differentiation with respect to time yields

$$\frac{d(\sigma_{ij})_e}{dt} = G_e \frac{d\gamma_e}{dt} \tag{11-51}$$

The model for rate effects is a dash pot containing a piston in a viscous Newtonian liquid. For greater clarity, and to conform with equation (11-51),

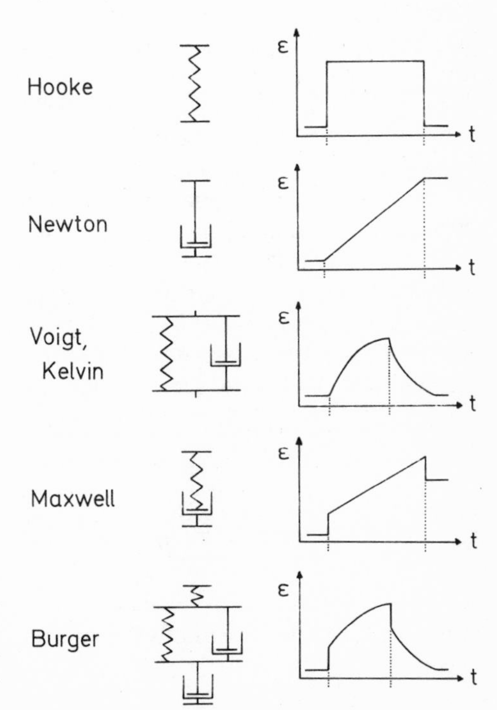

Figure 11-6. Deformation ε as a function of time, according to different models. The sample was loaded up to the time t_1 and the load was released at the time t_2 (represented by dotted lines).

the shear stress σ_{ij} in Newton's law will be written as $(\sigma_{ij})_\eta = \sigma_\eta$ and the shear rate dv/dy as the deformation rate $d\gamma_\eta/dt$:

$$\sigma_\eta = \eta \frac{d\gamma_\eta}{dt} \tag{11-52}$$

Integration leads to

$$\gamma_\eta = \frac{\sigma_\eta}{\eta} t \tag{11-53}$$

Maxwell bodies are obtained if Hookean and Newtonian bodies are connected in series (Figure 11-6). The Kelvin or Voigt model, on the other hand, contains Hookean and Newtonian bodies in a parallel arrangement (Figure 11-6). The Maxwell body is a model for relaxation phenomena and the Kelvin body is a model for retardation processes.

11.4.2. Relaxation Processes

A relaxation is defined in mechanics as the decrease in stress with constant deformation. That is, a stress must be applied during the deformation of a viscous liquid. When the deformation ceases, the stress will fall off (relax) as the molecules or molecular segments return to their rest state. The Maxwell model obviously describes this behavior very well. During rapid deformation, the spring will very quickly elongate since a viscous liquid responds only slowly to a rapidly applied stress. If deformation is kept constant, then, because of the relaxation of the spring to its equilibrium stress position, the piston subsequently begins to move slowly through the viscous liquid (see also Figure 11-6). If the stress is suddenly removed, the spring contracts immediately but the piston remains in the elongated state. Although originally intended only for energy-elastic bodies, the spring model can also be used to describe the shearing of entropy-elastic bodies, as exemplified by equation (11-43).

The rates of deformation $d\gamma/dt$ are additive in these processes. By combining the expressions for the rates of deformation according to Hooke's law [equation (11-51)] and according to Newton's law [equation (11-52)], we obtain the following for the total rate of deformation:

$$\frac{d\gamma}{dt} = \frac{d\gamma_e}{dt} + \frac{d\gamma_\eta}{dt} = G_e^{-1} \frac{d\sigma_e}{dt} + \frac{\sigma_\eta}{\eta} \tag{11-54}$$

The indices e and η can be neglected, since the stress is the same in both elements. When the deformation is constant, $d\gamma/dt = 0$, and equation (11-54) becomes

$$G^{-1}\frac{d\sigma}{dt} = -\frac{\sigma}{\eta} \tag{11-55}$$

or, when integrated,

$$\sigma = \sigma_0 \exp\left(\frac{-Gt}{\eta}\right) = \sigma_0 \exp\left(\frac{-t}{t_{\text{rel}}}\right) \tag{11-56}$$

$t_{\text{rel}} = \eta/G$ is the relaxation time. It indicates the time interval required for the stress to fall to a value $1/e$ times the original value.

With real polymers, however, there exists not only one, but a whole spectrum of relaxation times. In an ideal rubber, for example, all cross-linking points are separated by equal distances. With short periods of stress, the resulting stresses are compensated by the largely "free" rotation about the chain bonds within short relaxation periods of $\sim 10^{-5}$ s. With long stress periods, the cross-linking points can also shift in relation to one another. The long relaxation times characteristic of this process hinder the viscous flow of the material during short periods of stress. Between these two relaxation times there is a region in which the modulus of elasticity remains almost constant. With real rubbers, however, the distances between the cross-linking points are not equal, but vary over a wide range. Thus a whole spectrum of relaxation times is to be expected. This spectrum can be exemplified through the use of a model with a number of Maxwell bodies in a parallel arrangement.

11.4.3. Retardation Processes

Retardation is defined as the increase in deformation with time under constant stress. Retardation processes in a material can be recognized as "creep" or "strain softening." Since the phenomenon is also observed in apparently solid materials at room temperature, it is also called "cold flow." When the load is removed, a change in the deformation is often observed. In some cases, the sample readopts its original dimensions. Creep is therefore better described as retarded elasticity then as viscous flow. Although creep can be modeled by a series of coupled Maxwell elements, the corresponding mathematical equations are difficult to solve and a special model with parallel spring and dash-pot elements (Kelvin or Voigt element) is preferred. Since creep is a deformation at constant stress, it is only necessary to add the two expressions for the stress in Hookean bodies [equation (11-50)] and in Newtonian fluids [equation (11-52)]:

$$\sigma = \sigma_e + \sigma_\eta = G_e\gamma_e + \eta\frac{d\gamma_\eta}{dt} \tag{11-57}$$

From this it follows through integration that (index K characterizes the Kelvin element)

$$\gamma_K = \frac{\sigma}{G}\left[1 - \exp\left(\frac{-Gt}{\eta}\right)\right] = \gamma_\infty\left[1 - \exp\left(\frac{-t}{t_{ret}}\right)\right] \quad (11\text{-}58)$$

Here, the indices e and η are again omitted because the strain is the same in both elements. In equation (11-58), γ_∞ is a constant and t_{ret} the retardation time. As a rule, there is usually a whole spectrum of retardation times in the retardation processes also. Retardation and relaxation time distributions are similar, but they are not identical since they pertain to different models of the deformation behavior.

11.4.4. Combined Processes

Macromolecular materials usually possess entropy-elasticity together with viscous and energy-elastic components. Such behavior was only partly comprehensible by use of the models discussed up to now. It can be described very satisfactorily, however, by a four-parameter model in which a Hooke body, a Kelvin body, and a Newton body are combined (see the lowest figure in Figure 11-6). With this model, the deformation must again be added, i.e., with equations (11-50), (11-53), and (11-58),

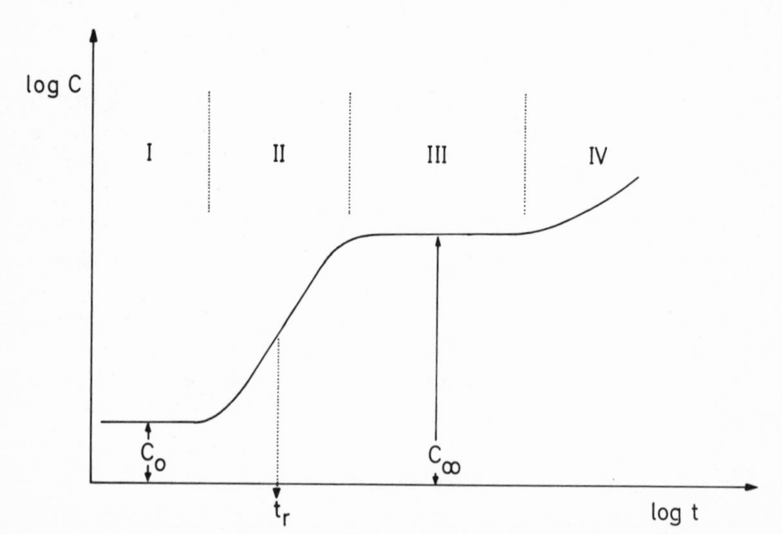

Figure 11-7. Schematic representation of the compliance C as a function of time. t_r is the orientation time. (I) Glassy state, (II) viscoelastic state, (III) entropy-elastic state, (IV) viscous flow.

$$\gamma = \gamma_e + \gamma_K + \gamma_\eta = \frac{\sigma}{G} + \gamma_\infty \left[1 - \exp\left(\frac{-t}{t_r} \right) \right] + \frac{\sigma}{\eta} t \tag{11-59}$$

or by introducing the compliance $C = \gamma/\sigma$,

$$C = C_0 + C_\infty \left[1 - \exp\left(\frac{-t}{t_r} \right) \right] + \frac{t}{\eta} \tag{11-60}$$

According to this equation, the mechanical behavior observed depends very much on the ratio of the test time to the orientation time t_r. With $t \gg t_r$, the exponential term in equation (11-60) makes almost no contribution to the total deformation. Conversely, the damping contribution as described by this term in the total deformation will become more noticeable the nearer the test and orientation times t and t_r approach each other (see Figure 11-7).

Similar reasoning applies to the temperature dependence. At low temperatures, $t_r = \eta/G$ tends toward infinity as the viscosity η becomes very large; only a Hookean elasticity is observed. At high temperatures, on the other hand, the third term (viscous flow) predominates. In between lies a range of temperatures at which the test and orientation times are comparable. A damping will then be observed at these temperatures.

11.4.5. Dynamic Loading

For dynamic measurements, the sample is subjected to a periodic stress. In the simplest case, the stress is applied sinusoidally. The applied stress σ then alters with the time t and the angular frequency ω according to

$$\sigma = \sigma_0 \sin \omega t \tag{11-61}$$

where σ_0 is the amplitude. In the linear stress/strain region, the resulting deformation γ exhibits the same frequency as the initiating oscillation, but with an angular phase shift of ϑ (Figure 11-8):

$$\gamma = \gamma_0 \sin(\omega t - \vartheta) \tag{11-62}$$

The tangent of the angle δ is termed the loss factor. The amplitude ratio $\gamma_0/\sigma_0 = C$ is the absolute compliance.

According to equation (11-62), the deformation can be separated into two parts:

$$\gamma = \gamma_0 [C'(\omega) \sin \omega t - C''(\omega) \cos \omega t] \tag{11-63}$$

The first part is in phase with the applied change in stress, and the second is out of phase by an angle of $\pi/2$. The amplitude C' of the first part is called the real or storage compliance, and the amplitude C'' of the second part is

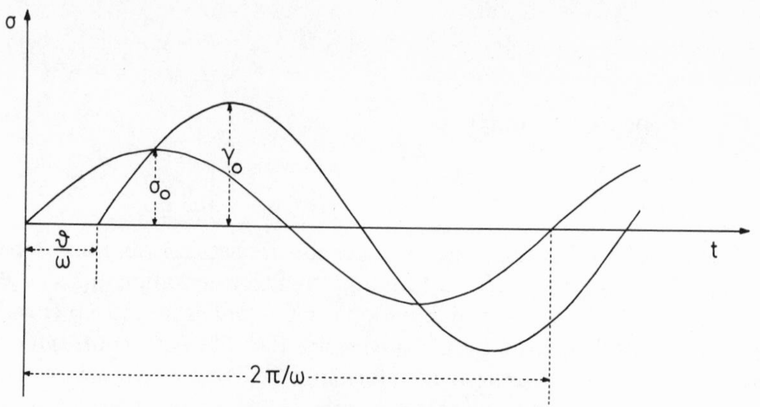

Figure 11-8. Schematic representation of the tensile stress σ as a function of the time t with dynamic (sinusoidal) loading (see text).

called the imaginary or loss compliance. The real (G') and imaginary moduli (G'') are not simply their countervalues. The real compliance measures the amount of reversible elastic deformation and the imaginary compliance measures the energy lost through conversion to heat. The relationships between these quantities are

$$C' = C \cos \vartheta, \qquad C'' = C \sin \vartheta, \qquad \tan \vartheta = C''/C'$$

$$G' = \frac{C'}{(C')^2 + (C'')^2}, \qquad G'' = \frac{C''}{(C')^2 + (C'')^2} \tag{11-64}$$

The behavior during dynamic stress can also be clarified, of course, by different models.

11.5. Deformation Processes

11.5.1. Tensile Tests

In a typical tensile-strength test, a standard rod of the sample is stretched in a tensile-strength machine and elongated at a constant rate while the stress σ_{ii} and strain ratio $\alpha = L/L_0$ (or time) are recorded. Strain, as used here, is only concerned with the change in length of the sample. It has no a priori meaning on the molecular level, and does not mean that molecules are in a strained or oriented state in the stretched sample. The elongation $\varepsilon = (L - L_0)/L_0$ is often used instead of the strain ratio. As normally expressed, a sample is said to be 150% elongated when it has been stretched to 2.5 times its original length. Figure 11-9 shows a typical stress/strain diagram.

Often, only a few points from the stress/strain diagram are given instead of the whole diagram.

Hooke's law is followed for small stresses (or strains) in the region from the origin to point I of Figure 11-9 [see equation (11-1)]. The point I is consequently called the *proportionality limit* or the *elastic limit*. The latter name is incorrect because of the entropy-elasticity remaining above the point I. According to definition, the proportionality limit is reached when the sample shows a remaining 0.1% strain after removal of the stress. The *technical elastic limit* is defined as an extension of 0.2%.

Point II is the maximum of the force/deformation curve and is called the *yield stress* σ_s. The *drawing stress* is at point III. Finally, the *tenacity at break* or the *tensile strength* σ_B and the *elongation at break* ε_B occur at the point IV. The region between the points Ia, II, and III is what is known as the ductile region. The decrease in tensile stress with increasing strain between II and III is known as stress softening. The increasing stress with increasing strain between III and IV is called stress hardening. The stress hardening or stress softening is only nominal, since the strain ratio, or elongation, is always in relation to the original cross-sectional area of the sample (see also below).

In many cases, a "specific tensile strength" or "elongation at break" is given as

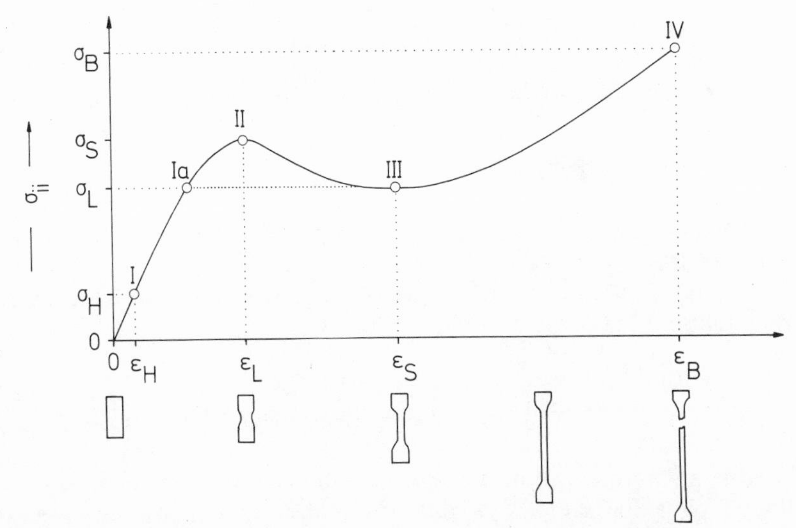

Figure 11-9. Tensile stress σ_{ii} as a function of strain ε of a rod at constant temperature. (I) Proportionality limit, (II) upper flow limit with yield stress limit and yield strain limit, (III) lower flow limit (in this region a deformation break is possible), (IV) break point with tensile strength and elongation at break. The ductile region is given by the surface Ia–II–III. Stress softening occurs along II–III, and stress hardening along III–IV. The stress/strain curve is typical of materials deforming with necking.

$$\text{specific tensile strength} = \frac{\text{stress}}{\text{density}} \tag{11-65}$$

Another term often used is

$$\text{specific rigidity} = \frac{\text{modulus of elasticity}}{\text{density}} \tag{11-66}$$

The rigidity of a body depends not only on the modulus of elasticity E, but also on the geometrical shape of the body. For example, the necessary flexural moment M required to bend a body with the radius R to a given extent can be approximately given by

$$M \approx \pi E R^2/4 \tag{11-67}$$

Of two bodies of the same shape and cross-sectional area, the body with the greater modulus of elasticity is the more rigid. The rigidity decreases with decreasing cross-sectional diameter.

According to the usual convention, the original cross-sectional area A_0 of the sample is used to calculate the stress $\sigma_{ii} = F/A_0$ from the stretching force F. The cross section is reduced, however, on deformation. Consequently, the actual or true stress σ'_{ii} operating at a given strain is greater than the nominal or engineering stress σ_{ii}, i.e., for constant volume

$$\sigma'_{ii} = \frac{F}{A} = \frac{F}{A_0}\frac{L}{L_0} = \sigma_{ii}\frac{L}{L_0} \qquad (V = \text{constant}) \tag{11-68}$$

The true strain ε' is also different from the nominal or engineering strain ε:

$$\varepsilon' = \int_{L_0}^{L} \frac{dL}{L} = \ln\frac{L}{L_0} = \ln\frac{A_0}{A} \tag{11-69}$$

Nominal and actual stress/strain diagrams differ characteristically (Figure 11-10).

An empirical relationship between the actual stress and the actual strain holds for many thermoplasts:

$$\left[\log\left(\frac{\sigma'}{\sigma^*}\right)\right]\left[\log\left(\frac{\varepsilon'}{\varepsilon^*}\right)\right] = -C \tag{11-70}$$

σ^* and ε^* are normalization parameters used to shift the curves of $\log \sigma' = f(\log \varepsilon')$ plots for samples of different molecular weights deformed at different rates and temperatures onto a single master curve.

The parameter σ^* is equal to the critical stress required to unfold chain folds in an ideal system with $C = 0$. The parameter ε^* is the extension required for the formation of extended-chain conformations. For example,

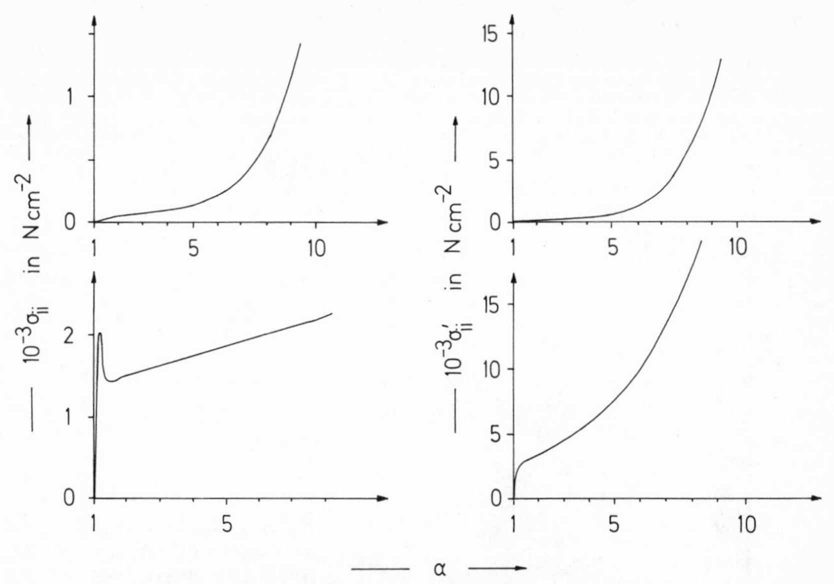

Figure 11-10. Stress/strain diagram of a natural rubber (above) and an it-poly(styrene) (below) at room temperature. Left: experimental curves; right: true stress curves. Numerical values were not given for the lower left figure in the original work. (After P. I. Vincent.)

$\varepsilon^* = 5.55$ for poly(ethylene), which corresponds to a strain ratio of 240. The constants C are typical of a given polymer; they are, for example, 0.175 for nylon, 0.230 for poly(propylene), and 0.384 for poly(ethylene).

11.5.2. Necking

The shape of a stress/strain plot for a sample depends on its chemical structure (constitution, configuration, molecular weight, molecular-weight distribution, cross-linking), on its physical structure (crystallinity, orientation), and consequently also on the processing conditions in producing the sample (annealing, drawing, injection molding, etc.), on additives (fillers, plasticizers), on the form of the sample (geometric shape, film thickness, etc.,) and on the deformation conditions (rate, temperature).

The same sample can behave as brittle, tough, or rubberlike (Figure 11-11) according to whether the experimental temperature lies above or below the glass-transition temperature. In some cases a telescoping effect or "necking" occurs (see also Figure 11-9). A necking is characterized by the occurrence of an increasing extension with a stress that decreases or at least remains constant instead of increasing.

The telescoping effect can be recognized by the formation of a neck

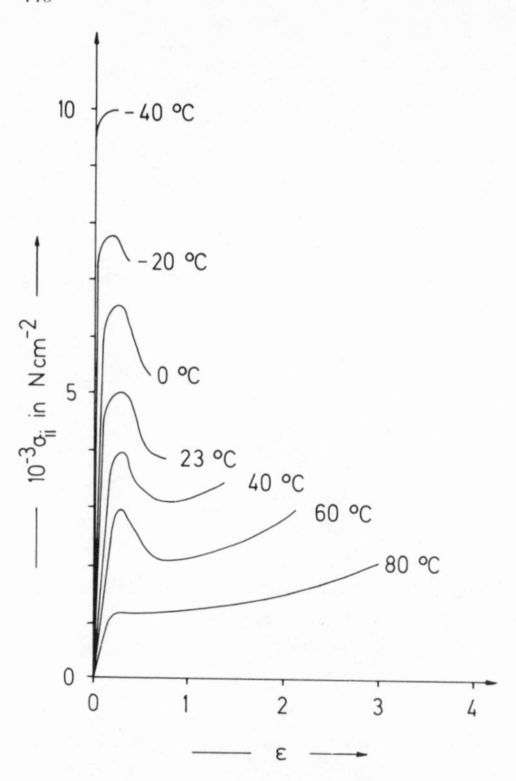

Figure 11-11. Stress/strain curves for a poly(vinyl chloride) at temperatures between −40°C and +80°C (after R. Nitsche and E. Salewski). The samples appear to be brittle at −40°C, to be ductile (tough) from −20 to 23°C, to show cold flow from 40 to 60°C, and are rubberlike at 80°C.

during deformation (Figure 11-9). A constriction begins to form above the upper flow limit. The cross section of the constriction decreases up to the lower flow limit. With continued elongation, the length of the constricted portion continues to increase at the expense of adjacent parts of the sample, although the constriction cross-sectional area remains practically constant. The flow zone moves along the test sample, and a kind of neck is formed.

The temperature can rise locally to 50°C above the surrounding temperature during neck formation. This causes a decrease in the viscosity, which, of course, leads to increased flow. The effect, however, is also observed under isothermal conditions, and so must be primarily caused by some other factor. There are, of course, locally microscopically small differences in cross section, which, for the same applied force, causes the stress to be greater at the smaller cross sections. A localized neck is formed and is stabilized by molecular orientation and/or heat released on stretching. The released heat causes a localized decrease in viscosity, etc.

The telescope effect is observed with practically all polymers, including X-ray amorphous polymers. One of the few exceptions is cellulose acetate. Necking can often be prevented by very slow stretching. The difference between the brittleness temperature and the experimental temperature is decisive for neck formation. At sufficiently low temperatures, every body behaves as a brittle body, and no drawing occurs.

The brittleness temperatures of amorphous polymers sometimes lie 100–150 K lower than the glass-transition temperatures, and so amorphous polymers are still tough (that is, they can be stretched) below their glass-transition temperatures. For the same reason, partially crystalline polymers can also be deformed below their melting temperatures. With increasing closeness to the glass-transition temperature, however, the viscous character of their behavior becomes increasingly important. The yield point occurs at lower stresses and the elongation at break increases (see Figure 11-9).

Thin films exhibit a greater tensile strength than thick films, since the statistical probability of local defects occurring is smaller with thin films.

The stress/strain plots of drawn films and fibers often differ significantly from those of undrawn samples (Figure 11-12).

With increasing width of the molecular weight distribution of the test sample, the viscosity is increasingly non-Newtonian and, as a result, the processibility is better. In contrast, samples with narrower molecular-weight distributions (especially when cross-linked) have better dynamic properties. The presence of a significant fraction of low-molecular-weight, highly crystalline material makes a sample brittle. In contrast, high-molecular-weight fractions lead to more elastic samples, but, because of the increased viscosity, the evenness of extruded samples is not so good; the surface finish is poorer.

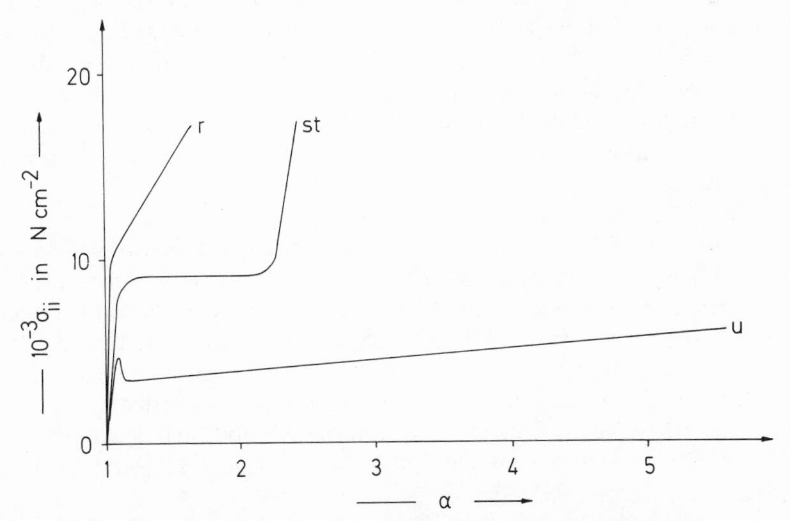

Figure 11-12. Stress/strain plot of a poly(ethylene terephthalate) after different pretreatments; u, undrawn; st, biaxially oriented; r, drawn and annealed. (After R. A. Hudson.)

11.5.3. Hardness

The hardness of a body is generally understood to be its mechanical resistance to the penetration of a needle, a sphere, etc. A definition of hardness which is generally valid and suitable for all materials does not yet exist, and, as a result, there is no generally applicable test method.

As a rule, the hardness measurements usually also reflect other material properties. For this reason, only similar materials can be compared with the same hardness test method.

In Brinell's method, a small steel sphere is pressed into the test body with a certain force. The depth of penetration, i.e., the retained plastic deformation, is measured. The measurement is only carried out, therefore, after the load is removed. The Brinell hardness test is especially suitable for hardness tests on metals, where the measurements are made above the flow limit in the plastic region.

Rockwell's hardness test works in a similar way to Brinell's hardness test, i.e., it uses the depth of penetration. Contrary to Brinell's method, however, it measures the penetration of a sphere while still under a load, and then measures the remaining elastic deformation. For this reason, the Rockwell method always gives lower degrees of hardness than the Brinell method. In addition, the degrees of hardness according to the Rockwell method are not measured in physical units, but in scale numbers of 0–120. Steel balls are used with soft materials, and diamond points with hard ones. The Vicker's hardness test uses a diamond pyramid. A modified Rockwell method is used for plastics. It should be noted that, with the Rockwell hardness thus determined, the plastic deformation contribution increases only gradually, because of creep. With metals, on the other hand, the deformation is always plastic, and therefore, also independent of time. Plastics, therefore, exhibit a relatively high Rockwell hardness compared to metals.

The so-called Shore hardnesses are measured differently for metals and plastics. With hard materials (metals), a scleroscope is used to measure the rebound of a small steel ball. This Shore hardness is thus measured by a dynamic method, which yields the rebound hardness (the "impact elasticity" of the rubber industry). Soft plastics, on the other hand, are tested with a Shore durometer. This measures the resistance to the penetration of the point of a cone through the contraction of a calibrated spring. The durometer thus works according to a static method, and yields the true Shore hardness as understood by the rubber industry. Like the Rockwell hardness, the Shore hardness is given in scale divisions.

The pendulum hardness is used to test painted steel surfaces. For this method, a duroscope is used. Here, a small hammer similar to a pendulum is made to fall on the sample. There are also many other standard test methods for pendulum hardness tests.

In the Mohs hardness test, the resistance of the sample to scratching is tested. The Mohs hardness scale is divided into 10 degrees of hardness. These are fixed arbitrarily (e.g., talc = 1, Iceland spar = 3, quartz = 7, diamond = 10). A similar hardness scale is based on the scratching power of pencils of different hardness.

In all methods of testing hardness, the thickness of the material and the type of substrate are very important because the elasticity is usually also measured. In addition, it should be noted that hardness tests always measure the hardness of the surface, and not that of the material within the sample. The surface of a sample can be plasticized, for example, by water vapor from the air. If a plastic that can be crystallized is injected into a cold mold, then in some cases the surface is less crystalline than the interior, etc.

In a certain sense, the abrasion tests can also be counted among the hardness tests. Abrasion is affected partly by the hardness and partly by the frictional properties of the sample. Among thermoplastics, the best abrasion resistance is shown by polyureas, followed by the polyamides and polyacetals.

11.6. Fracture

11.6.1. Concepts and Methods

A polymer can fracture in many different ways, according to the type, conditions, and stress. Many polymers fracture or break almost immediately as stress is applied. With others, no change can be seen even after days and months. The break can be "clean" or rough. The elongation at break can be less than 1% or more than several thousand percent.

In the limiting case, two types of breaking processes are possible, namely, a brittle fracture and a tough or ductile break. With a brittle fracture, the material fractures perpendicular to the direction of stress with little if any flowing processes. With a tough break, on the other hand, tearing takes place in the direction of the pressure stress through shearing processes, and by positional transitions in crystalline regions. Consequently, a body is defined as being brittle when the extension at break is less than 20%.

The fracture behavior of brittle substances is frequently tested by flexural strength tests (Figure 11-13). In a flexural strength test, the body is loaded slowly with a continually increasing force. For this, the test sample is either supported at two points or else clamped at one point. The flexural

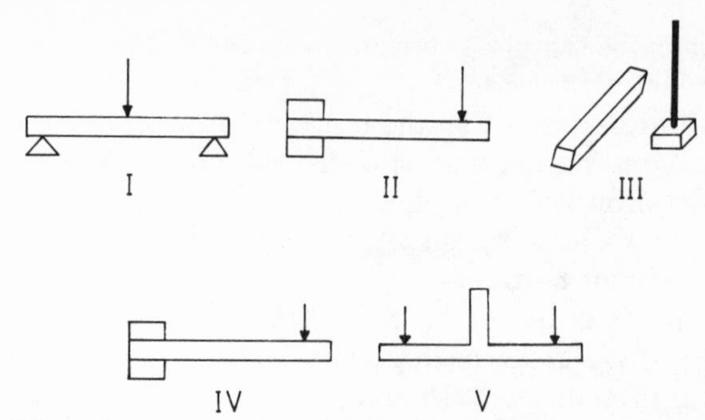

Figure 11-13. Schematic representation of different types of flexural tests. (I) Flexural test with sample supported at two points; (II) flexural test, sample clamped at one point; (III) pendulum impact test to determine the flexural impact strength; (IV) flexural impact test; (V) tensile impact test.

strength is an indication of the capacity of a body to change its form. Soft bodies can bend so much that the sample slips from the support.

To test the impact strength, the sample is quickly bent under stress up to the breaking point. The impact can be a pendulum or a flexural or a tensile impact. To test the notched-bar impact strength, the sample is first notched distinctly, and the subsequent tensile strength, so to speak, is measured. The ratio of the work done to the cross-sectional width is termed the impact strength.

11.6.2. Theory of Brittle Fracture

In principle, the force necessary to cause a brittle break $F = E/L$ can be calculated from the energy E required to separate chemical and physical bonds by an interbond-partner distance L. To break extended-chain poly(ethylene) crystals perpendicular to the chain direction (i.e., breaking covalent bonds), a force of about 20,000 N/mm^2 is necessary, whereas to cause a break parallel to the chain direction (i.e., working only against dispersion forces), only 200 N/mm^2 is required. Experimentally, however, a maximum tensile strength of 20 N/mm^2 is observed (the so-called crystal paradox). Consequently, the break must occur at inhomogeneities, since these lead to an inhomogeneous distribution of the tensile stress onto "disruption points" and thus lead to stress concentrations.

The break behavior of energy-elastic and entropy-elastic bodies is different. According to the break theory of Ingles for energy-elastic bodies, there is a relationship between the critical break stress $(\sigma_{ii})_{\text{crit}}$, the stress operating at the top of a crack σ_{ii}, the geometry of the crack, and the modulus of elasticity. In the simplest case of a crack of length L with a round tip of radius R, we have

$$(\sigma_{ii})_{\text{crit}} = \sigma_{ii}\left(\frac{R}{4L}\right)^{0.5} \tag{11-71}$$

The Ingles theory offers a very good description of the break behavior of silicate glass, since silicate glasses are practically solely energy-elastic and the crack propagation energies are of the same order of magnitude as the surface energies.

The break behavior of any desired elastic body is described by the Griffith theory. According to Griffith, a crack in an elastic body only propagates further when the elastically stored energy just exceeds the energy required to break chemical bonds. Combination of this with the Ingles concept leads to

$$(\sigma_{ii})_{\text{crit}} = \left(\frac{2E\gamma}{\pi L}\right)^{0.5} \tag{11-72}$$

where E is the modulus of elasticity and γ is the break surface energy, that is, the energy required to form a new, crack-free surface. The predicted dependence of the critical tensile stress on the reciprocal square root of the crack length has actually been observed (Figure 11-14). For small

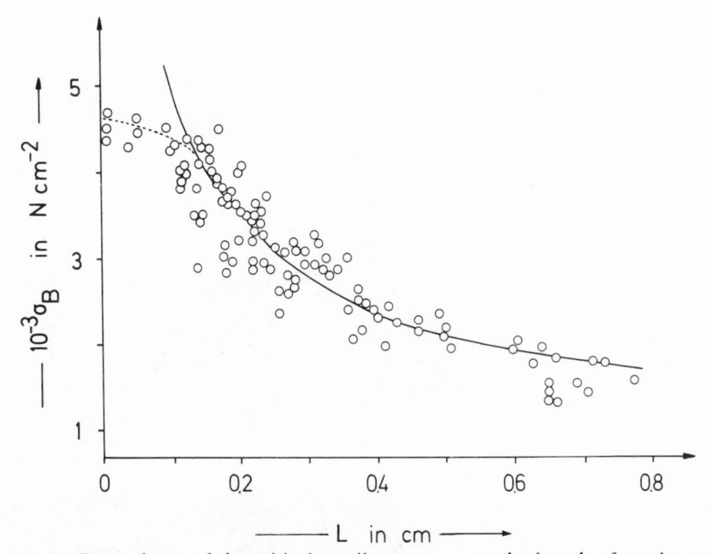

Figure 11-14. Dependence of the critical tensile stress σ_B on the length of cracks artificially introduced into poly(styrene) rods of cross sections between 0.3×0.5 and 2.8×0.5 cm^2 at elongation rates between 0.05 and 0.5 cm/min. The solid line gives the functionality predicted by the Griffith theory. (After J. P. Berry.)

crack lengths, measured data deviate from the Griffith theory and concentrate at a finite value of $(\sigma_{ii})_{crit}$ for $L = 0$. The crack length at which the deviating behavior is first observed results from crazing, and not from "naturally occurring" cracks.

Crazes occur perpendicular to the stress direction shortly before a destructive break. They may be up to 100 μm long and up to 10 μm wide. Crazes are not hairline cracks, that is, they are not totally void between the break surfaces. The spaces between the break surfaces in crazes mostly contain molecular bundles or lamellar material stretched in the stress direction. Consequently, in contrast to genuine breaks, crazes possess a structural and mechanical continuity. Because certain materials whiten on crazing, crazes are often called white breaks.

According to electron spin resonance data, free radicals are produced at chain ends even before a macroscopic break occurs. The free radical concentration depends only on the extension, and not on the tensile stress. Concentrations of 10^{14}–10^{17} free radicals/cm^2 are generally observed. Since free radical concentrations of only about 10^{13} free radicals/cm^2 occur

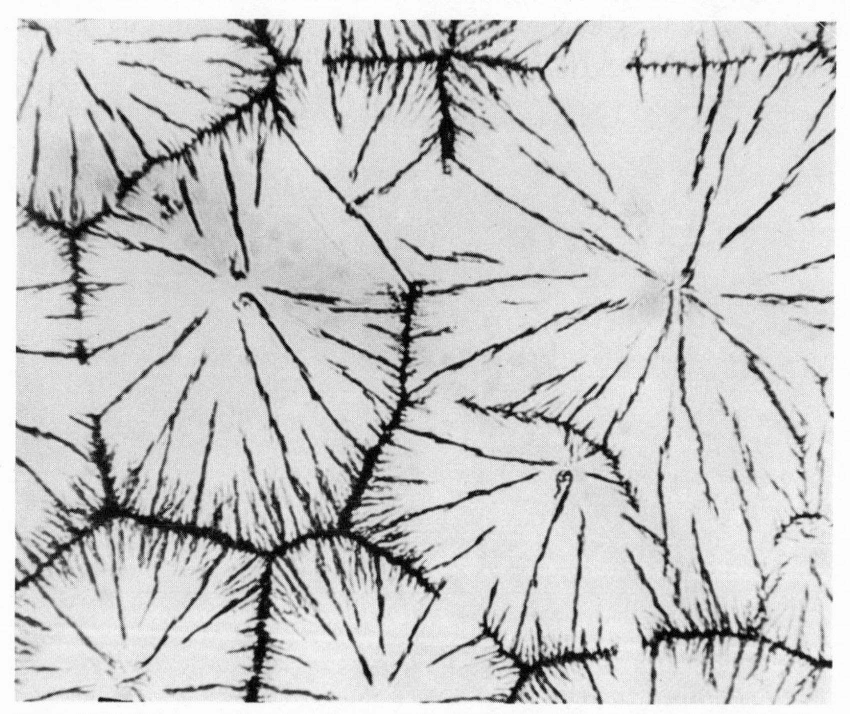

Figure 11-15. The onset of break at the amorphous positions in it-poly(propylene) spherulites, that is, between spherulites and radially within the spherulites (after H. D. Keith and F. J. Padden, Jr.).

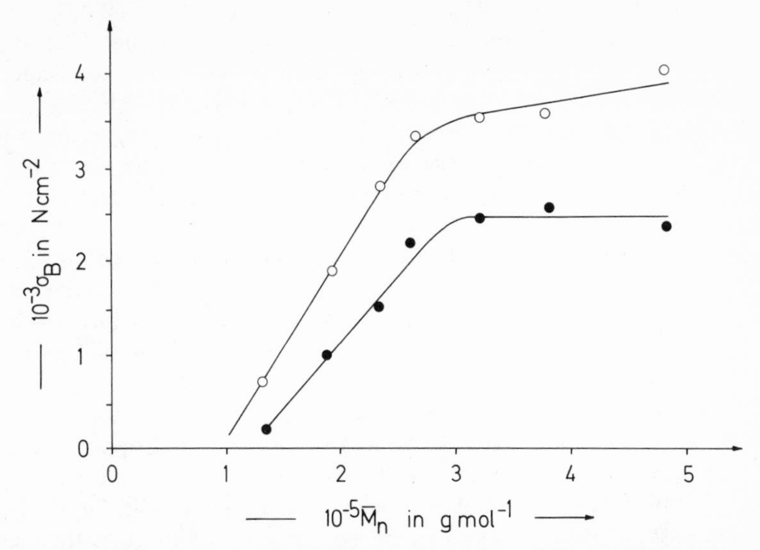

Figure 11-16. Tensile stress at break (tensile strength) σ_B of at-poly(styrenes) of narrow molecular-weight distribution and various number-average molecular weights $\overline{M}_n$. Measurements were made at 23°C and 50% relative humidity. Processing was by compression molding (●) or injection molding (○). Injection-molded samples are oriented. (After H. W. McCormick, F. M. Brower, and L. Kin.)

on the surface, free radicals must form in the test sample interior, that is, from the breaking of polymer chains.

In such break cases, break occurs generally in the amorphous regions, since the amorphous phase is set under stress on extension. Thus, break occurs at interlamellar bonds and at the surfaces of spherulites (Figure 11-15). Oriented samples generally exhibit a greater tensile strength in the orientation direction than do nonoriented samples (Figure 11-16). For each polymer a threshold tensile strength and molecular weight value exist above which the tensile strength either does not increase or increases very slowly.

11.6.3. Impact Strength

High-impact-strength thermoplastics always consist of a hard and a pliable component. The two components may be chemically or physically different. In partially crystalline polymers, the hard component is the crystalline component and the pliable component is the amorphous component. On the other hand, the hard component may be a material with a higher glass-transition temperature and the pliable component may be a rubber [for example, poly(styrene)–poly(butadiene)]. The two incompatible components may be combined in a block of graft copolymer. The important thing is that there be a phase separation between the hard and the pliable components, otherwise a plasticization effect would only be observed. Polymers consisting of a hard and a pliable component have high impact strength when the pliable component is discontinuously embedded in a matrix of the hard component. In the reverse case, one has a filled rubber.

Solvents determine which component will form the continuous or matrix phase. The component with the poorest interaction with the solvent always goes into the discontinuous phase. After this "emulsification of oil in oil," the solvent is removed. Description in terms of "oil in oil" emulsions allows an analogy to be drawn with emulsions of, for example, oil in water. The emulsifier in "oil in water" emulsions acts as a solvation agent between the water and the oil because of its hydrophilic and hydrophobic groups. Such an emulsifying role between "oil in oil" emulsifications of two incompatible polymers is played by block copolymers $(A)_m (B)_n$. The emulsifying effect of these block copolymers becomes stronger when their composition is close to $m = n$ and when the formula molecular weight of the blocks with respect to the molecular weights of the unipolymers A_q and B_p increases.

The pliable phase must be bound as strongly as possible to the hard phase, otherwise the phases will separate from each other on mechanically loading. Such a strong bond between the hard and the pliable components is automatically given by chemical bonds in block copolymers. The adhesion between the two phases of polymer mixtures can be improved by the addition of block and graft copolymers. The microconcentration of such "oil in oil" emulsifiers at the hard phase–pliable phase interface has been shown by UV-fluorescence spectroscopy on systems containing emulsifiers consisting of graft or block copolymers with a few fluorescing groups.

Good adhesion does not a priori lead to better transfer of the impact energy from the hard phase to the pliant phase. In fact, the pliant phase can only absorb significant amounts of energy in the region of the glass-transition. Furthermore, the energy absorption is so small that it has little effect on the stress ratios within the test sample. Equilibration of stress must therefore proceed from another mechanism.

On cooling a polymer mixture of a hard and a pliant component, the hard phase will try to desegregate from the pliant phase because of the different expansion coefficients of the demixing components. The two components are bound to each other, however, when the hard component is grafted onto the pliant component. But a non-cross-linked pliant phase further shrinks on cooling, which leads to stresses being set up in the pliant phase. Formation of holes within the pliant phase releases the stresses. If the pliant phase is cross-linked, it must dilate on cooling since it is under a tensile stress of about 6000–8000 N/cm^2 produced on all sides by the hard component. Since a compression stress on all sides of an elastomer leads to an increase in the glass-transition temperature of about 0.0024 $K/(N/cm^2)$, the glass-transition temperature of the rubber phase consequently sinks by 14–20 K through application of this tensile stress.

The particle size of the pliant phase required to produce a given impact strength depends on the hard phase. In high-impact poly(styrene), the diameters of the rubber particles must be at least 1000 nm. High-impact poly(vinyl chloride), however, requires rubber particles with diameters of less than 100 nm. Of course, poly(vinyl chloride) is a more ductile material than poly(styrene).

11.6.4. Reinforcement

Plastics can be "reinforced" through the mixing in of fibers, that is, their behavior is more fiberlike. The modulus of elasticity E_{mix} (and analogously, the tensile strength) of the mixture can be estimated from the moduli of elasticity of the fiber E_f and that of the matrix E_m with the aid of the volume fractions via the additivity rule:

$$E_{\text{mix}} = f E_f \phi_f + E_m(1 - \phi_f) \tag{11-73}$$

The orientation factor f is equal to one when all fibers lie in the tensile-stress direction. With three-dimensionally randomly distributed fibers, $f = 1/6$, and with two-dimensionally randomly distributed fibers, $f = 1/3$. If the fibers lie at the correct angle to each other, and if the tensile stress is applied along one of the fiber directions, then $f = 1/2$.

An optimum modulus ratio of $E_f/E_m \approx 50$ has been found empirically. Decreased rigidity is obtained with smaller ratios. At higher ratios, the fiber rigidity is not fully utilized. Processing of fibers that are too small leads to problems. The fibers are too short to be properly gripped by the matrix, and bundles, or knots, of fibers occur, with a decrease in the reinforcing effect.

The impact strength of reinforced plastics depends on the probabilities of crack onset and crack termination. In unnotched test samples, more voids are generated by increased fiber fractions and the impact strength decreases. In notched test samples, however, the notch is the largest void. In this case, increased fiber fractions impede the further progress of cracks. With longer fibers, there are fewer fiber ends, and, consequently, fewer defect sites and fewer possibilities for crack formation. A bond between fibers and matrix which is too good is not always an advantage. For example, if the fiber–matrix interface is not disrupted on impact, the crack proceeds further into the matrix. With poor bonding between the fiber and the matrix, on the other hand, the fiber separates from the matrix; the crack is redirected and the fiber serves to collect and focus the impact energy.

11.6.5. Plasticization

Plasticizers generally lower the glass-transition temperature, the elongation at break, and the tensile strength (Figure 11-17). In the case of plasticized poly(vinyl chloride), however, there are anomalies. Here, the tensile strength first rises with low concentrations of plasticizer and then falls again. This stiffening of the material cannot come from a stiffening of the molecules due to solvation. That is to say, when stiffening is caused by solvation, the glass-transition temperature must also rise, which is not found to be the case.

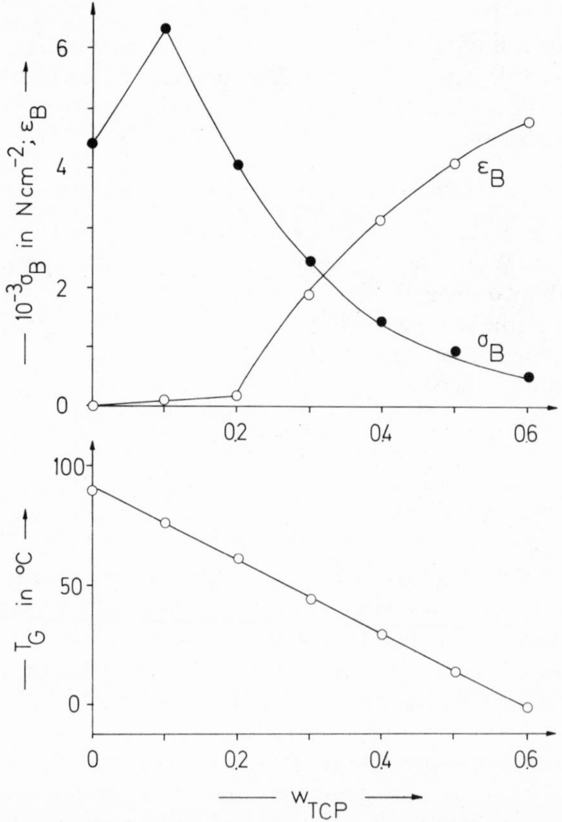

Figure 11-17. Tensile strength σ_B, elongation at break ε_B, and glass-transition temperature T_G poly(vinyl chloride) with different parts by weight of tricresyl phosphate w_{TCP} (after R. S. Spencer and R. F. Boyer).

11.6.6. Stress Cracking

The term stress cracking is used with metals and plastics when the material is damaged by the simultaneous action of chemical agents and mechanical forces. With plastics, a crazing is usually observed on the surface of the material under such conditions. It is also described, therefore, as stress corrosion cracking or crazing, or—since chemical reactions mostly play a minor role or none at all—stress crazing. Stress cracking plays an important role in bottles, tubing, cables, etc., that come into contact with chemical reagents.

The extent of the stress corrosion cracking varies according to the surrounding medium. The effects are usually small with nonwetting media. Here, stress corrosion cracking proceeds in three phases. In the case of stress corrosion cracking below the breaking point, weak points grow into visible hairline cracks or crazes perpendicular to the direction of stress. The crazes subsequently deepen and develop into cracks up to a limiting value, when the material is again strengthened. With wetting media, on the other hand, there is no limiting value to the development of the cracks.

The cause of stress cracking is not yet fully understood. It is established that there is still amorphous material in the cracks. This material can be deformed by cold flow. Of course, the extent of the cold flow is also determined by the diffusion and degree of swelling of the surrounding medium into the material. Wetting substances can build up a swelling pressure at the weak points.

The susceptibility of a material to stress cracking falls with increasing molecular weight and rises in the same material with increasing density. Stress cracking only occurs at the surface. Polymer softeners plasticize the surface, causing the stresses to compensate one another, so that the susceptibility to stress cracking falls. The mobility of the chain elements also ensures that no stress cracking occurs above the glass-transition (if amorphous) or melt temperature (if crystalline) of a material. The susceptibility to stress cracking is lowered by cross-linking.

11.6.7. Durability

In some cases, materials do not suffer damage immediately after a given stress is applied, but only after a certain time. Here, a distinction must be made between the fatigue limit and the fatigue strength. The fatigue limit is understood to be the stress under which a material suffers no damage even after an infinitely long time. Fatigue strength is understood to be the stress that destroys or damages the material after a given time.

The stress here can be either static or oscillating. In static tests (creep strength tests) the sample is, for example, subjected to a specific force, and

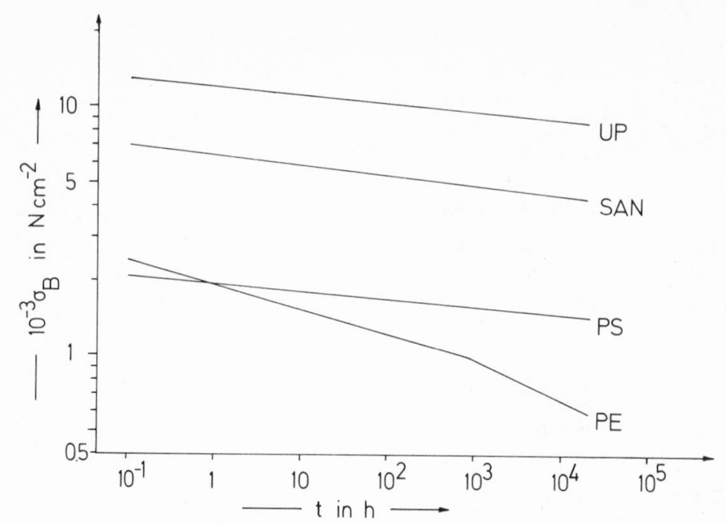

Figure 11-18. Fatigue strength under tensile stress (tensile strength) σ_B as a function of time for a glass-fiber-reinforced unsaturated polyester (UP), a high-impact-strength poly(styrene) (SAN), poly(styrene) (PS), and poly(ethylene) (PE) (after BASF).

the time up to the break is then measured. The same test is subsequently carried out with different forces. If the stress applied is a tensile stress, this is called the tensile creep strength. A static stress under compression would correspondingly be referred to as the compression creep strength. In order to determine the creep strength, the quantities that are proportional to the

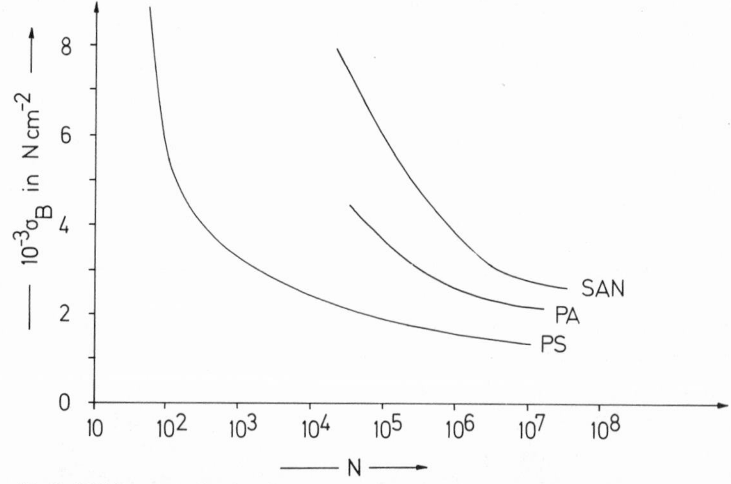

Figure 11-19. Wöhler curve for the alternating flexural stress applied to a high-impact-strength poly(styrene) (SAN), a humid polyamide (PA), and a poly(styrene) (PS). The flexural stress σ_B is measured as a function of the number N of alternations in the load.

force (for example, tensile stress) are usually plotted directly or as logarithms against the logarithm of the time (Figure 11-18). The creep strengths can vary considerably according to the polymer. At a load of 39 N/mm^2, for example, normal poly(styrene) samples have creep strengths of 0.01–10 h, whereas those of high-impact-strength poly(styrenes) are up to 10^4 h.

With alternating stresses, a distinction is made between those with alternating load and those with alternating torque. With the former, the fatigue strength under alternating flexing load is measured, and with the latter, fatigue strength under alternating torque. In analogy to the determination of creep strength, the quantities that are proportional to the force are again plotted against the logarithm of the number of changes in alternating load or torque to determine the "oscillation fatigue strength" (Figure 11-19). Curves of this type are called Wöhler curves. Here, again, large differences are found between normal and high-impact-strength poly(styrenes). With flexural stresses of 40 N/mm^2, normal poly(styrenes) break after only 300 changes of load, but high-impact poly(styrenes) only after one million.

Literature

General

A. V. Tobolsky, *Properties and Structures of Polymers,* Wiley, New York, 1960.

H. Oberst, *Elastische und viskose Eigenschaften von Werkstoffen,* Beuth-Vertrieb, Berlin, 1963.

L. E. Nielsen, Crosslinking—effect on physical properties of polymers, *Rev. Macromol. Chem.* **4,** 69 (1970).

I. M. Ward, *Mechanical Properties of Solid Polymers,* Wiley–Interscience, London, 1971.

A. Peterlin, Mechanical properties of polymeric solids, *Ann. Rev. Mater. Sci.* **2,** 349 (1972).

D. W. van Krevelen, *Properties of Polymers—Correlation with Chemical Structures,* Elsevier–North Holland, Amsterdam, 1972.

J. R. Martin, J. F. Johnson, and A. R. Cooper, Mechanical properties of polymers: the influence of molecular weight and molecular weight distribution, *J. Macromol. Sci. C (Rev. Macromol. Chem.)* **8**, 57 (1972).

E. A. Meinecke and R. C. Clark, *The Mechanical Properties of Polymeric Foams,* Technomic Publ. Co., Westport, Connecticut, 1972.

Journals

International Journal of Nondestructive Testing **1** (1970).

Polymer Mechanics **1** (1970) ff.

Testing of Polymers, J. V. Schmitz, ed., Interscience, New York, **1** (1965) ff.

Section 11.3. Entropy-Elasticity

L. R. G. Treloar, *Physics of Rubber Elasticity,* Oxford Univ. Press, Oxford, 1958.

M. Shen, W. F. Hall, and R. E. De Wames, Molecular theories of rubber-like elasticity and polymer viscoelasticity, *J. Macromol. Sci. C (Rev. Macromol. Chem.)* **2**, 183 (1968).

K. J. Smith, Jr., and R. J. Gaylord, Rubber elasticity, *ACS Polymer Div. Polymer Preprints* **14,** 708 (1973).

J. E. Mark, Thermoelastic properties of rubber-like networks and their thermodynamic and molecular interpretation, *Rubber Chem. Technol.* **46,** 593 (1973).

L. R. G. Treloar, The elasticity and related properties of rubber, *Rep. Prog. Phys.* **36,** 755 (1973); *Rubber Chem. Technol.* **47,** 625 (1974).

J. E. Mark, The constants $2C_1$ and $2C_2$ in phenomenological elasticity theory and their dependence on experimental variables, *Rubber Chem. Technol.* **48**, 495 (1975).

Section 11.4. Viscoelasticity

J. D. Ferry, *Viscoelastic Properties of Polymers,* second ed., Wiley, New York, 1970.

N. G. McCrum, B. E. Read, and G. Williams, *Anelastic and Dielectric Effects in Polymeric Solids,* Wiley, London, 1967.

R. M. Christensen, *Theory of Viscoelasticity: An Introduction,* Academic Press, New York, 1970.

J. J. Aklonis, W. J. MacKnight, and M. Shen, *Introduction into Polymer Viscoelasticity,* Wiley–Interscience, New York, 1972.

Section 11.5. Deformation Processes

A. J. Durelli, E. A. Phillips, and C. H. Tsao, *Introduction to the Theoretical and Experimental Analysis of Stress and Strain,* McGraw-Hill, New York, 1958.

J. W. Dally and W. F. Riley, *Experimental Stress Analysis,* McGraw-Hill, New York, 1965.

O. H. Varga, *Stress–Strain Behavior of Elastic Materials,* Wiley–Interscience, New York, 1966.

A. Peterlin, *Plastic Deformation of Polymers,* Dekker, New York, 1971.

J. G. Williams, *Stress Analysis of Polymers,* Longmans, Harlow, Essex, U.K., 1973.

A. R. Payne, Physics and physical testing of polymers, *Prog. High Polym.* **2,** 1 (1968).

H. J. Orthmann and H. J. Mair, *Die Prüfung thermoplastischer Kunststoffe,* Hanser, München, 1971.

G. C. Ives, J. A. Mead, and M. M. Riley, *Handbook of Plastic Test Methods,* Iliffe, London, 1971.

S. Turner, *Mechanical Testing of Plastics,* Butterworths, London, 1973.

J. K. Gillham, Torsional braid analysis, *Crit. Rev. Macromol. Sci.* **1,** 83 (1972).

Section 11.6. Fracture

E. H. Andrews, *Fracture in Polymers,* Oliver and Boyd, Edinburgh, 1968.

H. H. Kausch and J. Becht, Elektronenspinresonanz, eine molekulare Sonde bei der mechanischen Beanspruchung von Thermoplasten, *Kolloid-Z. Z. Polym.* **250,** 1048 (1972).

E. H. Andrews, Fracture of polymers, *in: Macromolecular Science* (Vol. 8 of Physical Chemistry Series 1) (C. E. H. Bawn, ed.), MTP International Review of Science, 1972.

H. Liebowitz, ed., *Fracture, Vol. 7, Fracture of Nonmetals and Composites,* Academic Press, New York, 1972.

G. H. Estes, S. L. Cooper, and A. V. Tobolsky, Block copolymers and related heterophase elastomers, *J. Macromol. Sci. C (Rev. Macromol. Chem.)* **4**, 313 (1970).

P. F. Bruins, ed., *Polyblends and Composites* (Appl. Polymer Symposia) Vol. 15, Interscience, New York, 1970.

S. Rabinowitz and P. Beardmore, Craze formation and fracture in glassy polymers, *Crit. Rev. Macromol. Sci.* **1,** 1 (1972).

R. P. Kambour, A review of crazing and fracture in thermoplastics, *J. Polym. Sci. D*7, 1 (1973).

J. A. Manson and R. W. Hertzberg, Fatigue failure in polymers, *Crit. Rev. Macromol. Sci.* **1,** 433 (1973).

Chapter 12

Compounding and Processing of Plastics

Most of the synthetically manufactured macromolecular substances are used as plastic materials in the materials science sense. In order to make a plastic material of a macromolecular substance, it must be equipped with antioxidants, fillers, lubricants, etc. Such protective additives are intended to improve the mechanical, electrical, and/or chemical properties of the plastic, simplify its processing, and give it a suitable, acceptable appearance or finish. Compounding is the controlled addition of such additives to the macromolecular substance.

The type and extent of the processing and the protection needed depend partly on the type of the plastic and partly on the intended application. Normally, plastics are classified according to their mechanical and thermal behavior as thermoplasts, elastomers, thermosets (duroplasts), and fibers (see also Section 11.1). According to the application, further distinctions are made among construction materials, insulating materials, adhesives, floor sealants, paints, films, soundproofiing media, etc.

12.1. Compounding

12.1.1. General

Batch polymerizations give products which often vary slightly from batch to batch despite the use of the same recipe. These batches are blended in order to be able to deliver products with the same specifications to the customer. This mixing process is also called microhomogenization. The product, complete with additives, is called a compound. Additionally, the compound is often converted into a form more suitable for handling and processing, e.g., by granulating. The mixing of granulates and similar prefabricates is called macrohomogenization. Granulates often have to be dried ("conditioned") before processing, otherwise voids or bubbles occur during manufacturing.

Compounds vary according to their preparation and the additives used. If poly(vinyl chloride), for example, is stirred into a mixture with the other additives, a heterogeneous compound results. Because of its heterogeneity, this so-called premix cannot be delivered as a ready-to-use product. On the other hand, if heavy-duty mixers are used, the frictional head developed is sufficient to cause encapsulation of the additive by the polymer, and "homogeneous" "ready-to-use" mixtures are obtained which can be processed directly with extruders and injection-molding machines (dry blends). If dry blends are treated at high temperatures, coarse-grained particles result (agglomerates).

When two polymers are intimately mixed, a polyblend (polymer alloy) is produced. Polyblends are usually thermodynamically unstable (see Section 6.6.6). Because of the very low diffusion coefficient, however, the demixing process is extremely slow, so that polybends appear kinetically stable.

To give suitable protection, it is often sufficient to work in just a small amount of an additive to a plastic. Since it is difficult to mix small amounts of additive efficiently with the polymer, use is made of what is called a master batch. This is a "concentrate," so to speak, of the additive in the polymer to be compounded. The true working-in is then effected by diluting the master batch with the required amount of polymer.

12.1.2. Fillers

Fillers are solid inorganic or organic materials. Inert fillers give increased bulk to the plastic, thus lowering its price. Active fillers improve certain mechanical properties, and are therefore often called reinforcing fillers or resins. The concept of reinforcement is rather poorly defined, since the increase in tensile strength and wear resistance, as well as that of the notched-bar impact strength and the fatigue strength under alternating flexural stress, are all termed reinforcement.

Inorganic materials such as chalk, flour, china, clay, mica, barytes, Fuller's earth, Aerosil (finely divided SiO_2), asbestos, glass-fiber, and metal or oxide single crystals (whiskers) are all used as fillers. Organic fillers include wood flour, cellulose flakes, foam-rubber chips. paper cuttings, paper streamers, woven ribbons, and chemical fibers. Plastics reinforced with glass or chemical fibers are termed GRPs or FRPs, respectively. Fillers are added in quantities of up to 30% in thermoplasts and 60% in thermosets.

Any one of three factors may be responsible for the reinforcement effect of active fillers. Some fillers can contract chemical bonds with the material to be reinforced. In the case of carbon black, this takes place by means of the radical reactions of the unpaired electrons, which are present in large numbers in carbon black. Carbon black particles cause cross-linking in elastomers.

Other fillers act purely through their volume requirements. The presence of the filler particles means that the molecular chains of the material to be reinforced cannot take up all the conformations that are theoretically possible. The molecular chains thus become less flexible and the tensile strength increases. The more finely distributed the filler, the greater is its effectiveness.

A third mode of action arises from the fact that the molecular chains under stress possess higher energies which enable them to slide along the surface of the filler, thus dissipating excess stress. This distributes impact energy better and raises the impact strength.

Two discrete phases are always found in reinforced plastics. The disperse phase consisting of filler should have a higher tensile strength and modulus of elasticity than the surrounding carrier material. The carrier phase, on the other hand, must have a higher elongation at break than the disperse phase. This is why fibers are such suitable reinforcing materials for plastics (e.g., unsaturated polyester resin, nylon, polycarbonates, polyethylene, etc.). With a tensile stress, the local stresses will be transported by shear forces to the plastic–fiber interface and distributed over the larger area of the fiber. For this, the fiber must adhere well to the plastic and have a certain length, otherwise it will slip out of the carrier material. The good adhesion of glass fibers to nylon, polycarbonate, or poly(styrene) is achieved by silanizing the fibers with, for example, vinyl triethoxysilane. The higher the E modulus of the carrier matrix, the lower is the minimum length of the fiber that is required. Fibers that are too long, however, are difficult to work into the necessary random distribution.

According to their length, the fibers are worked in differently. Short glass fibers with length of ~0.3–0.5 mm are mixed with the powdered plastics. The resulting material is then extruded and granulated. Long fibers are first impregnated continuously with the plastic, then cut to a length of 6–12 mm, and finally worked in.

Considerable reinforcement can be achieved by adding glass or chemical fibers. With glass-fiber reinforcement, the tensile, compression, and shear strengths, as well as the impact strength, increase to at least double the value of the carrier matrix material. The increase which can be obtained in this case depends on the plastic and on the type of reinforcing fiber. In these reinforced plastics, cracking is caused by the maximum shear stress which occurs at the fiber–plastic interface. Plastics with ductile deformation behavior, therefore, lead to better mechanical values than brittle ones (in comparison to glass-fiber-reinforced epoxy resins, glass-fiber-reinforced polyamides show a greater increase in tensile strength). With thermosets, it has been found that they are brittle when reinforced with glass fibers, but tough on reinforcing with chemical fibers. Therefore GFP (glass-fiber plastics) are particularly suitable for applications where a high tensile strength is needed, and CFP (chemical-fiber plastics) for those applications where a high impact strength and alternating load fatigue strength are needed.

With GFP, the notched-bar impact strength increases as the temperature rises; it falls with temperature in unfilled polymers. The moisture uptake and heat expansion are lowered by glass-fiber reinforcement. The thermal-threshold tensile strength increases, for a glass-fiber-reinforced nylon 6, for example, from 75 to 245°C (at 1.85 N/mm^2).

Even higher strength values are obtained when the glass fibers are not distributed randomly, but worked in in the form of mats or strands, i.e., oriented. In the same cured synthetic resin, for example, the tensile strength increases form 60 to 200 N/mm^2 after reinforcement with a glass-fiber mat, and even up to 1200 N/mm^2 after processing with what is called the filament-winding method (Section 12.2.2).

12.1.3. Dyes and Pigments

Plastics are colored with soluble or insoluble, inorganic or organic dyes. Insoluble dyes are called pigments. Dyes and pigments are added to the plastics in quantities of 0.001–5%.

High demands are made on pigments and dyes with reference to heat stability, dispersibility, migration, light and weather fastness, physiological safety, shade or nuance, and cost. Heat stability is necessary because of the normally high processing temperatures. Light, weather, and migration stability and physiological harmlessness are tested in special tests (see, for example, Chapter 24 for the weather and light fastness, Section 10.5.6 for the migration susceptibility).

The nuance (hue, tone, color strength, color purity) depends on the chemical constitution and crystal modifications of pigments, and also on the particle size. If the pigment particles are smaller than about half the wavelength of the incident light, the colored plastic will be transparent. Pigments

Table 12-1. Pigments Used to Color Plastics

Color	Pigment
White	Titanium dioxide (only the rutile modification), ZnO, ZnS, Lithopone (ZnS + $BaSO_4$).
Yellow	CdS (acid-sensitive), $Fe_2O_3 \cdot \times H_2O$, $PbCrO_4$ (chrome yellow), benzidine yellow, flavanthrone yellow
Orange	Pigments from the anthraquinone group
Red	CdSe, iron oxide red, molybdate red, and many organic pigments
Bordeaux	CdSe, thioindigo, chinacridone
Violet	Many organic pigments
Blue	Ultramarine blue, cobalt blue, manganese blue [$Ba(MnO_4)/BaSo_4$], phthalocyanine blue
Green	Chromium oxide green, chlorinated copper phthalocyanines
Metallic	Aluminum
Mother-of-pearl effect	Small platelets of lead carbonate
Black	Carbon black

should have diameters between roughly 0.3 and 0.8 μm. With pigments of this particle size, films and fibers down to a thickness of 20 μm can be colored. With thinner films or fibers, what is called the melt break then occurs because the pigment particles are comparable in size to the thickness of the film, and the material then breaks readily at the point where a pigment particle occurs. Lighter tones can be obtained by further grinding, although this also causes the swelling capacity to increase. The hiding power increases with the difference between the refractive indices of pigment and plastic.

The pigments can be introduced into the polymer by various methods. In the case of plasticized PVC, the pigment is usually added as a paste in a plasticizer. In many cases a master batch (color concentrate) is used, or the pigments are mixed with fillers, in order to obtain better color balance. Granulates accumulate pigment powder on the surfaces as a result of electrostatic charging of the pigment particles when granulates are mixed with pigments in granulate dry-blend mixers (tumblers); a total of up to 1% of pigment can be introduced in this way. Pigments used in lacquers and printing inks are often covered with what are called carrier resins. Copolymers of vinyl acetate and vinyl chloride, reduced rosin, or ethyl cellulose are added as carrier resins.

The aggregation (lumping) of pigments is often due to the air inclusions; this problem is solved by applying a vacuum. In addition, pigments must be easily wetted. A better wettability can be achieved, for example, by treatment with surface-active agents. The complete pretreatment process for pigments is called conditioning.

Table 12-1 lists the pigments most frequently used.

12.1.4. Plasticizers

Plasticizers lower the glass-transition temperature (see Section 10.5.6). In the range of up to ~40% by weight of plasticizer, the decrease in the glass-transition temperature ΔT is directly proportional to the mass fraction w_W of the plasticizer (see also Figure 10-22). According to Section 10.5.6, the smaller and more spherical the plasticizer molecules and the poorer the solvent power of the plasticizer, the greater will be the effectiveness $\Delta T/w_W$ of the plasticizer. All these properties, however, encourage migration, sweating out, and evaporation of the plasticizer. In technology, therefore, a compromise is sought between the effectiveness of the plasticizer and the resistance to phase separation and sweating out.

Typical commercial plasticizers, therefore, do not have too low a molecular weight (a slower rate of diffusion and, because of the danger of phase seperation, not too poor a solvent power or swelling ability. The solvent must also be able to enter into some kind of interaction with the polymer. For poly(vinyl chloride), therefore, one uses, for example, low-volatile esters of phthalic acid or phosphoric acid or oligomeric polyesters of glycols with adipic or sebacic acid. They are added in quantities of 10–40%. It is very often necessary for commercial plasticizers to be physiologically harmless.

12.1.5. Release Agents, Lubricants, Stabilizers, and Antistatics

Release agents are intended to prevent plastics from adhering to molds. Silicones are normally used for this purpose in the rubber industry.

Lubricants lower the coefficient of friction between polymer and machine parts, so that lower pressures can be used during processing. Stearic acid and zinc stearate are typical lubricants. With emulsion polymerizations, the emulsifier residues remaining in the polymer can act as a lubricant. Lubricants are added in amounts of 1–5%. Only a small amount of added lubricant really works by lubrication, however, since the greater part of the additive resides in the interior of the plastic. Lubricants often produce translucent or opaque films.

Stabilizers are intended to prevent degradation reactions (in the broadest sense) of plastics during processing (heat stabilizers against thermal decomposition) and in practical use (antioxidants and UV absorbers to protect against light and aging). They are added in amounts of ~1%. See Chapter 24 for their protection mechanisms.

Antistatics prevent electrostatic charges from forming on plastic surfaces. Electrostatic charging of the surface leads to such problems as easier soiling by dust pickup, etc. Antistatic agents are usually hygroscopic, surface-active substances (see Section 14.1.5). Carpets can be made antistatic by spinning-in, twisting-in, and texturizing-in fine steel wires for every 5–10 fibers.

12.2. Processing of Thermoplasts, Thermosets, and Elastomers

12.2.1. Classification

The choice of a processing technique is influenced technically by the rheological properties of the material and the form or shape of the desired product. Important economic features such as the cost of the processing machines and the number of pieces which can be produced per unit time also play a considerable role.

The processing methods can be classified according to process technology or according to the technique used to give the product its final shape. In the former, the methods can be subdivided according to the rheological states which play a predominant role in an given technique, as shown in Table 12-2. Here, elastoviscous is defined as rheological behavior with predominantly viscous and little elastic character, whereas viscoelastic is behavior with mainly elastic and little viscous character. Materials that behave elastoplastically show a marked flow limit. Naturally, all kinds of intermediate patterns of behavior are found between these two. Spinning into fibers can be considered a special case in the processing of plastics. In general, the processibility improves with increasing width of the molecular-weight distribution.

The methods can also be classified, as in Table 12-3, according to the type of form or shape which is produced.

Classification according to the pressure used, the continuity of the process (discontinuous, semicontinuous, continuous, or automated), or the

Table 12-2. Processing Methods for the Various Rheological States

Rheological state	Method
Viscous	Molding, compression molding, direct injection molding, coating
Elastoviscous	Injection molding, extruding, calendering, milling, internal mixing (kneading)
Elastoplastic	Drawing, blowing, foam forming
Viscoelastic	Sintering, welding
Solid	Chipping or granulating procedures, welding, adhesion

Table 12-3. Processing Methods for Producing Various Shapes or Forms

Shape or form	Method
Producing a completely new shape	Molding, dip-coating, compression molding, injection molding, extrusion, foam forming, sintering
Modifying a shape	Calendering, stamping, bend shaping, deep drawing
Combining shapes	Welding, adhesion, turning and rivetting, shrink coating, braiding
Coating	Laminating, painting, flame spraying, fluidized-bed sintering, cladding
Separating	Cutting, chipping
Finishing	Surface-finishing, texturing

degree of finish (intermediate component, i.e., section, or finished product, e.g., foam) is also possible (see *Plastics Handbook* or BS, ASTM, or DIN standards).

12.2.2. Processing via the Viscous State

Processing via the viscous state occurs by molding, compression molding, direct injection molding, and coating. The viscosity of the materials being processed must be low, e.g., melts of monomers or prepolymers, or polymer solutions or dispersions are used. In molding and compression molding, melts of prepolymers are used in the manufacture of intermediate or finished products from thermosets (thermoplast melts are less common in these processes). Compression molding also includes cavity molding and jet molding. Solutions or dispersions of thermoplasts or elastomers are processed by direct injection molding or coating. The latter includes dip-coating, painting, and laminating, as well as filament winding.

In *molding*, liquid materials are poured into a mold and there "hardened," i.e., polycondensed or polymerized (Figure 12-1). Phenol and epoxy resins are processed in this manner, as are monomers such as methyl methacrylate, styrene, vinyl carbazole, and caprolactam (reaction molding). The molding of materials that form gels is called setting [(plasticized poly(vinyl chloride)]. In molding, the cost of equipment is low. Metal parts can easily be incorporated into the product during processing. There are also two disadvantages, however: The rate of production is slow, so that the method is only economical for the production of up to ~3000 parts/yr. In addition, exothermic reactions are difficult to control. For this reason, the processing of polyester resins by molding has not really become established. Monomers polymerizing to thermoplasts are only processed by molding in the case of specialized products, e.g., methyl methacrylate for lenses or false teeth.

Casting and centrifugal casting are two variations of molding. Films produced by casting are more homogenous than those manufactured by calendering. In particular, cellulose acetate, polyamides, and polyesters are prepared by molding. Centrifugal casting is primarily used in the automated production of hollow bodies of plasticized PVC.

Several different processes can be used to produce a *laminate*. In what is called the hand application process, for example, glass-fiber mats are impregnated with unsaturated polyester resins. The impregnated mats are then removed from the mold by hand and pressed between rollers. The final molding is effected by cold pressing. The method is suitable for small numbers of objects with large surface areas (e.g., boat hulls).

To produce what are called prepregs, mats are repeatedly dipped, compressed, prehardened in an oven, and then processed by compression-cavity molding.

Filament winding is also a lamination method. For this method, what are called glass-fiber rovings are impregnated in resin, arranged in a geometrical pattern around a (removable) form, and then left to harden. Rovings consist of many fiber structures (called ends), which in turn contain $\sim$204 glass fibers. The method is used to manufacture hollow bodies of very high strength.

Particularly thin articles (e.g., rubber gloves) are produced by *dip-coating*. In this case, the mold negative is dipped into a latex (a dispersion) or a paste for as long and/or as often as is necessary to obtain the desired thickness. The latex viscosity should be less than 12 Pa s; the flow limit as low as possible. Latices of natural rubber poly(chloroprene), and silicones, as well as PVC pastes, are processed in this way.

For *spray-coating*, the viscosity should be less than 7 Pa s. The latices, plastisols, or solutions should have little or no dilatancy. One can describe this as injection casting.

In *spread coating* (*laquering*), solutions of film-forming polymers are applied and the solvent allowed to evaporate. Metallic and nonmetallic components and films of cellulose or aluminum are lacquered. Lacquered cellulose films are less permeable to water vapor, but poly(ethylene)-bonded films compete with these on the market (Section 12.2.6). Paints and lacquers

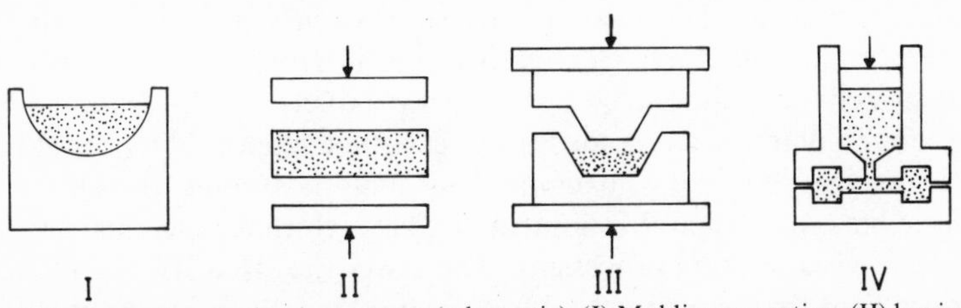

Figure 12.1. Processing via the viscous state (schematic). (I) Molding or casting, (II) laminating, (III) compression molding (cavity molding), (IV) direct injection molding or injection casting.

should have a relatively low viscosity on brushing, which produces shear rates of about 20,000 s^{-1} (see next chapter also).

In *compression molding*, powders or mold materials are usually preheated (possibly by high frequency) in the form of plates (laminates), put into a press, and then simultaneously pressed and hardened (Figure 12-1). Generally, only thermosets containing a great deal of filler are used as mold materials in compression molding, i.e., phenolic, urea, melamine, and unsaturated polyester resins. Inlays are also frequently used (fabrics, mats, etc.) in compression molding.

In *cavity-compression molding*, the cold powder or mold material is placed under pressure in a heated mold (Figure 12-1). Glass-fiber-reinforced, unsaturated polyester resins are processed by the heated-cavity-compression molding method. The vulcanization of rubber can also be carried out using this method. High-fidelity records are pressed out of the thermoplasts PVC or poly(ethylene-co-vinyl acetate), while cheaper records are injection molded.

In *injection compression molding*, a warm, compressed material is injected into a cold mold under pressure. Again, as in compression molding, only thermosets are generally used. Thermoplasts are only processed by injection compression if this offers economic advantages [e.g., poly(chlorotrifluoroethylene) or rigid PVC]. Injection compression molding is particularly suitable for the production of thick-sectioned parts or large numbers of parts with low bulk. Compared with compression molding, the process has the following advantages: Automation is possible because of the greater production rate; the products have higher dimensional stability than those produced by compression molding since in compression molding pressure varies with the amount of filler; finally, lower viscosities and pressures are needed when the mold material is preheated. Disadvantages compared to compression molding, on the other hand, are the higher material consumption, the orientation of filler particles by the injection process, and the high investment costs for very thin-walled sections or very large articles.

12.2.3. Processing via the Elastoviscous State

Processing via the elastoviscous state occurs in injection molding, extrusion, rolling (milling), calendering, and internal mixing (Figure 12-2). The intermediate components or finished products which result are—with the exception of injection molded products—not supported by a mold when they leave the processing machine. The materials being processed must therefore have, at least at this point, a viscosity which is much higher than in the case of viscous-state processing. The lower the viscosity, the higher the shear gradient that can be used during processing. With injection molding,

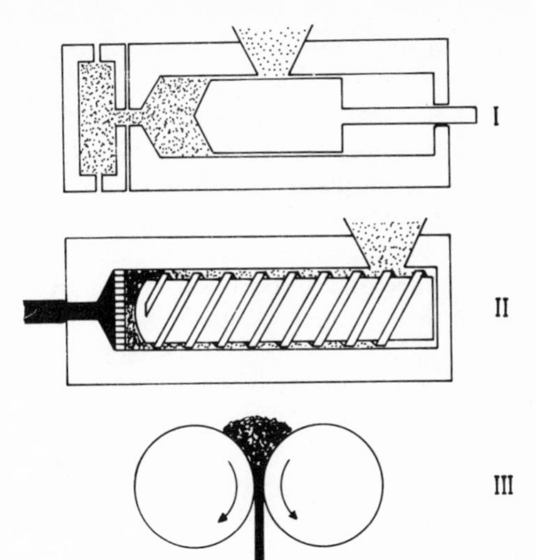

Figure 12-2. Processing via the elastoviscous state (schematic). (I) Injection molding (with torpedo), (II) extrusion, (III) rolling or milling. In injection molding, the granulate is plasticized by heat and then injected into the mold by a torpedo (or with a screw). When the molded article is cold, the mold is removed and the article ejected. In extrusion, the heated material passes through a metal screen (filter) placed in front of the nozzle.

the shear gradients are ~1000–10,000, in extrusion ~100–1000, in rolling and calendering ~10–100, and in internal mixers below ~10 s^{-1}. With rolling, calendering, and internal mixing, of course, the material behaves viscoelastically rather than elastoviscously.

In *injection molding* (Figure 12-2), the materials are first preheated ("plasticized") and then transported past a torpedo (see figure), a screw, or a double screw (varying pitch and varying number of threads per length) into the cold or possibly slightly preheated mold (Figure 12-2). The torpedo or screw acts simultaneously as a plasticizer (frictional heat), shot capacity dosing, and injection system. A higher production rate is obtained with screws than with torpedos. The material cools under a relatively low pressure in the mold. Then the mold is removed and the part is ejected. The whole procedure can be automated. A whole series of thermoplasts and a few thermosets are processed by the injection-molding method. These include, as thermoplasts, poly(styrene), poly(ethylene), poly(propylene), poly(vinyl chloride), polyamides, polyurethane, poly(oxymethylene), poly(carbonates), poly(trifluorochloroethylene), poly(acrylates), cellulose derivatives, and poly(methyl methacrylate), and, as thermosets, unsaturated polyesters and phenolic and aminoplast resins. The materials for processing should have relatively low melt viscosities. The melt viscosities can be adjusted via molecular weight, molecular-weight distribution, branching, and processing temperature.

In *sandwich injection molding*, two polymerizates in separate injection units are successively injected into a mold such that the second shot is fully enclosed in the first. The process is suitable, for example, for the enclosing of foams in a more rigid coating or enclosing cheap plastics in better quality materials.

In *extrusion*, the preheated material is forced out of the extruder with a screw (Figure 12-2) or double screw and allowed to cool in a bath or in the air. Thermoplasts, elastomers, and thermosets are extruded. As a rule, thermosets are processed in torpedo-containing extruders. With thermosets, most of the curing reaction must occur in a heated pressure chamber. Pressures can approach several hundred bars. The rate of extrusion is lowest with thick-walled bodies. Tubes are extruded at rates of up to 10 m/min, films up to 150 m/min, and telephone-cable insulating material or fibers at up to 1000 m/min. In extrusion, the Barus effect (Section 11.3.1) and melt fracture (Section 7.6.1) may be observed. Tubes, films, ducts, cable insulation, and knot-free nets are produced by extrusion.

Extrusion with wide exit slits is a special case. Planar films of 20–100 μm in thickness, for example, are produced with wide exit slits. The film is then chilled by cold rolling or in water baths (melt-molding or chill-rolling method). Wide exit slits are also used in what is called extrusion coating of paper or cardboard with poly(ethylene). Papers thus treated can then be heat-sealed.

In the extrusion of thermosets, it is really the monomer or a prepolymer which is extruded, whereupon the production of polymer and molding occur simultaneously. In theory, extrusion with simultaneous molding is also possible with thermoplasts, but is only used to a very limited extent, e.g., with methyl methacrylate.

In *rolling* and *calendering* (Figure 12-2), elastomers or plasticized thermoplasts are processed into film, or fabrics or films are coated with thermosets. The rollers are heated and run at different speeds. The resulting friction leads to drawing and frictional heating. Films with thicknesses of 60–600 μm are produced by calendering. Thicker films are obtained by extrusion. An exception is highly filled, plasticized PVC, which is produced on two-roller calenders with thicknesses of up to 1 mm. Calenders are constructed in a variety of ways. The material resides on the rollers for different times on the various four-roller calenders:

o o o o	≈	o o o o	<	o o o o	<	o o o o	≈	o o o o
Z calender		S calender		normal F calender		inverse F calender		L calender

Internal mixing (Banbury mixing) serves to produce intermediate products from elastomers and is used in particular to work in fillers, mix elastomers, etc.

12.2.4. Processing via the Elastoplastic State

In some methods of processing, the existence of a flow limit is utilized. Press forming, stretch forming, blowing, and foaming are among these methods. Stretch forming and blowing are known as "cold-forming" processes, since the material is not heated. Cold forming is only possible in the ductile region of the stress/strain curve (see Figure 11-9). Polymers with too narrow ductile regions, e.g., poly(4-methyl pentene-1), cannot be vacuum-formed.

Press forming is understood to be the deformation under pressure of films of thermoplasts with molds or templates (Figure 12-3). In this way, hollow bodies are produced from cellulose, and drinking mugs from poly(ethylene). Press forming is also sometimes called deep-drawing, heat forming, or form stamping.

Stretch forming or stretching is a type of press forming with simultaneous drawing or stretching of the films. The stretching can be carried out mechanically with a stamp, with compressed air, or under vacuum. In the last case, the term vacuum forming is also used. A deep drawing with spring-loaded fasteners is called stretch molding. With deep drawing, 4000–8000 parts can be manufactured per hour on machines with 4–12 forms, so that 4–12 parts are produced at a time; the method is used mainly for the packaging of fruit, eggs, chocolate, etc. ABS polymers, cellulose acetate, poly(carbonate), poly(olefins), poly(methyl methacrylate), poly(styrene), and rigid PVC are processed by this method.

Blowing can be considered as a special form of stretch forming from annular nozzles producing unbroken, continuous hollow bodies, which are used as such or else cut into film afterward. Blowing also enables hollow products of two components to be produced, e.g., toothpaste tubes, which consist of polyamide outside and poly(ethylene) inside.

Extrusion blow molding is really a special type of extrusion. The extruder

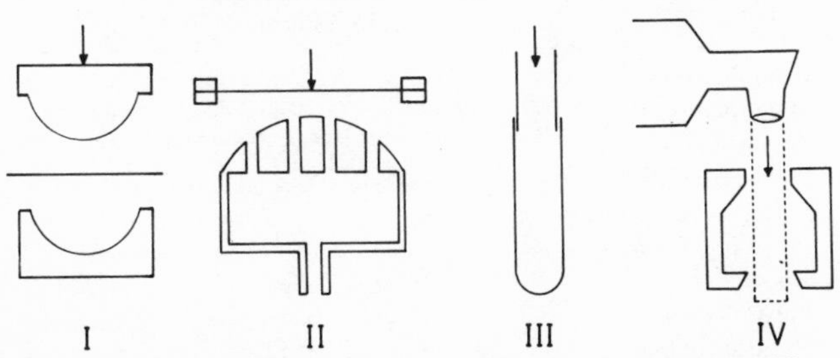

Figure 12.3. Processing via the elastoplastic state (schematic). (I) Pressure molding. (II) vacuum forming, (III) blow molding, (IV) extrusion blow molding with annular nozzle (see text).

head is directed vertically downward (Figure 12-3) and the continuous tubing which comes out of an annular nozzle is blown into a mold. In practice, closing the mold gives rise to a hollow, closed-bottomed component, which is otherwise only possible with the more laborious rotation molding. If, instead of being extruded, the product is produced with an injection molding machine, this is also called injection blowing or injection mold blowing. Poly(ethylene), rigid PVC, polyamides, high-impact-strength poly(styrene), and poly(carbonate) are processed by hollow-body blow molding.

Foam-forming is also a type of blowing. Foams are classified according to their hardness, their cellular structure, or the basic raw material. In the case of rigid foams, the glass-transition temperature is considerably above the application temperature, and with flexible foams, it is considerably below it. The cellular structure can be characterized in terms of open, closed, or mixed pores. Foams are produced mechanically, physically, or chemically with low-pressure (<10 bar) or high-pressure (up to 1000 bar) methods. In the mechanical production of foam, latices or prepolymers are beaten or stirred vigorously with added surface-active agents, and the resulting foam is then fixed by chemical cross-linking of the foam-forming material. This method yields open- or mixed-pore foams. In what is called physical foam-forming, gases which were previously incorporated under pressure are allowed to expand. The blowing of PVC plastisoles by nitrogen and the production of foam poly(styrene) through the vaporization of low-boiling hydrocarbons (vapor blow forming) are typical of this technique (see also Section 25.2.6.2). In chemical foaming, finished polymers are blown with gas-releasing materials [e.g., the elimination of N_2 from azo compounds, equation (20-3)], or the formation of NH_3, H_2O, and CO_2 from

Table 12-4. Important Materials Capable of Producing Plastic Foams and the Methods Used to Produce Them[a]

Polymer	Method of production		
	Mechanical	Physical	Chemical
Phenolic resin	r	r	r
Melamine resin	r	r	r
Polyurethane foam	—	r,f	r,f
Poly(styrene)	—	r	—
Poly(vinyl chloride)	—	r,f	r,f
Poly(vinyl formal)	f	—	f
Silicone	—	—	r,f
Poly(ethylene)	—	—	r,f
Natural rubber	—	r,f	r,f
Natural rubber latex	f	—	f

[a] r = rigid foam; f = flexible foam.

ammonium carbonate (NH_4HCO_3), or the production of polymer with simultaneous elimination of gas (e.g., the conversion of isocyanates with carboxylic acids; see Section 28.1). The most important foaming agents and the methods used are compiled in Table 12-4. Because of their low heat conductivity, foams are used for insulation, and because of their low density, as shockproof packing materials.

12.2.5. Processing via the Viscoelastic State

Welding and sintering can be cited as types of processing carried out in the viscoelastic state. Cutting, bracing, and adhesion presuppose the solid state.

In *welding*, thermoplasts are locally heated to a tacky consistency under nitrogen (or sometimes under air). The process serves to bond tubes and articles, mainly of poly(ethylene) or poly(vinyl chloride). The bond forms at the tacky points of contact. In autogeneous welding, the material being bonded forms its own seams; for this the tubing is usually arranged with the flanges in contact. In heterogeneous welding, the tubing or articles are butted; the joint is made with a welded seam of another material. Heating is usually produced by hot gases. In friction welding, the parts to be bonded are rapidly rotated against each other. The glass- or melt-transition temperature is exceeded by the resulting heat of friction, so that self-diffusion results. This "melting" is also encouraged by pressure. In induction welding, a metal band is laid in the groove, the parts are pressed together, and a 60-kHz frequency current is applied. The resulting heat of induction gives rise to welding of the parts.

Sintering is used for the treatment of surfaces, the production of porous materials, or the manufacture of large, hollow bodies. The material is compressed to a sinter under pressures of 500 MN/cm^2 and then carefully heated in such a way that only the surface layers begin to melt. The particles adhere, creating bodies with open, porous channels. These permanently porous bodies are used as filter supports, as bodies with large surface areas for physical processes such as thermal exchange, or for ventilation surfaces. Poly(ethylene), poly(propylene), poly(tetrafluoroethylene), poly(methyl methacrylate), and poly(styrene) are used as raw materials.

In the *double-rotation centrifuge* process, the heated grains are sintered together in a kind of centrifuge. Hollow bodies of poly(ethylene) with capacities of up to 10,000 liters are manufactured by this technique.

Fluidized bed sintering is not really a type of sintering. Here, heated metal parts which have previously been roughed by sandblasting or treated with primers are dipped into a fluidized bed of the plastic powder. The plastic powder, with particle sizes of $\sim 200\ \mu m$ melts on the warm surfaces and flows into a thick film of ~ 200–400 μm. In this way, for example, garden furniture can be coated with polymides. Poly(ethylene), poly(vinyl chlorides), and polyamides are often processed in this way, as is, less frequently, cellulose acetobutyrate.

The metal parts for treatment must also be roughed or pretreated with primer for *flame spraying*. The granulated thermoplasts are then melted in a flame-spraying gun and sprayed onto heated metal surfaces. The method is particularly suitable for small numbers of pieces. Poly(ethylene), PVC, cellulose esters, and epoxy resins are processed in this way.

Hot-blast sprinkling is a type of flame spraying in which the metal surfaces do not need to be heated. In many cases, heating of the metal surfaces is best avoided because it causes undesirable changes in the metal structure. In hot-blast sprinkling, the plastic powder is softened by an electrical arc at about 1600°C (under Ar, He, N_2) and blasted onto the worked article with a spray gun. In this way, the metal is only heated up to 50–60°C. Polyamides and epoxy resins can be processed by this method.

12.2.6. Processing via the Solid State

In *laminating* or coating, a preformed solid film of plastic is bonded with glue or adhesive directly onto the material to be laminated. Composite films of cellulose and poly(ethylene), for example, are produced in this way. In these composite films, the cellulose protects from foreign odors and the poly(ethylene) assures water-tightness. Composite films of poly(isobutylene) and poly(ethylene) are used in the construction of chemical apparatus made of steel. Poly(ethylene) takes care of the protection against corrosion, and poly(isobutylene), with adhesives, the adhesion to steel. Steel sheet laminated with PVC can be processed as normal, but is noncorrosive in the absence of any further treatment. Metal aircraft parts are often coated with films of poly(tetrafluoroethylene). Poly(tetrafluoroethylene) is difficult to wet, and this decreases the tendency to ice over.

Composites are often constructed by bonding two materials together as in the laminating technique. In the case of synthetic leathers, called poromers, layers of different plastics are bonded together. Corfam™, the first product of this type, consists, for example, of an upper layer of vapor-permeable polyurethane and a central layer of a mixed fabric consisting of porous 95% poly(ethylene terephthalate) fleece held together by an elastomeric polyurethane cord. The fleece is produced from fibrils (see Section 12.3.1).

Composites are particularly useful as construction materials. Car bodies, for example, can be manufactured according to the sandwich-construction technique: A hard foam of polyurethane is sandwiched between glass-fiber-reinforced epoxy resins. This mode of construction, however, requires a great deal of manual work, and is therefore uneconomical. It has recently become possible to produce car bodies much more economically by vacuum forming or deep drawing glass-fiber-reinforced plastics.

In the solid state, thermoplasts and thermosets are worked unrestrained by *stamping* or *cutting* (peeling), and under restraint (clamped) by *sawing*, *boring*, *turning*, or *milling*. Cutting is also used to some extent in the manufacture of films from intermediate products or components of celluloid or poly(tetrafluoroethylene). All the other methods are used merely to produce special parts. If possible, high cutting speeds should be avoided when working under strain, otherwise the plastic heats up too much, becomes viscoelastic, and "smears." At high cutting speeds, therefore, the degree of restraint and chip cross section must be small.

12.3. Fiber Processing

12.3.1. Introduction

Filaments can be drawn from the melts or concentrated solutions of most macromolecular substances, and also from those of a few low-molecular-weight compounds (honey, soap solutions) (see Section 12.3.3). Filament forming, then, is not a property which is peculiar to macromolecular substances. The filaments of low-molecular-weight compounds have only a low mechanical strength, however. If the filament material consists of long molecular chains, then these will be partly oriented on stretching or drawing. The resulting filament has increased strength. Only a few filaments, however, have a sufficiently high tensile strength to be suitable for use in textiles or industry.

The first provisions for filament formation, therefore, are long-chain macromolecules. Extensive branching decreases filament formation and affects the mechanical properties of the filament, since fewer contact points per unit length are possible between the chain molecules. For the same reason, there must be a certain minimum degree of polymerization for filament formation (Table 12-5). Below this minimum degree of polymerization, the tensile strength is virtually zero (Figure 11-16). The stronger the forces between the chains, i.e., the more polar the monomeric unit and/or the more readily the macromolecules crystallize, the lower will be the minimum degree of polymerization required for filament formation. Crystallization, however, is not necessarily an essential prerequisite for filament formation; for example, bristles for brushes, etc., are manufactured from noncrystalline poly(styrene). Too high a crystallinity, in fact, is hardly desirable, because then the filament will be too brittle.

Table 12-5. The Lowest Degree of Polymerization $\overline{X}_n$ of Various Polymers at Which it is Possible to Produce Filaments[a]

Substance	$\overline{X}_n$
Nylon 6	50
Poly(ethylene terephthalate)	70
Cellulose	130
Poly(acrylonitrile)	300
Poly(vinyl alcohol)	300
Poly(styrene)	600

[a] Data from H. Mark.

A distinction is made between filaments and fibers. Filaments are endless, i.e., they are more or less infinite in length. They consist of a single thread (monofilaments) or several threads (multifilaments). Fibers consist of filaments cut to 30–150 mm (staple fibers). Fibers and filaments are classified according to their diameter: Those of about 5–50 nm are also called filaments, fibrils are in the range of a few hundred nm, and fibers exceed this figure. The titer of a fiber or filament is a length-based mass. The titer is given in tex or decitex (1 tex = 1 g/km), and previously in denier (1 den = 1/9 tex). The ratio of the fiber length to the fiber diameter is called the slimness ratio.

In addition, fibers are classified according to their origin into natural and man-made fibers. Natural fibers can come from vegetable, animal, or mineral sources. Man-made fibers include regenerated and completely synthesized fibers.

All presently used vegetable fibers are cellulosic (cotton, flax, hemp, ramie, jute, sisal), and all presently used animal fibers are proteins (wool, natural silk, camel hair). Asbestos is a mineral fiber.

Regenerated fibers are fibers produced from material of natural origin. The material undergoes some chemical process or modification (viscose silk, acetate silk, nitrocellulose, alginate fibers, etc.). Synthesized fibers are synthetic fibers completely synthesized from other raw materials.

In addition, fibers are classified according to application into textile and industrial fibers (see also Section 12.3.6). Textile fibers are used for yarn, weaves, knitted fabrics, etc., and industrial fibers are used for filter cloths, ropes, etc.

A distinction is made between melt, wet, dry, gel extrusion, and dispersion spinning. In all cases, one can start with polymer granulate, or in principle, with the polymerizate direct from the reactor. Split fiber formation is completely different. Fleece is generally produced by a thermally or chemically generated interfiber adhesion.

12.3.2. Fiber Formation

The capacity to form long, continuous fibers varies with different materials. The fiber length which can be obtained with a given material depends on the viscosity η of the liquid and the rate v of spinning. With increasing product $v\eta$, the length L of the liquid fiber goes through a maximum (Figure 12-4). The maximum length of fiber which can be obtained is called the spinnability.

From the relation $L = f(v\eta)$ it is obvious that the spinnability is governed by two processes, namely the cohesive break (or the swell effect) and the melt break (capillary break, melt fracture). According to Section 11.3.1, a certain amount of elastic energy can be stored in all viscoelastic fluids. This phenomenon leads, among others, to the Barus effect.

The Barus effect or swell effect increases with stored energy (increasing with shear stress and decreasing with capillary length). If a certain stored energy value is exceeded, the fiber breaks. This break is called the cohesive break because it depends on the cohesive energy (or surface tension) of the material, as well as on the viscosity of the liquid and the fiber production rate and on the modulus of elasticity. Melt fracture is the same as the melt break discussed in Section 7.6.1, and it is a consequence of elastic surface waves and irregularity of flow appearing above a critical extrusion rate.

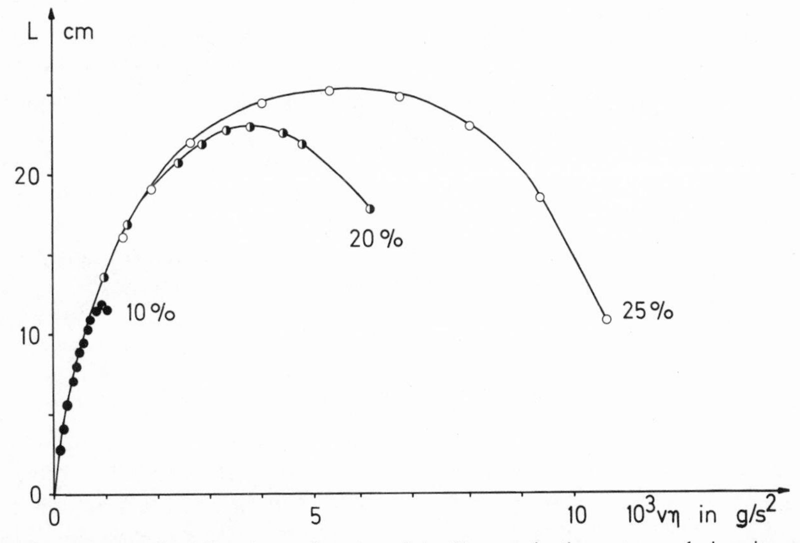

Figure 12-4. Length of fiber L as a function of the fiber production rate v and viscosity η in spinning solutions of cellulose acetate at different concentrations in an acetone–water mixture (85:15). L at the maximum is the spinnability. (From A. Ziabicki, according to data from Y. Oshima, H. Maeda, and T. Kawai.)

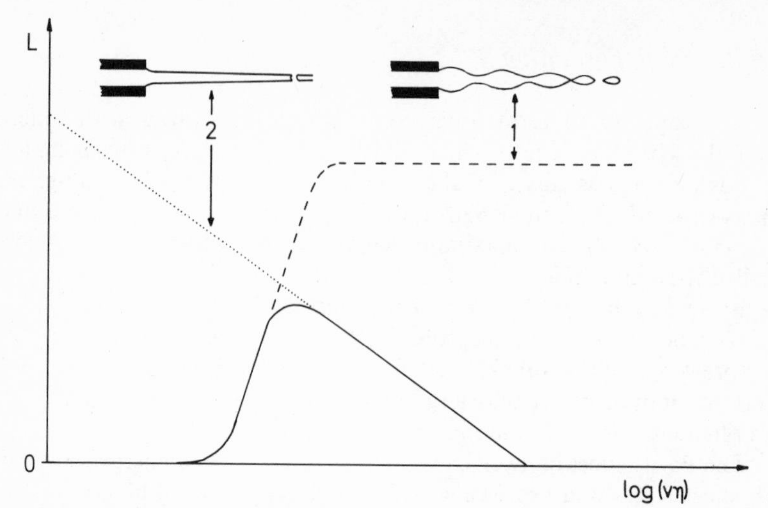

Figure 12.5. Influence of the capillary break (---) and cohesion break (···) on the spinnability. Length L of the fibers as a function of the fiber production rate v and viscosity η. (—) Effect observed when capillary and cohesion breaks overlap. (After A. Ziabicki.)

This results in a periodic swelling and breaking of the fiber. Sometimes this phenomenon is made use of commercially, since under carefully controlled conditions, periodic swelling leads to a fiber of decorative appearance and high bulk.

Because of capillary break and cohesive break, an optimum value is observed for L at a given value of $v\eta$ (Figure 12-5). The form of the function $L = f(v\eta)$ has the following consequence: If the rate and/or the viscosity during spinning is too low, then capillary break occurs, and the fibers disintegrate into individual drops. Too low a viscosity is obtained, for example, when the solution is too dilute or the spinning temperature too high. Conversely, a cohesive break is obtained when the relaxation times are too high, which can be caused by too high a molecular weight or too rapid gelling during wet spinning. The cohesive break is a brittle break.

12.3.3. Spinning Techniques

Melt spinning is the most economical method. The polymer granulate is fed into a heated reservoir. The melt is then pumped or extruded through spinnerets and the filaments are allowed to cool in the air. The filaments are produced at rates of up to 1200 m/min. Only thermally stable or stabilized polymers that give a melt can be melt spun. Polyamides, polyesters, poly(olefins), and glass are spun in this way (Figure 12-6).

The *wet-spinning* method is used for polymers that decompose on melting. Solutions of 5–20% are passed through the spinnerets by a spin pump. The filaments are coagulated in a precipitation bath and drawn in a drawing or stretching bath. The method gives much lower filament production rates than melt spinning; namely 50–100 m/min; it is less economical because of solvent recovery costs. Rayon, viscose rayon, and poly(vinyl alcohol) are spun out of aqueous solutions with this method.

In the *dry spinning* method, 20–45% solutions are used. After leaving the spinneret orifices, the filaments enter a 5–8-m-long chamber in which jets of warm air are directed toward the filaments, causing the solvent to evaporate. With filament production rates of 300–500 m/min, the method allows higher rates of spinning than the wet-spinning process. The capital cost of equipment is higher, but the running costs are lower than in wet spinning. In addition, this method can only be used to spin those polymers for which readily volatile solvents are known. The dry-spinning process is used for poly(acrylonitrile) (25% in dimethylformamide), chlorinated poly(vinyl chloride) (45% in acetone), and cellulose triacetate (~20% in CH_2Cl_2).

In *gel spinning, extrusion spinning*, or *gel-extrusion spinning*, 35–55% solutions are used. The resulting filaments have a greater dimensional stability because of their lower solvent content, so that the rates of filament production can be greater (~500 m/min). Poly(acrylonitrile) and poly(vinyl alcohol) are spun by this method.

Dispersion spinning is a special spinning process for insoluble and nonmelting polymers. The dispersions of the polymers for spinning have other organic polymers added to increase their viscosity and to stabilize the fibers. This polymer is then burned off after the filaments have been

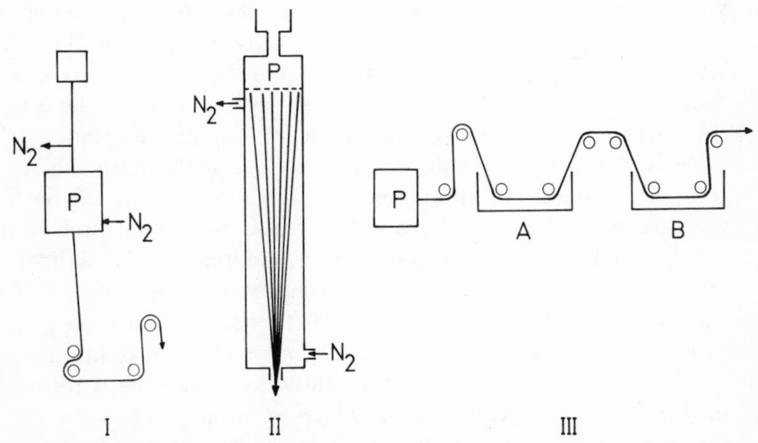

Figure 12-6. Spinning processes (schematic). (I) Melt spinning, (II) dry spinning, (III) wet spinning; P, spin pump; A, precipitation bath; B, drawing bath.

produced, and during this process the filament-forming polymer also sinters together. Poly(tetrafluoroethylene) fiber (Section 25.7.4.3) is produced by this process.

In polymerization spinning, the monomer is polymerized together with initiators, fillers, pigments, and flame retardants, or other desired additives. The polymerizate is not isolated, but directly spun at rates of about 4000 m/min. The process is only suitable for rapidly polymerizing monomers.

12.3.4. *Spin Processes*

All the spinning processes proceed roughly according to the same scheme in four stages. In the first stage, the liquid to be spun is extruded through the orifices of a spinneret. The length of filament which can be obtained is governed by the spinnability (Section 12.3.2). In the second stage, the actual filament begins to form. In this transition region, internal stresses equilibriate. In the first two stages, the filament retains its external shape. In third stage, the still semiliquid filament is drawn or stretches under its own weight, causing a slight orientation of the chains to occur. In the fourth state, the filament is drawn.

The length of time for which the liquid remains in the spinneret orifices is ~0.1–100 ms. The relaxation times for this process (see Section 11.4.2), on the other hand, lie between ~100 and 1000 ms. Relaxation processes are therefore important in spinning. They are particularly evident in the Barus effect (Section 11.3.1) and in elastic turbulence (Section 7.6.1).

In the spinning process, the molecular chains are oriented by three effects: flow orientation inside and outside the spinneret orifices, and orientation by deformation. For the orientation of the molecular chains that occurs in the spinneret to be effective in orienting the filament, the rate of stabilization of the filament must be greater than the reciprocal relaxation time. This requirement applies only to the surfaces and not to the interior of the filament. The orientation of the molecule within the spinneret thus has little influence on the orientation of the molecule in the finished filament.

Outside the spinneret, the molecules also become oriented by flow. As the distance from the spinneret mouth increases, the optical birefringence increases first slowly and then rapidly up to a limiting value (Figure 12-7). This limiting value is determined by the rigidity of the filaments and the resulting limited mobility of the molecule. This process produces the greatest observed proportion of orientation. Finally, a smaller contribution comes from yet a third process, namely, an orientation of the chains through a deformation of the physical network which is formed.

The crystallinity of the filaments produced by the spinning process

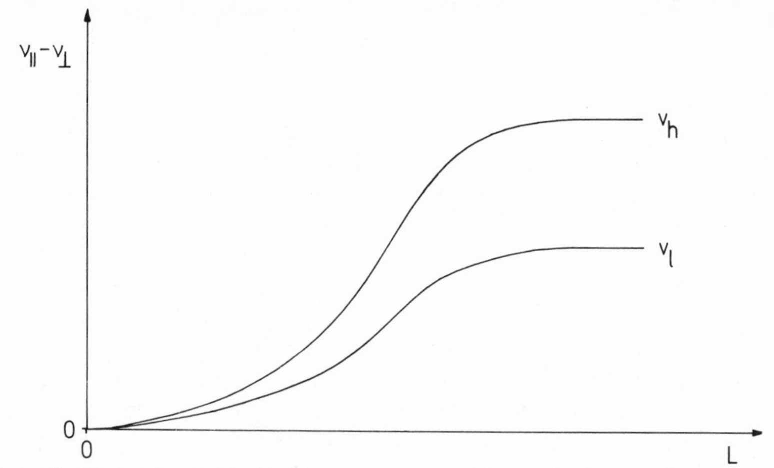

Figure 12-7. Dependence of the birefringence $n_{\parallel} - n_{\perp}$ in fibers on the distance L from the nozzles at low (v_l) and high (v_h) fiber production rates (schematic).

can vary greatly. It depends on how easily the polymer can be crystallized under the spinning conditions and the rate of stabilization of the filament. With slowly crystallizing, rapidly stabilizing polymers the crystallinity is almost zero [an example is the melt spinning of poly(ethylene terephthalate)]. Polymers that crystallize well and stabilize slowly show practically their maximum possible crystallinity [e.g., poly(vinyl alcohol) and cellulose on wet spinning]. Filaments of it-poly(propylene) and polyamides obtained by melt spinning exhibit crystallinity between these two extremes. In general, the highest degrees of crystallinity are obtained with wet spinning, because the mobility of the molecules is greatest in this case. With melt spinning, on the other hand, the mobility, and thence the degree of crystallinity, is small, because the filament is usually cooled very rapidly to below the glass-transition temperature. Spherulites are seldom found in melt spinning and not at all in wet or dry spinning.

The shape of the filament cross section depends on the spinneret shape and diffusion processes away from the spinneret. Circular spinneret orifices give cylindrical filaments. Because of diffusion processes, noncylindrically set filaments are obtained from dry or wet spinning with circular spinneret orifices. Noncylindrical filaments can also be obtained by using spinneret orifices of different shapes. In such cases, the filament thickening due to the Barus effect must be considered: For example, triangular orifices with concave sides give triangular filaments with convex sides. Noncylindrical filaments are desirable since the larger surface area gives better dyeability, and filaments, fibers, and weaves have a more acceptable appearance and a better hand.

12.3.5. Drawing

In drawing, the filament is irreversibly extended by many times its own length. With films and ribbons, the same process is called recking. The maximum elongation that can be achieved depends on many factors (see also Section 11.5.1). To elongate, for example, monofilaments to their maximum extent, it has been found empirically that the so-called capillary geometry (the ratio of the capillary length to the capillary cross section) must be about 3–7. Fibers, films, and reels obtain their great tensile strength only after drawing (Table 12-6). Less elongation is required with rigid molecules than with flexible ones. Fibers of cellulose derivatives, for example, need only to be drawn by 80–120% in order to obtain sufficient tensile strengths, whereas it-poly(propylene) must be elongated by up to 600–800%.

The molecular or supramolecular processes in the recking process are not completely understood. It is possible that whole crystallite blocks are drawn out from their original structure and freshly oriented.

With films, in many cases, an annealing process is coupled to the recking. Films of poly(ethylene terephthalate), for example, are first elongated biaxially at 80°C and then annealed under stress at 200°C (i.e., below the melting point). On annealing, the crystallinity of the sample and the density increase, i.e., the tensile strength is also improved. Stresses are often included in the processing of films (extrusion, stretching). On reheating, the stresses are relaxed and shrinking occurs. Such films are very important in the packaging industry, especially those films that shrink below 100°C, since here the shrinking can be induced in a hot-water bath.

Fibers can also be produced from films. The process is limited to polymers that give well-oriented films of high mechanical strength. The films are slit with knives into ribbons of 1–10 mm thickness or mechanically split (fibrillated) by passing the oriented film through spiked rollers. Such split fibers are predominantly produced from poly(propylene), but can also be made from polyamides or poly(vinyl chloride). Ropes, baling twines, packing or garden furniture fabrics, webbing, and carpet backing can be produced from split fibers.

Table 12-6. Tensile Strength σ_B of Oriented and Unoriented Polymers

Polymer	σ_B, N/mm^2	
	Unoriented	Oriented
it-Poly(propylene) (F fiber)	35	700
Poly(ethylene terephthalate)	35	500
Poly(styrene) (monofilament)	35	110

12.3.6. Fiber Properties

Properties of fibers and objects produced from them depend on the chemical nature of the fiber raw material, the physical structure of the fibers, the fiber shape, and how the fibers are joined together. The specifications for textile and industrial fibers are different because they serve different purposes (Table 12-7).

The mechanical properties of the fibers do not depend on the crystallite orientation, but they do depend on the tie molecules between the crystallites (see also Section 5.4.2). These bundles of tie molecules or strands can transport forces very effectively. It is important to note that laboratory-produced fibers often have defects, and in comparing the tensile strengths of these with industrially produced fibers, one should always take the highest tensile strength value of such laboratory produced fibers. Otherwise, one is comparing the faults and defects of the fibers and not the influence of structure.

When possible, synthetic fibers should have melting points in the range 240–270°C. Lower melting points exclude certain application areas and higher melting points make fast, even spinning more difficult. Even spinning is absolutely necessary, since no more than one defect per 2000 km (about 10 kg) is acceptable. The glass-transition temperature should, if possible, lie between 110 and 160°C. The recovery of the fibers is inadequate when the glass-transition temperature is too low.

Fiber properties are also strongly influenced by the fiber shape, especially the cross-sectional shape. Cotton is a round, hollow fiber. Silk has a triangular cross section, which gives it a fine lustre, hand, and rustle. Wool possesses a "scaly" surface, and also appears to be a two-component

Table 12-7. Fiber Specifications[a]

Property	Fiber: Textile	Fiber: Industrial
Tensile strength, g/den	3–5	>7–8
Elongation at break, %	>10	8–15
Elastic recovery, %	>5	—
E modulus in the conditioned state, g/den	30–60	>80
Temperature at which the tensile strength is zero, °C	>215	>250
Additional requirements	Dyeable, dry-cleanable, easy care	Resistant to chemical attack, nonflammable

[a] With the exception of elastic fibers. 1 den = 1 g/9000 m.

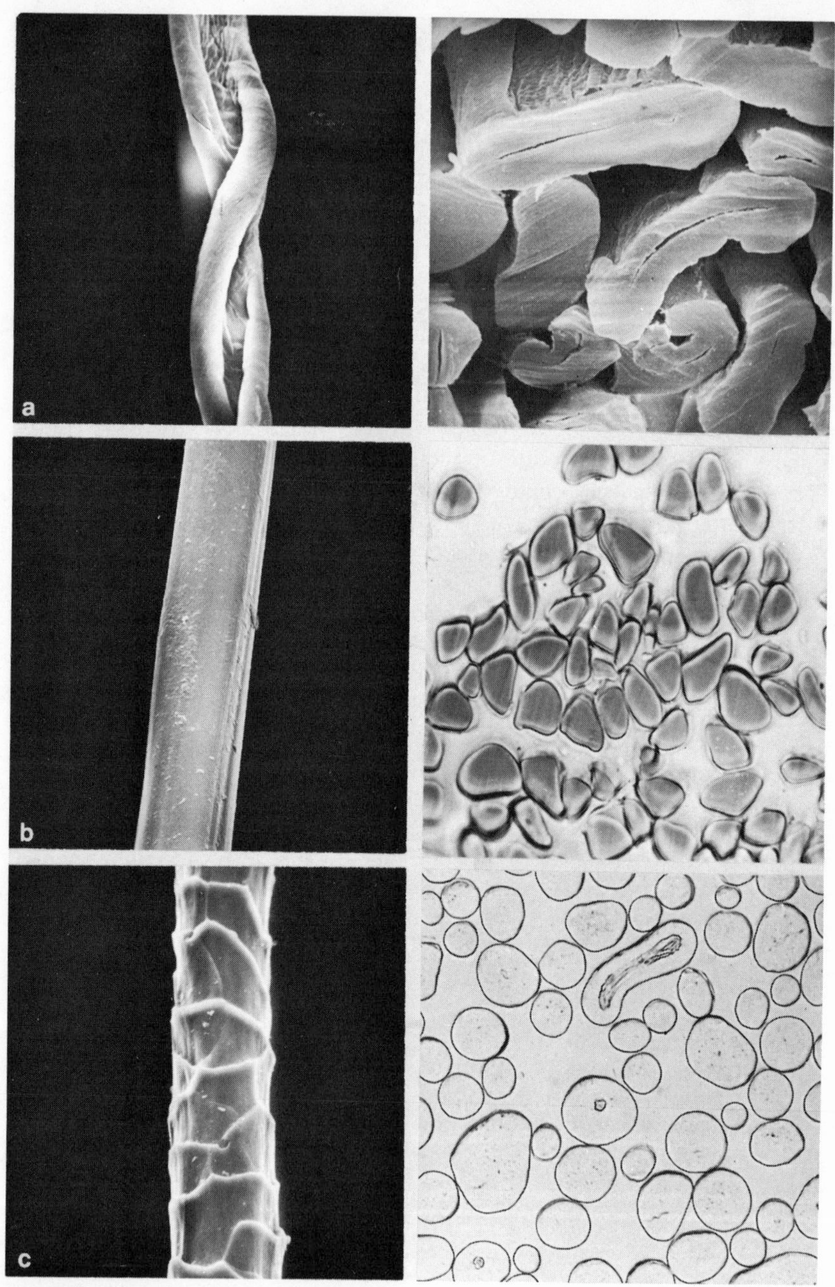

Figure 12-8. Lengthwise (left) and cross-sectional (right) shape of various natural fibers: (a) cotton; (b) silk; (c) wool. (Institut für angewandte Mikroskopie der Fraunhofer-Gesellschaft, Karlsruhe).

fiber, since each half has a different water absorption capacity. From this comes the crimpability and bulk of the wool. Synthetic fibers can be produced with various cross sections, and consequently, with various feels or hands. In addition, the thermoplasticity of synthetic fibers allows them to be fixed in various shapes; they can be "texturized" (Figures 12-8 and 12-9).

Good bulk and crimp stability can be obtained with bicomponent fibers (also known as conjugated fibers, twin fibers, or fibers of bilateral structure). The two components may lie side by side, and have a nucleus–mantle structure or a matrix–fibrillar (M/F) structure (Figure 12-10). The M/F fibers are known as matrix fibers in the United States and not as bicom-

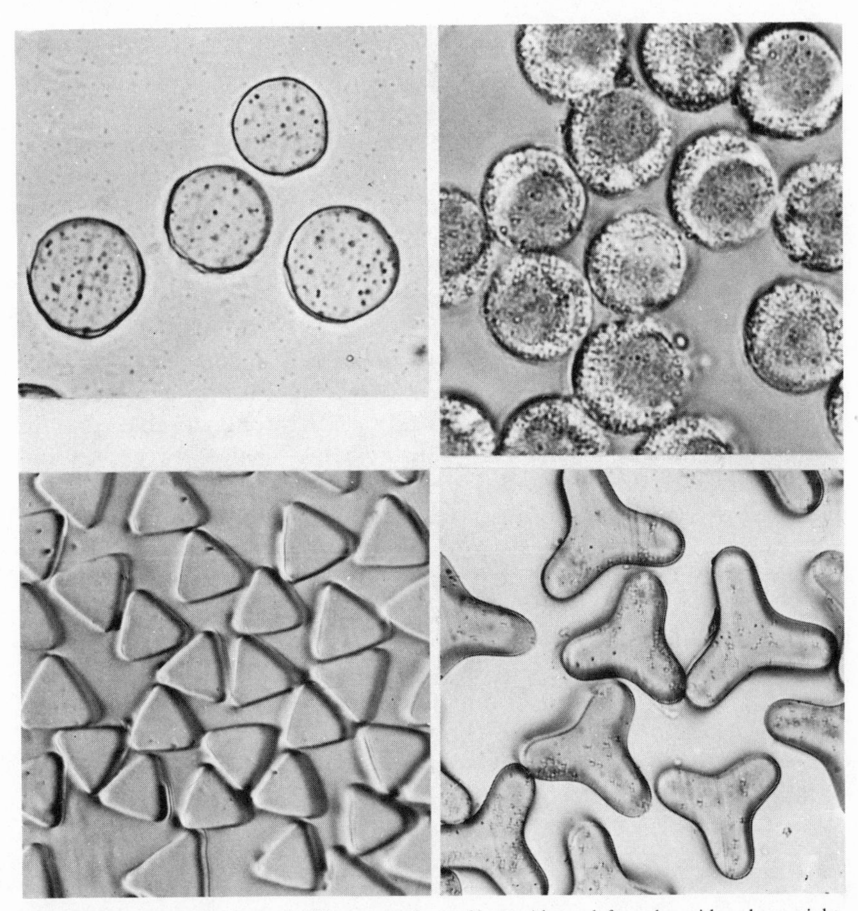

Figure 12-9. Cross sections of different synthetic fibers. Above left, polyamide; above right, bicomponent fiber with a nucleus of poly(hexamethylene diamine) and a mantle of poly(ε-caprolactam); below left, polyester; below right, polyamide with trilobal cross section (Institut für angewandte Mikroskopie der Fraunhofer-Gesellschaft, Karlsruhe).

Table 12-8. Production Rates for Textile Products

Process	Rate, m^2/h
Weaving	5
Machine knitting	
with needles	20
with hooks	80
Fleece formation	
dry	1,000–10,000
wet	10,000–50,000
Paper making	200,000–500,000

ponent fibers. Wet, dry, and emulsion spinning and, more rarely, melt spinning are suitable for producing bicomponent fibers. Two spin dopes of different composition are separately brought to the spin orifice and only together directly before the spin orifice. Examples of such two-component fibers are industrial fibers from cellulose with a poly(ethylene) mantle, or textile fibers such as elastic M/F fibers from two acrylonitrile copolymers. In the latter case, the "fiber" consists of a 70% acrylonitrile, 20% ethyl acrylate, and 10% methylol acrylamide copolymer, and the "matrix" is a 20% acrylonitrile, 70% ethyl acetate, and 10% methylol acrylamide copolymer.

The hand of the textile product is also strongly influenced by how the fibers are joined together, i.e., by twisting, weaving, knitting, etc. According to the process, the textile production rate varies strongly (Table 12-8). In fleece production (spun bonding), a distinction is made between wet and dry processes. In the dry process, the polymer melt is spun, and the fiber produced is continuously drawn, cut, irregularly set on a band or ribbon, and welded together by heating for short durations. For example, this process can be used to produce paper from the coarsest packing grades to the finest writing grades from poly(ethylene). The paper is suitable for writing because of capillary attraction and not because of any subsequent chemical modification of the surface. In the wet fleece-making process, fibers that are cut relatively short are suspended in water and allowed to settle on a water-impermeable bed.

Fleeces are also called nonwoven fabrics. They are used, e.g., as carpet backs or for inlets of suits.

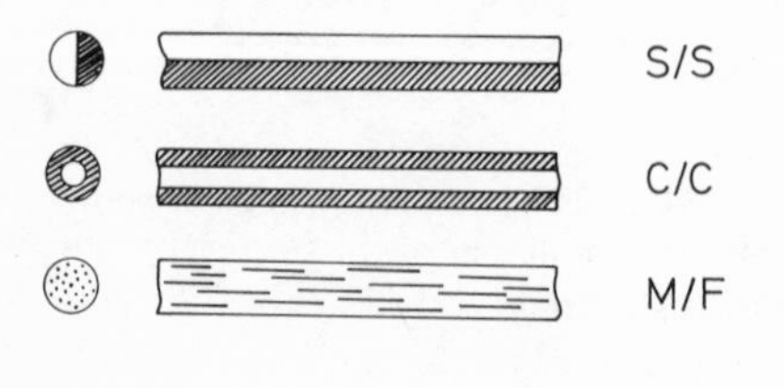

Figure 12-10. Diagram of the cross section and length section of bicomponent fibers. S/S, side by side; C/C, core/cover; M/F, matrix/fibril.

12.4. Finishing (Surface Treatment)

The surfaces of plastics are sometimes additionally treated or covered by a coating of metal or glass, for technical (surface hardness, friction) or aesthetic reasons (gloss).

12.4.1. Metallizing

Layers of metal up to thicknesses of $\sim \mu$m can be applied by chemical methods or by vacuum evaporation. The layers have a high gloss. For thicker layers, these methods are not economical, and as far as it is possible with the plastic, coating is performed with galvanizing methods. In all these methods, the plastic surfaces must first be thoroughly degassed, degreased, or dried.

Almost all plastics can be coated by metals under vacuum. Car hoods thus produced from high-impact plastics with a metal finish are cheaper, even today, than metal ones. The disadvantage of this method is the low adhesive strength, particularly with thicker metal layers.

Chemical metallization is, in practice, used only for silvering. The plastic surfaces are dipped in solutions of silver salts and the silver is then reduced chemically. Here again the adhesive strength is low.

So far, only ABS polymers have been galvanized on a commercial scale, since only in this case can a sufficiently high adhesive strength be obtained. When the plastic surfaces are pickled, the rubber elastic components are anodized. This produces pores and channels in which, for example, silver can be deposited chemically. The silver then forms the adhesive base for the copper layers subsequently deposited electrochemically, and these layers are then reinforced by the galvanized coating. Here, too, it is difficult to manufacture metal layer thicknesses of more than $\sim 10\ \mu$m because the different thermal coefficients of expansion of plastics and metals can easily lead to stresses, and thence to bubbles or cracks.

12.4.2. Glass Coating

Plastic surfaces can be made scratch resistant by a 3-μm-thick glass coating. Such coatings are produced by vaporizing certain borate–silica glasses in the heat of an electric arc. The process is economical because of the high rate of evaporation and the correspondingly short deposition times. Conversely, direct evaporation, or evaporation with the aid of cathode rays, is too slow. Of course, the deposition rate should not be too rapid or cracks will occur. SiO_2 coatings are not sufficiently thermally stable because of the large differences between the thermal expansion coefficients of coating material and polymer.

Coverings can be obtained by the use of alcoholic solutions of hydrolyzable alcoholates of polyvalent metals (e.g., Ti, Si, Al). Evaporation of the alcohol in the presence of air produces simultaneous hydrolysis and the formation of an insoluble gel or network. Gel formation at low temperatures gives a product still containing M—OH groups: The coating is hydrophilic and antistatic. Network formation at higher temperatures leads to metal oxides; the coating is scratch resistant.

Scratchproof coatings on poly(methyl methacrylate) can be produced by application of a 50 :50 mixture of poly(silicic acid) and poly(tetrafluoroethylene-co-hydroxyalkyl vinyl ether). It is not known why coatings as scratchproof as glass can be produced in this manner.

Literature

Reviews

Reviews in Polymer Technology, M. Dekker, Vol. 1 (1972).

Section 12.1.1. Preparation

O. Lauer and I. Engles, *Aufbereiten von Kunststoffen*, Hanser, Munich, 1971.

Section 12.1.2. Fillers

S. Oleesky and G. Mohr, *Handbook of Reinforced Plastics*, Reinhold, New York, 1964.
G. Kraus, E., *Reinforcement of Elastomers*, Interscience, New York, 1965.
W. S. Penn, *GFP Technology*, MacLaren, London, 1966.
P. H. Selden, ed., *Glasfaserverstärkte Kunststoffe*, Springer, Berlin, 1967 (third ed. of H. Hagen, *Glasfaserverstärkte Kunststoffe*).
R. T. Schwartz and H. S. Schwartz, *Fundamental Aspects of Fiber Reinforced Plastic Composites*, Interscience, New York, 1968.
W. C. Wake, ed., *Fillers for Plastics*, Iliffe, London, 1971.
P. D. Ritchie, ed. *Plasticizers, Stabilizers and Filler*, Butterworths, London 1972.
G. Kraus, Reinforcement of elastomers by carbon black, *Adv. Polym. Sci.—Fortschr. Hochpolym. Forschg.* **8**, 155 (1971).
J. Gilbert Mohr. ed., *SPI Handbook of Technology and Engineering of Reinforced Plastics/Composites*, second ed., Van Nostrand Rheinhold, New York, 1973.
L. Mascia, *The Role of Additives in Plastics*, Halsted, New York, 1975.

Section 12.1.3. Dyestuffs and Pigments

C. H. Giles, The coloration of synthetic polymers—a review of the chemistry of dyeing of hydrophobic fibers, *Br. Polym. J.* **3**, 279 (1971).
T. B. Reeve, Organic colorants for polymers, *J. Macromol. Sci. D (Rev. Polym. Technol.)* **1**, 217 (1972).
T. C. Patton, ed., *Pigment Handbook* (3 vol.), Wiley–Interscience, New York, 1973.
R. R. Myers and J. S. Long, *Pigments*, Part I, M. Dekker, New York, 1975.

Section 12.1.4. Plasticizers

K. Thinius, *Chemie, Physik und Technologie der Weichmacher*, Verlag Technik, Berlin, 1960.
I. Mellan, *Industrial Plasticizers*, Pergamon, Oxford, 1963.
P. D. Ritchie, ed., *Plasticizers, Stabilizers and Fillers*, Butterworths, London, 1971.

Section 12.2. Processing of Thermoplasts, Thermosets, and Elastomers

J. A. McKelvey, *Polymer Processing*, Wiley, New York, 1962.
T. H. Ferrigno, *Rigid Plastic Foams*, Reinhold, New York, 1963.
E. G. Fisher, *Extrusion of Plastics*, second ed., Wiley–Interscience, New York, 1964.
H. Götze, *Schaumkunststoffe*, Strassenbau, Chemie und Technik Verlagsges., Heidelberg, 1964.
J. S. Walker and E. R. Martin, *Injection Molding of Plastics*, Butterworths, London, 1966.
D. V. Rosato and C. S. Grove, Jr., *Filament Winding*, Wiley–Interscience, New York, 1968.
W. Schaaf and A, Hahnemann, *Verarbeitung von Plasten*, VEB Deutscher Verlag für Grundstoffindustrie, Leipzig, 1968.
O. J. Sweeting, *The Science and Technology of Polymeric Films*, 2 vols. Wiley–Interscience, New York, 1968.
B. S. Benjamin, *Structural Design with Plastics*, Van Nostrand Reinhold, New York, 1969.
H. Domininghaus, *Kunststoffe II (Kunststoffverarbeitung)*, VDI Taschenbücher, T 8, VDI-Verlag, Düsseldorf, 1969.
A. Kobayashi, *Machining of Plastics*, McGraw-Hill, New York, 1969.
L. I. Naturman, Cold forming, where to next? *Plast. Technol.* **15** (4), 39 (1969).
W. S. Penn, ed., *Injection Molding of Elastomers*, MacLaren, London, 1969.
J. V. Schmitz, *Bibliography on Polymer Testing, Processing and Applications*, Soc. Plastics Engineers. Inc., and Plastics Institute of America, Inc., New York, 1969.
K. Stoeckhert, *Formenbau für die Kunststoff-Verarbeitung*, C. Hanser, Munich, 1969.
W. R. R. Park, *Plastics Film Technology*, Van Nostrand Reinhold, London, 1970.
C. M. Blow, ed., *Rubber Technology and Manufacture*, Butterworths, London, 1971.
R. E. Elder and A. D. Swan, *Calendering of Plastics*, Iliffe, London, 1971.
E. G. Fisher, *Blow Molding of Plastics*, Butterworths, London, 1971.
A. Höger, *Warmformen von Kunststoffen*, C. Hanser, Munich, 1971.
E. Meinecke, *Mechanical Properties of Polymeric Foams*, Technometric Publ., Westport, Connecticut, 1973.
M. Morton, ed., *Rubber Technology*, second ed., Van Nostrand Reinhold, New York, 1973.
P. C. Powell, *Plastics for Industrial Designers*, The Plastics Institute, London, 1973.
I. I. Rubin, *Injection Molding-Theory and Practice*, Wiley–Interscience, New York, 1973.
W. A. Holmes-Walker, *Polymer Conversion*, Wiley, New York, 1975.

Section 12.3. Fiber Processing

J. W. S. Hearle and R. H. Peters, *Fiber Structure*, Butterworths, London, 1963.
R. H. Peters, *Textile Chemistry*, Elsevier, Amsterdam, 1963.
H. Rath, *Lehrbuch der Textilchemie*, second ed., Springer, Berlin, 1963.
L. R. McCreight, H. W. Rauch, Sr., and W. H. Hutton, *Ceramic and Graphic Fibers and Whiskers*, Academic Press, New York, 1965.
W. Bernard, *Praxis des Bleichens und Färbens von Textilen*, Springer, Berlin, 1966.
O. A. Battista, ed., *Synthetic Fibers in Papermaking*, Wiley–Interscience, New York, 1968.
H. F. Mark, S. M. Atlas, and E. Cernia, eds., *Man-Made Fibers*, 3 vols., Wiley–Interscience, New York, 1968.
H. Balk, Fibrillieren—Ein neues Verfahren zur Herstellung von Synthesefasern, *Kunststoff-Berater* **15**, 1091 (1970).
K. Meyer, *Chemiefasern (types, trade names, producers), VEB Fachbuchverlag*, second ed., Leipzig, 1970, Supplement volume, 1971.

H. Mark, N. S. Wooding, and S. M. Atlas, *Chemical After-treatment of Textiles,* Wiley, New York, 1971.
C. Placek, *Multicomponent Fibers,* Noyes Development, Pearl River, 1971.
R. Jeffries, *Bicomponent Fibers,* Merrow Publ., Watford, England, 1972.
R. W. Moncrieff, *Man-Made Fibers,* sixth ed., Halsted Press, New York, 1975.

Section 12.4. Surface Treatment

G. Kühne, *Bedrucken von Kunststoffen,* Hanser, Munich, 1967.
P. Schmidt, *Beschichten von Kunststoffen,* Hanser, Munich, 1967.
B. Rotrekl, K. Hudecek, J. Komarek, and J. Stanek, *Surface Treatment of Plastics,* Khimiya Publ, Leningrad, 1972 (Russian).

Construction with Plastics

E. Baer, ed., *Engineering Design for Plastics,* Reinhold, New York, 1974.
B. S. Benjamin, *Structural Design with Plastics,* Von Nostrand Reinhold, New York, 1969.
G. Schreyer, *Konstruieren mit Kunststoffen,* Hanser, Munich, 1972.
P. C. Powell, *Plastics for Industrial Designers,* The Plastics Institute, London 1973.

Chapter 13

Interfacial Phenomena

13.1. Spreading

Insoluble molecules spread on liquid surfaces. This behavior corresponds to that of an ideal gas if the coverage is small. Analogous to the ideal gas equation, the following is valid for the relationship between the surface pressure, given as $\gamma_0 - \gamma$, the difference between the surface tensions of the covering, and the covered surface, and the surface area per molecule of spreading material:

$$(\gamma_0 - \gamma) A = kT \qquad (\text{for } A \rightarrow \infty) \tag{13-1}$$

In principle, then, the molecular weight of the spreading material can be determined with equation (13-1). The surface pressure and the specific surface area are measured in a Langmuir trough. Here, a fixed quantity of material spreads out over a given surface, which is separated on the one side by an easily moved float. The pressure exerted on this float by a given surface area of a given quantity of material is then the surface pressure. These measurements are not simple to carry out, since only low pressures are found with low quantities of material, and the surface of the covering phase must be meticulously clean. Therefore the method has not become a routine method for determining molecular weights.

The plot of $\gamma_0 - \gamma = f(A)$ yields interesting conclusions concerning molecules of different shape or conformation. The more rigid poly(vinyl benzoate) molecules, for example, lead to a collapse of the surface at low values of A, but the more flexible poly (vinyl acetate) molecules do not (Figure 13-1). Isotactic and syndiotactic molecules show different behavior during spreading studies, but the curves are difficult to interpret quanti-

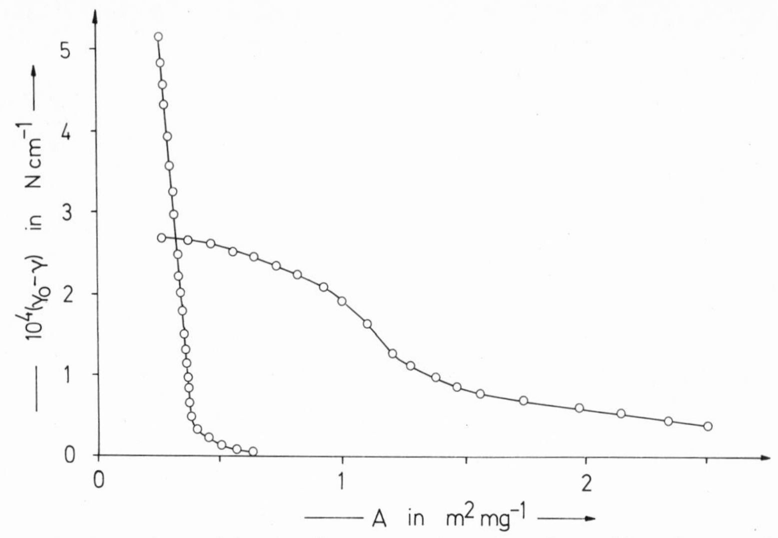

Figure 13-1. Dependence of the spreading pressure $(\gamma_0 - \gamma)$ on the specific surface area A in poly(vinyl acetate) (PVAC) and poly(vinyl benzoate) (PVBE) (after N. Berendjick).

tatively as a function of molecular quantities. In addition, there is the fact that the apparent molecular weight calculated at finite concentrations using equation (13-1) can also depend in some cases on the chemical nature of the hypophase. Such behavior is found during spreading studies with proteins, and indicates association–dissociation phenomena in the proteins.

13.2. Interfacial Tensions

13.2.1. Surface Tension of Liquid Polymers

Because of the high viscosities of polymer melts, the surface tension of liquid polymers can only be measured by some of the so-called static methods. The Wilhelmy-plate method and the suspended-drop method are suitable, but the capillary method and the wire-ring-detachment method are not suitable, since the measured surface tension depends on the speed at which the test is carried out. Varying the speed of the test alters the apparent viscosity of polymers with non-Newtonian behavior.

The Wilhelmy-plate method consists in partially immersing a plate in a wetting liquid. The surface tension of the liquid γ_{lv} acts downward on the plate. When the plate is wetted and its lower edge just resides on the liquid surface, the force acting on the plate is $\gamma_{lv} l_{per}$, where l_{per} is the perimeter of the plate. By measuring the restraining force on the plate in air and in

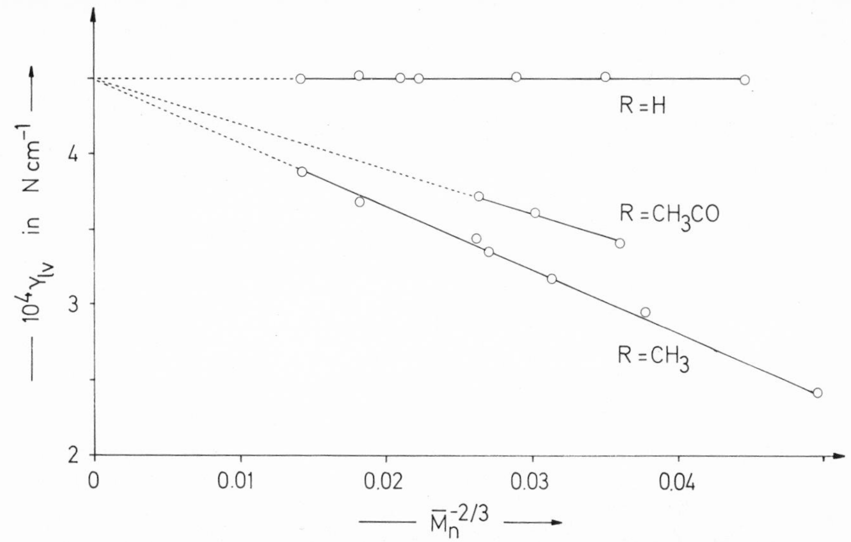

Figure 13-2. Dependence of the surface tension γ_{lv} of poly(oxyethylenes), $RO(CH_2CH_2O)_nR$, on the number-average molecular weight $\overline{M}_n$ at 24°C (data from various authors).

contact with the liquid surface, the surface tension can be calculated. The method has been used only to measure the surface tension (= liquid–air interfacial tension) and not the interfacial tension between two polymer liquids

The shape of a suspended drop depends on the surface tension as well as on the gravitational force. The drop is photographed and the diameter at various positions is measured. A consistent shape factor can be evaluated when hydrodynamic equilibrium is reached.

The surface tension of a polymer depends on its end groups, its molecular weight, and on the temperature. The dependence on molecular weight can be given empirically by

Table 13-1. Surface Tension γ_{lv}^{∞} of Polymers of Infinitely High Molecular Weight

Polymer	Temperature, K	$10^4\gamma_{lv}^{\infty}$, N/cm
Poly(dimethyl siloxane)	293.2	21.3
Poly(fluoroalkanes)	293.2	26.2
Poly(styrene)	449.2	30.0
Poly(isobutylene)	297.2	35.6
n-Alkanes	293.2	37.7
Poly(ethylene glycol)	297.2	45.0

[a]Data from various authors.

$$\gamma_{lv} = \gamma_{lv}^{\infty} - K\langle M_n \rangle^{-2/3} \qquad (13\text{-}2)$$

The slope constant K is influenced by the chemical nature of the end groups, as shown in Figure 13-2 for poly(ethylene glycols), $R{+}OCH_2CH_2{+}_{n-1}OR$. The term γ_{lv}^{∞} is independent of molecular weight and end groups and depends only on the temperature. Typical surface tensions are collected in Table 13-1. The temperature dependence of the surface tension is generally not very strong and is about $d\gamma/dT = (5\text{–}8) \times 10^{-4}$ N cm^{-1} K^{-1}.

13.2.2. *Interfacial Tension of Solid Polymers*

13.2.2.1. *Basic Principles*

On a solid, smooth surface a drop of liquid forms a certain contact angle ϑ (Figure 13-3). The value of the contact angle is determined by the resultant of the three interfacial tensions, liquid–vapor (ϑ_{lv}), solid–liquid (ϑ_{sl}), and solid–vapor (ϑ_{sv}):

$$\vartheta_{sv} = \vartheta_{sl} + \vartheta_{lv} \cos \vartheta \qquad (13\text{-}3)$$

At a contact angle $\vartheta = 0°$, there is complete spreading of the liquid on the surface, whereas at an angle of $\vartheta = 180°$ no spreading occurs. Real systems have contact angles of between 0° and 180°. Since the contact angle determines the spreadability, its cosine is a direct measure of the wettability of the surface. Equation (13-3) is valid for ideal surfaces in a vacuum. In real systems the equilibrium pressure p_e exerted by the absorbed liquid vapor on the solid must be considered (Young's equation):

$$\vartheta_{sv} = \vartheta_{sl} + \vartheta_{lv} \cos \vartheta + p_e \qquad (13\text{-}4)$$

Real surfaces are rough, not smooth. The ratio true surface/geometric surface is defined as the roughness r, and can only be equal to or greater than 1. Freshly cleaved mica has r values close to 1, polished surfaces have r values between 1.5 and 2.

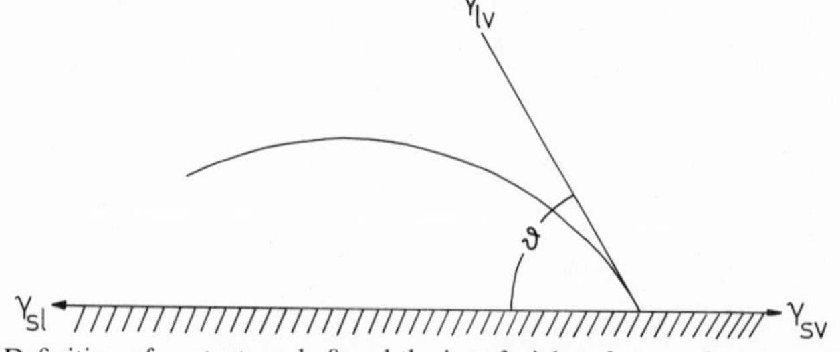

Figure 13-3. Definition of contact angle ϑ and the interfacial surface tensions γ_{sl} (solid/liquid), γ_{lv} (liquid/vapor), and γ_{sv} (solid/vapor).

Table 13.2. Interfacial Tension γ_{ll} between Two Liquid Polymers and Surface Tensions of the Pure Polymers at 150°C

		$10^5\gamma_{ll}$, N/cm							
Polymer[a]	$10^5\gamma_{lv}$, N/cm	PDMS	it-PP	PBMA	PVAc	PE	PS	PMMA	PEO
PDMS	13.6	0	3.0	3.8	7.4	5.4	6.0	—	9.8
it-PP	22.1	3.0	0	—	—	1.1	5.1	—	—
PBMA	23.5	3.8	—	0	2.8	5.2	—	1.8	—
PVAC	27.9	7.4	—	2.8	0	11.0	3.7	—	—
PE	28.1	5.4	1.1	5.2	11.0	0	5.7	9.5	9.5
PS	30.8	6.0	5.1	—	3.7	5.7	0	1.6	—
PMMA	31.2	—	—	1.8	—	9.5	1.6	0	—
PEO	33.0	9.8	—	—	—	9.5	—	—	0

[a]See Table A1 of the Appendix for polymer identification.

As a result of this roughness, an experimental average value ϑ_{exp} is measured instead of the theoretical contact angle ϑ. The roughness of the surface will tend to enlarge the liquid–polymer contact area. The opposed effects of cohesion and adhesion affect the response to an enlarged interface. In the case of liquids that spread poorly ($\vartheta > 90°$), cohesion predominates. The enlargement of the surface due to roughness is then counterbalanced by an increase in the contact angle ($\vartheta_{exp} > \vartheta$). With liquids that spread well ($\vartheta < 90°$), adhesion predominates. Therefore the liquid can cover a greater surface area on the roughened surface than on a smooth one, and the contact angle decreases ($\vartheta_{exp} < \vartheta$). The roughness can thus also be given as $r = \cos \vartheta_{exp}/\cos \vartheta$. The true contact angle ϑ can then be calculated from the roughness r, the true surface area, and the experimentally observed contact angle ϑ_{exp}.

The interfacial tensions between two liquid polymers of finite molecular weight are given in Table 13-2 and compared with the surface tensions of the individual polymers. It can be seen from Table 13-2 that the interfacial tensions are generally higher for increased polarity difference between the two polymers. However, the interfacial tensions are usually small. But the contact angles for polymer 1 on solid polymer 2 and vice versa can be very different. For example, the contact angle of poly(butyl methacrylate) on poly(vinyl acetate) is zero, but for poly(vinyl acetate) on poly(butyl methacrylate) it is 42°.

13.2.2.2. Critical Surface Tension

The contact angles in a homologous series of liquids vary systematically with regard to a given substrate. It was found empirically that the cosine of this contact angle varies linearly with the surface tension γ_{lv} of the liquid in contact with its saturated vapor (Figure 13-4). At a value of $\cos \vartheta = 1$, the limiting value of the surface tension corresponds to complete wetting of the substrate and is therefore termed the critical surface tension γ_{crit} of the substrate. This relationship between $\cos \vartheta$ and γ_{lv} applies, in the case of a given substrate, not only for a homologous series of liquids, but also quite well for liquids that are very different from one another. An example of this is found in the measurements on poly(ethylene) at 20°C with such different liquids as benzene ($\gamma_{lv} = 28.9 \times 10^{-5}$ N/cm), 1,1,2,2-tetrachloroethane (36.0), formamide (58.2), and water (72.0) (Figure 13-4). Thus, the critical surface tension γ_{crit} of the polymer appears to be almost constant for a given substrate.

The theoretical significance of the critical surface tension has not yet been established. Figure 13-4 defines γ_{crit} as the limiting value of γ_{lv} for $\cos \vartheta \to 1$. This corresponds to

$$\gamma_{\text{crit}} = \gamma_{sv} - (\gamma_{sl} + p_e) \tag{13-5}$$

according to the Young equation. The critical surface tension is only a true material constant when $\gamma_{\text{crit}} = \gamma_{sv}$.

The critical surface tension of all known solid polymers is lower than the surface tension of water at 72×10^{-5} N/cm (Table 13-3). All polymers are therefore relatively poorly wetted by water. The critical surface tension of polymers containing fluorine is particularly low, and they are poorly wetted by oils and fats as well as by water. Oils, fats, and glycerol esters

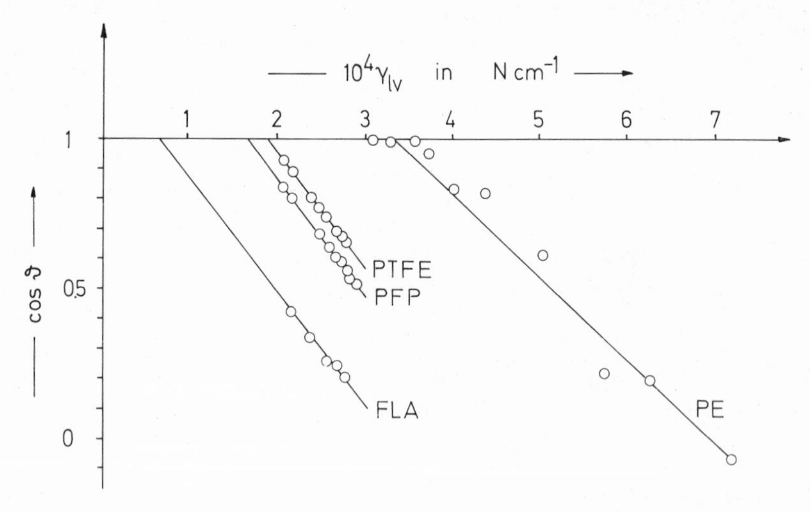

Figure 13-4. Dependence of the cosine of the contact angle ϑ on the interfacial surface tension γ_{lv} of the liquid used on different substrates in contact with air (20°C). PE, poly(ethylene); PTFE, poly(tetrafluorethylene); PFP, poly(hexafluoropropylene); FLA, perfluorolauric acid (monomolecular layer on platinum) (after R. C. Bowers and W. A. Zisman).

Table 13.3. Critical Surface Tension γ_{crit} of Clean, Smooth Polymers at 20°C

Surface	$10^5\gamma_{sv}$, N/cm	$10^5\gamma_{crit}$, N/cm
$—CF_3$	—	6
$—CF_2CF(CF_3)—$	—	16.2
$—CF_2—CF_2—$	19.0	18.5
$—CH_2—CF_2—$	30.3	25
$—CH_3$	—	22
$—CH_2—CHF—$	36.7	28
$—CH_2—CH_2—$	33.1	31
$—CH_2—CH(C_6H_5)—$	42.0	43
$—CH_2—CHOH$	—	37
$—CH_2—CHCl$	41.5	39
Poly(ethylene terephthalate)	41.3	43
Wool	—	45
Nylon 6,6	43.2	46
Urea/formaldehyde resin	—	61
Chlorinated wool	—	68
Poly(vinylidene chloride)	45	40
NaCl	—	150
Iron	—	1,200
Copper	—	2,700
Diamond	—	10,000

possess surface tensions of $\sim(20–30) \times 10^{-5}$ N/cm. Commercial use is made of this phenomenon, for example, in coating frying pans with poly(tetrafluoroethylene) to make them nonsticking.

13.3. Adsorption of Polymers

The adsorption, and hence the rate of adsorption, can be observed and determined, for example, by the increase in weight of the adsorbing material or the decrease in the concentration of solute remaining in the solution. A particularly elegant method of determining the rate of adsorption uses the change in buoyancy of the adsorbing material during the adsorption process. For this, the adsorbing material is hung on a sensitive balance in the form of films, and its change in weight is monitored continuously. The process of adsorption is not disturbed by this method. The rate of adsorption increases with increasing concentration and decreases with increasing solute molecular weight. The diffusion of the adsorbing material is therefore the rate-determining step of the adsorption process. If the system is stirred or agitated, then in some cases the deposition of a complete adsorption layer may be disturbed or the polymer may even be degraded.

The amount of adsorbed random-coil molecules increases steadily up to an ultimate value with increasing time according to this method. In the corresponding time, however, the viscosity of the remaining solution often goes through a minimum. The thickness of the layer (measured by ellipsometry) passes through a maximum. Ellipsometry is the measurement of the change in elliptically polarized light on reflection from a surface covered with an adsorbed layer.

These effects show that the adsorption equilibrium has not been reached when the adsorbed quantity has become constant. The reason for this is the high flexibility of long-chain macromolecules. An approaching macromolecule is first bound at many attachment points, and lies relatively smoothly on the surface. Some segments of a new, approaching macromolecule begin to crowd part of the region covered by segments previously adsorbed. As the area covered increases, therefore, the adsorbed macromolecules are raised more and more above the surface. The thickness of the adsorbed layer rises to a maximum, but the density of covering usually remains constant.

The quantity g_p of a polymer adsorbed per unit surface area usually increases sharply initially as the concentration c increases and reaches an ultimate value at higher concentrations. For technical reasons (e.g., high solution viscosity) this ultimate value is frequently not attained, as, for example, in the case of poly(isobutylene) with $\overline{M}_w = 50{,}000$ g/mol molecule in cyclohexane on aluminum at 22.5°C (Figure 13-5).

A marked stepwise adsorption is a characteristic observation with low-molecular-weight polymers (Figure 13-5). According to measurements

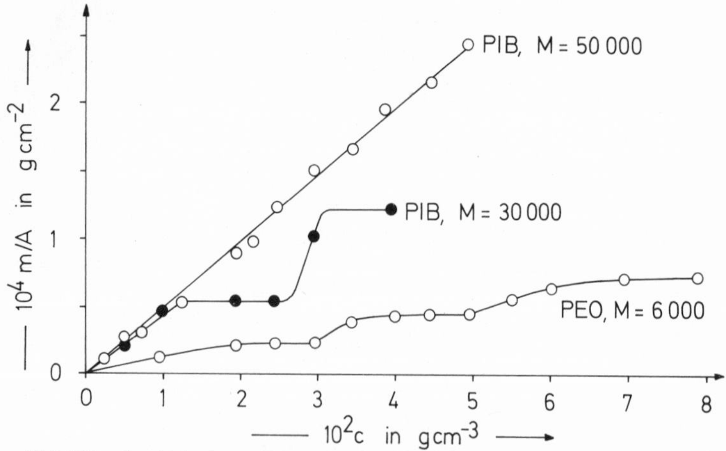

Figure 13-5. The adsorbed mass m/A per square centimeter of aluminum surface as a function of the concentration c of poly(isobutylene) (PIB) in cyclohexane (22.5°C) or poly(ethylene glycol) (PEO) in benzene (32.5°C) (after F. Patat).

on poly(ethylene glycols) in water on glass surfaces, the enthalpy of adsorption of the first layer is positive, but that of the second layer is negative. The entropy of adsorption of the first layer therefore must be positive. This increase in entropy can only come from desolvation of the macromolecule. The formation of layers is not caused by multiple layer adsorption, but is caused by some unknown ordering phenomenon.

It is difficult to conceive a molecular interpretation of the stepwise isotherms, because of a lack of systematic measurements on the thickness and density of the layers, the number of contact points, the structure of the layers, etc. It is observed, however, that the quantity adsorbed from a mixture is greater than that of the individual components.

13.4. Adhesives and Glues

13.4.1. Adhesion

In the strictly scientific sense, adhesion is the attraction between molecules across an interface. The receiving surface is termed the adherent and the material bound to it is the adhesive. According to this definition, the strength of the adhesion is determined by the number of contact points per unit surface area and the magnitude of the attraction forces at these points. In this definition, adsorption would be the decisive quantity, and it would only be necessary to allow for the forces between the adsorbent (the substrate) and the adsorbate (the substance bound to the surface).

In real systems, and particularly in technological processes, there are other important quantities besides adsorption, e.g., diffusion and/or chemical or electrical interactions. All these effects contribute to the observed adhesion. It is impossible to decide from measurements on adhesion alone which individual effect predominates or is present only as a contributing effect. If the adherent is completely covered with adhering groups, and every group occupies a site of 0.25 nm^2, then there are $\sim 5 \times 10^{14}$ groups/cm^2. With this number and the known bond strengths, strengths of 5000–25,000 kg/cm^2 are obtained for chemical bonds, 2000–8000 kg/cm^2 for hydrogen bonds, and 800–2000 kg/cm^2 for van der Waals bonds (dispersion forces, dipole forces). Experimentally, however, only up to 200 kg/cm^2 is found.

The type and extent of the interactions between adhesive and adherent are probably determined primarily by the physical state of the two materials. Here, three extreme cases can be distinguished in macromolecules, between which intermediate stages are, of course, possible. In the E/E type, both adherent and adhesive are in the viscoelastic state above the glass-transition temperature. In the G/G type, the two materials are below the glass-transition temperature. In the *G/E* type, the adherent is above, and the adhesive below, the glass-transition temperature. These physical states lead to the following consequences for the adhesion.

In the *E/E* type, the segments, and to some extent also the macromolecules of the adherent and the adhesive themselves, are able to move. Therefore, they can diffuse into one another. If adherent and adhesive are chemically equal, then self-diffusion is observed for this type. Self-diffusion leads to self-adhesion (autoadhesion).

The autohesion effect is especially good, therefore, when weak crystallization occurs on applying pressure or during annealing, as, for example, with natural rubber or with 1,5-*trans*-poly(pentenamers) (physical cross-linking). On the other hand, if the crystallization is too strong, the deformability of the adhesive is too small (see Section 7.4.2). If adherent and adhesive are chemically different, then in the *E/E* type this leads to interdiffusion and thence to heteroadhesion. Of course, marked interdiffusion is only possible when the different macromolecules are compatible with one another, and the strength of the autoadhesion or heteroadhesion depends on both diffusion and adsorption.

The *G/G* type is the other extreme. Since both materials are below the glass-transition temperature, the mobility of the segments is very restricted. The self-diffusion coefficients are estimated theoretically to be 10^{-21} cm^2/s, so that in the usual observation time diffusion effects would be very slight.

In the *G/E* type, similarly, there is only limited interdiffusion, as the adherent is below its glass-transition temperature. However, the chain ends of the adhesive have a certain mobility. They are able—particularly under pressure—to intimately cover the surface of the adherent so that a larger number of contact points are obtained. Adhesion in *G/E* types is therefore encouraged by roughening the adherent. Adsorption is very important in this type of adhesion. It is very difficult to decide whether diffusion or adsorption predominantly controls adhesion since both effects are approximately equally time and temperature dependent.

The bonding of a viscoelastic material (a film of glue) onto a solid surface can only be expected, then, when the surface tension γ_{lv} of the liquid is lower than the critical surface tension γ_{crit} of the solid body. According to equation (13-3), these two quantities are related to the contact angle ϑ and the interfacial surface tension ϑ_{sl} between solid and adhesive film. Since a chemical variation in the surface can also cause the surface tension to change, it is often possible to obtain better bonding through chemical modification of a surface. An example of this is the oxidation of the surface of polyolefins [see the critical surface tensions of poly(ethylene) and poly(vinyl alcohol) in Table 13-3].

13.4.2. Gluing

To achieve a good glue joint, the surfaces to be glued must be well wetted. The glue in the joint must then solidify. Finally, the glue joint must be sufficiently deformable so that stresses can equilibrate.

The strength of adhesion is usually measured through the yield stress of the bonded joint. Studies of this kind, however, only yield data on the strength of a bond when the overall deformation of the glue layer is equal. The material to be glued must not be deformable (Figure 13-6, II). When the material is readily deformed and the glue layer hardly deforms at all, then the bonded joint will be deformed much more readily at its extremities than in the middle. The points of weakness which then occur at the bond ends make the glue appear poor even when the adhesion is good. Thin films are often very difficult to glue, since they are readily deformed. To bond films, therefore, glues that are readily deformed must be used.

For the following discussion it will be assumed that the glue layer deforms more readily than the material to be glued. It must also be assumed that there are no chemical bonds between the material and the glue. The adhesion which is thus to be discussed depends mainly on adsorption and diffusion. The adhesive should be a clean material, and also it should be above its glass-transition temperature (if amorphous) or above its melting point (if partly crystalline). The lower the molecular weight of the adhesive, the more rapidly it can diffuse into the material to be bonded.

Adsorption, on the other hand, increases with an increasing number of contact points per adhesive molecule, i.e., with higher molecular weight. The adhesion should therefore exhibit an optimum value at a given molecular weight of the melt glue. A small number of branches per molecule of

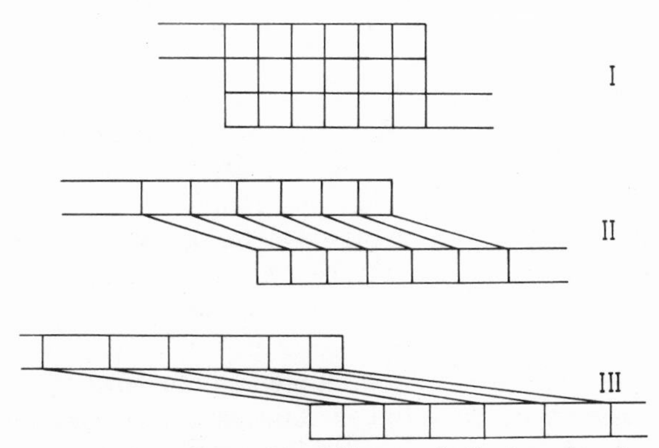

Figure 13-6. Schematic representation of the yield stress of adhesives when the adherent is more deformable than the glue layer.

adhesive lowers its melt viscosity and consequently increases the rate of diffusion. In the case of very highly branched molecules, on the other hand, fewer contact points can be formed per molecule of adhesive, so that adhesion should also pass through a maximum as branching increases.

Solutions of macromolecular materials are often used as glues. The solvent lowers the viscosity of the glue and simplifies its application. In addition, if the solvent is chosen correctly (conformity of the solubility parameters of material and solvent), it can swell or plasticize the adherent. The lowering of the glass-transition or melt temperatures which this produces encourages interdiffusion (transition from the G/E to the E/E type). After glueing has taken place, however, the solvent should no longer have any plasticizing effect. For this, it is necessary that the solvent diffuse out from the bond zone of the material–glue joint very easily; this can be achieved with low-molecular-weight, volatile materials. The plasticizing properties are also increased when polymerizable monomers are used as solvents. Naturally, a better bond will be formed between the adherent and the glue if it is possible for chemical bonds to be formed between them.

Melt glues are used above their melting or glass-transition temperatures. They are used, for example, in the textile industry to bond cover and support materials. Poly(ethylene), poly(ethylene-co-vinyl acetate), and terpolyamides are typical melt glues.

Glues can be classified as solid or soft glues according to the type and intended use. Solid glues are below the glass-transition or melt temperature after sticking; soft glues are above these temperatures. Among solid glues, cross-linked glues can be distinguished from non-cross-linked glues. Dispersions of poly(vinyl acetate) or starch solutions, for example, are used as non-cross-linked solid glues. Urea, phenolic and melamine resins, epoxy resins, unsaturated polyesters, and polyisocyanate glues act as cross-linked solid glues. In both types of solid glue, the action depends very much on the chemical nature of material and glue. In the case of soft glues, a distinction is made between contact glues and bonding glues. Contact glues are, for example, solutions of polar synthetic rubbers [such as poly(chlorobutadiene) or poly(butadiene-co-acrylonitrile)] or the polymers themselves (e.g., self-adhesive envelopes). Bonding glues are highly concentrated solutions of low-molecular-weight polymers, for example, of poly(isobutylene) or poly(vinyl ethers) or rubber decomposition products. They are used for adhesive tapes or sticking plaster.

In order to obtain good adhesion, it is usually necessary to prepare the surface of the material. The kind of preparation depends on the type of material and glue. With cross-linked solid glues, reactive groups can be produced on the material, for example, by oxidation with chromic acid or by a glow discharge. Since all glues are in a viscous or viscoelastic form when they are applied, roughening of the surface is always advisable.

Foreign-surface films must always be removed: adsorbed gases by evaporation, fats by means of organic solvents. Substances that facilitate bonding, known as wash primers, are used in the case of protective coatings. A typical wash primer consists of a dispersion of poly(vinylbutyral) (~40% butyral groups, 50% vinyl groups, 7% hydroxyl groups), to which a second resin (melamine resin, epoxide, etc.) is usually added. Wash primers have this name because they were "washed over" the iron decks of battleships before painting.

Literature

General

J. F. Danielli, K. G. A. Parthurst, and A. C. Biddiford, *Surface Phenomena in Chemistry and Biology*, Pergamon Press, New York, 1958.
I. R. Miller and D. Bach, Biopolymers at interfaces, *Surf. Colloid Sci.* **6**, 185 (1973).

Section 13.1. Spreading

W. D. Harkins, *The Physical Chemistry of Surface Films*, Reinhold, New York, 1952.
D. J. Crisp, Surface films of polymers, *in: Surface Phenomena in Chemistry and Biology* (J. F. Danielli, K. G. A. Pankhurst, and A. C. Riddifort, eds.), Pergamon Press, New York, 1958.
F. H. Müller, Monomolekulare Schichten, *in: Struktur und physikalisches Verhalten der Kunststoffe* Vol. 1 (R. Nitzsche and K. A. Wolf, eds.), Springer, Berlin, 1962.

Section 13.2.1. Surface Tension

G. L. Gaines, Jr., Surface and interfacial tension of polymer liquids, *Polym. Eng. Sci.* **12**, 1 (1972).
S. Wu, Interfacial and surface tension of polymers, *J. Macromol. Sci. C (Rev. Macromol. Chem.)* **10**, 1 (1974).

Section 13.2.2. Interfacial Tension of Solid Polymers

W. A. Zisman, Relation of the equilibrium contact angle to liquid and solid constitution, *Adv. Chem. Ser.* **43**, Am. Chem. Soc., Washington, 1964.

Section 13.3. Adsorption of Polymers

F. Patat, E. Killmann, and C. Schliebener, Die Adsorption von Makromolekülen aus Lösungen, *Fortschr. Hochpolym.-Forschg.* **3**, 332. (1961/64)
R. M. Screaton, Column fractionation of polymers, *in: Newer Methods of Polymer Characterization*, (B. Ke, ed.), Interscience, Wiley–New York, 1964.
Yu. S. Lipatov and L. M. Sergeeva, *Adsorption of Polymers* (in Russian), Naukova Dumka, Kiev, 1972; Wiley, New York, 1974.
S. G. Ash, Polymer adsorption of the solid/liquid interface, *in: Colloid Science*, Vol. I (D. H. Everett, ed.), Chem. Soc., London, 1973.
I. R. Miller and D. Bach, Biopolymers at interfaces, *Surface Colloid Sci.* **6**, 185 (1973).

Section 13.4. Adhesives and Glues

S. S. Voyutskii, *Autohesion and Adhesion of High Polymers*, Interscience, New York, 1963.

S. S. Voyutskii, Contact angle, wettability and adhesion, *Adv. Chem. Ser.* **43**, Am. Chem. Soc., Washington, 1964.

R. Houwink and G. Salomon, eds., *Adhesion and Adhesives*, Elsevier, Amsterdam, 1965.

R. S. R. Parker and P. Taylor, *Adhesion and Adhesives*, Pergamon Press, London, 1966.

R. L. Patrick, eds., *Treatise on Adhesion and Adhesives*, M. Dekker, New York, 1967–1969 (2 vols.).

J. J. Bikerman, *The Science of Adhesive Joints*, second ed., Academic Press, New York, 1968.

N. I. Moskvitin, *Physicochemical Principles of Glueing and Adhesion Processes*, Israel Program for Scientific Translations, Jerusalem, 1969.

D. H. Kaelble, *Physical Chemistry of Adhesion*, Wiley–Interscience, New York, 1971.

D. H. Kaelble, Rheology of adhesion, *J. Macromol. Sci. C (Rev. Macromol. Chem.)* **6**, 85 (1971).

P. E. Cassidy and W. J. Yager, Coupling agents as adhesion promoters, *J. Macromol. Sci. D (Rev. Polym. Technol.)* **1**, 1 (1972).

Journals

Adhäsion (Ullstein, Berlin)

Adhesives Age (Palmerton Publ., New York)

The Journal of Adhesion (Gordon and Breach)

Chapter 14
Electrical Properties

Matter can be classified according to its "specific" electrical conductivity σ into electrical insulators ($\sigma = 10^{-22}$–10^{-12} ohm^{-1} cm^{-1}), semiconductors ($\sigma = 10^{-12}$–10^{3} ohm^{-1} cm^{-1}), and conductors ($\sigma > 10^{3}$ ohm^{-1} cm^{-1}). The electrical conductivity is the reciprocal of the electrical resistance. Since electrical resistance is measured in ohms, the unit of conductivity is often written as mho instead of ohm^{-1} in American scientific literature.

Macromolecules with certain constitutional characteristics possess semiconductor properties (Section 14.2). The majority of the commercially used polymers, however, are insulators (Section 14.1). A consequence of their limited conductivity is that these polymers readily become electrostatically charged (Section 14.1.5). Specific conductivities are, for example, $\sim 10^{-17}$ ohm^{-1} cm^{-1} for poly(ethylene), 10^{-16} ohm^{-1} cm^{-1} for poly(styrene), and 10^{-12} ohm^{-1} cm^{-1} for polyamides (containing water?).

14.1. Dielectric Properties

When an electrical field is applied, the groups, atoms, or electrons of the insulator molecules are polarized. With stronger fields, electrons are displaced, giving rise to ions. With even stronger fields, the conductivity of the ions finally becomes so great that the material no longer shows any electrical resistance: It discharges. Electrical conduction need not only take place in the interior, it can also occur on the surface.

14.1.1. Polarizability

If a static electric field is applied to an nonconductor, then the electrons tend to be displaced relative to the atomic nuclei (electron polarizability). The corresponding displacement of atomic nuclei is called atomic polarizability. The electric moment $\boldsymbol{\mu}_i$ thus induced is directly proportional to the electric field $\mathbf{E}_i$, i.e., for the displacement polarizability (electronic and atomic polarizability)

$$\boldsymbol{\mu}_i = \alpha \, \mathbf{E}_i \tag{14-1}$$

Here α is the polarizability of the atom, group, or molecule. The greater α, the more energy will be adsorbed by the material.

Molecules with polar groups possess a permanent dipole moment $\boldsymbol{\mu}_p$. In these molecules, a static electric field produces an orientation polarizability, in addition to induced atomic or electron polarization; i.e., the most probable rest position for the permanent dipole lies preferentially in the direction of the field. Molecules with permanent dipoles thus often store more electrical energy than those with induced dipoles.

In general, polarizability is difficult to determine experimentally. However, the ratio of the capacity of a condenser in a vacuum to that in the medium under consideration, i.e., the dielectric constant of the medium, can be measured. At low frequencies, the dielectric constant of electrical nonconductors is almost independent of the frequency. At high frequencies, the dielectric constant depends on the frequency, since the permanent dipoles are no longer able to establish a preferred orientation, because of rapid alteration of the field.

14.1.2. Behavior in an Alternating Electric Field

In an alternating electric field, the dipoles of the dielectric medium attempt to align themselves in the direction of the field. The more rapidly the direction of the alternating field is changing, the less easily they are able to achieve this. The more the adjustment of the dipoles lags behind the applied alternating field, the greater is the electrical energy consumed in this effect (power loss). The available output power or voltage is thereby decreased, since power is lost by conversion into thermal energy.

The power loss depends on the phase difference between the alternating current produced by an applied alternating voltage. When the material behaves as a perfect dielectric, the phase difference between the alternating potential and the amplitude of the current is 90° and the power loss is zero. If current and voltage are in phase, then all of the electrical energy is converted into heat and the power output is zero. The ratio of power loss N_v to power output N_b is called the dielectric *dissipation factor*, $\tan \delta$:

$$\frac{N_v}{N_b} \equiv \frac{UI \cos (90 - \vartheta)}{UI \sin (90 - \vartheta)} \equiv \tan \delta \tag{14-2}$$

The real power output and the power loss can also be given in terms of the real ε' and imaginary ε'' dielectric constants (relative permittivities), respectively:

$$\varepsilon = \varepsilon' - i\varepsilon'' \tag{14-3}$$

when ε is the (complex) dielectric constant. The dissipation factor is

$$\tan \delta = \frac{\varepsilon''}{\varepsilon'} = \frac{\varepsilon \sin \delta}{\varepsilon \cos \delta} \tag{14-4}$$

ε' and ε'' may depend on the frequency ν. The function $\varepsilon' = f(\nu)$ corresponds to an energy storage and the function $\varepsilon'' = g(\nu)$ to an energy dissipation (Figure 14-1). The loss of energy per second, i.e., the power loss, is given as

$$N_v = \mathbf{E}^2 2\pi\nu\varepsilon \tan \delta \tag{14-5}$$

where $\mathbf{E}$ is the amplitude, ν the frequency of the alternating field, ε the relative permittivity (dielectric constant), and $\tan \delta$ the dissipation factor.

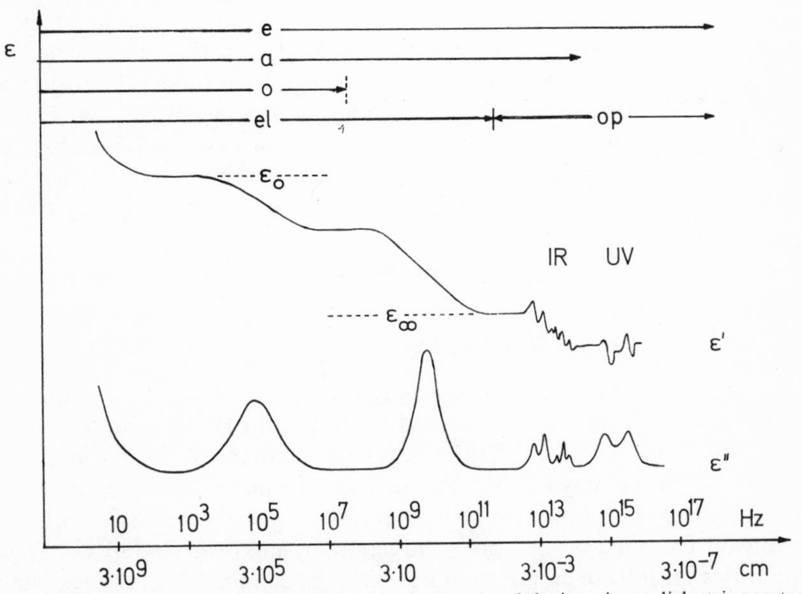

Figure 14-1. Dependence of the real dielectric constant ε' and the imaginary dielectric constant ε'' on the frequency or the wavelength (schematic); e, electron polarization; a, atomic polarization; o, orientation polarization; el, electrical region; op, optical region; IR, infrared region; UV, ultraviolet region.

The term $\varepsilon' \tan \delta$ is called the *loss factor*, and is not the same as the dissipation factor. Materials with a high $\varepsilon' \tan \delta$ are suitable for high-frequency-field heating, i.e., they can be welded in a high-frequency field. These materials are not suitable, on the other hand, as insulating materials for high-frequency conductors. Nonpolar plastics such as poly(ethylene), poly(styrene), poly(isobutylene), etc., have low dielectric constants ($\sim$2–3) and dielectric loss factors ($\tan \delta = 10^{-4}$ to 8×10^{-4}). As insulating materials they are of considerable importance in high-frequency-field technology. Polar materials such as poly(vinyl chloride), by contrast, have an $\varepsilon' \tan \delta$ that is at least 100 greater than the corresponding value of poly(styrene) or poly(ethylene). Therefore PVC can be welded extremely well using high-frequency fields.

The glass-transition temperature and other relaxation temperatures can be determined by investigating the behavior of polar macromolecules in an alternating electric field (see also Section 11.4.5). If the frequencies are low and the sample is above the glass-transition temperature, then the dipoles oscillate in phase with the alternating field. At high frequencies and below the glass-transition temperature this is no longer possible. The behavior of a polar macromolecule in an alternating field also depends on whether the dipoles are in the chain or in pendant side groups. In the case of poly(oxymethylene), $(\text{—}CH_2\text{—}O\text{—})_n$, the dipoles are in the chain. They are only capable of orienting themselves, therefore, when the segment mobility is high, i.e., only above the glass-transition temperature. In materials such as poly(vinyl ethers), $(\text{—}CH_2\text{—}CHOR\text{—})_n$, on the other hand, the dipoles are in pendant side groups. The orientation of these flexible groups, therefore, can take place either via segmental motion in the main chain or else by side-group movement. Thus, two areas of dissipation are observed here: at low frequencies, that due to segmental mobility, and at high frequencies, that arising from orientation of the side groups. If the experiment is carried out below the glass-transition temperature, then, of course, only the effect due to mobility of the side groups will be observed.

14.1.3. Dielectric Field Strength

Heat is developed within the polymer by the imaginary component of the dielectric constant. If the field is allowed to be effective for a very long time, then, because of the poor heat conductivity of the material, the heat produced may not be dissipated and the material may become hot. The imaginary component results from out-of-phase orientation of polar groups in the polymer or from conduction arising from impurities. These impurities must be of an ionic nature, since the conductivity of the polymer depends very much on the temperature. On the other hand, the electronic conductivity varies much less with temperature. Because of the strong dependence on temperature of the ionic conductivity, the accumulation of heat leads to increasingly better conductivity until, finally, a threshold (thermal breakdown) is reached. Thin films show a higher dielectric field strength than thick films, because of the better heat dissipation.

14.1.4. Tracking

Tracking is defined as a leakage current which flows across the surface of a material that is a good insulator in the dry, clean state, and occurs between charged particles consisting of conductive impurities. Since the surface resistance is two orders of magnitude lower than the specific resistance and is usually difficult to measure, the discharge resistance is tested under standard conditions. For this, a "normal impurity" is used, namely, a salt solution with an added detergent. This test solution is allowed to drip uniformly between electrodes at a standard potential difference and is separated by a standard distance on the surface of the test sample. The number of drops required before a discharge occurs is a measure of the discharge resistance. Initially, the surface becomes contaminated with conducting impurities, and when the contamination becomes sufficient, sparking occurs, which can cause decomposition or carbonization. Tracking then occurs.

Tracking does not occur in a material that does not carbonize under these conditions. If, for example, sparking produces monomer from the polymer by depolymerization, then the vaporizing monomer molecules prevent contaminating salt films from being deposited. Sparking occurs between gaps in the conducting film; an example of this is poly(methyl methacrylate). The same effect is shown when volatile degradation products are formed under the influence of sparking or arcing as, for example, in the case of poly(ethylene) or polyamides. Poly(*N*-vinylcarbazole), on the other hand, does not form volatile degradation products, and, therefore, has only a poor discharge resistance despite its good insulating properties.

14.1.5. Electrostatic Charging

Materials become charged electrostatically when the specific electrical conductivity is lower than $\sim 10^{-8}$ mho/cm. Thus, if two plastics are rubbed together or a plastic is rubbed against metal, different degrees of charging will be observed, depending on the materials rubbed together and the duration of rubbing. Poly(oxymethylene) rubbed against polyamide 6, for example, produces a charge of $+360$ V/cm on just one rubbing, a value of 1400 V/cm with 10 rubbings, and finally a limiting value of 3000 V/cm. If an ABS polymer with added antistatic agent is rubbed against poly(acrylonitrile), the extreme value is 120 V/cm. The same ABS polymer, however, has a limiting value of −1700 V/cm when rubbed against polyamide 6. The test methods are standardized.

The origin of this behavior is the transfer of electrical charges. Regions with both positive and negative charges are thus produced on the surface of the material, as has been established by dusting with differently charged dyes. As a rule, however, one kind of charge predominates, so that the surface appears to be positively or negatively charged. The sign of the overall charge depends on the position of the rubbing partner in the triboelectric series of nonmetallic materials (Table 14-1).

Because of the poor surface conduction of most macromolecular materials, the charge which is produced can only flow away slowly. The half-lives for such dissipation of the charge are usually different for positively and negatively charged substances (Table 14-2). The half-lives, which are often high, frequently have unpleasant effects in industry and in the household, for example, in the charging of godet rolls during spinning processes or in the accumulation of dust on household articles made from plastics.

The electrostatic charging effect can be prevented by various methods. One group of methods "grounds" the charges, e.g., by neutralizing them with ionized air (as in the textile industry) or by encasing rubber pipes in metal sheaths (as at gas stations). Alternatively, the materials can be protected externally or internally with antistatic agents. If, for example, up to 30% carbon black is worked into a copolymer of ethylene and vinylidene chloride, the material still retains practically all the good properties of the plastic, but this additive increases the specific conductivity up to about 10^{-2} mho/cm. The material no longer becomes electrostatically charged. In the case of an external antistatic treatment, materials that utilize the humidity of the air are applied to the surface. Contrary to internal addition, this form of treatment does not affect the specific conductivity, but does

Table 14-1. Triboelectric Series of Nonmetallic Materials

Material	Charge density, 10^{-6} C/g
Melamine resin	−14.70
Phenolic resin	−13.90
Graphite	−9.13
Epoxy resin	−2.13
Silicone rubber	−0.18
Poly(styrene)	+0.37
Poly(tetrahydrofurane)	+3.41
Poly(trifluorochloroethylene)	+8.22

Table 14-2. Half-Lives for Charge Loss from Charged Materials

Material	Half-lives, s	
	Positively charged	Negatively charged
Cellophane	0.30	0.30
Wool	2.50	1.55
Cotton	3.60	4.80
Poly(acrylonitrile)	670	690
Polyamide 66	940	720
Poly(vinyl alcohol)	8500	3800

change the surface resistance. Of course, external antistatic treatments are not permanent, and retreatments are required from time to time. The electrostatic charge can also be lessened if friction is lowered, e.g., by adding lubricants or by coating with poly(tetrafluoroethylene).

Conversely, the effects of the electrostatic charge can also be used commercially, namely, in electrostatic paint spraying and in the flocking of fabrics to produce velvety surfaces.

14.1.6. Electrets

Electrets are dielectric bodies that can retain an electric field for a certain time after it has been applied. They are only formed by polymers with poor electrical conductivity, for example, poly(styrene), poly(methyl methacrylate), poly(propylene), polyamides, or carnauba wax.

There are two procedures for the manufacture of electrets. In the first method, the polymer is heated to temperatures above the glass-transition temperature and then an electric field is applied (e.g., 25 kV/cm) and the polymer is allowed to form a glass while still under the influence of the field. An optimum working temperature seems to be at ~37°C above the glass-transition temperature T_G. In the second method, the polymer is allowed to glassify while flowing under pressure. Here, the optimum temperature appears to be at $(T_G + 57)$°C. When the electric field is removed, the bodies are positively charged on one side and negatively charged on the other. The difference in the charge diminishes only slowly, in a process that can extend over months.

As yet, the principles of electret formation are not fully understood. It is likely that both volume and surface polarizations can occur. A volume polarization is obtained with fields below ~10 kV/cm. That is, if an electret is parted parallel to the charged surface, two new capacitors result. With fields of more than ~10 kV/cm, an ionization, electronic failure, or breakdown due to the field takes place, giving a surface polarization. The polarizations at different field strengths also support this interpretation. At low field strengths, polarization opposes the electric field, which may be due to a charge migration by, for example, ionic impurities. At temperatures above the glass-transition temperature, the ion separation should occur readily, and then at $T < T_G$ the ion positions become frozen in. At high field strengths, air is ionized, and the surfaces of the electrets are polarized in the same sense as used in the case of electrodes.

14.2. Electronic Conductivity

14.2.1. Influence of Chemical Structure

Polymers with delocalized π electrons are usually semiconducting compounds. The specific conductivity depends on two factors: the transport of the individual charge within the macromolecule itself and the transport from molecule to molecule.

For good intramolecular electron transport to be achieved, the molecule must be as planar as possible. The more extended the delocalized π-electron system, the better will be the conductivity. The specific conductivity thus increases sharply in the series coronene, ovalene, circumanthracene, graphite (Table 14-3). The activation energy $E_0^{\ddagger}$ of the electronic conduction, which can be calculated using

$$\sigma = \sigma_0 \exp \frac{-E_\sigma^{\ddagger}}{2kT} \tag{14-6}$$

falls sharply in a similar fashion. This formula corresponds to the expression derived from Brillouin zone models. The formula is sometimes written without the factor 2, which also has to be considered in comparisons of activation energies.

Incorporated heteroatoms perturb the π resonance system, thus increasing conductivity (violanthrene as opposed to violanthrone). Poly(carbazene) and poly(azasulfene) also show considerably higher conductivity than poly(vinylene). It can be quite generally assumed that all conducting and semiconducting polymers are conjugated, but not all the (formally) conjugated polymers are semiconductors.

The intermolecular passage of electrons is made easier when the molecular chains are in a high state of order. Crystalline poly(acetylene) thus has a specific conductivity which is four orders of magnitude higher than that of amorphous poly(acetylene). For high levels of crystallization to be possible, the crystal unit cells must be constructed as simply as possible.

The passage of electrons occurs more readily when the macromolecules are cross-linked. Such cross-linking systems with conjugated double

Table 14.3. Specific Conductivity of Polymers and Low-Molecular-Weight Compounds[a]

Material		Temp., °C	Specific conductivity, ohm^{-1} cm^{-1}	Activation energy, eV
Name	Chemical constitution			
Cellulose, dry	—	25	10^{-18}	?
Gelatine, dry	—	130	2×10^{-14}	3.1
Tobacco mosaic virus	—	130	9×10^{-14}	2.9
Deoxyribonucleic acid	—	130	2×10^{-12}	2.4
Coronene	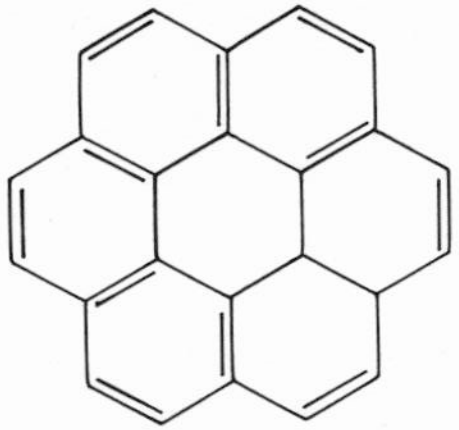	15	6×10^{-18}	0.85
Ovalene	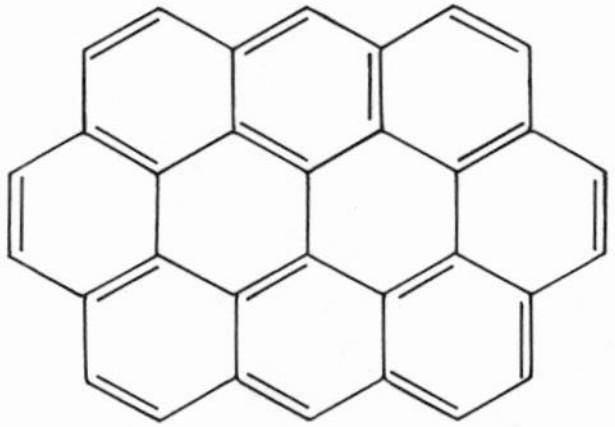	15	4×10^{-16}	0.55

continued overleaf

[a] T_p is the pyrolysis temperature; 1 eV = 1.6021×10^{-19} J.

Table 14-3 (Continued).

Material		Temp., °C	Specific conductivity, $ohm^{-1}\ cm^{-1}$	Activation energy, eV
Name	Chemical constitution			
Circumanthracene	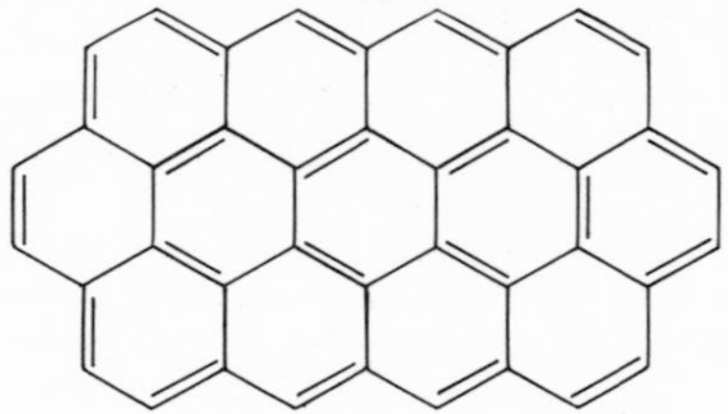	15	2×10^{-13}	?
Graphite	—	25	10^{5}	0.025
Violanthrene	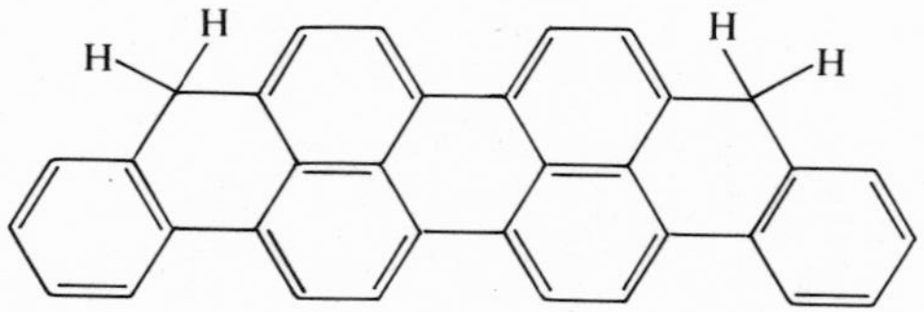 H H H H	15	5×10^{-15}	0.43
Violanthrone	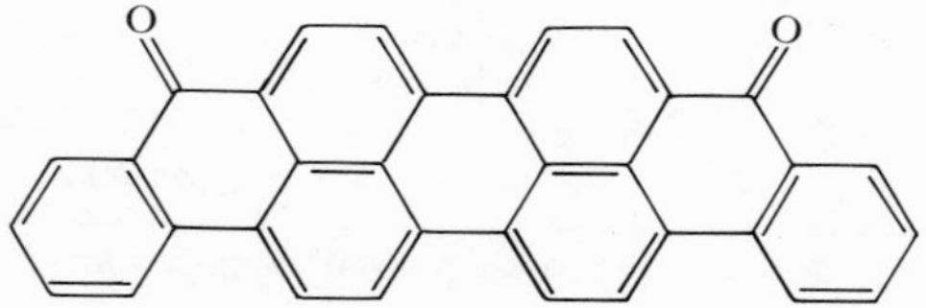 O O	15	4×10^{-11}	0.39

Poly(methylene)	$-(CH_2)_n-$	25	$<10^{-17}$	?
Poly(vinylene)	$-(CH{=}CH)_n-$	25	$<10^{-8}$	?
Poly(acetylene)	$-(C{\equiv}C)_n-$	25[b] 25[c]	$<10^{-8}$ $<10^{-4}$	0.83 ?
Poly(phenylene)	$-(C_6H_4)_n-$	25	10^{-11}	0.94
Poly(*p*-divinylbenzene)[d]	$-(CH_2-CH)_n-$ / C_6H_4 / $-(CH_2-CH)_n-$	25	10^{-15} 10^{-12} 10^{-6} 10^{2}	? ? ? ?
Poly(carbazene)	$-(N{=}CR)_n-$	25	$\sim10^{-5}$	~0.2
Poly(azasulfene)	$-(NS)_n-$	25	~8	~0.02

[b]Amorphous.
[c]Crystalline.
[d]Oxidized and pyrolized at 500, 600, 700, and 1000°C, respectively, for the four values listed for specific conductivity.

bonds are produced, for example, by the oxidation and subsequent pyrolysis of poly(*p*-divinylbenzene). Occasionally, however, cross-linked polymers are also prepared unintentionally in some syntheses. Linear poly(*p*-phenylenes), for example, occur when *p*-dihalogenbenzene is condensed with sodium. However, the reaction products have only a relatively low conductivity of $\sigma = 1.6 \times 10^{-11}$ mho/cm (25°C) and a relatively high activation energy of $E_\sigma^\ddagger = 0.94$ eV. On the other hand, when benzene is polymerized with Friedel–Crafts catalysts, it is more probable that cross-linked or highly branched polymers will be obtained, because at $\sigma = 0.1$ mho/cm ($E_\sigma^\ddagger = 0.025$ eV), the specific conductivity is almost as high as that of the condensation product from hexachlorobenzene with sodium ($\sigma < 5$ mho/cm).

The intermolecular transfer of electrons is also made easier by the formation of electron donor–electron acceptor complexes. The complex of poly[(styrene)$_{45\%}$-co-(1-butyl-2-vinyl pyridine)$_{55\%}$] and tetracyano-*p*-quinodimethane (with 15% tetracyano-*p*-quinodimethane), for example, possesses a specific conductivity of $\sigma = 1 \times 10^{-3}$ mho/cm. Contrary to the cross-linked, semiconducting polymers, these products are also soluble and can be cast into films.

14.2.2. *Measuring Techniques*

The specific conductivities of organic semiconductors extend into the range of those of metalloids or metals. The concentrations of charge carriers are also, at 10^9–10^{21} particles/cm^3, almost as high in some cases as those of metals (10^{21}–10^{22} particles/cm^3). The mobility of the charge carriers, on the other hand, is 10^{-6}–10^2 cm^2 V^{-1} s^{-1}, in general, considerably lower than that of metals and inorganic semiconductors (10–10^6 cm^2 V^{-1} s^{-1}). It is therefore doubtful whether the simple Brillouin zone model used for inorganic semiconductors can be used with organic semiconductors. With this zone model, it is assumed that electron clouds exist, which implicitly presumes a high charge-carrier mobility. For this reason, a modified zone model is discussed in which wave quanta made up of electrons and phonons* are considered to move. According to this model, there are lattice defects and deformations. The charge carriers trapped within the defects would then jump discontinuously from defect to defect. The relatively low activation energies of $E^\ddagger = 0.03$ eV at average specific conductivities of $\sigma = 10^{-8}$ mho/cm could confirm this defect-jump model.

To characterize the electrical properties of macromolecular semiconductors, the conductivity, the activation energy of the conductivity,

*The thermal oscillation of a lattice is considered as the movement of an elastic body with the energy $h\nu$. This is known as a phonon.

the concentration of free radicals, and the thermal electromotive force (EMF) are usually measured. In this, it is necessary to establish that the sample possesses no ionic conductivity (impurities!) and no surface conductivity. Even dampness can increase the conductivity by several orders of magnitude. Since the substances are usually in the form of amorphous powders, they are compressed into tablets. The contacts are either metal electrodes in contact under pressure with the surface or, for example, conductive pastes.

To determine the thermal EMF, the sample is placed between two plates at different temperatures. The thermal voltage that occurs with a difference in temperature of 1°C is called the Seebeck coefficient. The Seebeck coefficient is positive when the hotter pole is positive. A positive Seebeck coefficient originates from an excess of electron defects (*p*-type conductivity) and a negative coefficient from an excess of conducting electrons (*n*-type conductivity). The concentration of free radicals, measured using electron-spin resonance, need not be identical, or course, with the concentration of conducting electrons.

Literature

Section 14.1. Dielectric Properties

N. G. McCrum, B. E. Read, and G. Williams, *Anelastic and Dielectric Effects in Polymeric Solids,* Wiley, London, 1967.

M. E. Baird, *Electrical Properties of Polymeric Materials,* Plastics Institute, London, 1973.

P. Hedvig, *Dielectric Spectroscopy of Polymers,* Halsted, New York, 1975.

E. Fukada, Piezoelectric dispersion in polymers, *Prog. Polym. Sci. Jpn.* **2**, 329 (1971).

Section 14.2. Electronic Conductivity

B. A. Bolta, D. E. Weiss, and D. Willis, *in: Physics and Chemistry of the Organic Solid State,* Vol. II (D. Fox, ed.), Wiley–Interscience, New York, 1965, p. 67.

J. E. Katon, eds., *Organic Semiconducting Polymers,* M. Dekker, New York, 1968.

R. H. Norman, *Conductive Rubbers and Plastics,* Elsevier, Amsterdam, 1970.

E. Fukada, Piezoelectric dispersion in polymers, *Prog. Polym. Sci. Jpn.* **2**, 329 (1971).

H. Meier, Zum Mechanismus der organischen Photoleiter, *Chimia* **27**, 263 (1973).

Ya. M. Paushkin, T. P. Vishnyakova, A. F. Lunin, S. A. Nizova, *Organic Polymeric Semiconductors,* Wiley, New York, 1974.

E. P. Goodings, Polymeric conductors and superconductors, *Endeavour* **34,** 123 (1975).

Chapter 15

Optical Properties

The optical properties of a material depend on its interaction with the electromagnetic field of the incident light. A great many optical properties result from the fact that this interaction involves a whole series of molecular parameters. Two principal groups of optical properties can be distinguished: those resulting from molecular property averages and those caused by local deviations from these averages. Refraction, absorption, and diffraction phenomena belong to the first group; scattering effects belong to the second group. There are many relationships between the two groups; consequently the phenomena under consideration are often only limiting cases.

15.1. Light Refraction

If a ray of light is incident on a transparent body at an angle α with respect to the normal to its surface, it passes inside the body and is found to form a different angle β with respect to that normal (Figure 15-1): The light is "refracted." The refractive index n is a numerical measure of the refraction and depends on the angle of incidence α and the angle of refraction β:

$$n = \frac{\sin \alpha}{\sin \beta} = \frac{\sin \alpha'}{\sin \beta'} \qquad (15\text{-}1)$$

The refractive index varies with the wavelength of the incident light. The Abbé number v is given as a measure of this "dispersion." v is obtained from three refractive index measurements at the wavelengths 656.3, 589.3,

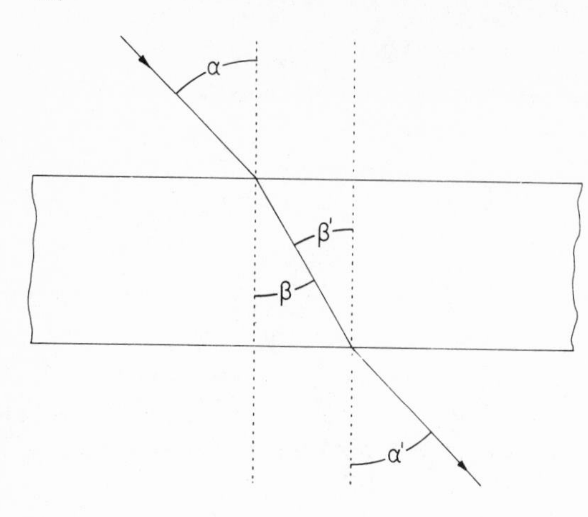

Figure 15-1. Definition of the angle of incidence α and angle of refraction β; for light incident on a homogeneous plate with plane parallel sides, $\beta = \beta'$, and if the two sides are in contact with the same medium, $\alpha = \alpha'$, i.e., the ray emerges parallel to its initial direction but displaced a distance that depends on the refractive index n.

and 486.1 nm:

$$v = \frac{n_{589} - 1}{n_{486} - n_{656}} \tag{15-2}$$

The capacity to separate the colors of white light increases as v decreases.

The refractive index n of a material depends, according to the Lorenz–Lorentz relationship, on the polarizability **P** of all the molecules residing

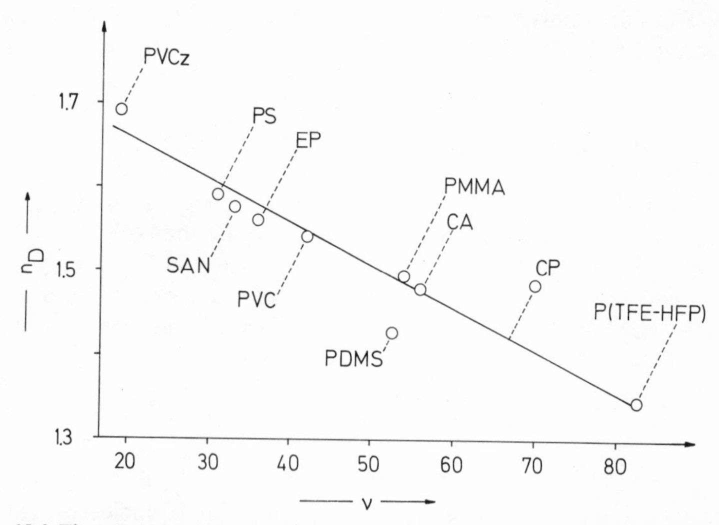

Figure 15-2. The relationship between the refractive index n_D for the sodium D line (589.3 nm) and the Abbé dispersion v for various polymers [see equation (15-2)]. PVCz. Poly(N-vinyl carbazole); P(TFE-HFP), copolymer of tetrafluoroethylene and hexafluoropropylene. For other abbreviations, see Table A1, Appendix.

in a uniform field:

$$\frac{n^2 - 1}{n^2 - 2} = \frac{4}{3}\pi P = \frac{4}{3}\pi N\alpha = \frac{4}{3}\pi N \frac{\mu}{E} \tag{15-3}$$

The polarizability P is given by the number N of molecules per unit volume and the polarizability α of the isolated molecule. The polarizability α in turn, depends on the dipole moment μ induced by an electric field of strength E. Consequently, both α and n increase with increasing number and mobility of electrons in the molecule. Thus, carbon has a much higher polarizability than hydrogen. Since for this reason the hydrogen contribution to the polarizability can, to a first approximation, be ignored, most carbon–carbon chain polymers have about the same refractive index (1.5). Deviations from this "normal value" only occur when there are large side groups [e.g., poly(N-vinyl carbazole)] or if highly polar groups are present (i.e., fluorine-containing polymers) (see also Figure 15-2). Further, on the basis of the molecular structure, one can estimate that the refractive indices of all organic polymers should lie within a range of only 1.33–1.73.

15.2. Light Interference and Color

15.2.1. Basic Principles

Some of the light incident on a homogeneous, transparent body is reflected from the surface (external reflection) and some passes inside, where it is reflected at an interior boundary of the body (internal reflection).

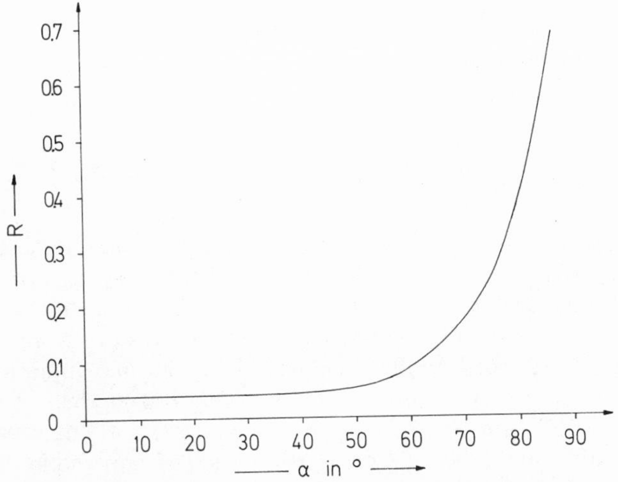

Figure 15-3. The reflectivity R as a function of the angle of incidence α for a material with $n = 1.5$.

According to Fresnel, the relationship between the intensity of the reflected light I_τ, and the intensity of the incident light I_0 involves the angle of incidence α and the angle of refraction β (for angle definitions, see Figure 15-1):

$$R = \frac{I_\tau}{I_0} = \frac{1}{2}\frac{\sin^2(\alpha - \beta)}{\sin^2(\alpha + \beta)} + \frac{\tan^2(\alpha - \beta)}{\tan^2(\alpha + \beta)} \tag{15-4}$$

The reflectivity R is small for small α and begins to increase sharply for high α (Figure 15-3).

15.2.2. Iridescent Colors

Iridescent colors can arise from the interference of light incident on films consisting of many layers. At each interface between layers, only a small portion of the incident light is reflected. The rest (neglecting absorption) passes through to the next interface, where it is again partly reflected and partly transmitted. If all the layers are of equal thickness, then for particular wavelengths (which depend on the optical densities and thickness of the films) the light reflected at the interfaces may be in phase, so that constructive interference occurs and the reflected light will have high intensity. If two polymers of refractive indices n_1 and n_2 make up alternate layers of thickness d_1 and d_2, then the wavelength λ_m of the $\bar{m}$th-order reflections for perpendicularly incident light is given by

$$\lambda_m = \frac{2}{m}(n_1 d_1 + n_2 d_2) \tag{15-5}$$

The relative intensities of the individual wavelengths depend on the optical density fractions of the two polymers, i.e.,

$$f_1 = \frac{n_1 d_1}{n_1 d_1 + n_2 d_2} \tag{15-6}$$

When the optical densities are equal ($f_1 = f_2 = 0.5$), the reflections will be suppressed for even-number orders and they will be of maximum intensity for odd orders. If $f_1 = 0.33$, on the other hand, third-order reflections are suppressed. The first-order reflections are still strong, and the second-, fourth-, etc., order reflections have less than maximum intensity. When the first-order reflection is $\lambda_1 = 1\ \mu\text{m}$ for $f_1 = 0.50$, there is a strong reflection for $(1.5/3)\ \mu\text{m} = 0.5\ \mu\text{m}$, no reflection for $(1.5/4)\ \mu\text{m} = 0.375\ \mu\text{m}$, etc. Such a film would reflect in the near-infrared (1.5 μm) and in the blue–green (0.5 μm).

Bandwidths are larger for variable layer thickness. In some circumstances, the whole visible spectrum is reflected when there is a suitable number of layers of two polymers with the right choice of refractive indices and the right choice of the various layer thicknesses. Such films have a metallic appearance.

15.3. Light Transmission and Reflection

15.3.1. Total Reflection

Total reflection occurs when the incident light is reflected without loss. This is especially important in the case of total internal reflection, since this principle is used in what is called fiber optics.

Total internal reflection only occurs above a quite specific minimum (critical) internal angle of incidence. The relationship $\sin \alpha \geqq 1/n_1$ is valid for a material of refractive index n_1 in air. Consequently, $\alpha_{crit} = 42°$ for $n_1 = 1.5$. The light is totally reflected on the interior interface and in a suitable array will be transmitted in a zigzag path through the system (Figure 15-4).

When the light transmitter is surrounded by air, the optically effective external surface is free. Surface scratches and dust deposits lead to light scattering and, consequently, loss of light intensity. Therefore, a smooth housing consisting of a transparent material of lower refractive index n_2 is used. The refractive index difference $n_1 - n_2$ should be as large as possible since it determines the light entry angle $2\alpha_0$ via

$$n_0 \sin \alpha_0 = (n_1^2 - n_2^2)^{0.5} \tag{15-7}$$

The entry angle $2\alpha_0$ is the maximum entry angle for the transmission of light through the light transmitter in a surrounding medium of refractive index n_0 (see Figure 15-4). For example, a core of poly(methyl methacrylate) and covering of partially fluorinated polymers, and a core of high-purity soda glass in a covering of poly(tetrafluoroethylene-co-hexafluoropropylene), have been introduced technologically for the transmission of, respectively, visible light and ultraviolet light.

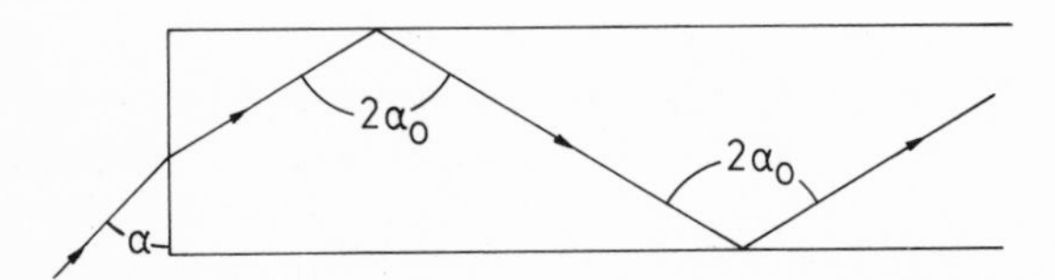

Figure 15-4. The principle of light transmission inside a body. $2\alpha_0$ is the entrance angle.

With flexible bundles of light transmitters, light can, for example, be transmitted "around corners," and one can even "look around corners." Light transmitters are used in medicine to illuminate or observe internal organs, in industry for rear lights on autos, and for the postmarking of postage stamps on letters and parcels, etc.

15.3.2. Transparency

When light is incident perpendicularly on an optically homogeneous sample, since the angles $\alpha = 0$ and $\beta = 0$, the Fresnel equation (15-4) reduces to

$$R_0 = \frac{(n - 1)^2}{(n + 1)^2} \tag{15-8}$$

The internal transmittance τ_i (transmittivity, transparency) is thus

$$\tau_i = 1 - R_0 \tag{15-9}$$

The refractive index is about $n \approx 1.5$ for most polymers. Consequently, the transparency can be a maximum of 96%, with at least 4% of the light being reflected at the polymer–air interface.

This ideal transparency is only rarely achieved, since the light is always absorbed and/or scattered to some extent. The most transparent polymer, poly(methyl methacrylate), has a maximum transparency of 92% (Figure 15-5) and this only in the range of about 430–1110 nm. On each side of this range, the transparency decreases because of absorption. Polymers generally absorb infrared radiation. Exceptions are the halogenated poly(ethylenes).

The "hiding power" of a paint can also be estimated with the Fresnel equation. The corresponding refractive indices n_1 of the pigment and n_2 of the polymer must be considered with pigmented paints:

$$R_0 = \frac{(n_1 - n_2)^2}{(n_1 + n_2)^2} \tag{15-10}$$

Consequently, the hiding power increases with increasing infractive index difference. For this reason, rutile (a TiO_2 modification with $n_D = 2.73$) is almost exclusively used when a high-quality white pigment is required, since it has the highest refractive index of all white pigments. The hiding power increases exponentially with the refractive index difference, so small variations in the polymer composition can lead to great changes in the hiding power.

Hiding power is not only influenced by the reflection; it is also influenced by light scattering. Light incident on a particle will be scattered

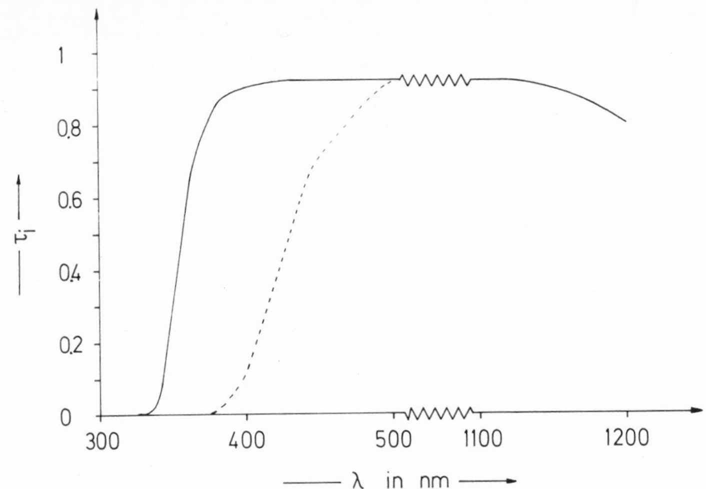

Figure 15-5. Internal transmittance as a function of wavelength for poly(methyl methacrylate). The maximum possible transparency of 96% is almost achieved in the region 430–1115 nm. The broken line indicates the shift obtained with a UV absorber.

in all directions (see Section 9.5). The scattering intensity increases with the particle size. Back-scattering decreases with particle size, and a large back-scattering is desirable for good hiding power. Consequently, the hiding power as a function of particle size passes through a maximum. The scattering intensity increases with increasing pigment concentration. If the pigment concentration is too high, the same light ray will be scattered many times. Multiple scattering lowers the relative scattering intensity, and the hiding power decreases. This loss in hiding power becomes marked when the interparticle distance becomes less than three times the particle diameter.

15.3.3. Gloss

Gloss is defined as the ratio of the reflection of the sample to that of a standard. In the paint industry, for example, the standard is a sample of refractive index $n_D = 1.567$. Consequently, the gloss as a ratio of two reflections depends, according to equation (15-4), on the refractive indices of the test sample and the standard, as well as on the angles of incidence and refraction of the light (Figure 15-6). The gloss of the polymer increases with increasing refractive index.

The theoretically maximum possible gloss calculated in this way is only rarely achieved in practice. The surfaces are always a little uneven. Also, optical inhomogeneities below the surface, that is, in the medium itself, cause marked light scattering. The relative fraction of light scattered

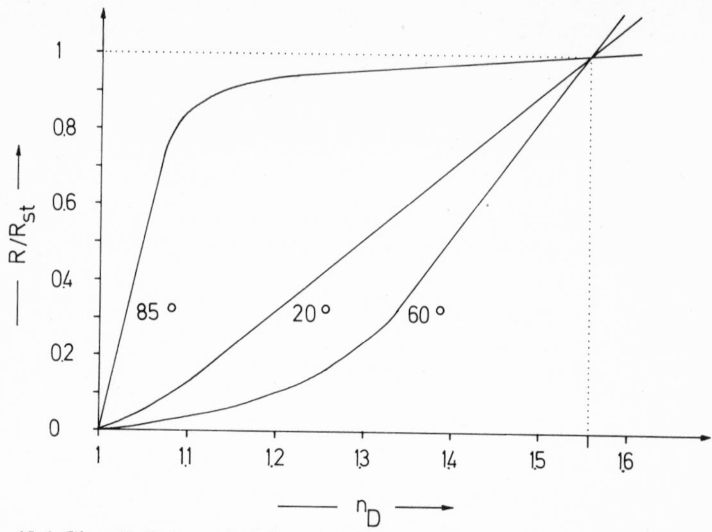

Figure 15-6. Gloss (R/R_{st}) as a function of refractive index n_D of the sample for various angles of incidence. A body with $n_D = 1.567$ is chosen as standard.

from the surface and from the medium depends on the angle of incident light. In general the two contributions can be separated by measuring the scattering in air and then, after immersion of the test sample, in a liquid having the same refractive index as the test sample. By subtraction, the light scattering fraction coming from the surface of the test sample can be obtained.

15.4. Light Scattering

15.4.1. Phenomena

All considerations so far discussed are only valid for optically homogeneous systems. In optically inhomogeneous systems, the medium acts in two ways on an electromagnetic wave traversing it. On the one hand, the amplitude and the phase of the wave are altered, that is, the wavefront is distorted. In optical terms, there is a loss in "resolving power"; in the plastics industry terminology, there is a loss in "clarity".

On the other hand, the electromagnetic wave loses some of its energy by scattering on encountering an inhomogeneity (see Section 9.4). The contrast lost because of forward scattering is called "haze". The combination of contrast loss by forward and backward scattering makes a sample "milky".

The internal transmittance of optically inhomogeneous materials is given by the reflected R, scattered f_{sc}, and absorbed f_{abs} fractions:

$$\tau_i = 1 - R - f_{sc} - f_{abs} \tag{15-11}$$

The reflected fraction can be eliminated by immersing the sample in a liquid of the same refractive index. In this case, the changes in both scattering and absorption are proportional to the thickness of the sample L. The proportionality coefficient is, consequently, given by the absorption coefficient K and the scattering coefficient, the so-called turbidity S:

$$\tau_i = 1 - (K + S)L \tag{15-12}$$

Equation (15-12) assumes a once-only scattering. For the general case, one starts from an infinitely thin sample, and integration gives

$$\tau_i = \exp[-(K + S)L] \tag{15-13}$$

Previously, the sum $K + S$ was known as the extinction coefficient.

The Kubelka–Munk theory relates the extinction coefficient to the reflection. In the simplest case, it is assumed that light is only scattered in two directions: in the incident and in the backward direction for an incident ray normal to the surface of the test sample. Also, both incident light and emitted light are diffuse. According to Kubelka and Munk, then,

$$\frac{K}{S} = \frac{(1 - R_\infty)^2}{2R_\infty} \tag{15-14}$$

where R_∞ is the reflection for an infinitely thick film.

15.4.2. Opacity

A light-scattering body appears opaque when there is a variation in refractive index or differences in the orientation of anisotropic volume elements, or both.

Local variations in the refractive index only lead to opacity when different structures are present whose dimensions are greater than the wavelength of incident light. On the other hand, these structures may not be too large, since an infinitely large single crystal does not scatter light.

Consequently, the clarity of a material can be considerably increased by decrease in the structure size. On the other hand, approach of the refractive index values of the two phases to each other leads to only a small increase in the clarity. If the refractive index of PVC is greater than that of the disperse phase in a PVC/ABS mixture, the material appears milky yellow in reflected light. It is milky blue in the reverse case.

Effects resulting from variations in the refractive index can be distinguished from those resulting from variation in orientation of anisotropic volume elements by the use of polarized light. The horizontally polarized scattering observed with incident vertically polarized light (H_v scattering) originates from anisotropy of the scattering elements. The V_v scattering depends on the anisotropy of the scattering elements as well as on the differences in refractive index.

An ordered spherulite is spherically symmetric. Consequently, the scattering in the interior of an H_v sample (see also Figure 5-25) should be zero. A finite scattering intensity in the center thus indicates disorder. The size of the spherulite can be determined from the angle at which maximum scattering occurs.

Lamellar structures with order over regions of dimensions greater than the wavelength of incident light are less optically heterogeneous than spherulitic structures. Thus, they scatter less light; they are more transparent. Consequently, poly(ethylene) films drawn and oriented under certain conditions are clear, although the samples are crystalline and even have superstructures of dimensions greater than the wavelength of incident light.

Literature

R. W. Burnham, R. M. Hanes, and C. J. Bertleson, *Color: A Guide to Basic Facts and Concepts*, Wiley, New York, 1963.

N. S. Capany, *Fiber Optics*, Academic Press, New York, 1967.

T. Alfrey, Jr., E. F. Gurnee, and W. J. Schrenk, Physical optics of iridescent multilayered plastic films, *Polym. Eng. Sci.* **9**, 400 (1969).

G. Ross and A. W. Birley, Optical properties of polymeric materials and their measurement, *J. Phys. D* **6**, 795 (1973).

Appendix

Table A1. International Abbreviations for Plastics, Rubbers, and Fibers[a]

Abbreviation	Plastic/fiber/rubber
ABR	Copolymers from acrylic esters and butadiene
ABS	Copolymers from acrylonitrile, butadiene, and styrene
ACM	Copolymers from acrylic ester and 2-chlorovinyl ether
AMMA	Copolymers from acrylonitrile and methyl methacrylate
ANM	Copolymers from acrylic ester and acrylonitrile
BR	Poly(butadiene)
BT	Poly(butene-1)
CA	Cellulose acetate
CAB	Cellulose acetobutyrate
CAP	Cellulose acetopropionate
CAR	Carbon fiber
CF	Cresol/formaldehyde resin
CFK	Man-made fiber-reinforced plastics
CHC	Chlorohydrin copolymers (from epichlorohydrin and ethylene oxide)
CHR	Chlorohydrin elastomer [poly(epichlorohydrin)]
CL	Poly(vinyl chloride) fiber
CMC	Carboxymethyl cellulose
CN	Cellulose nitrate
CNR	Carboxynitroso rubber
CP	Cellulose propionate
CPVC	Chlorinated poly(vinyl chloride)
CR	Poly(chloroprene)
CS	Casein
CSR	Chlorosulfonated poly(ethylene)
EA	Segmented polyurethane fiber

[a] From ISO/DR 1252 (International Standardization Organization), DIN 7728 (Deutsch Industrie-Norm = German industrial standard), and EEC (European Economic Community) abbreviations.

Table A1 (continued)

Abbreviation	Plastic/fiber/rubber
EEA	Copolymer from ethylene and ethyl acrylate
EC	Ethyl cellulose
EP	Epoxide resin
EPDM	Elastomer from ethylene, propylene, and a diene
EPM	Elastomer from ethylene and propylene
EVA	Copolymer from ethylene and vinyl acetate
FE	Fluorine-containing elastomers
GEP	Glass-fiber-reinforced epoxy resins
GFK	Glass-fiber-reinforced plastics
GUP	Glass-fiber-reinforced polyester resins
IIR	Butyl rubber
MA	Modacrylic fibers
MC	Methyl cellulose
MF	Melamine/formaldehyde resins
MOD	Modacrylic fibers (EEC abbreviation)
NBR	Elastomers from acrylonitrile and butadiene
NR	Natural rubber
PA	Polyamides
PAC	Poly(acrylonitrile) fiber
PAN	Poly(acrylonitrile)
PBMA	Poly(butyl methacrylate)
PC	Poly(acrylonitrile) fibers (EEC abbreviation)
PCF	Poly(trifluorochloroethylene) fiber
PCTFE	Poly(trifluorochloroethylene) fiber
PDAP	Poly(diallyl phthalate)
PDMS	Poly(dimethyl siloxane)
PE	Poly(ethylene)
PE	Polyester fiber (EEC abbreviation)
PEO	Poly(ethylene oxide)
PES	Polyester fiber
PETP	Poly(ethylene terephthalate)
PF	Phenol/formaldehyde resin
PFEP	Copolymer from tetrafluoroethylene and hexafluoropropylene
PIB	Poly(isobutylene)
PL	Poly(ethylene) (EEC abbreviation)
PMMA	Poly(methyl methacrylate)
PO	Phenoxy resin
POM	Poly(oxymethylene)
POR	Elastomer from propylene oxide and allyl glycidyl ether
PP	Poly(propylene)
PPO	Poly(phenylene oxide)
PS	Poly(styrene)
PSB	Copolymer from styrene and butadiene
PST	Poly(styrene) fiber
PS-TSG	Injection-molding foam poly(styrene)
PTF	Poly(tetrafluoroethylene) fiber
PTFE	Poly(tetrafluoroethylene)
PU	Polyurethane fiber

Table A1 (continued)

Abbreviation	Plastic/fiber/rubber
PUA	Polyurea fiber
PUE	Segmented polyurethane fiber
PVA	Poly(vinyl ether)
PVAC	Poly(vinyl acetate)
PVAL	Poly(vinyl alcohol)
PVB	Poly(vinyl butyral)
PVC	Poly(vinyl chloride)
PVCA	Copolymer from vinyl chloride and vinyl acetate
PVDC	Poly(vinylidene chloride)
PVDF	Poly(vinylidene fluoride)
PVF	Poly(vinyl fluoride)
PVFM	Poly(vinyl formal)
PVID	Poly(vinylidene cyanide)
PVM	Copolymer from vinyl ethers and vinyl chloride
SAN	Copolymer from styrene and acrylonitrile
SBR	Elastomer from styrene and butadiene
SMS	Copolymer from styrene and α-methyl styrene
SI	Silicones
TR	Thermoplastic elastomers
UF	Urea/formaldehyde resin
UP	Unsaturated polyester
VF	Vulcan fiber

Table A2. Trivial and Trade Names of Macromolecular Substances[a]

Name	Substance	Manufacturer
ABS polymers	Generic term for copolymers or polyblends from acrylonitrile, butadiene, styrene	—
Abson	ABS	Goodrich, U.S.
Acetate	Generic name for fibers from cellulose-$2\frac{1}{2}$-acetate	—
Aceta	Cellulose acetate	Bayer, Germany
Aclar	Fluorinated polycarbonate film	Allied Chemical, U.S.
Acrilan	Poly(acrylonitrile)	Chemstrand Corp. (Monsanto), U.S.
Acronal	Dispersions based on uni- and copolymers of acrylic esters	BASF, Germany
Acrylic	Generic name for fibers from at least 85% poly(acrylonitrile)	—
Acrylan rubber	Butyl acrylate/5–10% acrylonitrile copolymer	Monomer Corp., U.S.
Adiprene	Polyurethane elastomer	Du Pont, U.S.
Alathon	Ethylene/vinyl acetate copolymer	Du Pont, U.S.

[a] This list is not claimed to be complete. Register of trade names: J. B. Titus, *Trade Designations of Plastics and Related Materials*, Plastics Evaluation Center, Dover, New Jersey, 1970; *Deutsche Rhodiaceta, Chemiefasern auf dem Weltmarkt*, Deutsche Rhodiaceta, Freiburg/Br., 1966.

Table A2 (*continued*)

Name	Substance	Manufacturer
Albertol, Alberlat	Modified phenolic resins	Chem. Werke Albert, Germany
Alkathene	Poly(ethylene) (high pressure)	ICI, Great Britain
Alkydal	Polyester resin	Bayer, Germany
Algoflon	Poly(tetrafluoroethylene)	Montedison, Italy
Alloprene	Chlorinated rubber	ICI, Great Britain
Amberlite	Synthetic ion-exchange resins	Roehm & Haas, U.S.
Ameripol	Poly(isoprene)	Firestone, U.S.
Ameripol SM	*Cis*-1,4-poly(isoprene)	Firestone, U.S.
Aralac	Albumin fiber	National Dairy Prod., U.S.
Araldite	Epoxide resins	CIBA, Switzerland
Ardil	Fiber from peanut protein	ICI, Great Britain
Arnite	Poly(ethylene terephthalate) (as plastic)	AKU, The Netherlands
Atlac	Polyester resin	Atlas, U.S.
Asplit	Phenoplast	Hoechst, Germany
Balata	*Trans*-1,4-Poly(isoprene)	Natural product
Bakelite	Phenol–formaldehyde resins	Bakelite Inc., U.S.
Barex	Copolymer from acrylonitrile and methyl methacrylate (3:1)	Vistron, U.S.
Beckacite	Phenoplast	Reichhold, U.S.
Bodanyl	Poly(caprolactam)	Feldmühle, Rorschach, Switzerland
Boralloy	Boronitride	Union Carbide, U.S.
Buna N	Copolymer from butadiene and acrylonitrile.	Hüls, Germany
Buna S or SS	Butadiene/styrene copolymer (elastomer)	Hüls, Germany
Butacite	Poly(vinyl butyral)	Du Pont, U.S.
Buton	Cross-linkable plastic produced at high polymerization temperatures from butadiene and styrene	Esso, Great Britain
Butyl rubber	Poly(isobutylene) with 2% isoprene	Bayer, Germany
Carbowax	Poly(ethylene glycol)	Union Carbide, U.S.
Cariflex	Block copolymer of styrene/butadiene/styrene	Shell, The Netherlands
Carlona	Poly(ethylene)	Shell, The Netherlands
Carlona Pt	Poly(propylene)	Shell, The Netherlands
C 23	Ethylene/propylene copolymer	Montecatini, Italy
Celcon	Poly(formaldehyde) (from trioxane with some ethylene oxide)	Celanese, U.S.
Cellidor	Thermoplast based on Cellit	Bayer, Germany
Cellit	Cellulose acetate or acetobutyrate	Bayer, Germany
Cellon	Cellulose acetate	Dynamit Nobel, Germany
Cellophane	Hydrate cellulose from pulp	Kalle, Germany
Celluloid	Cellulose nitrate, plasticized with camphor	Dynamit Nobel, Germany
Chicle	Raw material for chewing gum (mixture of *trans*-1,4-poly(isoprene) + triterpenes	Natural product
Chinon	Graft copolymer of 70% acrylonitrile on 30% casein	Toyoba, Japan

Table A2 (continued)

Name	Substance	Manufacturer
Chlorinated rubber	Chlorinated natural rubber	Bayer, Germany
Cibanoid	Urea–formaldehyde resin	CIBA, Switzerland
Clarifoil	Cellulose acetate	British Celanese, Great Britain
Collacral K	Industrial poly(vinyl pyrrolidone)	BASF, Germany
Courlene	Poly(ethylene) (fiber)	Courtaulds, Great Britain
Courlene PY	Poly(propylene) (fiber)	Courtaulds, Great Britain
Courtelle	Poly(acrylonitrile)	Courtaulds, Great Britain
Corfam	Permeable artificial leather from polyurethane/polyester/polyester fleece	Du Pont, U.S.
Corvic	Poly(vinyl chloride)	ICI, Great Britain
Coral rubber	*Cis*-1,4-poly(isoprene)	Goodrich, U.S.
Crofon	Optical fibers from polymethyl methacrylate and polyethylene	Du Pont, U.S.
Cycolac	ABS	Marbon, U.S.
Cycolon	ABS	Marbon, U.S.
Dacron	Fiber from poly(ethylene terephthalate)	Du Pont, U.S.
Darvic	Poly(vinyl chloride)	ICI, Great Britain
Delrin	Poly(oxymethylene) (from formaldehyde)	Du Pont, U.S.
Densothene	Poly(ethylene)	Metal Box, Great Britain
Desmodur	Isocyanate grades for polyurethane	Bayer, Germany
Desmophen	Polyester for polyurethanes	Bayer, Germany
Dexel	Cellulose acetate	British Celanese, Great Britain
Dexsil	Poly(carborane siloxane)	Olin, U.S.
Diofan	Dispersion of copolymers of vinylidene chloride	BASF, Germany
Diolen	Fiber from poly(ethylene terephthalate)	Glanzstoff, Germany
Dralon	Poly(acrylonitrile)	Bayer, Germany
Drawinella	Cellulose triacetate	Wacker, Germany
Duranit	Butadiene/styrene copolymer	Hüls, Germany
Durethan	Polyamides or polyurethanes	Bayer, Germany
Durette	Fiber of isophthalic acid and *m*-phenylene diamine	Monsanto. U.S.
Dutral	Ethylene/propylene copolymer	Montecatini, Italy
Dynel	Vinyl chloride/acrylonitrile copolymer	Union Carbide, U.S.
Edistir	Poly(styrene)	Montedison, Italy
Elvanol	Poly(vinyl alcohol)	Du Pont, U.S.
Enkatherm	Poly(terephthaloyl oxamidrazone)	AKZO, The Netherlands
Enjay-Butyl	Isobutylene/isoprene copolymer	Enjay, U.S.
Epikote	Epoxide resin	Shell, The Netherlands
Epon	Epoxide resin	Shell, The Netherlands
Estane	Polyurethane	Goodrich, U.S.
Ethocel	Cellulose ether	Dow, U.S.

Table A2 (continued)

Name	Substance	Manufacturer
Exonol	Poly(*p*-hydroxy benzoate)	Carborundum, U.S.
Fiber AF	Dehydrogenated cyclic polyacrylonitrile	Du Pont, U.S.
Fluon	Poly(tetrafluoroethylene)	ICI, Great Britain
Gaflon	Poly(tetrafluoroethylene)	Gachot, France
Galalith	Plastic from milk albumin	Internationale Galalith, Germany
Geon	Poly(vinyl chloride)	Goodrich, U.S.
Glyptal	Alkyd resin	General Electric, U.S.
Grafoil	Pure graphite film	Union Carbide, U.S.
GR	"Government rubber," former term for polymers produced in U.S. government-owned factories during the second world war	—
GR-I	Copolymer from isobutylene with 2% isoprene	—
GR-N	Poly(chloroprene)	—
GR-P	Thiokol	—
GR-S	Copolymer from butadiene and styrene	—
Grex	Poly(ethylene) (low pressure)	W. R. Grace & Co., U.S.
Grilen	*p*-Hydroxybenzoic acid + terephthalic acid + glycol (fiber)	Emser Werke, Switzerland
Grilon	Polyamide 6	Emser Werke, Switzerland
Grilonit	Epoxide compounds	Emser Werke, Switzerland
Gutta Percha	*Trans*-1,4-poly(isoprene)	Natural product
H film	Polyimide (pyromellitic anhydride + *p,p'*-diaminodiphenylene oxide)	Du Pont, U.S.
Hi-fax	Poly(ethylene) (low pressure)	Hercules Powder, U.S.
Hostaflon	Poly(trifluoromonochloroethylene)	Hoechst, Germany
Hostaform	Poly(oxymethylene), containing cyclic acetals as comonomers	Hoechst, Germany
Hostalen	Poly(ethylene) (low pressure)	Hoechst, Germany
Hostalen PP	Poly(propylene)	Hoechst, Germany
Hostalit	Poly(vinyl chloride)	Hoechst, Germany
Hostaphan	Poly(ethylene terephthalate glycol ester)	Kalle, Germany
Hydron	Poly(hydroxy ethyl methacrylate)	Hydro-Dent, U.S.
Hycar	Group of elastomers (e.g., nitrile rubber, styrene/butadiene rubbers)	Goodrich, U.S.
Hygromull	Urea–formaldehyde resin (foam)	BASF, Germany
Hypalon	Chlorosulfonated poly(ethylene)	Du Pont, U.S.
Hystrel	Block copolymer from poly(butylene terephthalate) and poly(butylene glycol)	Du Pont, U.S.
Igelith	Poly(vinyl chloride)	BASF, Germany
Iolon	Ionomeric film	Du Pont, U.S.
Iporka	Urea–formaldehyde foam	BASF, Germany
Irrathene	Irradiated, cross-linked poly(ethylene)	General Electric, U.S.
Isonate	Cellular polyurethane	Upjohn, U.S.
Kapton H	Polyimide from pyromellitic dianhydride and *p,p'*-diaminodiphenyl ether	Du Pont, U.S.

Table A2 (continued)

Name	Substance	Manufacturer
Kaurit glue	Urea–formaldehyde resin	BASF, Germany
Kautex	Poly(vinyl chloride)	Kautex Werke, Germany
Kel-F	Poly(trifluoromonochloroethylene)	M. W. Kellog, U.S.
Kel-F elastomer	Copolymer from vinylidene fluoride and trifluorochloroethylene	M. W. Kellog, U.S.
Kinel	Polyimide	Rhone Poulenc, France
Kodar	Polyester	Eastman, U.S.
Kodel-2	Polyester from terephthalic acid and 1,4-dimethylol cyclohexane	Eastman, U.S.
Kodel-10	Poly(ethylene terephthalate)	Eastman, U.S.
Kralastic	ABS	US Rubber, U.S.
Kryston	Polyester	Goodrich, U.S.
Krytox	Perfluorinated polyether	Du Pont, U.S.
Kynol	Phenol/formaldehyde fiber	Carborundum, U.S.
Lanital	Fiber from milk albumin	Snia Viscosa, Italy
Lastrile	Fiber from copolymers with 10–15% acrylonitrile and an aliphatic diene	Generic name
Leacril	Poly(acrylonitrile)	ACSA, Italy
Leguval	Unsaturated polyester	Bayer, Germany
Levapren	Ethylene/vinyl acetate copolymer	Bayer, Germany
Lexan	Polycarbonate from bisphenol A and phosgene	General Electric, U.S.
Lignostone	Compressed wood	Röchling, Germany
Lopac	Copolymer from methacrylonitrile and styrene or α-methyl styrene (9:1)	Monsanto, U.S.
Lucite	Poly(methyl methacrylate)	Du Pont, U.S.
Luparen	Poly(propylene)	BASF, Germany
Luphen	Phenoplast	BASF, Germany
Lupolen	Poly(ethylene) (high pressure)	BASF, Germany
Luran	Copolymer from styrene/acrylonitrile	BASF, Germany
Lustrex	Poly(styrene)	Monsanto, U.S.
Lutofan	Vinyl chloride-containing copolymers in the form of solutions (*L*) or dispersions (*D*)	BASF, Germany
Lutonal	Poly(vinyl ether)	BASF, Germany
Luvican	Poly(vinyl carbazole) (no longer produced)	BASF, Germany
Luvitherm	Poly(vinyl chloride) (film)	BASF, Germany
Lycra	Elastomer from segments of polyether and polyurethane	Du Pont, U.S.
Makrolon	Polycarbonate from bisphenol A and phosgene base units	Bayer, Germany
Marlex	Poly(ethylene)	Phillips, U.S.
Melan, Melamin	Melamine/formaldehyde prepolymer	Henkel, Germany
Melbrite	Melamine/formaldehyde resin	Montedison, Italy
Meraklon	Poly(propylene)	Montecatini, Italy
Merinova	Casein fiber	Snia Viscosa, Italy
Methyl rubber	Poly(2,3-dimethylbutadiene)	Bayer, Germany (World War I)
Mipolam	Poly(vinyl chloride)	Dynamit Nobel, Germany

Table A2 (continued)

Name	Substance	Manufacturer
Mirlon	Polyamide	Viscose-Suisse, Switzerland
Modacrylic	Generic name for fibers with 35–85% poly(acrylonitrile), excluding rubbers	—
Modal	Generic name for fibers from regenerated cellulose of modified structure	—
Moltopren	Polyester or polyether + diisocyanate + water (foam)	Bayer, Germany
Moplen	Poly(ethylene) and poly(propylene)	Montedison, Italy
Mouldrite	Urea, phenol, and melamine/formaldehyde resins	ICI, Great Britain
Movil, Mowil	Poly(vinyl chloride)	Polymer Ind., Italy
Mowilith	Poly(vinyl acetate)	Hoechst, Germany
Moviol	Poly(vinyl alcohol)	Hoechst, Germany
Mylar	Polyester (film)	Du Pont, Germany
Neoprene	Poly(chloroprene)	Du Pont, Germany
Niax	Polyether from propylene and glycerine or 1,2,6-hexantriol	Union Carbide, U.S.
Nitron	Cellulose nitrate	Monsanto, U.S.
Nomex	Polyamide from isophthalic acid + *m*-phenylene diamine	Du Pont, U.S.
Noryl	Poly(phenylene oxide)	General Electric, U.S.
Novodur	ABS polymer	Bayer, Germany
Novolak	Phenol/formaldehyde condensate	Dynamit Nobel, Germany
Nylon	Generic name for polyamides	(Du Pont, U.S.)
Nylon 6-T	Terephthalic acid + hexamethylene diamine	Celanese, U.S.
Nylsuisse	Adipic acid + hexamethylene diamine	Viscose-Suisse, Switzerland
Olefin	Fibers from at least 85% other olefins, excluding rubber	Generic name
Oppanol B	Poly(isobutylene)	BASF, Germany
Oppanol C	Poly(vinylisobutyl ether)	BASF, Germany
Oppanol O	Copolymer from 90% isobutene and 10% styrene	BASF, Germany
Orlon	Poly(acrylonitrile)	Du Pont, U.S.
Pale crepe	Light, unsmoked natural rubber	—
Paralac	Polyester resin	ICI, Great Britain
Parlon	Chlorinated rubber	Hercules Powder, U.S.
Parylen N	Poly(*p*-xylylene)	Union Carbide, U.S.
Parylen C	Poly(monochloro-*p*-xylylene)	Union Carbide, U.S.
PBI	Poly(benzimidazole)	Celanese, U.S.
PE CE	Post-chlorinated poly(vinyl chloride) (fiber)	BASF, Germany
Penton	Poly(2,2-dichloromethyl trimethylene oxide)	Hercules Powder, U.S.
Perbunan C	Poly(chloroprene)	Bayer, Germany
Perbunan N	Butadiene/acrylonitrile copolymer	Bayer, Germany
Perduren	Thioplasts	Hoechst, Germany
Periston	Poly(vinyl pyrrolidone)	Bayer, Germany
Perlenka	Poly(caprolactam)	AKU, The Netherlands
Perlon	Generic name for polyamides from caprolactam (nylon 6)	—

Table A2 (continued)

Name	Substance	Manufacturer
Perlon U	Polyurethane	Bayer, Germany
Perspex	Poly(methyl methacrylate)	ICI, Great Britain
Phenoxy	Copolymer from bisphenol A + epichlorohydrin	Union Carbide, U.S.
Phenyl T	Polymerized phenylsesquisiloxane (ladder polymer)	General Electric, U.S.
Philprene	Butadiene/styrene copolymer	Philips Petrol, The Netherlands
Plaskon	Urea–formaldehyde resin	Allied Chemicals, U.S.
Plexidur	Poly(methyl methacrylate)	Roehm and Haas, U.S.
Plexiglas	Poly(methyl methacrylate)	Röhm & Haas, Germany
Plexol	Oil-soluble methacrylate copolymer (viscosity improver)	Röhm & Haas, Germany
Pliofilm	Rubber hydrochloride	Goodyear, U.S.
Pliolite NR	Cyclo rubber	Goodyear, U.S.
Pluronics	Ethylene oxide/propylene oxide copolymer	Wyandotte Chem., U.S.
Pollopas	Urea–formaldehyde resin	Dynamit Nobel, Germany
Polymin	Poly(ethylene imine)	BASF, Germany
Polyox	High-molecular-weight poly(ethylene oxide)	Union Carbide, U.S.
Polysar butyl	Isobutylene/isoprene copolymer	Sarnia, Canada
Polysulfone	Copolymer from bisphenol A + *p,p'*-dichlorodiphenyl sulfone	Shell, The Netherlands
Polythene	Poly(ethylene) (high pressure)	Du Pont, U.S.
PPO	Poly(2,6-dimethyl phenylene oxide)	General Electric, U.S.
Pro-fax	Poly(propylene)	Hercules Powder, U.S.
Propiofan	Poly(vinyl propionate)	BASF, Germany
Qiana	Fiber from *trans*-diamino dicyclohexyl methane + dodecane dicarboxylic acid	Du Pont, U.S.
Q2	Polyamide from 1,4-*bis*(aminomethyl) cyclohexane + suberic acid	Eastman, U.S.
Rayon, rayonne	Generic name for fibers from regenerated cellulose or cellulose derivatives	—
Rhodester	Cellulose acetate	Soc. Rhone Poulanc, France
Rhodia	Cellulose $2\frac{1}{2}$ acetate	Soc. Rhodiaceta, France
Rhodiaceta-nylon	Nylon 66	Soc. Rhodiaceta, France
Rhovil	Poly(vinyl chloride)	Soc. Rhovil, France
Ribbon straw	Cellulose $2\frac{1}{2}$ acetate (artificial straw)	British Celanese, Great Britain
Rilsan	Nylon 11	Acquitaine-Organico, France
Rovicella	Cellulose (viscose)	Feldmühle Rorsch., Switzerland
Royalene	Poly(ethylene) or poly(propylene)	U. S. Rubber Co., U.S.
RT 700	Cellulose (viscose)	Glanzstoff, Germany

Table A2 (continued)

Name	Substance	Manufacturer
Rubazote	Natural rubber	Expanded Rubber, Great Britain
Ruvea	Nylon 66 (artificial straw)	Du Pont, U.S.
Ryton	Poly(thio-1,4-phenylene)	Phillips Petroleum, U.S.
Ryton	Poly(thio-1,4-phenylene)	Phillips Petroleum, U.S.
Saflex	Poly(vinyl acetal)	Monsanto, U.S.
Saran	Generic name for fibers from polymers with not less than 80% vinylidene chloride	—
Scotchcast	Epoxide resin	Minnesota Mining & Mfg., U.S.
Silicone	Generic name for polymers with (—SiR_2—O—) links	Bayer, Dow, General Electric
Silopren	Polysiloxane rubber	Bayer, Germany
SKS	Copolymer from butadiene and styrene	USSR
Smoked sheet	Smoked natural rubber	—
Sov Pren	Poly(chloroprene)	USSR
Spandex	Generic name for fibers from elastic polyurethanes	—
Styroflex	Poly(styrene) (fibers)	Ndd. Seekabelwerke, Germany
Styrofoam	Poly(styrene) (foam)	Dow, U.S.
Styron	Poly(styrene), also copolymers	Dow, U.S.
Styropor P	Poly(styrene) (foam)	BASF, Germany
Supralen	Poly(ethylene) (pipes)	Mannesmann, Germany
Surlyn A	Ionomer (copolymer from ethylene + some acrylic acid or maleic anhydride)	Du Pont, U.S.
Tedlar	Poly(vinyl fluoride)	Du Pont, U.S.
Teflon	Poly(tetrafluoroethylene)	Du Pont, U.S.
Teflon FEP	Copolymer from tetrafluoroethylene and hexafluoropropylene	Du Pont, U.S.
Tego	Phenoplast	Resinous Products, U.S.
Tenax	Poly[oxy-1,4-(2,6-diphenyl)-phenylene]	AKZO, The Netherlands
Terital	Poly(ethylene terephthalate)	Soc. Rhodiadoce, Italy
Terlenka	Poly(ethylene terephthalate)	AKU, The Netherlands
Terluran	High-impact poly(styrene) (graft polymer of styrene and acrylonitrile on styrene/butadiene copolymer)	BASF, Germany
Terylene	Poly(ethylene terephthalate)	ICI, Great Britain
Thiokol	Polysulfide (rubber)	Thiokol, U.S.
Thornel	Graphite yarn	Union Carbide, U.S.
TPX	Poly(4-methyl-pentene-1)	ICI, Great Britain
Travis	Vinyl acetate/vinylidene cyanide copolymer	Hoechst/Celanese
Trevira	Polyester (fiber)	Hoechst, Germany
Triacetate	Generic name for fibers from cellulose triacetate	—
Tricel	Cellulose triacetate	Bayer, Germany
Trolit AE	Cellulose ether	Dynamit Nobel, Germany

Table A2 (continued)

Name	Substance	Manufacturer
Trolit F	Cellulose nitrate	Dynamit Nobel, Germany
Trolitan	Phenol–formaldehyde resin	Dynamit Nobel, Germany
Trolitul	Poly(styrene)	Dynamit Nobel, Germany
Tronal	High-impact poly(styrene)	Dynamit Nobel, Germany
Trovidur	Poly(vinyl chloride)	Dynamit Nobel, Germany
Trovitherm	Poly(vinyl chloride) (films)	Dynamit Nobel, Germany
Tylose	Cellulose ether	Kalle, Germany
Tynex	Nylon 6,6	Du Pont, U.S.
Ultramide A	Nylon 6,6	BASF, Germany
Ultramide B	Nylon 6	BASF, Germany
Ultramides	Nylon 6,10	BASF, Germany
Ultrapas	Melamine–formaldehyde resin	Dynamit Nobel, Germany
Urylon	Poly(nonamethylene urea)	Toya, Japan
Versamides	Group of "polymerized" vegetable oils whose ester groups are converted with di- and triamines	General Mills, U.S.
Vestamides	Various nylon grades	Hüls, Germany
Vestan	Polycondensate from terephthalic acid and 1,4-dimethylol cyclohexane	Hüls, Germany
Vestolen A	Low-pressure poly(ethylene)	Hüls, Germany
Vestolen P	Poly(propylene)	Hüls, Germany
Vestolit	Poly(vinyl chloride)	Hüls, Germany
Vestopal	Unsaturated polyester, dissolved in styrene	Hüls, Germany
Vestoran	Vinylchloride/vinyl acetate copolymer	Hüls, Germany
Vestyron	Poly(styrene)	Hüls, Germany
Vicara	Albumin fiber	Virginia-Carolina Chem., U.S.
Vibrathane	Polyurethane elastomer	Naugatuck, U.S.
Vinidur	Poly(vinyl chloride) film	BASF, Germany
Vinnipas	Poly(vinyl acetate)	Wacker, Germany
Vinnol	Poly(vinyl chloride)	Wacker, Germany
Vinoflex	Vinyl chloride/vinyl ether copolymer	BASF, Germany
Vinylite, Vinyon	Vinyl chloride/vinyl acetate copolymer	Carbide & Carbon Chem., U.S.
Vinylon	Poly(vinyl alcohol) (fiber)	Synthetic Fiber Mfrs. Group, Japan
Viscoplex	Oil-soluble methacrylate copolymer (viscosity improver)	Röhm & Haas, Germany
Viscose	Generic name for fibers from regenerated cellulose (prepared by the xanthate method)	—
Vistanex	Poly(isobutylene)	Standard Oil, U.S.

Table A2 (continued)

Name	Substance	Manufacturer
Viton A	Vinylidene fluoride/hexafluoropropylene copolymer	Du Pont, U.S.
Vulcollan	Polyurethane	Bayer, Germany
Worbaloid	Cellulose nitrate	Worbla AG, Switzerland
Zein	Generic name for fibers from vegetable albumin	—
Zetafin	Ethylene/vinyl acetate copolymer	Dow, U.S.
Zytel 31	Nylon 6,10	Du Pont, U.S.
Zytel 101	Nylon 6,6	Du Pont, U.S.

Table A3. SI Units

Symbol	Quantity	Unit	Abbreviation
Basic quantities			
l	Length	meter	m
m	Mass	kilogram	kg
t	Time	second	s
I	Electric current	ampere	A
T	Thermodynamic temperature	kelvin	K
I_v	Luminous intensity	candela	cd
n	Amount of substance	mole	mol
Additional quantities			
α, β, γ	Plane angle	radian	rad
ω, Ω	Solid angle	steradian	sr
Derived quantities			
F	Force	newton	$\mathrm{N = J\ m^{-1} = kg\ m\ s^{-2}}$
E	Energy	joule	$\mathrm{J = N\ m = kg\ m^2\ s^{-2}}$
P	Power	watt	$\mathrm{W = J\ s^{-1} = V\ A = kg\ m^2\ s^{-3}}$
p	Pressure	pascal	$\mathrm{Pa = N\ m^{-2} = J\ m^{-3} = kg\ m^{-1}\ s^{-2}}$
v	Frequency	hertz	$\mathrm{Hz = s^{-1}}$
Q	Electric charge	coulomb	$\mathrm{C = A\ s}$
U	Electrical potential difference	volt	$\mathrm{V = J\ C^{-1} = W\ A^{-1} = kg\ m^2\ s^{-3}\ A^{-1}}$
R	Electrical resistance	ohm	$\mathrm{\Omega = V\ A^{-1} = kg\ m^2\ s^{-3}\ A^{-2}}$
G	Electrical conductance	siemens	$\mathrm{S = A\ V^{-1} = s^3\ A^2\ kg^{-1}\ m^{-2}}$
C	Electrical capacitance	farad	$\mathrm{F = C\ V^{-1} = s^4\ A^2\ kg^{-1}\ m^{-2}}$
Φ	Magnetic flux	weber	$\mathrm{Wb = V\ s = kg\ m^2\ s^{-2}\ A^{-1}}$
L	Inductance	henry	$\mathrm{H = V\ s\ A^{-1} = kg\ m^2\ s^{-2}\ A^{-2}}$
B	Magnetic flux density	tesla	$\mathrm{T = V\ s\ m^{-2} = kg\ s^{-2}\ A^{-1}}$
Φ_v	Luminous flux	lumen	$\mathrm{lm = cd\ sr}$
E_v	Illumination	lux	$\mathrm{lx = lm\ m^{-2} = cd\ sr\ m^{-2}}$

Table A4. Prefixes for SI Units

Factor	Prefix	Symbol
10^{12}	tera	T
10^{9}	giga	G
10^{6}	mega	M
10^{3}	kilo	k
10^{2}	hecto	h
10^{1}	deca	da
10^{-1}	deci	d
10^{-2}	centi	c
10^{-3}	milli	m
10^{-6}	micro	μ
10^{-9}	nano	n
10^{-12}	pico	p
10^{-15}	femto	f
10^{-18}	atto	a

Table A5. Conversion of Old Units into New Units

Physical quantity	Name of old unit	Symbol for old unit	Definition of old unit
Length	Ångstrom	Å	10^{-10} m = 10^{-8} cm
Length	micron	μ	10^{-6} m = μm
Length	millimicron	mμ	10^{-9} m = nm
Length	inch	in.	0.0254 m
Mass	pound (avoirdupois)	lb	0.45359237 kg
Force	kilogram-force	kgf	9.80665 N
Force	dyne	dyne	10^{-5} N
Pressure	atmosphere	atm	101,325 N m^{-2}
Pressure	torr	Torr	(101,325/760) N m^{-2}
Pressure	conventional millimeter of mercury	mm Hg	$13.5951 \times 980.665 \times 10^{-2}$ N m^{-2}
Energy	kilowatt-hour	kWh	3.6×10^{6} J
Energy	thermochemical calorie	cal	4.184 J
Energy	electron volt	eV	$\sim 1.6021 \times 10^{-19}$ J
Energy	erg	erg	10^{-7} J
Titer	denier	den	$\frac{1}{9}$ tex = $\frac{1}{9}$ g km^{-1}
Viscosity	centipoise	cP	0.001 Pa s
Angle	degree	°	0.017453 rad

Table A6. Fundamental Constants

Quantity	Symbol, value, and unit
Speed of light *in vacuo*	$c = 2.997925 \times 10^{18}$ m s^{-1}
Charge of proton	$e = 1.60210 \times 10^{-19}$ C
Faraday constant	$F = 9.64870 \times 10^{4}$ C mol^{-1}
Planck constant	$h = 6.6256 \times 10^{-34}$ J s
Boltzmann constant	$k = 1.38054 \times 10^{-23}$ J K^{-1}
Avogadro constant (Loschmidt number)	$N_L = 6.02252 \times 10^{23}$ mol^{-1}
Gas constant	$R = 83.1433$ bar cm^3 K^{-1} mol^{-1} = 8.31433 J K^{-1} mol^{-1}
Permeability of vacuum	$\mu_0 = 4\pi \times 10^{-7}$ J s^2 C^{-2} m^{-1} (exactly)
Permittivity of vacuum	$\varepsilon_0 = \mu_0^{-1} c^{-2} = 8.854185 \times 10^{-12}$ J^{-1} C^2 m^{-1}

Index

Pages 1-532 will be found in Volume 1, pages 533-1131 in Volume 2.